T0260786

PHILOSOPHY IN AN AGE OF SCIENCE

PHILOSOPHY IN AN AGE OF SCIENCE

Physics, Mathematics, and Skepticism

———

Hilary Putnam

EDITED BY

Mario De Caro and
David Macarthur

HARVARD UNIVERSITY PRESS

Cambridge, Massachusetts
London, England
2012

Library of Congress Cataloging-in-Publication Data
Putnam, Hilary.
Philosophy in an age of science : physics, mathematics, and skepticism /
Hilary Putnam ; edited by Mario De Caro and David Macarthur.
p. cm.
Includes index.
ISBN 978-0-674-05013-6 (alk. paper)
1. Philosophy, American—20th century.   2. Philosophy, American—21st century.
I. De Caro, Mario.   II. Macarthur, David.   III. Title.
B945.P871 2012
191—dc23   2011035336

Contents

PART TWO
———————

Mathematics and Logic

PART THREE
———————

Values and Ethics

Preface

HILARY PUTNAM

For a number of years my old friend Lindsay Waters, the Executive Editor for the Humanities at Harvard University Press, urged me to publish another collection of papers. If I did not agree at once, it was because I was continuing to write new papers that I wanted to include in my next collection, but that seems to be a nonterminating state of affairs, and it is obviously no longer a good reason not to accept his suggestion. Another reason for hesitation was that the sheer amount of work involved in choosing papers, finding or creating electronic files, securing permissions, and so on, was a daunting prospect at my age, but two wonderful friends, Mario De Caro and David Macarthur, agreed to relieve me of the bulk of that work (the three of us together did the actual proofreading and, where necessary, revision of the papers). I am tremendously in their debt for this, as well as for the beautiful introductory essay they have supplied, which describes in detail the raison d'être of each of the parts of this volume. In addition to thanking them, I also want to say how much I have enjoyed our fellowship (which includes Lindsay Waters) as we worked to put this volume together.

Among the editorial decisions the four of us made early on was a decision to have a separate volume of reviews and replies, and we still plan to bring out such a volume in the not-too-distant future. Also, we have not included my papers on pragmatism because I plan to put those into a future joint volume of papers on pragmatism with my best friend and longtime collaborator (and critic), my wife Ruth Anna Putnam. As a result, the present volume consists almost entirely of fairly recent work; in fact, only four of the thirty-six papers in this volume were written before the year 2000 and only twelve were written before 2005. Even so, the papers (and some of the footnotes I have added to previously published papers) show that my views on some issues continue to change or, as I prefer to think, improve.

This fact will, of course, confirm a certain image of me, the image of the philosopher who "changes his mind." I am not ashamed of that image; I have never wanted to be the sort of philosopher who pretends to have the final answer to all the big questions. On the other hand, I know that there are also real continuities in my work, and the reader can find my description of those continuities in Chapter 2 of this volume. Those continuities include an interest in the nature of objectivity in at least three important areas of human knowledge and action: science, mathematics, and ethics. Those were the topics on which I focused in my first collection of papers, *Mathematics, Matter and Method*, published by Cambridge University Press in 1975, and they continue to be at the center of my reflections.

As just mentioned, Chapter 2 describes why and how I "changed my mind" about the nature of the sort of "realism" that I defend (and also says where some critics have thought mistakenly that I changed my mind). My "internal realist," that is, antirealist, period ended in 1990 (although some of my critics seem not to have caught up with that change even now, more than twenty years later). When it happened, I explained that change of mind in several places, one of them being the Dewey Lectures that I gave at Columbia University in 1994. But, under the influence of a certain side of Wittgenstein's thought, I did not simply come out and say that I am (again) a realist in metaphysics. That is not something I was willing to say in public until I delivered the lecture titled "From Quantum Mechanics to Ethics and Back Again" in Dublin at a conference (belatedly) celebrating my eightieth birthday. Instead, I said that I was now in favor of "commonsense realism," a term to which, not surprisingly, no one could attach a clear sense, and, as a result, one to which many people attached whatever sense they wanted, including "internal realism." The reason I did not want to use the word "metaphysics" is that although I felt uncomfortable with Wittgenstein's "therapeutic" conception of philosophy for a long time, it was only about 2006 that I definitely repudiated that conception, first in e-mails to friends (this is the e-mail age, after all), and now in Chapter 28 of the present volume. (However, that chapter also emphasizes that I believe that there is much in Wittgenstein we can learn from; nothing is to be gained, here or anywhere, by throwing out the baby with the bathwater.)

In the same period in which this collection was being put together with the aid of Mario and David, I was working with Hilla Jacobson on a volume on the subject of perception (an area in which I have moved away from the "disjunctivist" position I defended in *The Threefold Cord: Mind, Body, and World*). That volume is still in the process of being written, but the reader can get an idea of my current thinking on the subject from Chapter 4 and from Chapters 35 and 36.

In addition to Mario De Caro, David Macarthur, and Hilla Jacobson, I have profited over the years in which these papers were being written from long philosophical conversations with many other friends and former students, including David Albert, Yemima Ben-Menahem, Ned Block, Stanley Cavell, James Conant, Charles Travis, and Steven Wagner. And, looking back more than sixty years, I continue to cherish my conversations (which began in 1953) with Paul Benacerraf and Rudolf Carnap, and, before that, with my *Doktorvater,* Hans Reichenbach. I have long since moved away from both Carnap's and Reichenbach's (different) versions of logical empiricism, but I continue to draw inspiration from Reichenbach's belief that the philosophical examination of the best contemporary and past science is of great philosophical importance, and from the example of Carnap's continual reexamination and criticism of his own past views, as well as from both Carnap's and Reichenbach's moral and political commitment.

Hilary Putnam:
Artisanal Polymath of Philosophy

MARIO DE CARO

DAVID MACARTHUR

Let us not pretend to doubt in philosophy what we do not doubt in our hearts.

—C. S. Peirce

Even the hugest telescope has to have an eye-piece no larger than the human eye.

—Ludwig Wittgenstein

A Retrospective Overview

The present volume collects the recent philosophical papers of Hilary Putnam, who, in an age that increasingly tends toward specialization, is a genuine Renaissance man of philosophy combining conceptual imagination, mathematical genius, scientific erudition, humanistic concerns, and moral vision. In this respect he deserves to be compared with Aristotle, Gottfried Leibniz, Immanuel Kant, John Stuart Mill, and Bertrand Russell. Putnam has justly been called "the history of recent philosophy in outline,"[1] but he also represents its possible future, as the present volume attests.

A striking feature of this new collection, the first in eighteen years, is that we find Putnam returning to some of his very first enthusiasms in philosophy, such as mathematical logic, philosophy of mathematics, and philosophy

---

1. John Passmore, *Recent Philosophers* (La Salle, Ill.: Open Court, 1988), 97.

of quantum mechanics. This return is to be welcomed because, as Alfred North Whitehead said, "Fundamental progress has to do with the reinterpretation of basic ideas."[2] We also find Putnam's latest reflections on the perennial problems of realism, the fact/value divide, skepticism, and naturalism.

Because the present volume represents the sixth collection of Putnam's philosophical papers, it is a fitting occasion to cast a retrospective eye over the long and winding road that Putnam has traveled in his various rethinkings and reconceivings of the problems of philosophy and what he considers to be the most fruitful responses to them. Putnam's notorious "changes of mind" are, first and foremost, evidence of a powerfully imaginative philosophical intelligence that is more concerned to manifest the virtues of curiosity, imagination, and honesty than the questionable virtue of intellectual steadfastness.[3] Nonetheless, philosophers reading Putnam often see these "changes" as requiring some apology, as though it was agreed on all sides that the proper role of a philosopher is to fix on some position and then, ever after, to defend it against all objections. In contrast to this legalistic conception of philosophy, Putnam's tendency to constantly reconsider the motivations or grounds or intelligibility of the problems and the (present) best replies to them is to be celebrated as exemplifying an open-minded, inclusive vision of philosophy, a form of *fallible democratic experimentalism,* to be judged by its fruits on a trial-and-error basis.

This conception represents Putnam's absorption of the insights of Dewey's approach to epistemology, which sees inquiry at its best as characterized by various ethical or political virtues, such as fair-mindedness, openness to criticism, and toleration of a wide range of alternative points of view. These democratic virtues bring to light a Platonic analogy between self and society, applying equally to one's own intelligent self-reflections and to the social realm of one's debates, exchanges, and collaborations with others.

2. Quoted in W. H. Auden and Louis Kronenberger, *The Viking Book of Aphorisms* (New York: Viking, 1966).

3. Sidney Morgenbesser is reputed to have joked perceptively: "Putnam's the quantum philosopher. You can't understand him and his position at the same time." The suggestion is that Putnam is such an advanced thinker that no one else can keep up with him. By the time we have understood his position he's "leaped" to a different one! In other words, he is always one or two steps ahead of us. The moral is that Putnam's philosophy is not fundamentally a matter of his position at any one time, but the path he has taken. A related suggestion of Paul Franks: Putnam is not best understood in terms of his "position" or "positions" but in terms of the dialectical tensions and complexity that express themselves through his "positions."

A word of warning: Deweyan experimentalism in philosophy should not be confused with the movement called "experimental philosophy," which sees philosophy as a generalized form of empirical inquiry.[4] Putnam does not follow W. V. Quine's naturalism, which collapses philosophy as a whole into the sciences. In contrast to Quine, Putnam is happy to acknowledge the existence and importance of conceptual truths in the relatively a priori business of conceptual inquiry. Putnam's approach to philosophy can be better appreciated by seeing that, as he puts it, "philosophical tasks are never really completed," and that "there are no last words in philosophy";[5] that is, we must always allow for the possibility of there being new insights, new perspectives, and new conceptual and empirical possibilities to take account of.

Apart from following the relationships among the many "Putnams" as represented by the many doctrines and movements of thought he has defended at one time or another, another difficulty in reading a Putnam paper is to understand what its dialectical point is. The smoothness and beguiling conversational tone of Putnam's writing often belie the complexity and nuance of his philosophical moves. Consider, for example, the well-known model-theoretic argument. This is not uncommonly read as a skeptical attack on the notion of reference *as such*. It is as if the argument demonstrates that because "one can 'Skolemize' absolutely everything," as Putnam puts it, we must conclude that "it seems to be absolutely impossible to fix a determinate reference (without appeal to nonnatural mental powers) for any term at all."[6]

This reading then seems to gain further credibility from the apparent alignment of Putnam with Quine's notorious inscrutability-of-reference thesis. So we find a recent commentator in the authoritative *Stanford Encyclopedia of Philosophy* writing, "Putnam's *general* goal in the model-theoretic argument is to show that our language is semantically indeterminate—that there's no fact of the matter as to what the terms and predicates of our language refer to."[7]

---

4. Such work includes Stephen Stich's comparison of the epistemic intuitions of college students in Western and Asian countries. See, e.g., Shaun Nichols, Stephen Stich, and Jonathan Weinberg, "Normativity and Epistemic Intuitions," *Philosophical Topics* 29 (2001): 429–460

5. See Chapters 18 and 22 in this volume.

6. Hilary Putnam, *Philosophical Papers,* vol. 3, *Realism and Reason* (Cambridge: Cambridge University Press, 1983), 16.

7. Timothy Bays, "Skolem's Paradox," in E. N. Zalta, ed., *The Stanford Encyclopedia of Philosophy,* Spring 2009 ed., http://plato.stanford.edu/archives/spr2009/entries/paradox -skolem/.

This stunning misreading, which effectively turns the Putnam of *Reason, Truth and History* into a skeptic about reference, is paradoxically the very opposite of Putnam's own position at the time, which was that the determinacy of reference (e.g., the fact that "cow" refers to cows) is "a logical truth."[8] By missing Putnam's dialectical strategy, the author completely misses Putnam's philosophical point.

Putnam's target was a metaphysical realist who, like Putnam himself at that time, believed in a veil or "interface" between the "mind/brain" of the speaker/hearer of a natural language and the world,[9] and who, again like Putnam himself (then and now), believed that the success of the physical sciences is explained by the fact that the terms used in their theories typically refer to various sorts of physical "objects" with which we causally interact. Although Putnam no longer accepts the model-theoretic argument, its purpose was, first, to show that this metatheoretical "explanation" of scientific success (including its satisfaction of various operational and theoretical constraints) is consistent with *antirealist* theories of truth and reference, and further, that it is the metaphysical realist's picture, not the antirealist's, that fails to explain the determinacy of reference.[10]

Thus the point of this argument was not to argue against *any* determinate notion of reference, as this commentator and others suppose, but only to argue that metaphysical realism has not furnished one. Because Putnam took it for granted (and, for that matter, still takes it for granted) that our terms enjoy determinate reference to things in the world, he took the argument as a whole to provide us with a reductio ad absurdum of metaphysical realism. He believed (and still believes) that given our situated perspective in the world, which crucially includes our cognitively engaged perceptual relatedness to

8. Hilary Putnam, *Meaning and the Moral Sciences* (London: Routledge, 1978), 136–137.

9. See, e.g., Hilary Putnam, "Computational Psychology and Interpretation Theory," in Putnam, *Philosophical Papers,* vol. 3, *Realism and Reason,* 139–154. There Putnam wrote, "The brain's 'understanding' of its own 'medium of computation and representation' consists in its possession of a verificationist semantics for the medium, i.e. of a computable predicate which can represent acceptability, or warranted assertibility or credibility." Ibid., 142. Note that this "understanding" was supposed to take place entirely inside the brain; that is, the account of understanding excluded transactions between the brain and the environment as irrelevant.

10. The endgame of this argument concerns the viability of the causal realist's (e.g., Devitt's) claim that a determinate reference relation is fixed by causation itself. Putnam regards this move as metaphysical pie in the sky. Cf. Hilary Putnam, *Words and Life,* ed. James Conant (Cambridge, Mass.: Harvard University Press, 1994), chap. 14.

external things, there is, in general, a fact of the matter about what our terms refer to.

The reader may therefore wonder why Putnam gave up the model-theoretic argument. There were two principal reasons: first, Putnam realized that the verificationist semantics (also known as "internal realism") that the argument defended was just as incompatible with the idea that we can refer to "epistemically ideal" conditions as it is with the idea that we can refer to external things;[11] and, second, he saw that the idea that in perception we are aware of "sense data" (or corresponding brain states) made the problem of access unsolvable and had to be given up.[12]

We have presented Putnam's "changes of mind" as manifesting various intellectual virtues and exemplifying a pragmatist conception of philosophy, but there is a sense in which the usual response to these changes is, in any case, somewhat of an overreaction. It is important to see that Putnam's shifts in position are not arbitrary or temperamental but represent sustained efforts at self-criticism—a philosophical ideal—that manifest underlying patterns or commonalities of intention, concern, and commitment. Tracing the outlines of these patterns, we discern five themes that provide a useful map to help the reader negotiate the relations between Putnam past and Putnam present, between the technical and the nontechnical, and between the theoretical and the practical in this complex and wide-ranging field of evolving discussions.

The five themes are these: (1) the sympathetic critique of logical positivism (including the work of the "honorary" positivist, Quine); (2) the enduring aspiration to be realistic about rational (or conceptual) normativity in philosophy; (3) antiessentialism about a range of central philosophical notions; (4) the reconciliation of fact and value and, more broadly, the reconciliation of the scientific worldview and humanistic traditions of thought; and (5) the movement from reductive scientific naturalism (in early Putnam) to liberal naturalism (in recent Putnam).

---

11. "If there is a problem how, without postulating some form of magic, we can have referential access to external things, there is an equal problem as to how we can have referential access . . . to 'sufficiently good epistemic conditions.'" Hilary Putnam, *The Threefold Cord: Mind, Body, and World* (New York: Columbia University Press, 1999), 19.

12. "My picture of our mental functioning was just the 'Cartesian cum materialist' picture, a picture on which it has to seem magical that we can have access to anything outside our 'inputs.'" Ibid.

Let us make some remarks about each theme, the point of which is not to put Putnam in a nutshell—which, in any case, would be an impossible task.[13] Our aim is to cast light on certain aspects of Putnam's work that are often overlooked or poorly understood in order to help the reader find his or her way in these powerful but sometimes elusive writings. And part of their elusiveness is that Putnam has no single or simple philosophical vision. He is, as we want to put it, a philosophical artisan who, like a highly skilled craftsman (e.g., a Japanese carpenter), has no all-purpose tool for every problem but chooses from a large array of tools the right tool for the job at hand. Even when the problems are more or less the same, that does not imply that the same "tool" is appropriate: "The fact that the difficulties are in a sense the same does not mean that they do not require special treatment in each case."[14] Putnam's methods and tools in philosophy are always, to some extent, purpose-built. When one is looking at the craftsman at work, it can be hard to see "the method in the madness," but you can trust Putnam that there always is.

The Critique of Logical Positivism

Logical positivism, understood broadly to include the writings of Quine,[15] was arguably the most important philosophical movement of twentieth-century Anglo-American philosophy. Because Putnam was a student of Hans Reichenbach and a colleague of Quine's, it is hardly surprising that many strains of his thought can be best appreciated as a kind of Socratic dialogue between Putnam and representative positions of logical positivism or updated versions of them. Consider, for example, his work on the nature and limits of a priori and analytic truth, the viability of a fact/value distinction, the failure of the project to formalize inductive logic, the failure of the project to unify the sciences, conceptual pluralism, and the phenomenon of conceptual relativity in theoretical discourses. Putnam's attitude to this philosophical legacy is best characterized by appeal to Hegel's notion of *Aufhebung* (overcoming through

---

13. In his teaching Putnam would say, "Any philosopher that can be put in a nutshell belongs there!" [DM].

14. See Chapter 23 in this volume.

15. Putnam has written, "I am inclined to class Quine as the last and the greatest of the logical positivists." Hilary Putnam, *Realism with a Human Face*, ed. James Conant (Cambridge, Mass.: Harvard University Press, 1990), 269.

critical sublation), a concept that perfectly captures Putnam's attempt to engage sympathetically with logical positivism in order to both criticize its shortcomings and actively recover and inherit its deepest insights.

Putnam's treatment of metaphysical questions in the present volume provides a telling case study to assess this complex dialectical relationship with the positivist tradition. Putnam follows the positivists in rejecting incoherent and irresponsible metaphysics (that is, metaphysics with no or too few links to what has weight in our lives), but he does not agree with them that this spells the end of metaphysics as such. Far from it. Putnam is the first to insist that the life of philosophy depends on encouraging vigorous metaphysical discussion, which is, of course, consistent with finding some metaphysical disputes relatively fruitless (e.g., ontology, especially in its recent analytic revival).[16] Certain metaphysical issues, such as those raised by critical reflection on contemporary physics (e.g., concerning the reality of its latest theoretical posits, or the correct interpretation of quantum mechanics, or the status of string theory), are, in fact, unavoidable. And, as we shall see, he even finds insights in metaphysical realism—a position that he has long criticized. But for all his sympathy with metaphysical discussion, Putnam wants to resist the metaphysical impulse to demand final once-and-for-all answers to such questions.

Furthermore, in those cases where metaphysics is found wanting, Putnam's nuanced account of occasion-sensitive sense making allows for a more sympathetic approach to its potential insights than the positivists' verificationist theory of meaningfulness, which flat-footedly equates metaphysics with cognitively meaningless pseudostatements—allowing them only an "emotive meaning." Turning the tables on positivism, Putnam is of the view that the positivists' own theory of meaningfulness in terms of conditions of empirical verification is metaphysical in spite of itself. Paradoxically, the deniers of metaphysics often turn out to be unwitting metaphysicians themselves.[17]

16. See Hilary Putnam, *Ethics without Ontology* (Cambridge, Mass.: Harvard University Press, 2004).

17. To give another example of this Putnamian insight: despite Hume's antimetaphysical ambitions, some of his antimetaphysical arguments depend crucially on a disputable metaphysical construal of experience that problematically equates experience with sensory impressions (or images) in the mind.

Putnam, too, was a verificationist during his internal realist period (1976–1990),[18] but his notion of verification was never simply taken over from the positivist tradition. Putnam's notion of verification-in-principle made allowance for a very liberal conception of "ideal" (or, simply, good enough) epistemic conditions for warranted assertion. It is also worth pointing out that Putnam's motivation for verificationism was not, as it was for the positivists, to formulate an empiricist theory of meaningfulness. Rather, his concern was to avoid positing metaphysically mysterious conceptions of truth or truth makers that made no genuine contact with the human practice of employing the term "true."[19]

A Realistic Attitude to Rational Normativity

Putnam is famous for endorsing, at different stages in his philosophical career, a diversity of "realisms," including scientific realism, metaphysical realism, internal realism, and commonsense (or pragmatic or natural) realism. To make matters even more complex, Putnam now distinguishes between being a "metaphysical realist" in the sense he gave to that term in his internal realist days and "being a realist in one's metaphysics" in a broader sense:

> As I explained "metaphysical realism," what it came to was precisely the denial of conceptual relativity. My "metaphysical realist" believed that a given thing or system of things can be described in exactly one way if the description is complete and correct, and that way is supposed to fix exactly one "ontology" and one "ideology" in Quine's sense of those words, that is, exactly one domain of individuals and one domain of predicates of those individuals. Thus it cannot be a matter of convention, as I have argued that it is, whether there are such individuals as mereological sums; either the "true" ontology includes mereological sums or it does not. And it cannot be a matter of convention, as I have argued that it is, whether spacetime points are individuals or mere limits.

18. Putnam first defended "internal realism" in 1976; he renounced it in his reply to Simon Blackburn at the Gifford Conference in his honor at St. Andrews in late 1990. This reply was published in Hilary Putnam, "Comments and Replies," in Bob Hale and Peter Clark, eds., *Reading Putnam* (Oxford: Blackwell, 1994), 242–295.

19. For Putnam's view about the relation between truth and verification and that our grasp of empirical concepts depends on our perceptual verification abilities, see Hilary Putnam, "Pragmatism," *Proceedings of the Aristotelian Society* 95, no. 3 (1995): 291–306.

To be sure, this is one form that metaphysical realism can take. But if we understand "metaphysical realist" more broadly, as applying to all philosophers who reject all forms of verificationism and all talk of our "making" the world, then I believe that it is perfectly possible to be a metaphysical realist in that sense and to accept the phenomenon I am calling "conceptual relativity."[20]

Given Putnam's concern to endorse various forms of "realism," one might be forgiven for thinking that he is obsessed about how best to capture philosophically some intuitive idea of a mind-independent reality, but that would be to misunderstand his motivations. To see this, consider Putnam's reaction to Michael Devitt's purely ontological formulation of metaphysical realism in terms of the mind-independence of various commonsense and scientific entities.[21] Putnam's response is to emphasize that in considering the realism-antirealism debates there is no avoiding the semantic issues of truth and reference, as Devitt attempts to do. To maintain his semantic agnosticism, Devitt is forced to put his faith in an undefined notion of mind independence that cannot do the work required of it. The moral for Putnam (who echoes Frege here) is that these issues must be conducted in a semantic key.

We do more justice to Putnam's thought if we see these changes from one form of realism to another as reflecting an underlying commonality of purpose, a lifelong meditation on, and attempt to articulate, the indispensable *normative* dimension of our actual practices of rational criticism, including our use of central normative terms, such as "truth," "justification," "reference," and "meaning," as they function in one or another domain of discourse.[22] This commitment lies behind some of Putnam's most famous ideas, including his Twin Earth thought experiment for semantic externalism, his indispensability arguments, and his defense of the stability of reference across theory change (e.g., Niels Bohr's use of *das Elektron* to refer to the very same things, namely, electrons, despite large shifts in his embedding theory). What this shows is

20. See Chapter 2 in this volume.

21 Michael Devitt, *Realism and Truth* (Princeton, N.J.: Princeton University Press, 1991).

22. It seems to us that this theme takes precedence over the theme of thought's intentional directedness to reality, which Maximillian de Gaynesford, *Hilary Putnam* (Chesham: Acumen, 2006), treats as Putnam's leading concern. There is, of course, a close relationship between the two, but only in those cases where talk of a "reality" makes sense. In any case, it is a mistake to think that there is a single "key" that unlocks Putnam's thought in its entirety.

the extent to which Putnam is a post-linguistic-turn philosopher for whom issues of truth, meaning, reference, and understanding (or sense) are at the heart of, and so put significant constraints on, all philosophical reflection.

One important exception to this treatment of Putnam's "realisms" is scientific realism, which is an ontological doctrine concerning the real existence of the "unobservable" theoretical entities that pull their weight in successful scientific explanations, those of physics in particular. The no-miracles argument that Putnam defends in Chapter 4 (and of which more later) is a causal argument that does not depend on substantial semantic assumptions.[23] Indeed, Putnam's commitment to scientific realism has remained constant before, during, and after his famous internal realist period.[24] We might say that scientific realism is a local ontological realism and that it was always held in concert with, and sometimes not clearly distinguished from, other more general forms of semantically inspired realism that Putnam advanced at different times.

It is worth remarking that Putnam's concern with core normative phenomena is itself shaped by a deep problematic involving normativity, a problematic that animates the following remark:

> Our understanding of our concepts and our employment of them in our richly conceptually structured lives are not mystery transactions with intangible objects, transactions with something over and above the objects that make up our bodies and our environments; yet as soon as one tries to take a normative notion like the understanding of a concept or Wittgenstein's notion of the use of a word and equate that notion with some notion from stimulus-response psychology ("being disposed to make certain responses to certain stimuli"), or a notion from computational psychology, or a notion from the physiology of the brain, then the normativity disappears, and hence the concept itself disappears.[25]

The concern of this passage, as well as a significant part of the motivation for Putnam's focus on "realism" of one sort or another, is an attempt to explain the objectivity that attaches to central normative notions, such as truth, justification, and understanding. The shifts in Putnam's realist allegiances attest to the difficulty of steering clear of the twin threats of the hyperboliza-

---

23. Although it is worth noting that causation is an intentional notion on Putnam's view. See, e.g., Putnam, *Threefold Cord,* 137–150.

24. See Chapters 2 and 3 in this volume.

25. See Chapter 23 in this volume.

tion or sublimation of objectivity in metaphysical thinking (e.g., the Platonic or mathematical realist notion of "intangible objects") and the denial or denigration of objectivity in skeptical thought (e.g., psychological or physicalist reductionism). The criticism of this oscillation between subliming and ridiculousness, as we might put it, is played out again and again in Putnam's writings, from his attacks on the metaphysical realists' notion of truth or reference (subliming) to his critiques of postmodern nihilism and Rorty's relativist conception of justification (ridiculousness). The moral is: these are not our only options, nor are they our best.

Putnam's major achievement here is to show that we do not need to accept the appearance of a forced choice between an inhuman objectivity ("the view from nowhere") and no objectivity at all ("nihilism" or "skepticism"). It might be noted in passing that this attests to an important difference between Putnam's neopragmatism and that of Rorty, who effectively accepts the appearance of the forced choice and opts for the second alternative: no objectivity.

Putnam's vision of objectivity is one of his most important contributions to philosophy, although it has not received the attention it deserves. Once again, it is easy to misunderstand Putnam's position when he writes, "We have . . . *better and worse* versions, and that is objectivity enough."[26] This might just seem to raise the question of objectivity all over again: Better or worse for whom? Better or worse in what respect?

To explain Putnam's thought here, it is helpful to employ the Rawlsian distinction between "concept" and "conceptions."[27] The concept of objectivity is what all conceptions of objectivity share in common, namely, the abstract idea that there are better or worse answers or responses to our questions. But this tells us very little unless we know what sense we are to attach to these terms and how to apply them. This is what we have conceptions of objectivity for. In the history of philosophy there have been many conceptions (pictures or models) of objectivity to provide the ideas of better and worse with more or less specific content and to guide us in to how to apply them. Some of the more important conceptions include (1) the account of objectivity in terms of objects, e.g., Plato's forms, metaphysical realist "objects," empiricist sense

---

26. Hilary Putnam, *The Many Faces of Realism* (La Salle, Ill.: Open Court, 1987), 77.

27. This distinction is taken, somewhat adapted, from John Rawls: "Roughly, the concept is the meaning of the term, while a particular conception includes as well the principles required to apply it." John Rawls, *Political Liberalism* (New York: Columbia University Press, 1993), 14 n. 15.

data, perceptual objects; (2) the Kantian account of objectivity in terms of principles or rules of judgment; and (3) the conventionalist account of objectivity in terms of conventions or intersubjective agreement, e.g., linguistic conventions. In Putnam's way of thinking, our conceptions of objectivity are plural and ever expanding, but we make a profound mistake if we treat any one of them as what objectivity *really* is. Much of his effort has been directed against the object-based account of objectivity (in the philosophy of mathematics and ethics, in particular),[28] but his general lesson is that each of these different conceptions applies well to some, but not to all, aspects of our lives.

Regarding the concept of objectivity, Putnam is trying to get us to see that the core idea of a better or worse that transcends the speaker is presupposed in our everyday practices of adjudicating disputes even if we spell out the notion of better or worse in different ways on different occasions. This is as true in science and mathematics as in such notoriously problematic areas as ethics and aesthetics, where reasonableness and rational argument do not guarantee agreement. One important task for the philosopher, on Putnam's account, is to show what particular conception of objectivity, and hence what conception of reason and argument, is appropriate to each of the problematic situations we actually confront in our lives. Once again, this is an example of the artisanal approach to philosophy at work.

It is worth noting, however, that Putnam's interest in conforming to our actual practices of judgment, criticism, translation, and so on is not a matter of attempting to exhaustively describe the ordinary use of terms in the manner of ordinary-language philosophers. Putnam reveals an important aspect of the influence of his positivist teachers in aiming for "mild rational reconstructions"[29] of our normative notions, ones that might diverge for theoretical reasons from everyday practice even if practice remains an important constraint on such constructions.

Antiessentialism

Antiessentialism about truth is a familiar doctrine from the writings of Rorty, as is antiessentialism about language in the writings of Wittgenstein. But it is too little appreciated that Putnam has extended this Wittgensteinian and

---

28. Putnam, *Ethics without Ontology*.
29. Putnam used this expression in conversation [DM].

pragmatist move to a range of central philosophical notions, including, in addition to truth and language, meaning, reference, knowledge, reason, objectivity, and (moral) goodness. The consequences of this radical antiessentialism ramify very widely in Putnam's thought. It is this aspect of his thinking that is to the fore in his curious and oft-repeated insistence that he is not offering a "theory" of some topic of philosophical interest. This remark is somewhat mystifying unless we realize that he is talking about the traditional notion of an essentialist theory, one that assumes that the phenomenon being theorized is fixed and substantially unified. Putnam's work, then, is to help deepen our understanding of the "rough ground" of the variegated and ever-extendable phenomena under investigation.[30]

For example, although Putnam accepts that the references of empirical terms have causal constraints, he has denied, contrary to what many critics have thought, that he is offering a *theory* of reference. This does not mean that Putnam is adopting a "quietist" conception of the role of philosophy in this area. It can be better understood as saying that reference has no essence, and hence there is no single thing that is *the* relation of reference, although there is, of course, a single predicate "refers" that we employ in various ways.[31] On Putnam's alternative picture there is an extendable family of uses of the term "refers" that are different in different cases depending on the "object" in question: referring to tables and chairs is a different thing from referring to numbers, which is different again from referring to subatomic particles, and so on.

The central importance of this radically antiessentialist perspective, however, comes more clearly into focus if we consider that it is brought to bear on meaning itself. This is an important point of alignment between Putnam's thought and that of Wittgenstein (and also that of Austin).[32] On Putnam's view, the radical implications of which have largely been missed or ignored (Charles Travis is a notable exception),[33] what a grammatically

30. The reference is to Ludwig Wittgenstein's remark, "Back to the rough ground." Ludwig Wittgenstein, *Philosophical Investigations* (1953; repr., Oxford: Blackwell, 1958), §107.

31. Putnam follows Peirce in thinking of reference as a triadic predicate: person P refers to object O by symbol S.

32. Of major philosophers, Putnam is one of the few to continue to turn to Austin for philosophical insight. Another is Cavell, with whom Putnam also shares interesting commonalities of outlook and purpose. For example, Cavell's definition of philosophy as "the education of grownups" is Putnam's definition of choice. See Chapter 32 in this volume.

33. Charles Travis, *The Uses of Sense* (Oxford: Oxford University Press, 1989).

well-formed sentence composed of standard English words means is not settled simply by dictionary definitions and grammatically well-formed construction—although Putnam is concerned to point out that knowledge of standard dictionary definitions is linguistically obligatory for competent speakers, concerns individual words, and forms the basis of the important field of lexical semantics.

To determine the meaning as employed by a speaker in some real-life situation—which, to aid clarity, Putnam calls the "sense" in contrast to the dictionary "meaning" of a word—depends on all sorts of features of the occasion of use: Who is speaking? To whom? After what? Under what circumstances? Put otherwise, the sentence by itself has no single well-defined meaning (or truth-condition) but an indefinitely extendable range of occasion-sensitive senses (or reasonable understandings).

One immediate consequence of this outlook is that the project of giving a theory of meaning, insofar as this assumes a fixed and unified subject matter that can be considered independently of the messy business of pragmatics, is a fantasy. So, too, for similar reasons, is the project of attempting to analyze the concept of, say, knowledge into necessary and sufficient conditions. A key aspect of Putnam's recent work here has been to reject a facile contextualism that tries to domesticate this Wittgensteinian approach to meaning by reinstituting the idea of a fixed and stable subject matter (a core semantic meaning or truth-condition) based on the supposition that there is a fixed set of contextually varying parameters that can be filled in as needed.[34] Putnam's dismissal of this business-as-usual response is to demonstrate, through detailed consideration of many examples, that there is no fixed set of parameters of the sort imagined; the aspects of things we must take account of in order to determine what we understand by our words on occasions of their use are variable, open ended, and ever extendable.

Reconciling the Scientific Worldview and the Humanistic Tradition

The attempt to overcome the positivist fact/value dichotomy and to reconcile the scientific image(s) of the world with those aspects of our lives denigrated

---

34. For example, this is the position of Michael Williams regarding the use of the term "knowledge." See Chapter 30 in this volume.

by the positivists as noncognitive or by Quine as second-class (e.g., ethics, aesthetics, religion, intentionality) marks Putnam as an inheritor of central teachings of classical pragmatism. Putnam has recently published works on these themes, so we will comment only briefly on them here.[35]

One point, however, is worth repeating. Putnam's denial that there is a unique and complete description of the world in some metaphysically privileged vocabulary (say, the language of the natural sciences) reflects his commitment to conceptual pluralism. For example, a chair can be usefully and truthfully described in the language of physics, of carpentry, of furniture design, or of etiquette without it being the case that these vocabularies are reducible to some favored or fundamental vocabulary.

One of Putnam's most important insights regarding the question of fact and value is to see that if one has a subjectivist attitude toward moral values, according to which they are incapable of genuine truth and justification,[36] then consistency dictates that one must adopt the same subjectivist attitude toward the cognitive values of consistency, reasonableness, simplicity, and the like. These values are presupposed by reason in the areas of science, epistemology, and logic that the metaphysician takes for granted. So if all values were subjective, then so, too, would be all the "facts." Putnam thus confronts his scientistic opponents with an acute dilemma: either concede his point or treat these paradigmatically cognitivist domains as noncognitive, effectively sawing off the branch on which they are sitting, because surely no one will argue that all discourses are noncognitive, incapable of genuine truth and justification.

The present volume presents Putnam's latest reflections in his long-standing attempt not to make an idol out of science and not to impose scientific (or mathematical or logical) modes of reason or clarity or knowledge on every area of our lives. It is significant also in having Putnam's most sustained reflections on contemporary physical theory, as well as more abstract issues concerning the crucial interpretive role of philosophy in the sciences. A key concern of Putnam's throughout his work is to argue for the unique importance of philosophy to the sciences to counter the baleful influence of leading scientists who themselves often betray a dismissive attitude toward

---

35. Hilary Putnam, *The Collapse of the Fact/Value Dichotomy and Other Essays* (Cambridge, Mass.: Harvard University Press, 2002), and Putnam, *Ethics without Ontology*.

36. As Putnam explains, this will typically be held on disputable metaphysical grounds.

philosophy.[37] Another key concern is to argue for the indispensable importance of philosophy to modern society in a time of academic specialization, which often loses touch with the broader human and cultural concerns that have always driven thinking people toward philosophical reflection in the first place.

The Movement from Scientific Naturalism to Liberal Naturalism

Putnam's earliest papers already reveal a commitment to scientific realism that remains in force today (more on this later). But they also reveal a commitment to a reductive or scientistic form of naturalism. Indeed, in a 1958 paper titled "Unity of Science as a Working Hypothesis" (cowritten with Paul Oppenheim),[38] Putnam famously supposed that all sciences could be reduced to the language of physics. But for several decades since then Putnam has been a staunch critic of this kind of scientism and its various expressions, such as the project to naturalize reason or intentionality and the procrustean imposition of scientific or mathematical models of rationality across the board. The tendency of scientism to attack normativity by denial or attempted elimination or reduction has led Putnam to say that "scientism is . . . one of the most dangerous contemporary intellectual tendencies."[39]

The present volume is the clearest and fullest expression of Putnam's commitment to a *liberal naturalism* that combats the scientistic tendency that often finds expression under the banner of the term "naturalism."[40] Liberal naturalism contests scientific naturalism, especially those narrow or reductive forms that recognize only the natural sciences (or sometimes only physics)

37. For example, in a newspaper article Richard Feynman—echoing Barnett Newman's quip about aesthetics—is reported by Steven Weinberg to have said, "Philosophy of science is to scientists what ornithology is to birds." Dennis Overbye, "Laws of Nature, Source Unknown," *New York Times,* December 18, 2007.

38. Hilary Putnam and Paul Oppenheim, "Unity of Science as a Working Hypothesis," in Herbert Feigl, Michael Scriven, and Grover Maxwell, eds., *Concepts, Theories, and the Mind-Body Problem,* Minnesota Studies in the Philosophy of Science, vol. 2 (Minneapolis: University of Minnesota Press, 1958), 3–36.

39. Putnam, *Philosophical Papers,* vol. 3, *Realism and Reason,* 211.

40. For elucidation and discussion of the philosophical promise of liberal naturalism, see the two collections edited by Mario De Caro and David Macarthur: *Naturalism in Question* (Cambridge, Mass.: Harvard University Press, 2004) and *Naturalism and Normativity* (New York: Columbia University Press, 2010).

as legitimate and irreducible. Liberal naturalism wants to allow for the conceptual possibility of nonscientific understanding and knowledge, but from a perspective that still earns the right to the title of naturalism insofar as it rejects all forms of supernaturalism and is centrally concerned to take the sciences, including the social sciences, seriously (which, for Putnam, involves extensive familiarity with the methods and results of contemporary scientific practice, e.g., quantum mechanics).[41]

Moral or aesthetic or religious understanding would all count as nonscientific on Putnam's view because they are not in the business of prediction and control on the basis of discerning causal patterns or laws in the world. Such forms of understanding also make ineliminable reference to a subjective or agential point of view for which there is no plausible scientific account in the offing. A slogan for Putnam's position might be: scientific realism without scientific imperialism.

Liberal naturalism also makes room for at least the possibility of admitting a nonscientific, nonsupernatural ontology, such as "values" or the "abstract objects" of mathematics. Many think that Putnam is committed to just such "entities" because of his well-known sympathy for Quine's indispensability argument, which reasons from the indispensability of mathematics in the physical sciences to a realist interpretation of mathematics. But there are two importantly different ways to interpret this realism: (1) ontologically, as a commitment to there being a domain of "objects" that provide referents for logically regimented terms in the discourse (this is Quine's way); or (2) semantically, as a commitment to there being full-fledged truths of the relevant sort (in this case, mathematical truths) without this automatically committing one to "intangible objects" (this is Putnam's way).

As has become clear in recent work, including that represented in the present volume, Putnam does not think that the objectivity of mathematics and ethics, to take two prime examples, requires positing a special domain of "objects" for each of these regions of thought and talk to be about.[42] It is enough that they are rational discourses that involve their own normative practices for establishing and justifying claims to truth even if it is not easy to say what it is to "think in the right way" in discourses such as ethics.

41. See Part I of this volume.
42. See, in particular, Putnam, *Ethics without Ontology.*

Another important aspect of Putnam's liberal naturalism is his advocacy of a realist attitude toward the sciences themselves; that is, he wants philosophers to adopt a broadly empirical attitude to the evident plurality and disunity of the sciences (including the human or social sciences) rather than being happy to rely on largely armchair speculation about what the sciences are or must be. Again, a large part of Putnam's mission here has been to try to overcome the hidden metaphysical legacy of the positivist attitudes to science, such as reductionism, antirealism, and monism.

The Present Volume

The Theoretical Face

One of Wittgenstein's key insights is that we typically (and often unwittingly) think about the complex phenomena of the world—such as people, science, language, politics, art, and religion—by comparison with schematic pictures or models. We want to apply this insight to Putnam himself and say that we can see many of his past and present works as responding to a picture of the successful sciences as providing an absolute, value-free conception of the world—a metaphysically sanctified set of "objects"—to underwrite privileged notions of truth and reference. We can fill this picture out a little more by seeing these "objects" as impinging on the mind causally, not cognitively. Criticisms of taking this picture as the literal truth (i.e., as a science-inspired metaphysics telling us how things must be) underlie a good many of Putnam's positions. Hence the appropriateness of our title, *Philosophy in an Age of Science.*

There are two further motivations for the title. One is the liberal naturalism canvassed earlier. Another is the acknowledgment that the greatest challenge to philosophy today is to explain what role philosophy, understood as a relatively distinct discipline of inquiry, can plausibly have in light of the great successes of modern science and the technologically dependent consumer society it has made possible. Logical positivists held that philosophy is essentially philosophy of the logic of scientific discourse. Putnam's more complex and less restrictive vision of philosophy sees philosophy as overlapping with the sciences but allows that there are areas of philosophy, such as its interpretive and critical dimensions, that distinguish it. Moreover, in addition to its theoretical face, Putnam recognizes a moral face of the subject. But the importance of science to philosophy today can be gleaned from this: that science and reflections on science still retain a fundamental role in discussions of both our theoretical and practical lives.

Surveying Putnamian themes—an artisanal approach to philosophy; a fallibilist antiskeptical epistemology; conceptual pluralism; conceptual relativity; direct realism in the philosophy of perception (which is part of a broader commonsense realism); the interdependency of facts and values and of facts and conventions; the reconciliation of (the philosophy of) science and ethics, including an attempt to do justice to the world of everyday life; a liberal naturalism; and so on—it is hard to escape the conclusion that Putnam should properly be regarded as one of the two founding fathers of neopragmatism. The other is Richard Rorty. Putnam's passionate debates with Rorty,[43] particularly over the right way to think about the normative dimension of, and connection between, truth and justification, should not obscure the great importance of their joint efforts in articulating a rejuvenated form of pragmatism for our time that has the power to absorb and build on the best insights of both analytic and continental traditions of philosophy—including, of course, the lessons of logical positivism. An early editorial decision was to leave aside Putnam's papers on the classical pragmatists for a separate volume, but it is important to see that all the papers collected here bear the stamp of Putnam's distinctive brand of neopragmatism.

The main reason Putnam refuses to call himself a pragmatist is that he considers classical pragmatism to have been a "form of verificationism."[44] A further reason is that this label tends to be associated with a view he has no sympathy with, a so-called pragmatist analysis of truth in terms of practical benefits; so-called because, as Putnam has helped make clear, none of the classical pragmatists took themselves to be offering an analysis of truth in terms of a set of necessary and sufficient conditions, despite what most commentators seem to think.[45] The classical pragmatists, like Putnam himself, were antiessentialists.

---

43. See, e.g., Hilary Putnam, "Richard Rorty on Reality and Justification," and Richard Rorty, "Reply to Putnam," in Robert Brandom, ed., *Rorty and His Critics* (Oxford: Blackwell, 2000), 81–87 and 87–89. Also, Hilary Putnam, "Realism with a Human Face," in *Realism with a Human Face*, 3–29, and Richard Rorty, "Putnam and the Relativist Menace," *Journal of Philosophy* 90, no. 9 (1993): 443–461.

44. See Putnam, "Pragmatism," 291.

45. For example, William James distinguishes his account from an analysis by calling it "a *genetic* theory of what is meant by 'truth'" (emphasis added). William James, *Pragmatism:*

To avoid the pejorative connotations of the term "pragmatism," the term "neopragmatism" is particularly fitting. It also reflects the fact that Putnam tends to see philosophical issues through the lens of a sophisticated practical or use-based philosophy of language that, despite treating semantic notions (e.g., truth, reference, content) as ineliminable and irreducible, does not see them as explaining the meaningfulness of our utterances independently of their use. Putnam also focuses philosophical attention on the agent point of view and the primacy of our practices in approaching intellectual conundrums. It is plausibly this very neopragmatism that answers Putnam's call for a renewal of philosophy after a retrograde period of speculative analytic ontology, paradigmatically represented by David Lewis's work on possible worlds and David Armstrong's work on universals.[46]

Putnam may celebrate the interpenetration of philosophy and metaphysical questions, but, like Peirce, he is a longtime critic of ontology understood as a metaphysical theory of the supposedly fixed, explanatorily basic categories of being. The artisanal approach to philosophy sets itself against allegedly all-purpose explanatory tools and ready-made answers (e.g., possible-world metaphysics, truth-maker theory, the absolute conception of the world). Sensitivity to the issues at hand and a certain unteachable good judgment learned through experience and practice are required by philosophy at its best no less than by doctors, carpenters, or concert pianists.

The Moral Face

In *The Critique of Pure Reason* Kant famously proposes that the interests of reason can be distilled into three questions: "What can I know?" "What must I do?" and "What may I hope?"[47] Putnam is almost unique in the analytic and continental traditions of philosophy in having an imaginative vision of philosophy large enough, and a range of detailed and technically informed responses to philosophical problems wide enough, to encompass and provide a guide in approaching these questions, as the present volume ably demon-

---

New Name for Some Old Ways of Thinking (1907; repr. Cambridge, Mass.: Harvard University Press, 1978), 29.

46. Hilary Putnam, "Replies," in "The Philosophy of Hilary Putnam," special issue, *Philosophical Topics* 20, no. 1 (1992): 347–408.

47. Immanuel Kant, *The Critique of Pure Reason,* transl. Norman Kent Smith (Houndmills: Macmillan, 1965), B237/A192, A805/B833.

strates. But without Kant's substantial but no longer credible notions of synthetic a priori truth and an autonomous transcendental subject, Putnam sees his relation to these fundamental questions differently than Kant does. Putnam does not attempt to answer these questions directly so much as help free our responses to them from popular misconceptions, such as skepticism, e.g., postmodernism, scientism; ethical monism in all its forms, e.g., dogmatic religion, reductive naturalism; and political cynicism (arising from, e.g., the modern corporate-driven media and traumatic historical events such as Auschwitz and the failure of communism).

Put otherwise, Putnam's conception of philosophy embodies an ethos of hope and an orientation toward the good. It is to Dewey that Putnam turns to help articulate this vision. In a remark Putnam applauds, Dewey writes:

> [Philosophy's] primary concern is to clarify, liberate and extend the goods which inhere in the naturally generated functions of experience. It has no call to create a world of "reality" *de novo,* nor to delve into secrets of Being hidden from common sense and science. It has no stock of information or body of knowledge peculiarly its own; if it does not always become ridiculous when it sets up as a rival of science, it is only because a particular philosopher happens to be also, as a human being, a prophetic man of science. Its business is to accept and to utilize for a purpose the best available knowledge of its own time and place. And this purpose is criticism of beliefs, institutions, customs, policies with respect to their bearing upon good.[48]

Here is the moral face of philosophy that Putnam endorses. Philosophy is not a higher authority of reason sitting in judgment on ordinary men and women. Nor it is a special esotericism with secret knowledge of a deeper reality beyond that of "common sense and science." Its job is, as we want to put it, the artisanal one of skillfully deploying and extending all that we have learned from past experience and practice to new problematic situations involving, among other things, "beliefs, institutions, customs, policies," in order to further the cause of the "good": not *the* Good—a fixed, unified, supernatural(?) value—but the many and various real-world goods, including the bringing into being of new goods.

---

48. John Dewey, *The Later Works: 1925–1953,* vol. 1, *1925: Experience and Nature,* ed. J. A. Boydston (originally published in 1925); citation is from the Collected Works edition (Carbondale: Southern Illinois University Press, 1981), 305. Putnam quotes this passage approvingly in Hilary Putnam, *Renewing Philosophy* (Cambridge, Mass.: Harvard University Press, 1992), 188.

One might wonder about the deference to Dewey in this context. James Conant has aptly remarked on Putnam's tendency, especially evident in his teaching practice, to present his latest views through the mouths of a changing pantheon of philosophical heroes.[49] This could strike one as false modesty, but it is not that. Philosophers of a certain persuasion say that philosophy is a matter of reading and responding to a particular set of texts, but what Putnam's lecturing and writing offer his readers is a sense of philosophy as an engagement with actual flesh-and-blood people, with all their actual cares, concerns, and hopes. The problems are brought down to earth, humanized, we might say, and the imaginative bedding of the problems is acknowledged: in Putnam's hands the temperament, tastes, and cultural outlook of the philosopher shine through the particular problems and the particular philosophers that he is interested in and how he approaches or reads them. This is a perfect illustration of Putnam's sense that philosophy is a way of life as much as it is a theoretical discipline, and that if he or she practices it properly, the philosopher can be an inspiration to others where it is the whole person that inspires us. That is demonstrably true of Putnam himself.

We shall now present brief descriptions of each of the parts of the present volume: (1) "On the Relations between Philosophy and Science"; (2) "Mathematics and Logic"; (3) "Values and Ethics"; (4) "Wittgenstein: Pro and Con"; (5) "The Problems and Pathos of Skepticism"; and (6) "Experience and Mind."

On the Relations between Philosophy and Science

This part involves further articulations of such important themes as the criticism of scientism in all its multifarious forms, the interpenetration of science and philosophy without the collapse of philosophy into science, the characterization of scientific realism, and the problems quantum mechanics raises for ontological commitment. Of particular interest is Putnam's continued endorsement of scientific realism because it reveals the importance of being clear about the potentially confusing ways in which the term "realism" is used across different philosophical debates.

Putnam has *always* subscribed to the doctrine of scientific realism, understood as the affirmation of the reality of the theoretical or "unobservable"

---

49. See James Conant, introduction to Putnam, *Realism with a Human Face,* xvi–xvii.

entities of our successful sciences, especially physics, e.g., the reality of electrons. Realism in this context means the denial of fictionalism about theoretical entities. In particular, Putnam's realism about the "unobservable" theoretical entities posited by science is a rejection of a positivistic instrumentalist
construal of, say, electron talk as highly derived talk about the behavior of
various scientific instruments, e.g., vapor trails in cloud chambers. However,
it is important to note that Putnam does not follow Quine in treating ordinary objects like tables and chairs as scientific "posits" akin to the theoretical
entities of physics. Commonsense objects like tables and books do not await
the blessing of science for their reality, nor can science denigrate their existence, although it may reveal surprising facts about them.[50]

What importantly changed over the long period of Putnam's allegiance
to scientific realism was his conception of the truth of statements about such
entities as electrons, quarks, and fields and the relation of such truths to matters of justification: from a correspondence theory of truth[51] to the identification of truth with idealized warranted assertability and to his present position that truth in some areas (e.g., physics),[52] but not all areas (e.g., not in
ethics),[53] can sometimes outrun what we have epistemic access to. Nonetheless, throughout all these changes, including the internal realist period, Putnam always accepted that the reality of, say, electrons was causally independent of minds. And he always accepted a thesis of logical independence: the
existence of electrons neither entails nor is entailed by the existence of minds.

Truths, even on the internal realist picture, were understood in terms of
idealized conditions of verification, which allowed for truths not actually
verified and for which there may be no humanly possible verification (e.g.,
truths about volcanic eruptions in the Mesozoic period). Even for the internal

50. In one of Putnam's thought experiments, what we call "cats" turn out, upon investigation, to be automata controlled by a man on Mars by tiny radio transmitters located in their
pineal glands. See Hilary Putnam, *Philosophical Papers,* vol. 1, *Mathematics, Matter and Method*
(Cambridge: Cambridge University Press, 1975), 238–239.

51. In Chapter 2 of this volume, Putnam remarks that the paper "Do True Assertions Correspond to Reality?" (the third chapter of *Philosophical Papers,* vol. 2, *Mind, Language and
Reality*, (Cambridge: Cambridge University Press, 1975), 70–84) "is one of the few places . . .
where I defended Metaphysical Realism."

52. Cf. Hilary Putnam, "When 'Evidence Transcendence' Is Not Malign: A Reply to
Crispin Wright," *Journal of Philosophy* 98, no. 11 (2001): 594–600.

53. Cf. Putnam, *Ethics without Ontology,* lecture 1.

realist, then, the world was supposed to be mind independent on a reasonable interpretation of that expression.[54]

At this point it is important to register a significant change in Putnam's outlook. Throughout the 1990s and for several years after that Putnam held that the metaphysical realist notion of "mind independence" had not been made sufficiently clear to render it coherent. On this recently outdated view, metaphysical realism is not so much false as senseless and thus not even a candidate for truth or falsity. But since about 2006 Putnam has come round to the view that the metaphysical realist's notion of mind independence can be made sufficiently clear with the aid of semantic vocabulary. Let us explain.

The verificationist (including the internal realist) thinks that statements that human beings could not possibly verify—such as "There are no intelligent extraterrestrials"—must be, if not false, lacking in truth-value altogether.[55] If there are no intelligent extraterrestrials, then, according to internal realism, the truth of that statement depends on whether *Homo sapiens* could know that there are none (if sufficiently well positioned in space-time and not systematically deceived in some way), and it is not true if it would not be verifiable under sufficiently good epistemic conditions.

Against this, the metaphysical realist believes that the statement is bivalently true or false.[56] According to Putnam today, that is a coherent view, and one he endorses. It is true that the metaphysical realist (as defined in "Realism and Reason"),[57] in addition to believing that truth is nonepistemic in this sense, denies the existence of equivalent descriptions (i.e., conceptual relativity), but Putnam now prefers to describe that as a mistake rather than an "incoherent view." What Putnam described as "incoherent" in "Realism and Reason" was precisely the idea that an epistemically ideal theory might be

---

54. However, in Chapter 3 of this volume, Putnam goes so far as to say, "However, if verificationist semantics is solipsistic, as I just argued it is, then this move turns the 'externalism' and 'anti-individualism' of 'The Meaning of "Meaning"' into a pseudoexternalism and a pseudo-anti-individualism, just as construing other people as logical constructs out of *my Elementarerlebnisse,* as Carnap did in the *Aufbau,* turns the social dimension of language into a pseudosocial dimension, a 'sociality' within a solipsistic world." But Putnam did not see, at the time, that that was what he was doing, because he believed that his reference to "epistemically ideal" conditions bestowed a realist aspect on internal realism.

55. See Putnam, "When 'Evidence Transcendence' Is Not Malign."

56. Here we are ignoring the problem of ordinary vagueness, e.g., the case where there are "borderline" intelligent extraterrestrials but no clear cases.

57. Cf. Putnam, *Meaning and the Moral Sciences.*

false. But in hindsight Putnam agrees with the metaphysical realist (in both senses) that an epistemically ideal theory might indeed be false.

What has remained unchanged throughout is Putnam's claim that the notion of mind independence that verificationists and antiverificationists quarrel about cannot be made clear without entering into issues belonging to philosophy of language and philosophy of mind, in particular into the question of what constitutes understanding a statement.

In this part Putnam also defends his no-miracles argument for scientific realism against various sympathetic critics, including Yemima Ben-Menahem, Arthur Fine, and Axel Mueller. To shed some valuable light on this doctrine, we will briefly discuss a recent criticism advanced by Jack Ritchie.[58] Ritchie reads both the indispensability argument and the no-miracles argument as forms of inference to the best explanation for the existence of certain problematic entities: abstract (nonspatiotemporal) entities and theoretical ("unobservable") entities, respectively. Ritchie claims that inference to the best explanation is a metaphysical argument that misrepresents how ontological commitment actually works in science, and that any plausible naturalist philosophy must hold to the ontological standards of scientific practitioners themselves.

On Ritchie's view, neither the no-miracles argument nor the indispensability argument has any force (at least for a serious methodological naturalist), because scientists do not in fact accept realist interpretations of theoretical or abstract entities on the sole basis of inference to the best explanation.[59] Some specific experimental evidence must also be available. According to Ritchie, for example, it was only after J. B. Perrin experimentally verified Einstein's views on the Brownian motion that the scientific community came to believe in the real existence of atoms.[60]

However, it is important to distinguish inference to the best explanation as used in science and as used in metaphysics. Because Ritchie runs these together, he ends up denying any role for inference to the best explanation in

58. Jack Ritchie, *Understanding Naturalism* (Durham: Acumen, 2009).

59. Ritchie tends to think of "best explanations" as making no reference to experience or experimental evidence. This is a symptom of his misguidedly treating inference to the best explanation as a purely metaphysical affair.

60. On this issue Ritchie refers to Penelope Maddy, *Naturalism in Mathematics* (Oxford: Clarendon Press, 1997).

science—but that is simply a mistake, as Putnam has long argued.[61] First, important scientists, such as Ludwig Boltzmann and Einstein (not to mention Galileo and Newton), posited atoms as the best explanation of phenomena before Perrin's experiments. Second, the history of science is filled with examples of appealing to inference to the best explanation to argue for the existence of some disputed entities, from the discovery of Neptune to Darwin's acceptance of some (at the time unknown) causal mechanisms and entities implied by natural selection.

It is also important to see that Ritchie assimilates two different arguments of Putnam's involving two different kinds of realism, namely, (1) the indispensability argument for the truth of mathematical statements ("realism" in one sense) and (2) the no-miracles argument for the reality—the real existence—of certain theoretical entities in science ("realism" in another sense). In other words, Ritchie fails to distinguish semantic realism from ontological realism. The no-miracles argument is a causal argument: its central claim is that if there were not these theoretical ("unobservable")[62] entities causing these phenomena, then that would be a miracle. There is no similar argument for the reality of mathematical entities. The considerations of the indispensability argument are not causal at all. Indeed, Putnam actually agrees with Ritchie's point that the best explanation of mathematical discourse and practice does not (contra Quine) require positing mathematical entities ("intangible objects").

Again the Putnamian lesson is: do not think that inference to the best explanation is an all-purpose tool with a single function (say, ontological commitment). There are too many different kinds of explanation (and hence too many notions of what "best explanation" comes to) to make that at all plausible.

Mathematics and Logic

In this part of the volume Putnam reviews the current discussion regarding the indispensability argument as it pertains to mathematics and shows how

61. For example, see Hilary Putnam, "What Theories Are Not" (1962), in Putnam, *Philosophical Papers,* vol. 1, *Mathematics, Matter and Method,* chap. 13. For a useful general discussion of the role of inference to the best explanation in science, see Peter Lipton, *Inference to the Best Explanation* (London: Routledge, 1991).

62. A Putnamian theme is that the instruments of science can enlarge our idea of what is observable; e.g., we can literally see charged particles in a cloud chamber. So the term "unobservable" in these debates is, at the very least, misleading.

the conclusion of the argument has often been misinterpreted. We are also witness to a significant change in Putnam's outlook regarding the question whether it is possible to make progress with problems in the philosophy of mathematics. The pessimistic flavor of Putnam's previous position can be gleaned from the title of a paper he published, "Philosophy of Mathematics: Why Nothing Works."[63] His thinking in this area is now characterized by newfound hopefulness, a breath of fresh air. In the papers collected here a crucial methodological step forward is that Putnam now sees significant continuities between the problems in the philosophy of mathematics and problems in other areas of philosophy, such as ethics; so the approaches to the latter problems, or analogues of them, become relevant to the philosophy of mathematics as well. Again, it is part of the artisanal approach that one needs mathematical know-how and experience to apply these strategies to the philosophy of mathematics.

Early in his career Putnam presented an argument for the thesis that mathematics should be taken as true under some interpretation (against, above all, formalists); an argument that was dependent on considerations internal to mathematics, for example, its coherence, fertility, and success. He then developed an argument based on considerations external to mathematics for the conclusion that mathematics is true on a realist interpretation (against intuitionists, operationalists, if-then-ists, and others). The latter argument is his famous indispensability argument:

1. Physics is true in a realist sense.
2. Mathematics is indispensable to physics.
3. The only way of accounting for the indispensable applications of mathematics to physics requires that mathematics be interpreted as true in a realist sense.
4. Thus mathematics is true in a realist sense.

"Realism" is to be understood throughout as implying that the statements of physics and mathematical statements are either true or false, where these notions are not to be understood in antirealist terms of verifiability or provability. Contra Mark Colyvan, the argument is not an ontological one and, in particular, does not imply any form of Platonism in the philosophy of

---

63. Putnam, *Words and Life,* chap. 28.

mathematics.[64] Consequently, many theoretical issues that are currently debated in connection with Colyvan's formulation of the argument miss their intended target.

Putnam also addresses several other objections raised against his argument, including the claims that no physical theory is a good candidate for truth, that this argument at most proves that *some* mathematical statements should be considered true, and that the argument has no force against Benacerraf's challenge: how can we have epistemic contact with causally inert mathematical entities? Additionally, there are Hartry Field's argument that mathematics is not indispensable for physics and Penelope Maddy's claim that working scientists do not in fact believe in the existence of all the entities posited by their best scientific theories.

Furthermore, in this part Putnam addresses the liar paradox (by discussing some difficulties in Tarski's and Parsons's proposed solutions), presents some theorems concerning set existence, gives a new proof of Craig's

---

64. See Mark Colyvan, "Indispensability Arguments in the Philosophy of Mathematics," in E. N. Zalta, ed., *The Stanford Encyclopedia of Philosophy* (Fall 2004 edition), http://Plato. stanford.edu/archives/fall2004/entries/mathphil-indis/. Colyvan's misunderstanding of Putnam's version of the indispensability argument may, to some extent, be explained by the fact that Putnam has not always been concerned to distinguish his version from Quine's. In his 1971 book *Philosophy of Logic* (New York: Harper and Row, 1971; reprinted in Putnam, *Philosophical Papers*, vol. 1, *Mathematics, Method and Mind,* chap. 20), Putnam appears to accept a Quinean way of putting the indispensability argument: that because mathematics is indispensable for physics, we must take with ontological seriousness our quantification over "abstract entities" (see especially 347). Indeed, as late as 1994, in "Rethinking Mathematical Necessity," in *Words and Life,* 245–263, Putnam writes:

> As I read this and similar passages in Quine's writings, the message seems to be that in the last analysis it is the utility of statements about mathematical entities for the prediction of sensory stimuli that justifies belief in their existence. The existence of numbers or sets becomes a hypothesis on Quine's view, one not dissimilar in kind from the existence of electrons, even if far, far better entrenched.
>
> It follows from this view that certain questions that can be raised about the existence of physical entities can also be raised about the existence of mathematical entities—questions of indispensability and questions of parsimony, in particular. These views of Quine's are views that I shared ever since I was a student (for a year) at Harvard in 1948–9, but, I must confess, they are views that I now want to criticize. (245–246)

The reader is advised to see Chapter 9 in this volume for a retrospective assessment of this complicated issue.

theorem about Ramsey sentences, and discusses Quine's underdetermination thesis. There is also a discussion of some interesting implications of Gödel's incompleteness theorems, including a criticism of Noam Chomsky's suggestion that all human linguistic and scientific competence can be represented by a Turing machine and a proof due to Kripke (but never published by him) of the Gödel theorem, as well as the theorem that Peano arithmetic has no consistent finitely axiomatizable extensions.

Values and Ethics

It is fair to say that Putnam's ethical writings are dominated by his many discussions and criticisms of the fact/value dichotomy. And there can be no doubt that this emphasis is well placed because the idea of a metaphysical gulf between "facts" and "values" is a critical assumption of many contemporary philosophers, as well as having almost acquired the status of being a matter of conventional wisdom. It is a striking example of the way metaphysical ideas can shape and even dictate our ethical, aesthetic, and political outlooks. And in our world today it is a powerful idea behind various forms of skepticism about the good of reason, argument, and intelligent criticism, from the crippling subjectivism behind such familiar "truisms" as "There is no disputing matters of taste" to the apparent paradigm of good sense, "Just the facts, Ma'am!" (as if to reason well is to make a point of leaving evaluative matters aside) and to the argument stopper "That is just a question of values" (as if pointing out the evaluative nature of an issue is enough to stop all intelligent discussion of it). No wonder, then, that Putnam is on a campaign. But as in any campaign, there are casualties along the way.

Criticism of the fact/value dichotomy, as vitally important as it is, can sometimes seem remote from current ethical debate within philosophy. A virtue of this part is to provide some balance to Putnam's moral philosophy through detailed engagement with the positions of prominent contemporary ethical theorists, such as Simon Blackburn, Ronald Dworkin, Jürgen Habermas, Ruth Anna Putnam, and Thomas Scanlon. As evidenced here, some of Putnam's finest writing is in criticism of his contemporaries.

For example, let us briefly consider Putnam's response to Scanlon's "contractarian" theory of ethics, which puts great weight on a collaborative desire to be governed by principles that are not reasonably rejectable. Putnam argues that this is but one of a number of basic interests of morality. On Putnam's view, it is a mistake to try to reconstruct ethics rationally on any single "foundation," as Scanlon is not alone in trying to do. Apart from the moral interest on which Scanlon focuses, there are others no less important, including respect for the humanity of others, e.g., Kant; equality of moral rights and responsibilities, e.g., Locke; compassion for the suffering of others, e.g., Levinas; and a concern to promote human well-being, e.g., Aristotle. Putnam is no moral foundationalist, nor is he morally utopian.[65] He is acutely aware that the moral interests he discerns can come into conflict, but he hopefully remarks, "I believe that on the whole and over time, promoting any one of them will require promoting the others."[66]

One might usefully compare Putnam's work in ethics with Wittgenstein's method of trying to achieve perspicuous overviews *(Übersichten)* or useful models of our concepts to help draw attention to aspects of the highly complex flux of our concepts in action. Putnam's theorizing here wants to do justice to the highly complex nature of our concepts of ethics and morality, which have inherited a vast and not readily integrable range of influences and ideas over the course of history.

In this part, within a Dewey-inspired application of experimentalist problem-solving techniques to our practical problems, we also have Putnam's most recent thoughts about such real-world problems as the potential of human cloning, the picture of the human in economic theorizing, the idea of a just war, and a sympathetic criticism of the pros and cons of Kantian ethics. An important background idea shaping these discussions is Amartya Sen's and Martha Nussbaum's capabilities approach to ethics, which is further elaborated here. One of the concerns of which Putnam wishes to persuade his readers is that "the activity of putting forward and discussing . . . 'moral images of the world' . . . seems to me *the* indispensable task of philosophy"[67]—a task that includes moral images of our community and its internal structure, for example, the family.

---

65. Another sense in which Putnam is not utopian in his ethical thought is that he does not put too much weight on the notion of the "reasonable." Not only can there be reasonable disagreements in ethics or morality (as Cavell thinks), but our best attempts to say what is morally at stake in real-life situations are open ended and capable of endless improvement—in part because of their depth and complexity. Putnam's essay "The Depths and Shallows of Experience" (Chapter 33 of this volume) appropriates the Kantian notion of "indeterminate concepts" to characterize the ever-not-quite aspect of our moral concepts and descriptions of the moral life.

66. See Chapter 19 of this volume.

67. Putnam, "Replies," 377.

Wittgenstein: Pro and Con

One of the most interesting recent developments in Putnam's thought, as evidenced by the papers in this part, is a movement beyond a period of close alignment with the philosophy of Wittgenstein and the "new Wittgenstein"[68] reading of it. In particular, Putnam today is much more tolerant of metaphysics than the new Wittgenstein; is not happy to say that philosophers often talk nonsense in any interesting "grammatical" sense (although what they say may still lack full intelligibility); and is not convinced that traditional philosophy is an illness for which the right treatment is a kind of therapy.

To approach Putnam's complex relation to Wittgenstein, it is important to keep clearly in mind that it has never been helpful to regard Putnam as a Wittgensteinian,[69] a Kantian,[70] or, for that matter, a follower of any philosophical movement that takes a famous philosopher for its title. Indeed, given what we have called his artisanal approach to philosophy, there is a sense in which it is quixotic (except for local or polemic reasons, as in the case of neopragmatism) to expect to gain much understanding from attempting to fit Putnam under any philosophical rubric ending in "ian" or "ism."

Even if he is not a Wittgensteinian, Putnam retains strong sympathies for Wittgensteinian themes concerning language and mind, especially Wittgenstein's occasion-sensitive conception of sense making and his socially rooted world-involving conception of mindedness. But these essays also reveal that Putnam has serious reservations about Wittgenstein's philosophy of mathematics and his thesis-free vision of philosophy.

Putnam wants, first of all, to follow Cora Diamond's pathbreaking work in helping free Wittgenstein from a misguided antirealist reading that is widely

68. See Chapter 28 in this volume.

69. It is worth noting that some of Putnam's early papers were devoted to criticizing once-fashionable neo-Wittgensteinian views of meaning, e.g., Norman Malcolm's on "dreams." Putnam has also remarked, "I do not describe myself as a 'Wittgensteinian.' In part this is because I do not like sects in philosophy, and I do not like treating mere mortals as divinities." "Wittgenstein and the Real Numbers," Chapter 26 in this volume.

70. Putnam has written: "I believe there is much insight in Kant's critical philosophy, insight that we can inherit and restate; but Kant's 'transcendental idealism' is no part of that insight" ("Replies," 366). This is an important comment, especially in light of McDowell's recent sympathetic interpretation and appropriation of transcendental idealism. Cf. John McDowell, *Having the World in View* (Cambridge, Mass: Harvard University Press, 2004).

popular in the secondary literature, one that often simply takes for granted that the relationship between the *Tractatus* ("early Wittgenstein") and the *Philosophical Investigations* ("later Wittgenstein") is to be explained as a movement from a realist to an antirealist attitude to semantics.[71] This is consonant with a good deal of Putnam's writing in the 1990s, which is involved in the attempt to rescue Wittgenstein from his interpreters.[72]

Notwithstanding, Putnam cannot rescue Wittgenstein entirely from a taint of antirealism in the philosophy of mathematics. For example, Wittgenstein in the late 1930s and into the mid-1940s held that a mathematical proposition cannot be true unless we can decide its truth on the basis of a proof or calculation. But, Putnam argues, this seems to conflict with our ordinary commonsense realism with respect to number theory. Putnam is also troubled by Wittgenstein's rejection of set theory and his remarks on the law of the excluded middle. In these and other cases Wittgenstein's antitheoretical stance lapses, and Putnam finds him making disputable theoretical claims rather than following his professed aim of exploring the grammar of mathematical statements—although it should be noted that Putnam now is less convinced that we have a clear sense of what "grammar" is supposed to be. Such lapses, Putnam argues, are ultimately traceable to Wittgenstein's relative unfamiliarity with advanced mathematical practice and mathematical physics.

Another significant interpretive issue is Putnam's attempt to rebut a widespread view that Wittgenstein is an end-of-philosophy philosopher. Wittgenstein is not uncommonly read (by, say, Burton Dreben,[73] Paul Horwich,[74] and Rupert Read[75]) as simply engaged in the negative task of showing philosophers that they have, in their philosophizing, strayed into nonsensical expressions, primarily by attempting to give their expressions a metaphysical employment. His aim, on this view, is exhausted by showing the fly the way

71. Well-known versions of this traditional reading can be found in Michael Dummett, *The Logical Basis of Metaphysics* (Cambridge, Mass.: Harvard University Press, 1991), and Saul Kripke, *Wittgenstein on Rules and Private Language* (Cambridge, Mass.: Harvard University Press, 1982).

72. Putnam, *Renewing Philosophy* and *Words and Life.*

73. Dreben's view was presented in his legendary "Dreben-ars" at Boston University in the 1990s [DM].

74. Paul Horwich, *From a Deflationary Point of View* (Oxford: Oxford University Press, 2005).

75. Rupert Read, "'Discussion': A No-Theory Theory? Dan Hutto, *Wittgenstein: Neither Theory nor Therapy*," *Philosophical Investigations* 29 (2006): 73–81.

out of the fly bottle. Here Putnam demurs. Wittgenstein is not just a negative philosopher who equates philosophizing with various pathologies of reason; but to the extent that he is, and insofar as he thinks of metaphysics as an intellectual disease, Putnam is unsympathetic.

Fortunately, on the positive side, Wittgenstein "calls our attention to the plurality of our language games and the plurality of the forms of life that are interwoven with those games."[76] Furthermore, he is concerned to clear up confusions outside philosophy, for example, in science and mathematics. That is, he is interested in giving us a clearer sense of the life we lead with our concepts in philosophical and everyday settings.

The Problems and Pathos of Skepticism

Putnam's response to skepticism is significant for several reasons, not the least of which is that it provides an interesting vantage point from which to assess his relation to the pragmatist tradition. Putnam follows James, Peirce, and Dewey in thinking that if we are to take doubt seriously, we require an answer to the question, "What positive reason do we have to doubt in this case?" The mere possibility of doubt does not produce a genuine or living doubt, as Peirce famously argued. But Putnam goes well beyond this familiar pragmatist line of thinking in developing a sophisticated semantic criticism of philosophical skepticism.

Putnam's discussions of the writings of Stanley Cavell, P. F. Strawson, and Barry Stroud on skepticism reveal, once more, the power of the artisanal approach, which is on display in Putnam's careful and penetrating diagnostic thinking. For Putnam, as for Cavell, skepticism is seen as an attack on the everyday lifeworld;[77] and, at the same time, it is a deeply human tendency of thought that can never be eliminated once and for all. That is part of its pathos. But even if there is no eradicating skepticism in general, Putnam shows that particular

---

76. Chapter 28 in this volume.

77. Putnam's conception of "common sense" and Cavell's conception of the "ordinary" should not be assumed to be the same, or to be responding to precisely the same concerns. To give a preliminary characterization, we might say that Putnam is more concerned with our cognitively charged perceptual relation to middle-sized objects (e.g., tables, houses), whereas Cavell focuses attention on our general non-epistemic relation to things and on articulating the overlooked realm of everyday actions like eating breakfast or walking down the street—actions that are neither free nor unfree, voluntary nor involuntary, justified nor unjustified.

formulations of skepticism can be undermined on a case-by-case basis. This involves a painstaking demonstration of the subtle incoherence of skeptical doubt that must, if it is to be convincing, do justice to the specific words used in the formulation of doubt and our prima facie sense of their initial intelligibility.

One of the overarching themes of these essays is that skepticism depends on an untenable self-standing (or internalist) conception of semantics according to which words (and the thoughts that they are used to express) have their meanings quite independently of external conditions (e.g., features of the world or social norms of linguistic practice). Following the path blazed by Wittgenstein, Austin, Cavell, and Travis, Putnam demonstrates the philosophical power of a world-involving, occasion-sensitive approach to meaning and understanding that repeatedly finds the skeptic subtly stripping his or her words of the conditions of their full intelligibility or making conflicting demands on language that cannot be brought into a coherent harmony.

Stroud's derivation of external-world skepticism and Hume's derivation of the problem of induction are revealed to be not unanswerable problems but senseless conundrums that have no solution of the sort philosophers have long craved.

Experience and Mind

The papers collected in this part are representative of Putnam's recent thinking about the mind after having given up computational functionalism, the view that models the mind as the software running on the hardware of a computer—one of his most influential ideas and one that still has many defenders worldwide.[78] Of course, Putnam remains a functionalist in a broad sense because he continues to understand the mind in terms of its "functions," both internal and external, although these are not now characterized in computational terms. Putnam's current conception of the mind as a "structured system of object-involving abilities,"[79] however, builds on his long-standing commitment to semantic externalism by taking seriously that there is no interface between the mind and the world in perception or conception. Putnam's slogan "Meanings ain't in the head" now widens to become "The mind ain't in the head."

In this part Putnam presents arguments for the enduring importance of direct realism in the philosophy of perception. In part, Putnam's claim is his-

---

78. Cf. William Lycan, *Mind and Cognition* (Oxford: Blackwell, 1990), pt. 1.
79. Hilary Putnam, "Replies," in "The Philosophy of Hilary Putnam," 356.

torical, arguing that Aristotle should be regarded as a direct realist against the contrary view of Victor Caston.[80] In part, it is a matter of arguing for a better interpretation of direct realism under the name of "transactionalism," which involves discerning the insights and oversights of disjunctivism (e.g., John McDowell), intentionalism (e.g., Fred Dretske), and phenomenism (e.g., Ned Block). Both intentionalists and disjunctivists fail to do justice to the subjective character of perceptual experience (e.g., there are differences in what we regard as "pure" colors, and objects may have different "looks" depending on our visual system), although they are right not to think of such experience as nothing but "mental paint."

Coda

Iris Murdoch once wrote, "It is a *task* to come to see the world as it is."[81] Hilary Putnam has been carrying on this task courageously, longer, and with more philosophical tools and abilities at his disposal than any other contemporary philosopher. These are the fruits of a great mind. Enjoy!

For assistance with various aspects of the Introduction we would like to thank Massimo Dell'Utri, Paul Franks, Andrea Sereni, and Stephen White. Stephen, once again, deserves our gratitude for his encouragement and advice. The book was a collaborative effort, with Hilary Putnam playing a leading role. He generously made himself available to review our editorial decisions, offer suggestions, and provide whatever help we needed. We are especially grateful for his penetrating remarks on early drafts of the Introduction and, of course, for his good-humored enthusiasm for the project, and our work, throughout. We also want to warmly acknowledge Lindsay Waters and his editorial assistants at Harvard University Press for their friendly efficiency and skill in helping negotiate the thousand and one issues that arise in seeing a work such as this into print in its present form.

Warm greetings to our children, Gaia, Tulia and Thea who, without knowing it, played a vital role in this project!

80. Victor Caston, "Why Aristotle Needs Imagination," *Phronesis* 41, no. 1 (1996): 20–55.
81. Iris Murdoch, *The Sovereignty of Good* (London: Routledge, 1970), 89.

PART ONE

On the Relations between Philosophy and Science

I

Science and Philosophy

Why Do We Need Philosophy?

I want to begin by considering why and how the very need for philosophy became a question (in our times, at least).

A reason often given for the contemporary question concerning the need of or role for philosophy is that philosophy for so long—from the Middle Ages until the end of the nineteenth century, in fact—was heavily invested in two "ontotheological" (i.e., metaphysical-cum-theological) ideas, namely, (1) the idea of God (although the "God of the philosophers" was always very different from the God of the celebrated "man—or woman—on the street"),[1] and (2) the idea of the immateriality of the soul.[2] Although most nineteenth-century scientists were still churchgoers, the posture of twentieth-century (and now twenty-first-century) science has been decidedly secular, and analytic

1. The God of the philosopher is very different from the God of the ordinary believer, whether the philosopher be, for example, Aquinas, or Spinoza, or Kant, or Hegel.

2. Spinoza was an exception here. For Kant, however, the immaterial soul existed only in the noumenal realm, as the "transcendental unity of apperception"; and Hegel characteristically wanted matter and soul to be in some sense identical and also nonidentical.

philosophy, for the most part, has the same posture. That is not to say that there are not scientists or analytic philosophers who are religious. But the idea of God as an entity we need to postulate in order to account for the existence of the natural world or as a foundation for morality is no longer widely accepted. Indeed, although I am a practicing Jew, I myself do not believe either in "onto-theology" or in the idea that ethics requires a religious "foundation." In fact, in my *Jewish Philosophy as a Guide to Life: Rosenzweig, Buber, Levinas, Wittgenstein,* I describe my current religious standpoint as "somewhere between John Dewey in *A Common Faith* and Martin Buber."[3] I understand Dewey to be saying that the kind of reality God has is the reality of an ideal, and that is the way I, too, conceive of God. But that is *not* the subject of this chapter.

For the large number of people, including many religious people, who have lost the belief in the God of ontotheology, the metaphysical First Cause and *ens necessarium,* and now find themselves unable to believe in the God of the unreflective believer or in life after death, it *can* seem that the whole of philosophy between, say, Plato and Hegel was a vast mistake. That is what Heidegger's proclamation of the end of "ontotheology" seemed to amount to; and Wittgenstein's description of the new task of philosophy as "showing the fly the way out of the fly-bottle" has often been read as pronouncing a similar verdict on traditional philosophy.[4] Moreover, not only has belief in God, as God was traditionally conceived, ceased to be something that educated people take for granted in the West, but the successes of evolutionary biology, genetics, computer science, and brain science have demolished the idea that the "seat" of our mental faculties must be an immaterial soul. In addition, many natural scientists are hostile to traditional philosophy and are glad to dismiss it as obsolete. (Nor are they particularly friendly to analytic philosophy, which they often regard as scientifically uninformed hairsplitting.)

In sum, even if the enormous prestige and the enormous success of science have not jeopardized the position of the humanities as a whole as important disciplines, they have called into question the raison d'être of philosophy. Is philosophy really just a relic from past ages that we need either to discard or to replace with something else, even if we disguise the fact that the latter is what we are doing by retaining the *word* "philosophy"?

3. Hilary Putnam, *Jewish Philosophy as a Guide to Life: Rosenzweig, Buber, Levinas, Wittgenstein* (Bloomington: Indiana University Press, 2008), 100.

4. Ludwig Wittgenstein, *Philosophical Investigations* (Oxford: Blackwell, 1953), § 309.

Logical Positivism as a Failed Response to the Question of Philosophy's Function

Logical positivism was a product of the combined influence of the great physicist Ernst Mach's version of empiricism and of Bertrand Russell's belief that in the new logic that he and Whitehead did so much to perfect, a tool had been discovered that would either solve or dissolve the traditional problems of philosophy once and for all. It was thus a decidedly twentieth-century movement. In fact, Carnap's first great book, *The Logical Construction of the World,* was not published until 1928. In that book there is still an interest in epistemology, and Husserl is even cited as a forerunner.[5] But in 1934 Carnap wrote the following words: "All statements belonging to Metaphysics, regulative Ethics, and (metaphysical) Epistemology have this defect, are in fact unverifiable and, therefore, unscientific. In the Viennese Circle, we are accustomed to describe such statements as nonsense."[6]

In these words we have the essence of logical positivism. In fact, they express in an extremely condensed way the two chief principles of the movement: (1) The meaningful statements in our language are exactly the statements that can be tested and "verified" (established to be true or false by the methods of science).[7] All other statements are *nonsense*—totally devoid of discussable content. (2) Every one of the fields into which traditional philosophy was divided: metaphysics, ethics, and even epistemology—referred to in this passage as "(metaphysical) Epistemology"—must be abandoned because they consist entirely of such "nonsense."

So what is left for philosophers to do? How could the logical positivists continue to teach in philosophy departments, train graduate students, publish

---

5. There are references to Husserl and Husserl's student Becker in Rudolf Carnap's *Der logische Aufbau der Welt* (Leipzig: Felix Meiner Verlag, 1928), as well as uses of the Husserlian term "epoché," in the original edition, most of which were removed by the translator in the English (Berkeley: California University Press, 1967) edition. See Pawel Przywara, "Husserl's and Carnap's Theories of Space" (2006), http://philsci-archive.pitt.edu/archive/00002858/.

6. Rudolf Carnap, *The Unity of Science* (London: Kegan Paul, Trench, Trubner and Co., 1934), 26–27.

7. Later this was softened to the requirement that meaningful statements must be "confirmable or disconfirmable," and still later by still more complicated requirements. For a brief history, see Hilary Putnam, *The Collapse of the Fact/Value Dichotomy* (Cambridge, Mass.: Harvard University Press, 2002), chapter 1.

in philosophy journals, and even create new philosophy journals? The answer Carnap went on to give, especially in his 1935 book *The Logical Syntax of Language,* was that they were not really doing what had traditionally been *called* "philosophy" at all; they were engaged in studying the "logic of science."

This might lead one to expect that Carnap must then have proceeded to prove a number of results in a scientific field called by that name. But one would be wrong. In all the books and papers that Carnap wrote—books and papers that are rightly prized as brilliant contributions to important philosophical debates—there is, to be blunt, not one contribution to anything that anyone today recognizes as logic of science. There are attempts to "reduce" nonobservational terms in science (e.g., "atom," "gene," "gravitational field tensor") to observational terms (an attempt that was a total failure, as Carnap came to recognize). There was an attempt in *The Logical Syntax of Language* itself to provide a syntactic characterization of "analytic" sentences (which category was supposed to include all the sentences of mathematics)—another failure. There were, as mentioned, brilliant contributions to philosophical debates, particularly debates in the philosophy of language and the philosophy of mathematics. Finally, there was an attempt to formalize inductive logic, with results universally regarded as disappointing. In sum, to the extent that Carnap achieved anything important and impressive, it was in philosophy and not in logic of science, and to the extent that he tried to obtain impressive and important results in logic of science, he met with complete failure. Carnap and his followers did not succeed in folding philosophy into science, as they hoped to do.

Postmodernism as Another Failed Response

If logical positivism sought to counter the perceived danger to philosophy from science, the danger that science had rendered philosophy *obsolete*, postmodernism sought to restore the prestige of philosophy by retorting that science itself, and, indeed, everything we think of as a description of "facts," is just a form of fiction—useful fiction, to be sure, but in the end just more of the many webs of ideology Western culture keeps spinning. It is not, I hasten to add, that the postmodernists thought that there is some other kind of discourse that is free of deceit. Rather, at least in the hands of Jacques Derrida and his followers (but to a greater or lesser extent in the hands of the other gurus of the movement as well), discourse itself is seen as *inherently* deceptive, if only because it tempts us to believe that there is such a thing as a truthful representation of reality, and there is not and cannot be any such thing. Or so postmodernists claim.

One might think that such a revolutionary claim would be backed by strong arguments. In fact, even Richard Rorty, the most intelligent analytically trained philosopher to be won over to the movement, described "a lot" of Derrida's arguments as "just awful."[8] When he defended postmodernist views, Rorty did so by adapting arguments from analytic philosophers (not one of whom subscribed to Rorty's use of their arguments, however!), namely, Sellars, Quine, Davidson, and myself. But, rather than discuss this can of worms here, let me simply say that "representationalism"—the view that we can and often do succeed in *representing* parts and aspects of reality in language—is just what used to be called "realism," and realism does not become a terrible fallacy to be looked down on with scorn just because a number of professors have given it the new name "representationalism" and *declared* it to be such.

The Importance and Value of Philosophy

If neither the positivist response (fold philosophy into science) nor the "postmodernist" response (declare science to be fiction) is tenable, what remains? Looking back over my six decades of "doing philosophy," I find that two definitions of philosophy appeal to me the most, and also that each definition must be supplemented by the other.

The first definition comes from Stanley Cavell's famous book, *The Claim of Reason,* and I should like to quote from the passage in which it occurs at some length:[9]

> But if the child, little or big, asks me: "Why do we eat animals? or Why are some people poor and others rich? or What is God? or Why do I have to go to school? or Do you love black people as much as white people? or Who owns the land? or Why is there anything at all? or How did God get here?"

8. "Searle is, I think, right in saying that a lot of Derrida's arguments (not to mention some of Nietzsche's) are just awful." Richard Rorty, *Philosophical Papers,* vol. 2, *Essays on Heidegger and Others* (Cambridge: Cambridge University Press, 1991), 93n.

9. Stanley Cavell, *The Claim of Reason* (Oxford: Clarendon Press, 1979), 125.

I may feel my answers thin, I may feel run out of reasons without being willing to say, "This is what I do" . . . and honor that.

Then I may feel that my foregone conclusions were never conclusions *I* had arrived at, but were merely imbibed by me, merely conventional. I may blunt that realization through hypocrisy or cynicism or bullying. But I may take the occasion to throw myself back on my culture, and ask why we do what we do, judge as we judge, how we have arrived at these crossroads. . . . In philosophizing I have to bring my own language and life into imagination. What I require is a convening of my culture's criteria, in order to confront them with my words and life as I pursue them and as I may imagine them; and at the same time to confront my words and life with the life my culture's words may imagine for me: to confront the culture with itself along the lines in which it meets in me.

This seems to me a task that warrants the name of philosophy. . . . In this light, philosophy becomes the education of grownups.[10]

The second definition comes from Wilfrid Sellars's essay "Philosophy and the Scientific Image of Man."[11] "The aim of philosophy," he wrote, "is to understand how things in the broadest possible sense of the term hang together in the broadest possible sense of the term."

Cavell's definition, "the education of grownups," and the examples he gives of the "child's" questions that may show the "grownup" that he or she needs education, point to what I will call the *moral* face of philosophy: the face that interrogates our lives and our cultures as they have been up to now, and that challenges us to reform both. Sellars's definition points to what I will call the *theoretical* face of philosophy, the face that asks us to clarify what we think we know and to work out how it all "hangs together." Indeed, Cavell's definition comprehends Sellars's, because a grownup who does not care whether his view of things hangs together hardly counts as "educated."

---

10. For a discussion of this view, see also Hilary Putnam, "Philosophy as the Education of Grownups: Stanley Cavell and Skepticism," in Alice Crary and Sanford Shieh, eds., *Reading Cavell* (London and New York: Routledge, 2006), 117–128; reprinted in this volume as Chapter 32.

11. Wilfrid Sellars, "Philosophy and the Scientific Image of Man," in Robert Colodny, ed., *Frontiers of Science and Philosophy* (Pittsburgh, Pa.: University of Pittsburgh Press, 1962); reprinted in *Science, Perception and Reality* (London: Routledge and Kegan Paul; New York: Humanities Press, 1963; reissued, Atascadero, Calif.: Ridgeview Publishing Co., 1991), 37.

Logical positivism, in a limited way, sought to preserve the theoretical face of philosophy (albeit dismissing many important theoretical issues as "metaphysical") while banishing the moral face entirely, while postmodernism wants to preserve the moral face of philosophy—albeit often reducing it to what Richard Rorty once described as "the hallucinatory effects of Marxism, and of the post-Marxist combination of De Man and Foucault currently being smoked by the American Cultural Left"[12]—at the expense of the theoretical face.

My own view is that philosophy at its best has always, in every period, included some philosophers who brilliantly represent the moral face of the subject and some philosophers who brilliantly represent the theoretical face, as well as some geniuses whose insights span and unite both sides of the subject. To renounce either the moral ambitions of philosophy or its theoretical ambitions is not just to kill the subject of philosophy; it is to commit intellectual and spiritual suicide.[13]

Does Science Need Philosophy?

That philosophy is not to be identified with science is not to deny the intimate relation between science and philosophy. The positivist idea that all science does is to predict the observable results of experiments is still popular with some scientists, but it always leads to the *evasion* of important foundational questions. For example, the recognition that there is a problem of *understanding* quantum mechanics, that is, a problem of figuring out just how physical reality must be in order for our most fundamental physical theory to work as successfully as it does, is becoming more widespread, but that recognition was *delayed* for decades by the claim that something called "the Copenhagen interpretation" of Niels Bohr had solved all the problems. Yet the "Copenhagen interpretation," in Bohr's version, amounted only to the vague philosophical thesis that the human mind could not possibly understand how the quantum universe was in itself; so the mind should just confine itself to

---

12. Rorty, *Philosophical Papers,* vol. 2, *Essays on Heidegger and Others,* 137.

13. Philosophy, when it conceives of itself as a purely theoretical subject, first, tends to become overambitious (e.g., it then tries to realize the dream of a grand metaphysical system); and, second, it is held up by the public to the standard of cumulative knowledge. But because, unlike science, philosophy is not cumulative knowledge when it pretends to be a science, people see right through that pretense.

telling us how to use quantum mechanics to make predictions *statable in the language of classical, that is to say, non-quantum-mechanical, physics*! (In my lifetime, I first realized that the "mood" had changed when I heard Murray Gell-Mann say in a public lecture sometime around 1975, "There is no Copenhagen Interpretation of quantum mechanics. Bohr brainwashed a generation of physicists!")

Only after physicists stopped being content to regard quantum mechanics as a mere machine for making predictions and started taking seriously the question, but what does this theory actually *mean*? could real progress be made. Today many new paths for research have opened as a result:[14] string theory, various theories of quantum gravity, and "spontaneous-collapse" theory are only the beginning of quite a long list. And Bell's famous theorem, which has transformed our understanding of the "measurement problem," would never have been proved if Bell had not had a deep but at the time highly unpopular interest in the meaning of quantum mechanics.[15]

In cosmology, however, there has unfortunately been somewhat of a revival of the positivist contempt for the question of the meaning of general relativity in recent years, but, owing to the influence of Einstein, who always recognized that physical theories are not mere formal systems, the great majority of astrophysicists continue to try to *understand* the nature of cosmic space-time and of the forces that shape the destinies of astronomical objects (including black holes), and not simply to say, as Steven Weinberg now appears to urge, that sometimes it is more convenient to use one theory, and sometimes it is more convenient to use another, and there is no reason to ask which is really true.[16] It is precisely at the level of fundamental physical science, in fact,

14. For more details, see Hilary Putnam, "A Philosopher Looks at Quantum Mechanics (Again)," *British Journal for the Philosophy of Science* 56, no. 4 (2005): 615–634; reprinted in this volume as Chapter 6.

15. For a simple statement and proof of Bell's theorem, see Gary Felder, "Spooky Action at a Distance" (1999), http://www4.ncsu.edu/unity/lockers/users/f/felder/public/kenny/papers /bell.html.

16. For an account of Weinberg's views, see Yemima Ben-Menahem, *Conventionalism* (Cambridge: Cambridge University Press, 2006), 93–94. Although Weinberg claims that his version of general relativity (which employs a flat space-time) is equivalent to Einstein's version (with its curved space-time), this is not the case, because the topologies of space are different in the Einstein version and the Weinberg version. The three-dimensional spatial cross sections of the universe were finite epsilon seconds after the big bang, and although inflation may have produced a situation in which the curvature of three-dimensional space came to be undetectably small later, a change from a "finite universe" to an "infinite universe" (or from a

that it becomes clear that the sharp separation that the positivists thought they saw, and that our culture often takes for granted, between metaphysics and physics is most untenable. Both physics and metaphysics flourish most when they interact and interpenetrate, that is, when they push Sellars's question, "How things in the broadest possible sense of the term hang together in the broadest possible sense of the term."[17]

Science and Normativity

Part of the difficulty of seeing the relation between science and philosophy clearly is that false ideas about science abound in ordinary life and philosophy, including the influential idea that there is something like an "algorithm" for doing science. Yet this idea is not one that any scientist I know believes in. A few years ago, speaking to an audience that contained at least fifty Nobel Prize winners, I said the following:

> I have argued that even when the judgments of reasonableness are left tacit, such judgments are presupposed by scientific inquiry. (Indeed, judgments of *coherence* are essential even at the observational level: we have to decide *which* observations to trust, which scientists to trust—sometimes even which of our *memories* to trust). I have argued that judgments of reasonableness can be objective. And I have argued that they have all of the typical properties of "value-judgments." In short, I have argued that my pragmatist teachers were right: "knowledge of facts presupposes knowledge of values."[18] But the history of the philosophy of science in the last half

---

curved finite cross section to a flat one) would involve a discontinuity in the universe's evolution that no astrophysicist has postulated. The fact is that Weinberg simply ignores the difference between "indetectably small" and "zero" curvature—a typical verificationist move. In addition, he ignores the fact that, as he has confirmed to me in correspondence, his "flat spacetime" version has no description of black holes.

17. Sometimes conceptual crises arise in science such that in order to make progress, science has to find a reasonable way to resolve them—one example being the time of Newton and the great debate between Newton and Leibniz about whether space is absolute or not; another more recent example being the birth of modern set theory and the debates among Cantor, Frege, Poincaré, Hilbert, Brouwer, and others. Today there are raging debates about string theory (is it physics at all? or metaphysics?). It is characteristic of these crises that both scientists and philosophers contribute to the discussion.

18. I heard this in my undergraduate years at the University of Pennsylvania from my teacher C. West Churchman, who attributed it to *his* teacher E. A. Singer Jr., who was in turn a student of William James.

century has largely been a history of attempts—some of which would be amusing, if the suspicion of the very idea of justifying a value judgment which underlies them were not so serious in its implications—to *evade* this issue. Apparently any fantasy—the fantasy of doing science using only deductive logic (Popper), the fantasy of vindicating induction deductively (Reichenbach), the fantasy of reducing science to a simple sampling algorithm (Carnap), the fantasy of selecting theories given a mysteriously available set of "true observation conditionals," or, alternatively, "settling for psychology" (both are Quine's)—is regarded as preferable to rethinking the whole dogma—the last dogma of empiricism?—that facts are objective and values are subjective and never the twain shall meet.

Not one of those scientists disagreed with me!

Nevertheless, even if what we might call "the myth of inductive logic" is not one to which scientists themselves subscribe, it has had a powerful influence on the way laypersons and philosophers (including philosophers of science) think about science—an influence that, I would argue, it is one of the functions of philosophy, at its best, to combat. In fact, the idea that beliefs about "facts" must be fundamentally different from "attitudes" toward values was supported by the most influential defender of the idea that values are "noncognitive," the father of "emotivism" (as the doctrine that value judgments are mere modes of emotional "persuasion" was called), Charles Stevenson, precisely by employing the claim that value judgments cannot be verified by "induction and deduction" (while "beliefs about facts" supposedly could be).[19] Thus what began as a technical issue in epistemology, the issue about "induction," came to play a crucial role in discussions of the possibility of the rational discussion of value issues—an issue of immense importance to our culture.

The case for the idea that facts and values are deeply entangled[20] draws on some of the best philosophical work of the last hundred years: on Quine's attack on the positivists' analytic/synthetic distinction, on the work done by some of Wittgenstein's best followers (notably Iris Murdoch, Philippa Foot, and John McDowell) on the way "thick" ethical concepts such as "cruel" resist "factorization" into a "purely descriptive" component and an "expressive" or "emotive" component, and on my own observations on the way in which the epistemic values that inform science are, after all, *value* concepts too. For a number of years, in fact, I have argued that in science, and particularly in

19. On Stevenson, see Chapter 18 in this volume.
20. Cf. Putnam, *Collapse of the Fact/Value Dichotomy.*

the social sciences, we are unavoidably dealing with an entanglement of facts, theories, and values. It is like a three-legged stool—all three legs are needed, or it falls over. The all-too popular idea that if something is a "value judgment," then it must be wholly "subjective" rests not just on shaky foundations but on foundations that have completely collapsed. In my opinion, understanding this is vital if we are to regain faith in the possibility and the importance of *rational debate about values*.

The Scientific and Manifest Images

The need to "save the appearances"—that is, to do justice to the manifest image—in our theorizing about the world has been an important aspiration of philosophy from its inception. Aristotle's insistence that metaphysics must "save the appearances" can be viewed as a form of respect for ordinary language as against Platonic speculation. If "ordinary language" is just something to be sneered at, then is the whole vocabulary we have for describing the word of human agents to be either despised or else replaced by the "Newspeak" of some social science?

To me it seems clear that the descriptions of human life we find in the novels of Tolstoy or George Eliot are not mere *entertainment*; they teach us to perceive what goes on in social and individual life. And such descriptions require the many subtle distinctions that ordinary language has made available to us. The question of the relevance or irrelevance of "how we speak" is not just a question for philosophers, although it is that too. It is a question for philosophers because once ordinary language is laughed out of the room, philosophical theories are no longer held responsible at all to the ways we actually speak and actually live; but it is a question for more than just philosophers because, at bottom, contempt for ordinary language is contempt for all the humanities.[21]

Another Look at the History

I began this paper with a particular account of the "crisis" of philosophy, a view made popular in Europe by Heidegger's *Sein und Zeit*. On that view, a natural one for an ex-seminarian like Heidegger, traditional philosophy was seen as

---

21. I do not think that philosophy can be turned into a science because there are areas of philosophy that are essentially humanistic, and I think that turning the humanities into science is a fantasy, and a dangerous fantasy at that. But there are parts of philosophy that overlap with science.

"ontotheology." (Heidegger's solution to the crisis was that we should all become Heideggerians.) Another account of the "crisis," one I might have begun with instead of the one I gave, can be found in the writings of Bertrand Russell and also in the logical positivists. On that account, it was the realization that philosophy led to debates that are never settled, coupled with the supposed fact that "the new logic" could resolve the old unsettleable problems, that required that traditional philosophy be entirely replaced.

"Progress" in philosophy need not consist in "settling" issues once and for all. Indeed, it does not consist of that in any serious area of human endeavor or inquiry that I know of. And those problems have quite old roots, which suggests that philosophy was never just a handmaiden to theology, even if in the Middle Ages it was often urged to be that. To give only one of many possible examples: The issue about whether talk of unobservables in physical science is really "representational" already appeared at the time of Berkeley and Hume and is thus centuries old.[22] How to interpret quantum mechanics is simply a contemporary spin-off occasioned by the puzzling character of that theory. Philosophy was never just ontotheology, and even when philosophers were concerned with ontotheology, they were concerned with much more than that. That is the first reason that the idea of a fundamental "crisis" in philosophy and of the "end of philosophy" is deeply mistaken. And if the questions of philosophy are indeed "unsettleable," in the sense that they will always be with us, that is a wonderful thing, not something to be regretted.

22. Think, for example, of the debates between Samuel Clarke and Leibniz concerning the reality or lack of reality of (1) action at a distance and (2) absolute space.

2

From Quantum Mechanics to Ethics and Back Again

As you will doubtless have guessed from the title of this chapter, I am going to write about my philosophical career as a whole (up to the ninth decade, anyway). "Back Again," by the way, does not mean that I have abandoned my interest in ethics! Not at all. Rather, it refers to the fact that in December 2005 I published a paper titled "A Philosopher Looks at Quantum Mechanics (Again)."[1] I could also have titled these reflections "From Philosophy of Mathematics to Ethics and Back Again" because in 2006 I delivered a lecture titled "Indispensability Arguments in the Philosophy of Mathematics."[2]

My reason for pointing this out is that I am tired of being described as a philosopher who was interested only in philosophy of science until the late 1970s, and who then turned to "soft" philosophy. A simultaneous interest in *both* science and ethics has characterized my thinking for decades; in fact,

---

1. Hilary Putnam, "A Philosopher Looks at Quantum Mechanics (Again)," *British Journal for the Philosophy of Science* 56, no. 4 (December 2005): 615–634; reprinted in this volume as Chapter 6.

2. This lecture was delivered at the 40th Chapel Hill Colloquium in Philosophy, October 6–8, 2006, and is reprinted in this volume as Chapter 9.

I gave a seminar on metaethics in 1964, when I was professor of philosophy of science at MIT. But I have also *never* stopped thinking about and writing about philosophy of physics and philosophy of mathematics.

In the introduction (dated September 1974) to *Mathematics, Matter and Method,* the first volume of my first collection of papers, I wrote:

> It will be obvious that I take science seriously and that I regard science as an important part of man's knowledge of reality; but there is a tradition with which I would not wish to be identified, which would say that scientific knowledge is all of man's knowledge. I do not believe that ethical statements are expressions of scientific knowledge; but neither do I agree that they are not knowledge at all. The idea that the concepts of truth, falsity, explanation, and even understanding are all concepts which belong exclusively to science seems to me to be a perversion. That Adolf Hitler was a monster seems to me to be a true statement (and even a "description" in any ordinary sense of "description"), but the term "monster" is neither reducible to nor eliminable in favor of "scientific" vocabulary. (This is not something discussed in the present volume. It is a subject on which I hope to write in the future.)
>
> If the importance of science does not lie in its constituting the whole of human knowledge, even less does it lie, in my view, in its technological applications. Science at its best is a way of coming to know, and hopefully a way of acquiring some reverence for, the wonders of nature. The philosophical study of science, at the best, has always been a way of coming to understand both some of the nature and some of the limitations of human reason. These seem to me to be sufficient grounds for taking science and philosophy of science seriously; they do not justify science worship.[3]

I still stand by every one of those words.

Although the misdescription of the evolution of my thinking I just referred to usually comes from people who regret my interest in ethics, in American pragmatism, and in Wittgenstein—people who think that, as Quine famously put it, "philosophy of science is philosophy enough"—there is a related but more subtle misunderstanding that I find even in the writings of some philosophers who are certainly sympathetic to the attitude expressed in those words I wrote in 1974. Today I often find myself divided into three "Putnams"

---

3. Hilary Putnam, *Philosophical Papers,* vol. 1, *Mathematics, Matter and Method* (Cambridge: Cambridge University Press, 1975), xiii–xiv.

(somewhat like "early, middle, and late Wittgenstein"). "Early Putnam" is said to have been a staunch scientific realist; "interim Putnam" is said to have repudiated scientific realism in favor of internal realism; and "more recent Putnam" is often described as defending "commonsense realism" (which is not usually described in any detail at all). But that account is confused (although the confusion is partly my own fault, as I shall explain). To explain why it is confused, I need to say something about scientific realism.

Realism: Scientific, Metaphysical, and Internal

The so-called interim period in my career, that is, the period from 1976 to 1989 during which I espoused what I described as the "picture" of truth as "idealized rational acceptability," or verification under ideal conditions, opened with my reading an address titled "Realism and Reason" to the American Philosophical Association (Eastern Division) in December 1976.[4] That address was, in a way, the manifesto of what I called "internal realism." But the opening lines of that "manifesto" contained an unfortunate slip that was the cause of much of the confusion in subsequent years about just what I meant to be defending. In my own mind, the purpose of those opening lines was to make it clear that I was *not* repudiating *scientific* realism. I should like to quote part of those lines for you:

> In one way of conceiving it, realism is an empirical theory. One of the facts that this theory explains is the fact that scientific theories tend to "converge" in the sense that earlier theories are very often limiting cases of later theories (which is why it is possible to regard theoretical terms as preserving their reference across most changes of theory). Another of the facts it explains is the more mundane fact that language-using contributes to getting our goals, achieving satisfaction, or what have you.
>
> The realist explanation, in a nutshell, is not that language mirrors the world but that *speakers* mirror the world—i.e. their environment—in the sense of *constructing a symbolic representation* of their environment . . . —let me refer to realism in this sense—acceptance of this sort of scientific picture of the relation of speakers to their environment, and of the role of language— as *internal* realism. [I should have written "scientific realism." That was the slip.]

4. That address was collected in Hilary Putnam, *Meaning and the Moral Sciences* (London: Routledge and Kegan Paul, 1978), 123–138.

Metaphysical realism, on the other hand, is less an empirical theory than a model. . . . It is, or purports to be, a model of the relations of *any* correct theory to all or parts of THE WORLD. I have come to the conclusion that this model is incoherent.[5]

The slip might not have been important if I had continued to use "internal realism" as I explained it in this paragraph (although "scientific realism" would have been better), but later in that same lecture I gave it a very different sense. In a section titled "Why All This Doesn't Refute Internal Realism," I identified "internal realism" with the view that whether a theory has a unique intended interpretation "has no absolute sense."[6] At that point it is clear that "internal realism" was *now* a name for the view I had developed in the lecture as a whole, a view on which truth and idealized verifiability were supposed to coincide, and *not* a name for the view described in the opening quotation, the view on which *both* metaphysical realists and holders of the antirealist view I now advanced were supposed to be able to agree.

Although I have been aware for many years that I used "internal realism" in these two inconsistent ways in "Realism and Reason," it was only upon reading Maximilian de Gaynesford's book about the evolution of my views[7] that I appreciated how much confusion that inconsistency had occasioned. In that book, with much of which I agree, and many of whose philosophical attitudes I share, "scientific realism," the term I had used repeatedly in the first two volumes of my *Philosophical Papers* published in 1975,[8] is identified with what I called "metaphysical realism" in "Realism and Reason," with the result that that lecture, and in fact everything written by "interim Putnam," is seen as a repudiation of almost everything that I had written in the papers collected in those retrospective volumes. The truth is that over the years I have changed my mind with respect to only *three* of the nineteen papers in *Mathematics, Matter and Method,* namely, chapter 2 (which defended a Russellian "if-thenist" position in philosophy of mathematics) and chapters 9 and 10 (which dealt with quantum logic). Similarly, I have changed my mind with respect to only *four* of the twenty-two papers in *Mind, Language and*

---

5. Ibid., 123.
6. Ibid., 136.
7. Maximilian de Gaynesford, *Hilary Putnam* (Chesham: Acumen, 2006).
8. Vol. 1, *Mathematics, Matter and Method,* and vol. 2, *Mind, Language and Reality* (Cambridge: Cambridge University Press, 1975).

Reality, namely, chapter 3 (which is one of the few places in those two volumes where I defended metaphysical realism) and chapters 18, 20, and 21 (which proposed "functionalism" in the philosophy of mind). Of course, there are individual arguments in those papers I would now improve, and formulations I regard as needing qualification, but at no time did I regard myself as repudiating them. Yet those volumes are now supposed to represent a "metaphysical realist" period in my development! Although a confusion of this kind never seems to disappear from the literature, let me nevertheless repeat and emphasize: I have *always* regarded myself as a scientific realist, although of course not *only* a scientific realist, as the introduction to the first of those volumes made clear.

One more bit of history. I still owe you an explanation of *why* I called scientific realism "internal realism" in those opening paragraphs of "Realism and Reason." The reason has to do with a conversation I had in the late 1970s with my sorely missed and much-admired friend and onetime colleague Rogers Albritton. At that time I defended—as I still do, to the disgust of some of my friends and former students—the claim that scientific realism is the only philosophy of science that does not make the success of science a miracle. As an argument against a sophisticated antirealism such as Michael Dummett's, or Crispin Wright's, or my own former "internal realism," this was not a good argument. Indeed, as I pointed out in the paragraphs I quoted above, an "internal realist" will agree that "scientific theories tend to 'converge' in the sense that earlier theories are very often limiting cases of later theories (which is why it is possible to regard theoretical terms as preserving their reference across most changes of theory)."[9] "Antirealism" as a *theory of truth* needs to be rebutted in other ways, as I tried to do with my own "internal realism" in my 1994 Dewey Lectures.[10] But as an argument against such positions as logical positivism and van Fraassen's "constructive empiricism," I still believe it.

---

9. Putnam, *Meaning and the Moral Sciences,* 123. When I wrote those words, I was thinking of Richard Boyd's characterization of scientific realism, according to which terms in a mature science typically refer and theories accepted in a mature science are typically approximately true—which I still like, by the way.

10. Hilary Putnam, "The Dewey Lectures 1994: Sense, Nonsense, and the Senses; An Inquiry into the Powers of the Human Mind," *Journal of Philosophy* 91, no. 9 (September 1994): 445–517; reprinted in Putnam, *The Threefold Cord; Mind, Body, and World* (New York: Columbia University Press, 1999), 3–70.

Anyway, Rogers and I were discussing this, and he said, "Well of course, scientific realism is *science's* philosophy of science." Coming from other lips, I would probably have heard that as praising scientific realism. But with the particular intonation that Rogers gave to those words, I understood at once that he meant something else. What he meant was that if "scientific realism" was, as Boyd and I claimed, "an empirical theory," then it could be shown to be incompatible with antirealism only if it could be shown that an antirealist had to reject at least some empirical theories. To just assume *that* was to beg the case against antirealism. In short, one could accept scientific realism as a *part* of science, as something *internal* to science, while adopting an antirealist view of language as a whole, including scientific language (e.g., one could accept Dummett's view, or the view I went on to develop in *Reason, Truth and History*). And that is what I did, which is how I could refer to scientific realism as "internal realism," meaning in that sentence "the realism internal to science."

Conceptual Relativity

The idea of "conceptual relativity" has been a controversial feature of my work. Indeed, this idea has often been seen as in itself an antirealist one, which is another reason that many commentators continue to describe me as an "internal realist," although I rejected the "verificationist semantics" that was the essential idea of my so-called internal realism in 1990.[11] In the next section of this chapter I shall try to clear up that confusion by explaining why I believe that conceptual relativity is fully compatible with realism in metaphysics. But first I want to review briefly what conceptual relativity amounts to.

11. I publicly renounced that idea in my reply to Simon Blackburn at the conference on my philosophy at the University of St. Andrews in November 1990. A "written-up" version of that reply is published in Peter Clark and Bob Hale, eds., *Reading Putnam* (Oxford: Blackwell, 1994), 242–254. The reasons that I gave up my "antirealism" are stated in the first three of my replies in "The Philosophy of Hilary Putnam," an issue of the journal *Philosophical Topics* 20 (Spring 1992), 347–369, where I give a history of my use(s) of the unfortunate term "internal realism," and, at more length, in my Dewey Lectures. See also my exchange with Crispin Wright: Crispin Wright, "Truth as Sort of Epistemic: Putnam's Peregrinations," *Journal of Philosophy* 97, no. 6 (2000): 335–364; Hilary Putnam, "When 'Evidence Transcendence' Is Not Malign: A Reply to Crispin Wright," *Journal of Philosophy* 98, no. 11 (2001): 594–600.

First, a story. I taught for a quarter at the University of Washington in 2002, and one day when I stopped for a cup of coffee at one of the little cafeterias on the campus, I struck up a conversation with Matt Strassler, a physicist who was there to give a lecture. We started talking about quantum mechanics, and I became quite excited as I realized that he was describing a phenomenon I had written about for a long time under the name "conceptual relativity." Of course, he did not call it that (physicists call it "duality"), nor did he know of my writing, and I did not know the papers he referred to, although he was kind enough to send me a list. In brief, what he told me was that what an analytic philosopher would probably call the "ontology" of a conceptual scheme (specifically, a quantum-mechanical theory of a particular system) is not regarded as, so to speak, the load-bearing aspect of the scheme, because such a scheme has many different "representations" (Strassler's term), which are regarded as perfectly equivalent. For example, these representations may differ in how many dimensions they treat space (or space-time) as having, and over whether the particles in the system are or are not bosons![12]

My own notion of "conceptual relativity" (which I originally called "cognitive equivalence") is beautifully illustrated by the case Strassler described. The different "representations" are perfectly intertranslatable; it is just that the translations do not preserve "ontology."

What do they preserve? Well, they do not merely preserve macro-observables. They also preserve *explanations*. An explanation of a phenomenon goes over into another perfectly good explanation of the same phenomenon under these translations.

But who *is* to say what is a phenomenon? And who is to say what is a perfectly good explanation? My answer has always been: *physicists* are; not linguists and not philosophers.

Of course, in my writing I have sought simple examples, and the simplest, one that is now well known, involves mereological sums. The question "Are there really mereological sums?" is, in my view, a pseudoquestion. Although it is obvious that we could not do with only particles (in, for example, classical physics), it does not follow that we have to accept all the axioms/theorems of mereology, including the statement that for every X and Y there

12. See, for example, C. P. Burgess and Fernando Quevedo, "Bosonization as Duality," *Nuclear Physics* B 421 (1994): 373–390; e-Print Archive: hep-th/9401105; http://arxiv.org/abs/hep-th/9401105.

is a smallest object having both *X* and *Y* as parts. We can perfectly well do classical physics using sets rather than mereological sums, and we need the axioms of set theory anyway. Given that all scientific and nonscientific discourse can be formulated perfectly successfully both with and without the "assumption" of mereological sums, their "existence" is, in my view, best regarded as a matter of convention. The positivists got that one right (although for the wrong reasons).

Similarly, I regard the question whether *points* (in space or in space-time) are real "individuals" or simply logical constructs as a pseudoquestion.

A moment ago I spoke of the existence of mereological sums as a matter of convention. To be more precise: saying that there are seven objects on a certain table, namely, three billiard balls and four additional mereological sums of billiard balls, is a matter of fact *as opposed to saying* that there are three such objects (the second statement is true if there are two billiard balls on the table), while saying that there are seven mereological sums that can be formed of the objects on that table *as opposed to saying* that there are three individual objects on the table and eight *sets* of those objects (mereological sums are not sets, and there is an empty set but no empty object) is a matter of convention. The fact that we say *X* rather than *Y* may be a matter of convention, in whole or in part, while the fact that we say *X* rather than *Z* is not at all conventional.

What This All Has to Do with Realism

Because I criticized Maximilian de Gaynesford for portraying me (or at least "interim" me) as renouncing scientific realism, let me balance the account with some sincere praise: he rightly sees that the problem of intentionality has been a lifelong preoccupation of mine, and that various changes in my position were occasioned by the realization that one or another assumption about the nature of reference led to deep difficulties. As de Gaynesford rightly explains, I had to give up "functionalism," for example, that is, the doctrine that our mental states are just our *computational* states (as implicitly defined by a "program" that our brains are hard-wired to "run"), because that view is incompatible with the semantic externalism that years of thinking about the topic of reference had eventually led me to develop. If, as I said in "The Meaning of 'Meaning,'" our intentional mental states are not in our heads, but are rather to be thought of *as world-involving abilities,* abilities identified by the sorts of transactions with our environment that they facilitate, then they are not identified simply by the "software" of the brain.[13] (A further reason for giving up computational functionalism was that mental states are not only compositionally plastic, that is, capable in principle of being realized in different sorts of hardware, but *computationally plastic,* that is, capable of being realized in different sorts of software. But I shall not go into further detail about my philosophy of mind here.)

Because at least one friend has expressed regret that I gave up the slogan that "the mind and the world together make up the mind and the world," let me say how reflections on reference led me into and then out of that idea. (De Gaynesford is excellent on how I was led into the position, but I would have liked him to say more about how I was led out of it.)

The heart of internal realism was what I called "verificationist semantics" (I even tentatively endorsed Michael Dummett's idea that we should adopt intuitionist logic in one of my publications).[14] In my version of that semantics, truth was identified with verifiability under epistemically ideal conditions. But I did not, as some mistakenly took me to be doing, adopt the Peircean view that such conditions involve infinitely prolonged scientific inquiry, or the corollary that truth about the past is determined by what we can or will find out in the future. I also recognized that it is possible that for a variety of empirical reasons we may be unable to attain epistemically ideal conditions with respect to some of our inquiries; thus the truth may outrun what we can as a matter of fact verify. But this recognition reinstated the very problem of "access" that internal realism was designed to block! If there is a problem about how, without postulating "noetic rays," we can have access to external things, there is an equal problem about how we can have referential access to

---

13. Hilary Putnam, "The Meaning of 'Meaning,'" in Keith Gunderson, ed., *Language, Mind and Knowledge. Minnesota Studies in the Philosophy of Science,* vol. 7 (Minneapolis: University of Minnesota Press, 1975), 131–193; reprinted in Putnam, *Philosophical Papers,* vol. 2, *Mind, Language and Reality,* 215–271. I still believe that our so-called mental states are best thought of as capacities to function, but not in the strongly reductionist sense that went with the model of those states as "the brain's software." They are, so to speak, "long-armed" functional states—their "arms" reach out to the environment, and their identity depends, as Ruth Millikan has stressed, on their evolutionary history.

14. Hilary Putnam, introduction to Putnam, *Philosophical Papers,* vol. 3, *Realism and Reason* (Cambridge: Cambridge University Press, 1983), vii–xviii, which has the best statement of the position.

"sufficiently good epistemic conditions."[15] On my "internal realist" picture, which differed in this respect from Dummett's "antirealist" one, the world was allowed to determine whether I am in a sufficiently good epistemic situation or only seem to myself to be in one—thus retaining an essential idea from commonsense realism—but the conception of our epistemic situation was the traditional "Cartesian" one, on which our sensations are an interface "between" us and the "external objects." By the time of the Gifford Conference on my philosophy at St. Andrews in 1990, I had decided that this just would not do.

The alternative was to return to the Cartesian predicament itself and see whether it could be avoided. I had accepted that predicament because I had rejected Austin's criticisms of "sense-data" talk. I believed, as I wrote in "Models and Reality,"[16] that

> although the philosopher John Austin and the psychologist Fred Skinner both tried to drive sense data out of existence, most philosophers and psychologists think that there are such things as *sensations* or *qualia*. They may not be objects of perception, as was once thought (it is becoming increasingly fashionable to view them as states or conditions of the sentient subject, as Reichenbach long ago urged we should); we may not have incorrigible knowledge concerning them; they may be somewhat ill-defined entities rather than the perfectly sharp particulars they were once taken to be; but it seems reasonable to hold that they are part of the legitimate subject matter of cognitive psychology and philosophy, and not mere pseudo-entities invented by bad psychology and bad philosophy.

By 1990, however, I had begun to think that Austin's attack on the conception of our experiences as "sense data" that somehow are directly perceived, while the "external objects" are only "indirectly perceived," was right. In addition, I had begun to read William James and was impressed by his insistence that what he called "natural realism"[17] could be defended.[18] In

15. For a fuller discussion, see Hilary Putnam, "Between Scylla and Charybdis: Does Dummett Have a Way Through?" in R. E. Auxier and L. E. Hahn, eds., *The Philosophy of Michael Dummett* (Chicago: Open Court, 2007), 155–167.

16. Hilary Putnam, "Models and Reality," *Journal of Symbolic Logic* 4, no. 3 (September 1980): 464–482; reprinted in Putnam, *Philosophical Papers*, vol. 3, *Realism and Reason*, 1–25; see p. 15.

17. William James, "A World of Pure Experience," in James, *Essays in Radical Empiricism*, ed. Frederick Burckhardt and Fredson Bowers (Cambridge, Mass.: Harvard University Press, 1976), 21–44. The term "natural realism" appears on p. 21.

18. See Hilary Putnam, "James's Theory of Perception," in Putnam, *Realism with a Human Face* (Cambridge, Mass.: Harvard University Press, 1990), 232–251.

sum, I began to think that the problem of "access" to external objects that I had elaborated with the aid of devices from mathematical logic (model theory) was a replay of the older problem of epistemological dualism, even if the dualism was no longer a dualism of mental substance and physical substance, but one of brain states and everything outside the head. I concluded, and still conclude, that "natural realism" with respect to perception can indeed be defended, and that with natural realism with respect to perception back in place, the fear (or the bugaboo) that we may have no "access" to reality outside our heads can be dismissed as a bad dream.

I can now respond to the regret that I gave up the slogan "The mind and the world together make up the mind and the world." I called that slogan a "metaphor,"[19] and what it came to was this: we make up conceptual schemes, and those schemes and the entities that appear to be talked about (that "intensionally inexist," in Brentano's jargon) when we think or say those sentences are the only "world" of which we can speak. No wonder Rorty was fond of my writing in that period! It was not that he missed the realist elements in *Reason, Truth and History,* but that he regarded them as vestiges that he hoped to talk me out of. (Later he despaired of that hope.)

In sum, on the "internal realist" picture it is not only our experiences (conceived of as "sense data") that are an interface *between* us and the world; our "conceptual schemes" are likewise conceived of as an interface. And the two "interfaces" are related: I saw our ways of conceptualizing, our language games, as controlled by "operational constraints" that ultimately reduce to our sense data. And to give up "internal realism" and return to "natural realism," or to a reasoned philosophical defense of natural realism, involves giving up thinking of either experiences *or* concepts as "between" us and external realities. Or so I argued in my Dewey Lectures, and so I still believe.

Let me add that the problem with *Reason, Truth and History* is not that I failed to realize that many experiences are *conceptualized.* On the contrary, even then, in 1981, I emphasized that they are conceptualized.[20] It was that I failed to see that one could think of both experiences and concepts as forms—to use McDowell's language of *openness to the world.* I should not have seen us as "making up" the world (not even with the world's help); I should have seen us as *open* to the world, as interacting with the world in

19. Hilary Putnam, *Reason, Truth and History* (Cambridge: Cambridge University Press, 1981), xi.

20. Ibid., 54.

ways that permit aspects of it to reveal themselves to us. Of course, we need to invent concepts to do that. There is plenty of constructive activity here. But we do not construct reality itself.

In "Realism and Reason" (the lecture I described as an internal realist "manifesto") and subsequently I also used the term "metaphysical realism." I now think that that was a mistake. The mistake was not a simple slip, like using "internal realism" to mean two different things, as I recounted earlier. But it *was* a mistake nonetheless.

It was a mistake because, although I repeatedly explained what I meant by the term, there is a natural understanding of the phrase "metaphysical realism" in which it refers to a broad family of positions, and not just to the *one* position I used it to refer to. Although I was often impatient with critics who said, "But you haven't refuted *my* form of metaphysical realism," when "their" form was not the one I was talking about, I now sympathize with them. In effect I *was* saying that by refuting the one philosophical view I called by that name I was ipso facto refuting anything that deserved to be called "metaphysical realism," and that was not something I had shown.

As I explained "metaphysical realism,"[21] what it came to was precisely the denial of conceptual relativity. My "metaphysical realist" believed that a given thing or system of things can be described in exactly one way if the description is complete and correct, and that way is supposed to fix exactly one "ontology" and one "ideology" in Quine's sense of those words, that is, exactly one domain of individuals and one domain of predicates of those individuals. Thus it cannot be a matter of convention, as I have argued that it is, whether there are such individuals as mereological sums; either the "true" ontology includes mereological sums or it does not. And it cannot be a matter of convention, as I have argued that it is, whether space-time points are individuals or mere limits.

To be sure, this is *one form* that metaphysical realism can take. But if we understand "metaphysical realist" more broadly, as applying to all philosophers who reject all forms of verificationism and all talk of our "making" the world, then I believe that it is perfectly possible to be a metaphysical realist in *that* sense and to accept the phenomenon I am calling "conceptual relativity."

21. See particularly Hilary Putnam, "A Defense of Internal Realism," in Putnam, *Realism with a Human Face*, 30–42.

The Compatibility of Conceptual Relativity
with Realism

One of the many contributions of both John Austin and Ludwig Wittgenstein was to stress how many different sorts of things we do with language. A further insight, which they shared with the earlier American pragmatists, was that these uses and speech acts are not simply discrete and separate (something that the metaphor of a "toolbox" that Josiah Royce used in *The Problem of Christianity*[22] does not bring out) but interdependent. John McDowell has written about the impossibility of "disentangling" factual description and evaluation, and Quine famously spoke of the impossibility of disentangling fact and convention. I shall say more about the importance that the triple entanglement of fact, value, and convention has had in my philosophy in a few moments. For the moment, however, let us focus on descriptions, though without forgetting that description can both depend on and determine evaluations and conventions, or that descriptions can be of many different kinds and serve many different interests.

So how *should* realists who recognize the existence of cases of genuine conceptual relativity formulate their realism? Imagine a situation in which there are exactly three billiard balls on a certain table and no other objects (i.e., the atoms and other entities of which the billiard balls consist do not count as "objects" in that context). Consider the two descriptions, "There are only seven objects on that table: three billiard balls and four mereological sums containing more than one billiard ball" and "There are only three objects on the table, but there are seven *sets* of individuals that can be formed of those objects." What it means to be a *realist who recognizes conceptual relativity* with respect to *this* case is to believe that there is an aspect of reality that is independent of what we think at the moment (although we could, of course, change it by adding or subtracting objects from the table), and that is *correctly describable either way.*

The example is artificial because no one except a philosopher, to my knowledge, *ever* talks about "mereological sums." But in mathematical physics conceptual relativity is a ubiquitous phenomenon, and there the correct attitude is the same (or so I maintain). To take an example from a paper with

22. Josiah Royce, *The Problem of Christianity* (Washington, D.C.: Catholic University of America Press, 1913).

the title "Bosonization as Duality" that appeared in *Nuclear Physics* B some years ago,[23] there are quantum-mechanical schemes, some of whose representations depict the particles in a system as bosons, while others depict them as fermions. As their use of the term "representations" indicates, real live physicists—not philosophers with any particular philosophical axe to grind—do not regard this as a case of ignorance. In their view, the "bosons" and "fermions" are simply artifacts of the representation used. But the system is mind-independently real, for all that, and each of its states is a mind-independently real condition that can be represented in each of these different ways. And that is exactly the conclusion I advocate. (A still-unpublished paper by David Albert titled "Physics and Narrative" describes a much simpler case that needs no field theory, in which two different stories about the history of a two-particle electron paramagnetic resonance system are representations of the same reality, even though in one of them there is a period of time in which the spin of *A* in a certain direction is correlated with the spin of *B* in that same direction, rather than anticorrelated, and in the other "story" [frame-dependent history] there is no such period of time.)

To accept that these descriptions are both answerable to the very same aspect of reality, that they are "equivalent descriptions" in that sense, is to be a metaphysical realist without capital letters, a realist in one's "metaphysics," but not a "metaphysical realist" in the technical sense I gave to that phrase in "Realism and Reason" and related publications. And if I have long repented of having once said that "the mind and the world make up the mind and the world," that is because what we actually make up is not the world but language games, concepts, uses, and conceptual schemes. To confuse making up the *notion* of a boson, which is something the scientific community did over time, with making up real quantum-mechanical systems is to slide into idealism, it seems to me. And that is a bad thing to slide into.

A Caution

It is important not to confuse conceptual relativity with the very different phenomenon of conceptual pluralism. You may have noticed that all my examples of conceptual relativity come from science, which is where the phenomenon seems to occur. At one time I confused it with simple conceptual

23. Burgess and Quevedo, "Bosonization as Duality."

pluralism, and that led me to mistakenly give the fact that I can (depending on my interests) describe the contents of a room *either* by saying that the room contains a table and two chairs *or* by saying that it contains such and such fields and particles as an example of conceptual relativity. I do indeed deny that the world can be completely described in the language game of theoretical physics, not because there are regions in which physics is *false,* but because, to use Aristotelian language, the world has many levels of form, and there is no realistic possibility of reducing them all to the level of fundamental physics. For example, it is a true description of one aspect of reality to say that Immanuel Kant wrote some passages in *The Critique of Pure Reason* that are difficult to interpret, but that statement cannot be "translated" into the language of physics in any reasonable sense of "translated." It is a true description of another aspect of reality to say that Andrew Wiles and Richard Taylor gave a correct proof of Fermat's last theorem, and that statement too cannot be translated into the language of *physics.* And it is a true description of a third aspect of reality to say that the reason I took a certain route to Harvard Square on a certain day was that I mistakenly believed that it would be quicker. And both of the above descriptions of the room could well be correct. *But* the very fact that these descriptions do not belong to "schemes" that can be systematically translated into each other means that they are *not* "equivalent" in the technical sense of "mutually relatively interpretable." That is why they illustrate *conceptual pluralism* but not conceptual relativity in my sense.[24]

Some Objections

I have spoken of descriptions as answerable to "aspects of reality," and I know that such talk will raise some hackles. Let me say a word about some of those hackles.

One set of hackles I will not try to smooth down here is those of one sort of "Wittgensteinian." I say "one sort" because the Harvard Department of Philosophy was for years the scene of an intense disagreement about whether Wittgenstein was an "end-of-philosophy" philosopher. (My own view is that he was *sometimes* such a philosopher, but that is not what made him a *great*

---

24. The mistake was first pointed out to me by Jennifer Case. See Jennifer Case, "The Heart of Putnam's Pluralistic Realism," *Revue Internationale de Philosophie* 55, no. 4 (2001): 417–430; and Hilary Putnam, *Ethics without Ontology* (Cambridge, Mass.: Harvard University Press, 2004).

philosopher.) On the "end-of-philosophy" view, my whole lifelong preoccu-pation with "realism" was a *mistake.* For example, my dear personal friend and strong philosophical opponent, Burton Dreben, used to call me a "Girondist" (as opposed to the true Wittgensteinians like himself, who were "Jacobins"). According to him, realism and idealism are both nonsense, *period.*

Another dear personal friend and strong philosophical opponent who agreed with Dreben, although not for Wittgensteinian reasons, was Richard Rorty. In a paper that will appear in the forthcoming volume on me in the Library of Living Philosophers, he even accuses me of being a throwback, not to the Girondists, but to Parmenides! According to him, talk of reality com-mits me to thinking of reality as a "superthing," and thereby as opening the pseudoquestion, as he regards it, of the nature of reality.

Let me say for the record that I utterly and totally reject all versions of the "end-of-philosophy" story, whether they come from Wittgensteinians, Rortians, Heideggerians, Derridians, or whomever. Philosophy was not a mistake—not in Parmenides' time, not in Plato's time, not in Aristotle's or Descartes's or Hume's or Kant's times, and not in our time. As Etienne Gilson put it, "Philosophy always buries its undertakers."[25]

But let me turn to hackles whose lack of smoothness I take more seriously. Instead of speaking of "aspects of reality," I could obviously (if I had not wanted to raise the hackles I will not discuss here) have said "states of affairs." And have not Quine and Davidson convinced us, or at least convinced those of us who take them—as I certainly do—to have made great contributions to our subject, that such talk is useless and needs to be discarded?

For both of them, the objection stemmed partly from the belief that to quantify over so-called abstract entities such as states of affairs is to reify them, to be guilty of hypostatizing them. Even quantifying over *sets,* which Quine defended as necessary for science, was something that he accepted with explicit reluctance.[26] That is a reluctance that I do not share.[27] But over and beyond that reason, Quine argued that there are no clear (i.e., no precise) criteria of *identity* and *nonidentity* in the case of states of affairs, propositions, and the

---

25. Etienne Gilson, *The Unity of Philosophical Experience* (New York: Charles Scribner's Sons, 1937), 306.

26. See, for example, W. V. Quine, *Theories and Things* (Cambridge, Mass.: Harvard University Press, 1981), 100.

27. Cf. Chapter 11 in this volume.

like, and that *this* disqualifies them from serious discourse (which Quine equated with *scientific* discourse).

However, as Quine himself recognized, to disqualify quantification over every sort of thing, in the widest sense of "thing," for which we lack precise criteria of identity has an enormous price. In particular, he observed (correctly) that there are not precise criteria of identity for beliefs, meanings (semantic contents) of words and sentences, hopes, fears, and all the host of propositional attitudes. And Quine recognized that talk of all of these is indispensable in practice. At this point, however, he violated his own best pragmatic impulses by writing, "Propositional and attributary attitudes belong to the daily discourse of hopes, fears, and purposes; causal science gets on well without them. . . . A reasonable if less ambitious alternative [to attempting to make them "scienceworthy"] would be to keep a relatively simple and austere conceptual scheme, free of half-entities [*sic*] for official scientific business and then accommodate the half-entities in a second-grade system."[28] And he went on to tell us that only the first-grade system need be taken seriously in philosophy. Apparently the idea that there are not really any such things as meaning, hopes, fears, and purposes "gilt in der Philosophie aber in der Praxis nicht" (is valid in philosophy but not in practice)—an attitude that I find wrong both philosophically, because it frees philosophy from the control of relevance to practice, and practically, because it frees practice from the control of critical reflection.

In what was perhaps the first of my explicitly scientific realist papers, "What Theories Are Not"—the paper I read to an international congress on logic, methodology, and philosophy of science on August 27, 1960, the day Ruth Anna and I met—I illustrated the indispensable scientific role of terms that are not "precise" when I wrote,

> Usually a scientist introduces [a new technical term into the language] via some kind of paraphrase. For example, one might explain "mass" as "that physical magnitude which determines how strongly a body resists being accelerated, e.g. if a body has twice the mass, it will be twice as hard to accelerate." (Instead of "physical magnitude" one might say in ordinary language "that property of the body" or "that *in* the body which . . ."). Such "broad-spectrum" notions occur in every natural language; and

28. W. V. Quine, *Ontological Relativity and Other Essays* (New York: Columbia University Press, 1969), 24.

our present notion of a "physical magnitude" is already an extreme refinement.[29]

I also counted "thing" as a "broad-spectrum notion." Well, I think that "state of affairs" is also one of the "broad-spectrum notions" that we need in contexts in which we do not have a more precise notion available, and without which we could not intelligibly explain those more precise notions when we do introduce them. In short, I think that we need to stop apologizing for talk of "states of affairs."

In addition to sharing Quine's worries, Donald Davidson had an additional reason for shunning talk of states of affairs, namely, that it might lead to the resurrection of the so-called correspondence theory of truth.[30]

One problem with that "theory" is that it is hopelessly vague. As James and Kant both remarked, phrases like "agrees with reality" are, if anything, *less* clear than "true." But there are further problems as well. I will mention them only briefly because this is not a paper on the theory of truth. It would be much too long if I took on *that* topic here. But although there is a sense in which a true description "corresponds" to an aspect of reality, or to an actual state of affairs, or a way things actually happen to be (take your pick!), not all true sentences are *descriptions*. My own view of truth ever since my Dewey Lectures has been what I described as a "disquotational one" in a sense of that term I connected with Frege. *To make a statement is to assert something, and to say that that something is true is to assert the same thing.* But that does not commit me to inventing states of affairs to correspond to all the things that can be correctly asserted. And if one does invent such "states of affairs," then nothing is thereby added to the account of truth.

A further problem with "correspondence" talk, one also seen by Frege, I believe, and certainly seen by Wittgenstein, is that it makes it sound as if there is just *one* sort of correspondence, *one* single relation, that somehow

29. Hilary Putnam, "What Theories Are Not," in Ernest Nagel, Patrick Suppes, and Alfred Tarski, eds., *Logic, Methodology and Philosophy of Science* (Stanford: Stanford University Press, 1962), 240–252; reprinted in Putnam, *Philosophical Papers*, vol. 1, *Mathematics, Matter and Method*, 225.

30. Donald Davidson, "True to the Facts," *Journal of Philosophy*, 66 (1969): 748–764; reprinted in Davidson, *Inquiries into Truth and Interpretation* (Oxford: Clarendon Press, 1984), 37–54.

stands behind all our talk. But if one agrees with Wittgenstein that there are many uses even of so-called descriptive language, and further agrees with him in thinking of those uses as exercises of world-involving and context-sensitive abilities, then whether given words (even, say, "The cat is on the mat") correspond to a particular state of affairs will depend on *which* world-involving ability to use those words is being exercised and in *what context* that ability is being exercised. One might say that when a true statement does "correspond" to an aspect of reality, what *sort* of "correspondence" is involved depends both on the meaning of the statement and on the particular extra-linguistic context. It is certainly not a matter of mere "marks and noises" standing in a *fixed* relation R to something (to a state of affairs, the world as a whole, or whatever).

To sum up: to say, as I do, that when we describe things, in Sellars's "broadest possible sense of the term," we are *answerable* to those things, and that when we describe them correctly, there is an aspect of reality that is as we assert it to be, is obviously to be a realist in one's view of "how things, in the broadest possible sense of the term, hang together."[31] My sort of realism is compatible with conceptual relativity (or the cognitive equivalence of some theories that are incompatible at the level of surface grammar), and it does not presuppose or require a "correspondence theory of truth" of the kind that Davidson feared.

Pragmatism

My first exposure to pragmatism occurred during my undergraduate years at the University of Pennsylvania. My first teacher in the philosophy of science, C. West Churchman, was a pragmatist (a student of E. A. Singer Jr., who was in turn a student of William James). And the idea of the triple entanglement of fact, theory, and convention that was to be so important to me when I finally turned to the topic of valuation was very much at the center of attention in Churchman's courses, although that early influence lay dormant in my soul for a long time as I concentrated on work in mathematical logic, especially in the 1950s. (That work, together with my writing "The Analytic and the

31. Wilfrid Sellars, "Philosophy and the Scientific Image of Man," in Robert Colodny, ed., *Frontiers of Science and Philosophy* (Pittsburgh: University of Pittsburgh Press, 1962), 35–78; reprinted in Sellars, *Science, Perception and Reality* (London: Routledge, 1963), 1–40.

Synthetic"[32] in 1957, resulted in my being awarded tenure in both the Department of Mathematics and the Department of Philosophy at Princeton in 1960.) Likewise, I took a course with Abraham Kaplan, who was at that time influenced by John Dewey, at UCLA in 1950, as well as a seminar on Dewey's *Logic* taught by Donald Piatt the following year. But my serious reading of William James only began forty years later.

Although I do not normally call myself a pragmatist or, for that matter, any kind of "ist," I am not unhappy when I am so described, except when people assimilate my views to those of my aforementioned dear friend and philosophical opponent Richard Rorty. I do not share his reading of the classical pragmatists at all, nor do I know of any James or Dewey scholar who does. Nor do I share his desire to bring philosophy to an end, or his contempt for talk of experience, reality, and truth. But I do not wish to repeat myself.

I also want to stress the fact that my admiration for the classical pragmatists does not extend to any of the different theories of truth that Pierce, James, and Dewey advanced. But one can learn from any of the great philosophers without sharing all of their cherished beliefs. If that were not the case, how could we still be learning, as I think we are and should be, from Aristotle or from Kant?

There are many things that I think we can learn from the pragmatists. I have already gone on longer than I planned, so I shall close by mentioning only the following two:

The first is the realization that description and evaluation must not be regarded as two separate watertight boxes in which statements or uses of statements can be put. All description presupposes evaluation (although not necessarily moral evaluation), and all evaluation presupposes description. Although this idea of interdependence or mutual presupposition has been developed more carefully and extended by post-Quinean and post-Wittgensteinian philosophers, notably by Stanley Cavell, Philippa Foot, Iris Murdoch, John McDowell, Vivian Walsh, Morton White, and (beginning with *Reason, Truth and History*) myself, it was ubiquitous in the writing of James and Dewey (and even present in Peirce, although not ubiquitous).

---

32. Hilary Putnam, "The Analytic and the Synthetic," in Herbert Feigl and Grover Maxwell, eds., *Scientific Explanation, Space, and Time. Minnesota Studies in the Philosophy of Science*, vol. 3, (Minneapolis: University of Minnesota Press, 1962), 358–397; reprinted in Putnam, *Philosophical Papers*, vol. 2, *Mind, Language and Reality*, 33–69.

Second, I heartily approve of the pragmatists' insistence that philosophy can and should matter to our moral and spiritual lives. Unfortunately, nowadays it is often claimed that analytic philosophy is a good thing precisely because it eschews thinking about that sort of stuff. If any philosophy that does not eschew the question "How to live?" is not automatically convicted of being a bad thing, then at least it is shown to be bad philosophy, according to this point of view. (Fortunately, not all analytic philosophers agree! Paul Grice, for example, once devoted a highly analytic seminar to precisely that three-word question.) But reflection on our ways of living, and especially on what is wrong with those ways of living, has always been a vital function of philosophy. If philosophy abandons that aspiration, what is supposed to replace it? The aspiration that philosophy will become a cumulative body of knowledge? That is a pipe dream.

The only argument that I have seen for rejecting the idea that such writers as Emerson, Kierkegaard, Sartre, Marx, and Thoreau belong in philosophy is that they do not give "arguments" (meaning: their writings do not look like logic texts). In addition, I have heard it said that there is the danger that such writings may convince someone by "irrational persuasion."

Well, of course there is that danger! And it is the responsibility of each reader to avoid irrationality—but not by relying on a "method." What the objection overlooks is that philosophy need not have the sole function of establishing theses, pieces of propositional knowledge, via more or less rigorous argument. When that is the function, the standards of "analytic philosophy" are appropriate. But to read the writings of any of the philosophers I just listed is not to encounter a series of theses; it is to encounter texts that anger, provoke, inspire, transform, repulse, or all of these at once. Such encounters are important because a life that does not encounter such issues and such reactions is not worth living.

What the pragmatists saw—and with this remark I shall close—is that philosophy can have both of these functions, and yet others besides. We do not need to erect firm boundaries around philosophy to keep the aliens out. We need their contribution.

3

Corresponding with Reality

In 1959 Martin Davis and I proved a theorem that, together with results by Julia Robinson and a result by Yuri Matiyasevich, added up to a "negative solution" to Hilbert's tenth problem.[1] As Davis describes our way of working in George Csicsery's film *Julia Robinson and Hilbert's 10th Problem*,[2] I was a "fountain of ideas," and Martin's role was to say, "That's ridiculous . . . that's ridiculous . . . that's ridiculous" until every so often one of my ideas was not ridiculous, and then we would both set to work on it. There are many who would say that that is also my method of working in philosophy! Needless to say, I hope that the ideas I lay before you will not turn out to be among the "ridiculous" ones.

I shall not, in any case, propose any wholly new ideas here. Instead, I shall reexamine some ideas that I defended in the past and tell you what in them I

1. A good account of the Hilbert problems and their solvers is Ben Yandell, *The Honors Class* (Natick, Mass.: A. K. Peters, 2002). Hilbert's tenth problem was to give a decision method for determining whether an arbitrary Diophantine equation has a solution. The "negative solution" referred to was a proof that the decision method Hilbert asked for does not exist.

2. The film (a one-hour documentary) is available from Zala Films, www.zalafilms.com /films/juliarobinson.html.

think was insightful and what in them I now think was mistaken. Because those ideas are still influential, I hope that this may be of more than merely autobiographical interest. The first of them is one that has influenced both philosophy and cognitive science, namely, *functionalism*.

Liberalizing Functionalism

In "Minds and Machines" I assumed that the brains of both robots (pretend that there are intelligent ones!) and humans can be described as computers. I suggested, but did not commit myself to, the idea that the mental states of those robots and those humans could be identified with what I called the "logical states" of their brains, meaning by that the states described by their programs. I called them "logical states" to emphasize that the *physical* description of those states was irrelevant; if a robot "brain" and a human brain have the same program, then the human and the robot have the same mental states, if that idea is right.

Subsequently[3] I committed myself to this identification of mental states and computational states[4] as an "empirical hypothesis."[5] Very soon I found difficulties with the identification of mental states and computational states—difficulties that led me to various reformulations.[6] Eventually, I found that I could not reconcile this identification with my advocacy of externalist and anti-individualist semantics in "The Meaning of 'Meaning,'" and I finally discarded it as "science fiction."[7]

---

3. See, in particular, Hilary Putnam, "The Mental Life of Some Machines" and "Psychological Predicates" (both published in 1967); "Psychological Predicates" was retitled "The Nature of Mental States" when I collected these papers in Putnam, *Philosophical Papers*, vol. 2, *Mind, Language and Reality* (Cambridge: Cambridge University Press, 1975), 429–440.

4. I had originally thought of these computational states as those of a Turing machine or, more precisely, a finite automaton.

5. In those subsequent papers I replaced the term "logical states" with "computational states," but with no change of meaning.

6. For an account of those difficulties and the reformulations I attempted, see the essay "Putnam, Hilary," in Samuel Guttenplan, ed., *A Companion to the Philosophy of Mind* (Oxford: Blackwell, 1993), 507–513.

7. See Hilary Putnam, "Functionalism: Cognitive Science or Science Fiction?," reprinted here as Chapter 35.

But I knew of a position related to functionalism, but minus the science fiction, in Aristotle. In a paper written when I was still wedded to computational functionalism, I wrote, "What we are really interested in, as Aristotle saw, is form and not matter. *What is our intellectual form?* is the question, not what the matter is."[8] Part of what I believed Aristotle was saying was that our mental capacities are *capacities* to function in certain ways, and that the question for psychology is what those functions are, and not what the physical constitution of this or that organism is. What I ascribed to Aristotle was a form of "functionalism" that identifies mental activities and capacities with ways of functioning, and that is compatible with the idea that those ways of functioning can, in principle, be realized in different physical systems, but minus the "science fiction."

I am still wedded to this liberal version of functionalism. It also appealed to Martha Nussbaum, and in a joint paper we described the position we ascribed to Aristotle as "a tenable position even in the context of a modern theory of matter."[9] What I will argue today is that it can provide us with a powerful approach to central questions in metaphysics and cognitive science. Here follow some examples.

Realism and Antirealism

In Boston in 1976 I delivered the lecture in which I declared that metaphysical realism was "incoherent."[10] Although it is now many years since I realized that this was an error, the issues involved still fascinate me. As I now see it, there are three main ones:

1. What to make of the "model-theoretic argument" that I gave as my reason for abandoning metaphysical realism.

8. Hilary Putnam, "Philosophy and Our Mental Life," in Putnam, *Philosophical Papers,* vol. 2, *Mind, Language and Reality,* 302.

9. Hilary Putnam and Martha Nussbaum, "Changing Aristotle's Mind," in *Essays on Aristotle's "De Anima,"* ed. Martha C. Nussbaum and Amelie Rorty (Oxford: Clarendon Press, 1992), 27–56; reprinted in Hilary Putnam, *Words and Life* (Cambridge, Mass.: Harvard University Press, 1994), 23.

10. Hilary Putnam, "Realism and Reason," *Proceedings and Addresses of the American Philosophical Association,* 50 (1977): 483–498; reprinted in Hilary Putnam, *Meaning and the Moral Sciences* (London: Routledge and Kegan Paul, 1978), 123–138.

2. What to make of the "verificationist semantics" that was the core of the "internal realism" that I proposed as the alternative to metaphysical realism.

3. Is verificationism, either in the form of my "internal realism" or in Michael Dummett's form (which I acknowledged as inspiring "internal realism") itself coherent?

What to Make of the Model-Theoretic Argument

To explain briefly what the "model-theoretic argument" was, let us imagine a "normal" realist, who not only believes that there are mind-independent truths[11] (that is what makes her a "realist"), but also believes that there are rationally acceptable statements concerning the mind-independent world. (It is possible to be a metaphysical realist and an epistemological skeptic, by the way.)[12] She believes that there are theories that imply no false observation sentences[13] and are rationally acceptable. Call them "good theories." I assume (certainly she should not object to this assumption) that we can imagine a series of such good theories whose limit is an "ideal theory," that is, a theory that answers all the empirical questions it is possible for a human scientist to answer, and that meets all the methodological constraints it is rational to impose on theories, such as elegance, simplicity, and the like. What makes my imagined realist a metaphysical realist is that she conceives of truth as non-epistemic; that is, she believes that even an ideal theory might be false.

For example, an ideal theory might say that there are intelligent extraterrestrials somewhere in space-time, although in fact there are not any. There might be *overwhelming evidence* that there are intelligent extraterrestrials (somewhere, some time), *evidence* for laws according to which the probability that such never did, do not, and never will exist is less than one in a trillion, let us say (which would certainly justify *believing* that intelligent extraterrestrials

11. Or better, verification-independent truths.

12. That is how I read van Fraassen, although Steve Wagner disagrees.

13. It was a problematic feature of the model-theoretic argument that it took over from logical positivism the "observational/theoretical dichotomy" that I myself had criticized in "What Theories Are Not," in Ernest Nagel, Patrick Suppes, and Alfred Tarski, eds., *Logic, Methodology, and Philosophy of Science* (Stanford, Calif.: Stanford University Press, 1962), 240–252; reprinted in Hilary Putnam, *Mathematics, Matter and Method* (Cambridge: Cambridge University Press, 1975), 215–227.

exist in space-time), when, in fact, ours is a universe in which the one in a trillion chance that they do not exist is realized.[14] This is an example of the way in which "correspondence truth" can differ from even idealized verifiability. The purpose of the model-theoretic argument was to cast doubt on the very intelligibility of this very plausible set of beliefs.

The argument proceeded thus: It is a theorem of model theory that if the number of individuals is large, a consistent theory must have a very large number of different possible "interpretations" (ways of assigning denotations to its non-logical symbols).[15] In *Reason, Truth and History* I showed that this remains the case even if we specify what the truth-values of all observation sentences are to be.[16] Now, our imagined realist thinks that there is a unique (or, perhaps, close to unique) interpretation of the predicates and individual constants of her language that is the "correct" one. Suppose that she is right, and also suppose that the ideal theory is, in fact, false. Then this theorem tells us that in addition to the "correct" interpretation, there are other interpretations that make the ideal theory *true* and *assign the correct truth-values to all observation sentences.* Because these "nonstandard" interpretations differ from the "correct" one, it is obvious that they must be "incorrect." But what on earth can that mean? Here the Putnam of 1976 to 1990 and his critics disagreed. I thought that it could not mean *anything,* and thus I felt forced to believe that truth could not be radically nonepistemic. The critics saw nothing wrong with the supposition that the world determines a "correct" interpretation of our symbols, even if our "theoretical constraints" and our "observational constraints" are not what single out that "correct" interpretation. (Most of them said that it was "causal connection" that did the trick, a reply that I found merely empty noise.)

14. Observe that it is logically impossible to verify that the conjunction $P \wedge N \wedge S$ is true, where S is the statement that an observer cannot verify whether there is intelligent life in a region of space-time that that observer is unable to receive causal signals from and P is the possible empirical theory that tells us that (1) causal signals do not travel faster than light, (2) it is physically possible (and overwhelmingly probable) that there are intelligent extraterrestrials, (3) it is also physically possible that there are not, and (4) there are large regions of space-time that any particular physically possible observer is unable to receive causal signals from. It is logically impossible to verify $P \wedge N \wedge S$ if it is true, and yet $P \wedge N \wedge S$ is a statement that P itself tells us could be true (and even assigns a probability to).

15. Such a theory must have an infinite number of interpretations if the number of individuals is infinite, and a number of interpretations that grows exponentially with the number of individuals if that number is finite.

16. *Reason, Truth and History* (Cambridge: Cambridge University Press, 1981), chapter 2 and Appendix.

I now think that although it is true that there is nothing incoherent about the idea that truth is nonepistemic, a satisfying response to the model-theoretic puzzle requires more than a foot-stamping declaration that it is. That is why the other two issues that I listed need to be addressed as well.

What to Make of "Verificationist Semantics"

Most versions of antirealism, including the one I called "internal realism," are versions of verificationism; and strange as it may sound, it was functionalism that pushed me in the direction of verificationism. Here are some passages from an essay I published the same year as *Reason, Truth and History*:[17]

> The brain's "understanding" of its own "medium of computation and representation" consists in its possession of a verificationist semantics for the medium, i.e. of a computable predicate which can represent acceptability, or warranted assertibility or credibility.

(This is, of course, the "functionalist" picture of what I called "the mind/brain" that I had at the time.) From this I concluded that "such rules must be 'computable'; and their 'possession' by the mind/brain/machine consists in its being 'wired' to follow them or having come to follow them as a result of learning."

In her insightful criticism of this essay, Louise Antony remarked that I specifically reject the alternative she advocates, "namely, a purely truth-conditional semantics for Mentalese, on the grounds that such a semantic theory will yield no account of comprehension."[18] She goes on, "He points out that, on the one hand, a truth-conditional theory, by itself, says nothing about what understanding of Mentalese sentences consists in." In fact, I wrote:

> If we interpret Mentalese as a "system of representation" we do ascribe extensions to *predicate-analogs* and truth conditions to *sentence-analogs*. But the "meaning theory" which represents a particular interpretation of Mentalese is not *psychology*. [A meaning theory in Davidson's sense] yields only

---

17. All these passages can be found in Hilary Putnam, "Computational Psychology and Interpretation Theory," in Putnam, *Philosophical Papers*, vol. 3, *Realism and Reason* (Cambridge: Cambridge University Press, 1983), 142–143. They are all discussed by Louise Antony in "Semantic Anorexia: On the Notion of Content in Cognitive Science," in George Boolos, ed., *Meaning and Method: Essays in Honor of Hilary Putnam* (Cambridge: Cambridge University Press, 1990), 105–135.

18. Antony, "Semantic Anorexia," 124.

such theorems ["T-sentences"] as " 'Snow is white' is true in Mentalese if and only if snow is white." This contains no psychological vocabulary at all.

The reason we cannot say that the brain's understanding consists in "knowing" such "T-sentences" is that the notion of knowing cannot be a *primitive* notion in subpersonal cognitive psychology. And if we try to explain the brain's "knowing the T-sentences" in terms of the brain's *use* of those sentences, we are still left with the task of explicating "use"; and that, I said, "is what a verificationist semantics gives (and as far as I can see), what *only* a verificationist semantics gives."

In short, if functionalism is the right philosophy of mind, and the "natural semantics for functionalist psychology" is verificationist, then verificationism must be the right account of the "mind/brain's" understanding of its own internal language ("Mentalese"). Unlike Antony, I rejected the idea that nature (more specifically, evolution) fixes the interpretation of our words. I rejected this proposal because the model-theoretic argument seemed to show that evolution *cannot* assign to our words a *determinate* reference.[19] In addition, this idea seemed to me to ascribe magical powers to nature. Again, we were at a standoff; I did not see how nature could fix the reference of our terms, and many other philosophers felt that there was no problem. Let us now turn to the third of the issues I listed.

Is Verificationism Itself "Coherent"?

What neither my critics nor I questioned was the coherence of the verificationist picture that I had painted. What the critics argued was that I had not *proved* that nature could not fix what our words refer to without regard to our intentions, and what I replied was that the idea of the world imposing reference on our terms or truth-conditions on our predicates makes no sense. But I did not consider the possibility that it was internal realism that made no sense, or, more precisely, that it made sense only if one was prepared to retreat all the way to solipsism (and perhaps not even then).[20]

---

19. See Hilary Putnam, "Das modeltheoretische Argument und die Suche nach dem Realismus des Common Sense," in Marcus Willaschek, ed., *Realismus* (Paderborn, Germany: Ferdinand Schöningh, 2000), 125–142.

20. See "Wittgenstein and Realism," Chapter 22 of this volume, for a discussion of the coherence/incoherence of solipsism.

What I mean is this: On a verificationist account of *understanding*,[21] whether it be early Carnap's, or Dummett's (antiphenomenalist) view, or my "internal realist" story, the only substantive notion of correctness available to a thinker is that of *being verified*. If that is the only notion of correctness that my "mind/brain" is supposed to be able to use, or if the ability to tell that a statement is verified is the only ability that passes Dummett's requirement that my ability to distinguish truths from falsehoods be capable of being behaviorally "manifested," then my talk about other people is only intelligible to *me* as a device for making statements that are or will be verified by *my* experiences.

In *Reason, Truth and History* I was aware of the danger that my position could lead to solipsism, but I thought that I could define an intersubjective notion of truth *in terms of* verification ("justification"), thus:

> *S* is true if and only if believing *S* would be justified if epistemic conditions were good enough.

The problem I faced (but was not aware that I faced at that time) was this: Let us suppose, as seems reasonable, that whatever makes it rational to believe that *S* makes it rational to believe that *S* would be justified were conditions good enough. If my understanding of the counterfactual "*S* would be justified if conditions were good enough" is *exhausted* by my capacity to tell to what degree it is justified to assert it, as my "verificationist semantics" claimed, and that is always the same as the degree to which it is justified to assert *S* itself, then I might as well have simply said that my understanding of *S* is just my capacity to tell what confirms *S* (and to what degree it does so). The talk of epistemically "ideal" connections must either be understood

---

21. Verificationism becomes solipsistic when it purports to be an account of how the individual thinker/speaker *understands* his own language (and/or "Mentalese"). Of course, one could hold the position that the only normative property of statements is being confirmed/disconfirmed to this or that degree, although not necessarily by oneself; i.e., one could be a "social verificationist" or, better, an "intersubjective verificationist." But if one allows an unreduced notion of a statement's *being verified by someone who may not be able to communicate with one* (perhaps because she lived long ago, or is too far away, or even outside one's light cone), then why not allow the notion of referring to objects that I cannot myself confirm/infirm statements about, without going through the epicycle of saying that someone else must be able to confirm/infirm statements about them? It was the form of verificationism that seemed to unify theory of understanding with an account of the sort of rightness that our statements about the world aim at that appealed to me.

outside the framework of internal realism, or it too must be understood in a solipsistic manner.

Moral: My "internal realism," far from being an intelligible alternative to a supposedly unintelligible *metaphysical realism*, can itself possess no *public* intelligibility. And the situation may be worse: Dummett is right to worry whether a verificationist account of understanding does not commit one to antirealism about the past. If it does, then, as Yuval Dolev has shown,[22] even methodological solipsism collapses into hopeless paradoxes. The best way to show that the realist position is not just one of two equally tenable positions is to show that the verificationist account entails solipsism (and probably even entails a self-refuting antirealism about the past). If this is right, then it clearly becomes vital to find an account of our capacity to understand and use language that "fits" with realism.

The Meaning of "Meaning"

Some of my readers may be wondering, did I not myself *propose* a semantics that fits with realism in "The Meaning of 'Meaning' "? I think that in large measure I did, and in a paper I wrote about the same time, I even emphasized my realist intentions.[23] So why did I backslide into a verificationist

22. Yuval Dolev, *Time and Realism: Metaphysical and Antimetaphysical Perspectives* (Cambridge, Mass.: MIT Press, 2007); and Dolev, "Dummett's Antirealism and Time," *European Journal of Philosophy* 8, no. 3 (December 2000), 253–276. Dummett himself pointed out, in *Truth and the Past* (New York: Columbia University Press, 2004), ix, that if one holds the antirealist view that statements about the past, if true at all, "must be true in virtue of the traces past events have left in the present," then it must be that the meaning of every empirical statement about the past changes every time the reference of "the present" changes—that is to say, it must change at every moment. But, as Dolev points out, any attempt to say what a statement about the past meant a moment ago must then likewise be semantically unstable—indeed, the change in question will be indescribable.

23. The paper I refer to is Hilary Putnam, "Explanation and Reference," in Glenn Pearce and Patrick Maynard, eds., *Conceptual Change* (Dordrecht: D. Reidel, 1973), 199–221; reprinted in Putnam, *Philosophical Papers*, vol. 2, *Mind, Language and Reality* (Cambridge: Cambridge University Press, 1975), 196–214. I wrote "The Meaning of 'Meaning' " in December 1972 and "Explanation and Reference" shortly thereafter. However, "The Meaning of 'Meaning' " was not published until 1975, in Keith Gunderson, ed., *Language, Mind, and Knowledge*, Minnesota Studies in the Philosophy of Science, vol. 7 (Minneapolis: University of Minnesota Press, 1975), 131–193; reprinted in Putnam, *Philosophical Papers*, vol. 2, *Mind, Language and Reality*, 215–271.

position in "Realism and Reason" and the book and papers that followed it?[24]

The answer is that in "The Meaning of 'Meaning'" I employed the notion of "reference."[25] And as I just explained, by the time I wrote "Realism and Reason," I had convinced myself that verificationist semantics was "the natural semantics for functionalist (or 'cognitive') psychology." And the (computational) "functionalism" I had invented and still subscribed to at that time was totally *internalist*. Exercises of *external-environment-involving capacities* were certainly not among the "functions" that I considered to lie within the province of "functionalist (or 'cognitive') psychology." And "verificationist semantics" does not employ the notion of "reference!"

I did not, however, repudiate "The Meaning of 'Meaning.'" Instead, I adopted what amounts to a "deflationary" account of reference. In effect, I thought, "Sure we can *say* that 'cat' refers to cats; but that is just a tautology. It says nothing about how we *understand* 'cat'. The latter is what verificationist semantics tells us."[26]

However, if verificationist semantics is *solipsistic,* as I just argued it is, then this move turns the "externalism" and "anti-individualism" of "The Meaning of 'Meaning'" into a *pseudoexternalism* and a *pseudo-anti-individualism,* just as construing other people as logical constructs out of *my Elementarerlebnisse,* as Carnap did in the *Aufbau,* turns the social dimension of language into a *pseudosocial* dimension, a "sociality" within a solipsistic world.[27] But I did not

24. In particular, Hilary Putnam, *Reason, Truth and History,* and Putnam, "Models and Reality," *Journal of Symbolic Logic* 45, no. 3 (September 1980): 464–482, reprinted in Putnam, *Philosophical Papers,* vol. 3, *Realism and Reason,* 1–25, and in Paul Benacerraf and Hilary Putnam, eds., *Philosophy of Mathematics: Selected Readings,* 2nd ed. (Cambridge: Cambridge University Press, 1983), 421–445.

25. More precisely, I employed the notion of "intending to refer," but that notion clearly presupposes the notion of reference.

26. See "Models and Reality."

27. Today there are philosophers (e.g., Hartry Field, Paul Horwich, and some of the "Sellarsians") who consider themselves "realists," but who advocate a deflationist account of "refers to" and thus give up the idea that reference involves any sort of correspondence to extra-linguistic realities; sometimes they claim that saying that our words are connected to those realities "causally" suffices to avoid metaphysical antirealism. Because what our words are connected to causally frequently differs from what they refer to semantically, I find this a very odd position. (For example, is it at all clear what the words "causally connected to" are causally connected to?)

see that that was what I was doing because I believed that my reference to "epistemically ideal" conditions bestowed a realist aspect on "internal realism." As I said previously, that was a mistake. For fifteen years that mistake enabled me to believe that I could defend *both* semantic externalism and "internal realism."

But what *should* we do after we have seen that "verificationist semantics" is a nonstarter? I think that we should do precisely what I said we should do in "The Meaning of 'Meaning.'" But what about the objections that the externalism of that essay (like the externalism of Kripke's *Naming and Necessity,* which has many similarities to mine)[28] presupposes the notion of reference? How can a naturalist accept an unreduced semantic notion? (Horror!)

Antireductionism and Liberal Functionalism

In a word, the answer is that an up-to-date liberal functionalist should not think that she has to *reduce* all the notions she uses to nonintentional notions. It is true that we have no idea how to reduce the predicate "refers to" to nonintentional predicates, but that does not mean that talk of certain organisms using signs to refer to certain sorts of things and certain sorts of events should be considered talk of "occult" entities or properties. Psychologists, anthropologists, sociologists, and linguists have long been investigating sign behavior, and most of them do not eschew talk of what signs *refer to*.[29] The identification of naturalism with such "reduction programs" as the program of reducing the intentional to the nonintentional or dispensing with intentional and normative notions entirely is a mistake, and I have been explaining how that mistake led me, at one time, to abandon the very realist intuitions with which I started. Some naturalists are reductionists, to be sure, and reduction programs *have* sometimes succeeded, but counting oneself as a naturalist does not require one to subscribe to reduction programs that are, as far as we can now tell, utterly unrealistic. The liberalized functionalism I advocate

---

28. Saul Kripke, *Naming and Necessity* (Cambridge, Mass.: Harvard University Press, 1980).

29. Noam Chomsky is an exception. He thinks that talk of reference has no place in serious study of language, and he adds that we can have no intuitions about reference because "the terms *extension, reference, true of, denote,* and others related to them are technical innovations which mean exactly what their inventors tell us they mean." Noam Chomsky, *New Horizons in the Study of Language and Mind* (Cambridge: Cambridge University Press, 2000), 148.

is an antireductionist but naturalist successor to the original, reductionist, functionalist program. For a liberalized functionalist, there is no difficulty in conceiving of ourselves as organisms whose functions are, as Dewey might have put it, "transactional," that is, environment involving, from the start.

What I have in mind in speaking of a "liberal functionalist" is someone who, like me (or like me today), accepts the basic functionalist idea that what matters for consciousness and for mental properties generally is the right sort of *functional capacities* and not the particular matter that subserves those capacities, but (1) does not insist that those functions be "internal," that is, completely describable without going outside the organism's "brain" (thus Gibsonian "affordances" and Millikan's "normal biological functioning" in an environment can all be involved in the description of the "functional organization" of an organism);[30] (2) does not insist that those capacities be described as capacities to *compute* (although she is naturally happy when computer science sheds light on some part of our functioning); and (3) does not even eschew intentional idioms, if they are needed, in describing our functioning, although she naturally wants an account of how intentional capacities grow out of protointentional capacities in our evolutionary history.

The first of these three characteristics of my liberalized functionalism implies that the capacities to function that are relevant to mentality have "long arms"; they reach out to the environment, as it were, instead of being just the programs of a computer in the skull. The second characteristic implies that the psychological, biological, and neurological vocabulary needed to describe those functions and capacities to function need not be described in a vocabulary drawn from one science (e.g., computer science) or in any way fixed in advance; and the third characteristic implies that, in particular, intentional idioms (e.g., "refers to") are not taboo. It should not surprise you that I see the details as largely something to be worked out by scientists in a number of different fields, but with philosophers playing the necessary, if often unappreciated, role of critics and "gadflies."

At one time Jerry Fodor, who was my student many years ago, and whose work, let me say, I admire and always find stimulating even when we disagree, made an interesting but, I believe, ultimately unsuccessful attempt to

---

30. James J. Gibson, *The Ecological Approach to Visual Perception* (Boston: Houghton Mifflin, 1979); Ruth Millikan, *Language, Thought and Other Biological Categories* (Cambridge, Mass.: MIT Press, 1987).

reduce the semantic to the nonsemantic using complicated counterfactuals.[31] I do not know whether he still believes that this can be done, but I do not believe that it can. In addition, I have argued that most counterfactuals, including the ones Fodor himself used, assume what David Lewis called a "similarity metric,"[32] that is, a set of judgments about which similarities are "relevant" and which are "irrelevant," that is determined by human interests and by judgments of what is *reasonable* given those interests, and that such judgments presuppose our conceptual abilities; it is a fantasy to think that the similarity metric is just "out there" anyway.[33] Indeed, ordinary talk of "causing" has the same presuppositions.[34] Reductionists are forced in practice to use notions they have no hope of reducing and that do not belong to the privileged vocabulary of mathematical physics. (It has, by the way, become so common to regard "causes" as a notion the reductionist is "obviously" entitled to use that hardly anyone notices this anymore!) Likewise, saying, as Quine did, that notions that are admittedly indispensable, such as the notion of "sameness of meaning," are without significance when we are engaged in "limning the most general traits of reality"[35] (on the ground that they cannot be accounted for in some supposedly "scientific" vocabulary) is a form of reductionism that the liberal functionalist repudiates once and for all.

Truth

The position I have just taken with respect to the problem of *reference* resembles the one Hartry Field took in 1972 in his justly famous paper "Tarski's Theory of Truth,"[36] with one important difference. Field rightly pointed out that Tarski's *philosophical* claims for his theory were somewhat exaggerated.[37]

31. Jerry A. Fodor, *A Theory of Content* (Cambridge, Mass.: MIT Press, 1990).

32. David Lewis, *Counterfactuals* (Oxford: Blackwell, 1973).

33. This is argued in Hilary Putnam, *Renewing Philosophy* (Cambridge, Mass.: Harvard University Press, 1992), chapter 3.

34. Ibid., 47–48; see also the example of "the cause of the forest fire" on pp. 212–214 of the reprint of "Why There Isn't a Ready-Made World," *Synthese* 51 (May 1982): 141–168; reprinted in Putnam, *Philosophical Papers,* vol. 3, *Realism and Reason,* 205–228.

35. W. V. Quine, *Word and Object* (Cambridge, Mass.: MIT Press, 1960), 161.

36. Hartry Field, "Tarski's Theory of Truth," *Journal of Philosophy* 69, no. 13 (1972): 347–375.

37. But its mathematical significance was not, which is great.

What makes them exaggerated is that a notion that Field called "primitive reference" (that is, reference for the primitive terms of the formal language to which the theory is applied) is not *explained* when a Tarskian "truth-definition" for a particular language is constructed,[38] but is merely described by means of a *list,* for example, the list

(R) <"cat," The set of all cats>, <"longer than," the set of all ordered pairs *x, y* such that *x* is longer than *y*>, <"red," the set of all red things>, "<apple," the set of all apples>," . . .

This list tells us that in the language for which this particular "truth-definition" is constructed, "longer than" refers to the ordered pair <*x, y*> if and only if *x* is longer than *y*, and so on,[39] but this hardly counts as a clarification of *what it is* for a sign to "refer" to something.

When he published the paper I refer to, Field thought that what we need to do is to supplement Tarski's theory with a further theory—one yet to be discovered—that would explicate the notion of primitive reference in "physicalist" terms. Despairing of that, he now advocates a deflationary account of reference and truth. What he should have done is to recognize that it is perfectly acceptable that the notion of primitive reference is explained informally using our *unreduced* notion of reference (which is, in fact, more or less what Tarski in fact did!). But this, it is important to add, does not preclude either empirical or conceptual inquiry into reference and into the practice of referring. Field's trajectory illustrates the way in which, when liberal functionalism goes missing as a position, various untenable positions—among which I include deflationism as well as verificationism—may seduce a philosopher.

I have time to mention just one more example of an area in which liberal functionalism often goes missing. John McDowell is one of the deepest contemporary philosophers of mind, and I have been thinking about and commenting on his work ever since his *Mind and World* first appeared. However, like many other philosophers of mind, I have trouble with his claim that all perceptual experience (all "impressions," in the language of *Mind and World*)

---

38. The reader should think of a language whose primitive terms include "cat," "longer than," and "red." Because a formalized language has only finitely many primitive predicates and individual constants, the list in question will always be finite.

39. A pair <*x, y*> is in the extension of the predicate "longer than" if and only if *x* is longer than *y;* an object *x* is in the extension of the predicate "red" if and only if *x* is red; an object *x* is in the extension of the predicate "apple" if and only if *x* is an apple; and so on.

are *conceptualized*. But McDowell is unmoved by all this dissent; the idea that impressions have *nonconceptual* content is one he strongly opposes. It would take a book and not a few remarks to explore this controversy at length (and in fact Hilla Jacobson and I are currently writing such a book), but I will conclude by indicating some of the lines of thought we are pursuing.

To do that, I need to say a word about McDowell's argument. McDowell believes that to deny that sense impressions are conceptualized is just as bad as denying that they have content altogether; he feels that both denials give game, set, and match to the skeptic. But it is important to understand what sort of skepticism McDowell has in mind.

James Conant has distinguished between two varieties of skepticism, which he calls "Cartesian skepticism" and "Kantian skepticism."[40] "Cartesian skepticism," in Conant's sense, is skepticism about the possibility of knowledge of things and events "outside" the mind; it seems to be just what most philosophers identify as "Humean" skepticism. "Kantian skepticism" is a puzzle about the very possibility that one's thoughts, whether seemingly about an external world or even about one's own sense impressions, can have content at all. It is the Kantian variety that McDowell was concerned to exorcize in *Mind and World*.

Exorcizing it in the way McDowell proposes to do it in *Mind and World* requires our accepting three claims. *First,* we have to accept "minimal empiricism." "Minimal empiricism," in McDowell's sense, requires that sense impressions *justify,* and not merely cause, our beliefs about observable things.[41] *The second claim* is that understanding any empirical concept involves knowing what experiences justify beliefs that have that concept as a constituent.[42] And *the third claim,* the claim that rules out "nonconceptual content," is that

---

40. James Conant, "Varieties of Skepticism," in Denis McManus, ed., *Wittgenstein and Skepticism* (London: Routledge, 2004). Conant claims that in *The Threefold Cord: Mind, Body, and World* (New York: Columbia University Press, 1999), I failed to see that Kantian skepticism was McDowell's target, and not Cartesian (or Humean) skepticism.

41. McDowell is concerned to provide a picture of our relation to reality and of the nature of knowledge that does not rule out the possibility of knowledge, but I believe that he recognizes that there is no such thing as a "proof the skeptic must accept" that we are not massively deceived. In this respect he follows Stanley Cavell, who in turn follows Wittgenstein. It is an open question whether McDowell has an account of the justification of beliefs about *unobservables*.

42. This is a claim that seems to imply some form of verificationism, by the way, but McDowell does not discuss this issue, as far as I know.

experiences (or sense impressions—McDowell uses both terms) can justify beliefs only if they are conceptualized in the sense of being "articulated" like propositions. (In a short paper published since *Mind and World*,[43] McDowell gives up the claim that the experiences of language-using human beings, which he refers to by the Kantian term "intuitions" in this paper, are articulated like propositions, but he insists that they are "conceptualized" in a different sense: he still holds that "we should center our idea of the conceptual on the content of discursive activity."[44] That is, McDowell takes concept use to be a linguistic achievement, so the learning of a language is a prerequisite for having experiences.)[45]

The second claim seems to imply some form of verificationism, but I shall not discuss that issue here. Instead, I shall discus the first and third claims.

I believe that it is quite possible that the best scientific account of the nature of, say, visual impressions may be *incompatible* with the claim that they are articulated like propositions (and also incompatible with the revised claim that "capacities that belong to the higher cognitive faculty" are needed to have visual impressions), but I do not see this as having the disastrous consequences McDowell fears. McDowell's argument is that if our experiences are not "conceptualized," they cannot justify accepting and rejecting perceptual and other beliefs; and then those beliefs must be seen as mere *responses* to those experiences. But that is not how I see things.

I would draw a distinction that McDowell never makes, between a full-fledged apperception, that is, an awareness that something is the case, and a bare sensation or bunch of sensations. (No doubt McDowell does not draw this distinction because he does not believe that there are any bare sensations. But I believe that there are unconceptualized sensations, or, to use Ned Block's term, "qualia.") Apperceptions are genuine experiences, and they are neither the same as sensations (in fact, there are "amodal" apperceptions, that is, apperceptions with no distinctive phenomenal character—e.g., the apperception that one is looking at an object that has a side one does not see), nor are they the same as perceptual beliefs (it can seem to one as if one apperceives

43. John McDowell, "Avoiding the Myth of the Given," in McDowell, *Having the World in View* (Cambridge, Mass.: Harvard University Press, 2009), 256–272.

44. Ibid., 262.

45. Let me repeat: I plan to discuss this revised version of the thesis that experiences are "conceptualized" in a future publication with Hilla Jacobson.

something one knows one does not apperceive, e.g., in the Müller-Lyer illusion, but it does not then seem to one that one believes something one does not believe). Much that McDowell says about experiences is, in fact, true of *apperceptions* (in the case of mature normal humans). Mature human apperceptions are conceptualized (and animal apperceptions involve what I have elsewhere called "protoconcepts"). But qualia, although they may be affected by our conceptual activity and even at times may be "fused" with "apperceptive ideas," as William James claimed, are not conceptualized in McDowell's sense.

As a liberal functionalist sees things, under normal conditions *neither* our perceptual experiences *nor* sentences we accept are the beginning of the process of forming a perceptual belief. The beginning is outside our heads; the process of forming a perceptual judgment to the effect that there is a notepad on this table is an exercise of a "function," in fact, a whole system of functions, some shaped by evolution, and some shaped by cultural processes, that connect me to objects and goings-on in my environment (in this case, to the notepad and to the table). Forming beliefs in accordance with our normal biological functions and our linguistic upbringing is not just uttering noises that are mere responses to sense impressions, although those sense impressions are a *part* of the causal chain that constitutes the normal formation of a particular perceptual belief on the basis of seeing something in one's visual field. On an externalist-functionalist story, for either our beliefs or the protobeliefs[46] of animals and prelinguistic children to have content is just for them to function as representations of external states of affairs. If we think of psychological states (including, importantly, apperceptions) as including objects, inputs, or items that are otherwise nonpropositional and in important cases not themselves mental, then we will have to come up with a new account of representation and of transitions from inputs to representations (obviously a demanding project), but we will not insist that the nonpropositional/nonmental items must be "conceptualized." As Steve Wagner has put it,[47] "Once we understand rational functions ecologically, we can put the state transitions inside them without facing a paradox about what *objects* could be doing there."

Whether such an account of content makes our beliefs "justified" or only "exculpated" is a different question: it connects with "Kantian skepticism"

---

46. On "protobeliefs," see my notion of a "protoconcept" in *Renewing Philosophy*, chapter 1.
47. In an e-mail commenting on a draft of this chapter.

only if one accepts McDowell's insistence that to have content, they have to be connected in a *justificatory* way with impressions, and his further insistence that only what is conceptualized can be justificatory.[48] I repeat, given an *externalist*-cum-functionalist notion of what content is, *of course* perceptual beliefs (of both children and animals, as well as human adults) have content. It is not the case that by rejecting McDowell's picture of experience as a "tribunal," I have left myself vulnerable to *that* form of skepticism.[49]

What I am saying is that bare sensations are never a "tribunal"; it is apperceptions that are the tribunal. (If McDowell's view is "minimal empiricism," then perhaps mine is minimal Kantianism.) Nevertheless, if we reach the level of thinking about thinking, we know that if such and such of my beliefs are justified, I probably should have had certain sorts of sensations when I did or underwent certain things. And knowing *that* presupposes the ability to *refer* to my sensations. But referring to a sensation no more requires the sensation to be "articulated like a proposition" than referring to a *cliff* requires the cliff to be articulated like a proposition. Within my total web of belief, I can ask whether my sensations are such as one would expect them to be if those beliefs were true. But that question does not presuppose that the sensations themselves are articulated like propositions. No doubt McDowell would say that by treating this as a question within my "web of belief," I am guilty of "coherentism." If so, it is not the "coherentism" he finds in Neurath and Davidson, according to which the only constraints on belief are *sentences* the believer accepts.

What I am suggesting does, however, flout the sharp line McDowell finds between questions of justification and questions of fact, between the so-called "space of reasons" and the so-called "realm of law." Discussing that dichotomy would require a much longer chapter, but I will just make two comments on it.

First, even if the dichotomy were absolute, which I do not think that it is, it would not imply that only what is shaped like a proposition can justify a proposition. That it does not is shown, I believe, by the phenomenon of justification

48. Re "justification": I follow Stanley Cavell in thinking that for most of our beliefs about the quotidian objects and goings-on around us, the question "Is that justified?" does not arise.

49. With respect to Cartesian skepticism, I also agree with Stanley Cavell that there is a "truth" in skepticism that must be acknowledged, but that *is not* to say that I "do not know" that there is a notepad on this table.

by what Wittgenstein called "imponderable evidence"[50] (his example was a sensitive observer's judgment that someone is *feigning* an emotion). I expect that McDowell will claim that this is not really a counterexample to his claim that evidence needs to be propositionally structured, but I am skeptical. (And I certainly do not think that the possible findings of neuroscience can be ruled out as irrelevant to whether it is or not.)

Second, the line between questions of natural law and matters that lie in "the space of reasons" is indeed sharp when "justification" has the sense of *deductive* justification; questions of what follows from what deductively are not decided by experiment. But the whole history of science suggests that when it comes to *nondeductive justification,* the line between "empirical fact" and "method" is blurry: we are often led by the success of novel sorts of theories to reconsider our canons of justification themselves. (The whole history of science since the Enlightenment could be described as a sequence of such reconsiderations.) We did not perform *experiments* to decide that, for example, it is not a principle required by reason itself that every event must have a cause, and that a difference in the effects must be traceable to some difference in the causes, but neither did we discover that independently of being impressed by the success of quantum-mechanical theories that denied this ancient principle. And it is not only when questions of fundamental physical theory are involved that we are forced to reconsider "what justifies what": at one time people thought that they knew very well that certain experiences confirm judgments that so-and-so is a witch, to cite a familiar example, but that "knowledge" had to be revised, and much more was involved in revising it than confronting judgments with sense impressions. The idea that evidence must be "conceptualized" is a belief that we need to reconsider in the light of neuroscience and ecological psychology.

Of course, none of these criticisms of McDowell's views is meant to be final. But it seems to me that the point of view suggested by liberal functionalism is quite different from McDowell's. The overarching methodological question is whether the findings of the natural sciences can really be declared irrelevant to philosophical issues about perception and cognition, as McDowell in effect does, or whether progress with regard to them requires constant attention to those findings, as I believe. The course we take in philosophy depends on our answer to that question.

---

50. Ludwig Wittgenstein, *Philosophical Investigations* (Oxford: Blackwell, 1953), II, 227–228.

4

On Not Writing Off
Scientific Realism

In 1975 I published an essay in which I claimed that "the positive argument for realism is that it is the only philosophy that doesn't make the success of science a miracle."[1] I have been aware for some time that a number of my close friends, philosophers I admire, think that this "no-miracles" argument for realism was a bad one. The list includes Yemima Ben-Menahem, Arthur Fine, and Axel Mueller. But I have not been persuaded. Of course, that might just mean that I am a stubborn old philosopher. However (although that does not prove anything, of course), I do not think that that is the case. Rather, it seems to me that when their criticisms of the no-miracles argument do not simply rest on misunderstandings of my position, they turn on a serious confusion between scientific realism and metaphysical realism. That is what I will be arguing, but I will also pay attention to the "moderate constructionist" stance defended by Alfred Tauber (in whose honor the paper from which this chapter derives was written), a stance that is incompatible

---

1. Hilary Putnam, "What Is Mathematical Truth?" in Putnam, *Philosophical Papers,* vol. 1, *Mathematics, Matter and Method* (Cambridge: Cambridge University Press, 1975), 60–78. The quotation is from p. 73.

with the scientific realism that the no-miracles argument was meant to defend.

Yemima Ben-Menahem's List of Criticisms

I will begin with Yemima Ben-Menahem's account of the no-miracles argument (which she calls "the argument from the success of science" in her paper "Putnam on Skepticism"),[2] and a list that she gives of some common criticisms of the argument. She writes:

> "Realism," Putnam used to say in the 1970s, "is the only philosophy that does not make the success of science a miracle." Briefly, the idea is as follows. Electrons figure in our explanation of the workings of electrical equipment, and genes figure in our explanation of hereditary diseases. Realists understand the success of predictions derived from theories employing these notions straightforwardly in terms of the existence of electrons and genes. By contrast, nonrealists, suspicious of "theoretical entities," maintain that such notions are merely fictitious constructs that happen to work. But how can such "as-if" notions, or the theories employing them, be of any explanatory value? And how can the phenomenon of success, the fact that such fictions yield successful predictions, be explained? Realism, Putnam therefore concludes, provides the only explanation for the success of science.[3]

The paragraph in which she lists (and seems to endorse) the criticisms I mentioned runs as follows:

> As a number of writers have noted, the argument from success has several (somewhat interdependent) limitations. First, it is precisely the inference to the best explanation of the kind in question that opponents of realism find unconvincing. They are unlikely to be persuaded by another, albeit more general, argument of the same kind. Second, the claim that scientific practice rests on realist assumptions, and would be inexplicable without them, has been challenged by adducing nonrealist grounds for the same procedures. Third, despite the formal analogy with hypothetico-deductive explanations within science, the more general argument for realism does

2. Yenima Ben-Menahem, "Putnam on Skepticism," in Yemima Ben-Menahem, ed., *Hilary Putnam* (Cambridge: Cambridge University Press, 2005), 125–155.

3. Ibid., 127–128.

not meet the standards of scientific explanation in terms of its putative empirical import: it does not yield new scientific predictions unavailable to its opponents.[4]

Let us consider these criticisms in turn.

1. *It is precisely the inference to the best explanation of the kind in question that opponents of realism find unconvincing. They are unlikely to be persuaded by another, albeit more general, argument of the same kind.*

Well, what argument for a philosophical position is "likely to persuade" committed opponents of that position? My efforts in philosophy have always been intended to provide intellectual and moral support to those who have realistic sensibilities in science and "cognitivist" sensibilities in ethics; I did not expect that "What Theories Are Not"[5] would convert Carnap to scientific realism, for example. (Even in my "internal realist" period—which I now regard as misguided—the presence of the noun "realism" in the phrase "internal realism" was no accident.) In particular, the no-miracles argument brings out just how strange it is to suppose that a bunch of equations involving various parameters should give us successful predictions if not a single one of those parameters corresponds to anything real. (Although I have learned a lot from James and Dewey, the fact that both of them were fictionalists about theoretical entities has always seemed to me regrettable.)

2. *That scientific practice rests on realist assumptions, and would be inexplicable without them, has been challenged by adducing nonrealist grounds for the same procedures.*

Well, what "nonrealist grounds"? Bas van Fraassen, for example, simply "writes in by hand" the idea that scientists should try to construct theories that not only lead to successful prediction individually, but whose conjunction also forms a coherent system.[6] But if theories are only prediction devices, why should we care whether the various prediction devices that we use have the form of a *theory,* that is, a set of statements that *appear* to describe the world, rather than the form of a bunch of algorithms or even a bunch of mutually

---

4. Ibid., 128.

5. Hilary Putnam, "What Theories Are Not," in Ernest Nagel, Patrick Suppes, and Alfred Tarski, eds., *Logic, Methodology, and Philosophy of Science* (Stanford, Calif.: Stanford University Press, 1962), 240–252; reprinted in Putnam, *Philosophical Papers,* vol. 1, *Mathematics, Matter and Method,* 215–227.

6. Baas van Fraassen, *The Scientific Image* (New York: Oxford University Press 1980).

incompatible theories, as long as we know which to use when? If this last idea seems absurd, recall that Bohr believed that this is precisely what we do in quantum mechanics—according to him, we sometimes use (classical) wave forms of thinking and sometimes use (classical) particle ways of thinking and supplement both with the mathematics of quantum mechanics, which he regarded as a mere algorithm. (It is often forgotten that this is all that the famous "Copenhagen interpretation" of quantum mechanics amounted to in Bohr's version. As Murray Gell-Mann once remarked, "Niels Bohr brainwashed a whole generation of physicists into believing that the problem [of interpreting quantum theory] was solved fifty years ago.")[7]

3. *Despite the formal analogy with hypothetico-deductive explanations within science, the more general argument for realism does not meet the standards of scientific explanation in terms of its putative empirical import: it does not yield new scientific predictions unavailable to its opponents.*

To unpack this can of worms, let us examine this talk of "the standards of scientific explanation." I argued in "The Idea of Science" and in "The Diversity of the Sciences"[8] that (1) different sciences exemplify very different methodologies; and (2) unless we want to declare that history, including historical linguistics, and many other human sciences are not "science,"[9] it is wrong to posit that only what resembles *physics* is really "science." In fact, even in physics the idea that a theory must imply novel predictions to be "explanatory" is highly problematic for the following reasons.

First, consider the case of two theories that imply the same predictions but are not intertranslatable, for example, von Neumann's version of quantum mechanics and Bohm's. Because von Neumann's version was invented first,

7. Murray Gell-Mann, "What Are the Building Blocks of Matter?" in Douglas Huff and Omer Prewett, eds., *The Nature of the Physical Universe: The 1976 Nobel Conference* (New York: John Wiley and Sons, 1979), 27–46.

8. Hilary Putnam, "The Idea of Science," in P. A. French, T. E. Uehling, and H. K. Wettstein, eds., *The Philosophy of the Human Sciences,* Midwest Studies in Philosophy, vol. 15 (Notre Dame, Ind.: University of Notre Dame Press, 1990), 57–64; reprinted in Putnam, *Words and Life* (Cambridge, Mass.: Harvard University Press, 1994), 481–491; Hilary Putnam, "The Diversity of the Sciences: Global versus Local Methodological Approaches," in Philip Pettit, Richard Sylvan, and Jean Norman, eds., *Metaphysics and Morality: Essays in Honor of J. J. C. Smart* (Oxford: Basil Blackwell, 1987), 137–153; reprinted as "The Diversity of the Sciences", in Putnam, *Words and Life* (Cambridge, Mass.: Harvard University Press, 1990), 463–480.

9. And if we do declare this, what is supposed to follow? That there is no well-confirmed knowledge in these areas? That would be absurd.

it would then be Bohm's that failed the test of implying novel predictions and would thus be declared not to "meet the standards of scientific explanation." But if Bohm's had been thought of first, then it would be von Neumann's version that failed to meet those standards. In my view, the late J. S. Bell was right to say that the Bohm interpretation deserves much more attention than it has received, precisely because it makes sense of the phenomena predicted by quantum mechanics, while the von Neumann version posits an unacceptable dependence on the "observer."[10] Perhaps Bohm's theory is wrong, but the simple question "Does it lead to novel predictions?" fails to get to the real problems with it, which have to do with whether it can be reconciled with relativity.

Second, in fact not all successful scientific theories imply novel predictions. For example, it is doubtful that Darwin's theory of natural selection "implies predictions," although it has led to a host of more specific explanatory theories that do imply predictions. Natural selection is, indeed, a paradigm case of "an inference to the best explanation" that is accepted for its simplicity, plausibility, fruitfulness, and ability to unify specific explanations, even though it does not by itself lead to novel predictions.

Third, even in physics, but even more in the historical sciences, hypotheses are often regarded as confirmed if they unify the explanations of phenomena that previously had to be simply accepted on the basis of observation. The right question to put to the scientific realist explanation of the success of science is whether it does something of this sort. And Ben-Menahem herself notes that

> Putnam elaborates on the argument from success to highlight its analogy with hypothetico-deductive arguments as they figure in science. The claim here is that scientific practice is based on assumptions that make sense from a realist point of view but are unjustifiable from a nonrealist perspective. For example, scientists will typically conjoin several theories to derive new predictions. This procedure is understandable if each of the conjoined theories is considered to be true, for truth is preserved under conjunction. But utility, simplicity, economy, beauty and other attributes cited by nonrealists as surrogates for truth in the evaluation of theories lack this characteristic, that is, are not preserved under conjunction. Thus even the commonplace practice of conjoining theories cannot be justified on the nonrealist premise.[11]

10. J. S. Bell, *Speakable and Unspeakable in Quantum Mechanics* (Cambridge: Cambridge University Press, 1987, revised edition 2004).

11. Ben-Menahem, "Putnam on Skepticism," 128.

This is indeed an example of the way in which the supposition that well-confirmed theories are in general approximately true *explains* a feature of scientific method that otherwise has to simply be written in by hand, as it is by van Fraassen, which makes it all the more surprising that Ben-Menahem endorses this criticism.

In the rest of her paper, however, Ben-Menahem turns to a very different line of criticism. This is that later views of mine, with which she is in greater sympathy, supposedly make the appeal to the no-miracles argument otiose. I now turn to this part of her argument.

Yemima Ben-Menahem's Arguments from Internal Realism

In my closing lecture to my "80th Birthday Conference" in Dublin,[12] I confessed that in "Realism and Reason," the lecture I described as the "manifesto" of my internal realist period (my presidential address to the 1976 meeting of the Eastern Division of the American Philosophical Association),[13] I used "internal realism" in two different senses. At the beginning of that lecture the phrase was explained as synonymous with "scientific realism," and I explained the presence of the adjective "internal" thus: scientific realism is (I claimed) *science's* explanation of the success of science. It is "internal" in the sense of being internal to *science*. But, as I said in the Dublin lecture, "In a section titled 'Why All This Doesn't Refute Internal Realism,' I identified 'internal realism' with the view that whether a theory has a unique intended interpretation 'has no absolute sense.' At that point it is clear that 'internal realism' was *now* a name for the view I had developed in the lecture as a whole, a view on which truth and idealized verifiability were supposed to coincide, and *not* a name for the view described in the opening paragraphs that I just mentioned, the view on which *both* metaphysical realists and holders of the antirealist view I now advanced were supposed to be able to agree." I mention all this because the fact that "internal realism" has both of these very different senses in that "manifesto," a document from which Ben-Menahem quotes, explains some of the things she says.

---

12. The lecture was titled "From Quantum Mechanics to Ethics and Back Again"; it is Chapter 2 of the present volume.

13. Hilary Putnam, "Realism and Reason," reprinted in Putnam, *Meaning and the Moral Sciences* (London: Routledge and Kegan Paul, 1978), 123–138.

Here is what she writes:

At this point Putnam is obviously still committed to the argument from success. It seems to me, however, that this commitment is but a vestige of his earlier understanding of realism, and is bound to clash with the internal perspective. If my description of the transition to the internal perspective is an accurate reconstruction, internal realism should no longer be seen as an explanatory theory. If, in particular, in view of the model-theoretic argument, the concept of an independent reality to which true representations correspond is paradoxical, it remains paradoxical when used as a quasi-scientific explanation of success. Certainly, insofar as the best theory of the world is concerned, the inconceivability of its turning out to be false also implies that its truth cannot and need not be explanatory or explained, at least not in the ordinary sense of scientific explanation. As indicated above, I see the idea that truth and reference are to be taken at face value as the main thrust of internal realism. Truth and reference can neither explain the superphenomenon of human success, nor be explained by a superscientific theory. Taken at face value, they are irreducible to other notions and do not play the explanatory role assigned to the theoretical concepts of science.[14]

This would be perfectly correct if the target of the no-miracles argument were the same as the target of "internal realism" in the sense of the main body of "Realism and Reason," but it was not. Of course, because of the unfortunate slide on my part from one sense to another, I can certainly excuse Ben-Menahem for supposing the contrary. But let me say what I intended (I explain this in more detail in Chapter 2): the target of the no-miracles argument (what Ben-Menahem calls "the argument from success") is *not* antirealism in the sense of *antirealism about truth*. It is operationalism with respect to scientific theories. And one can perfectly well be a realist about truth, indeed, the rather special sort of realist about truth I called a "metaphysical realist" in "Realism and Reason," and an instrumentalist about scientific theories (today, Bas van Fraassen is an example of such a position), or an antirealist about truth and an anti-instrumentalist about scientific theories (my position in my "internal realist" period). The role that the "argument from success" played in my "internal realist" period was to argue that empirical explanations of what I called "the contribution of language using to success in achieving our goals," and, in particular, explanations of the success of the scientific method

14. Ben-Menahem, "Putnam on Skepticism," 133.

as I described it, are not rendered meaningless or otiose by the "verificationist semantics" I defended then.[15] Ben-Menahem assumes that I attempted to explain human success by appealing to the existence of an independent reality in a sense of "independent" that *excludes* verificationism, but that is a mistake. (That is a queer sense of "independent," by the way; the idea that verificationists deny the causal independence of external reality has always seemed to me a howler.) Whether reality is "independent" or not in that problematic sense has nothing to do with, for example, whether we should regard quantum mechanics the way in which Bohr wanted us to regard it or the way in which scientific realists like J. S. Bell want us to regard it; but the no-miracles argument has everything to do with the latter question.

Where Ben-Menahem is right is that when it comes to combating antirealism about truth, as opposed to instrumentalism about scientific theories, the "argument from success" does not work. But that is what I myself pointed out in the opening paragraphs of "Realism and Reason," and that is why I thought that "internal realism" in the sense of acceptance of the no-miracles argument was *compatible* with "internal realism" in the sense of a (would-be) sophisticated verificationism.

Ben-Menahem also appeals to my more recent writings about skepticism when she writes:

> The skeptic . . . is not merely portrayed as violating the canons of scientific method, but as purporting to doubt that which cannot reasonably be doubted. Putnam responds[16] to Strawson's query about "whether the nonexistence of the external world is really a coherent idea" as follows: "I have not been so far enabled (by the skeptic or, for that matter, by his familiar opponent, the traditional epistemologist) to give it a coherent sense." This formulation is typical of Putnam's recent writings, and illustrates the difference between his earlier and later responses to skepticism.[17]

---

15. I find it strange that Ben-Menahem writes, "I see the idea that truth and reference are to be taken at face value as the main thrust of internal realism." In my internal realist period I identified truth with "idealized rational acceptability," a metaphysical picture of a Kantian kind, as I pointed out, and insofar as I discussed reference at all, I gave a deflationary account. I am not sure I know what it is to take truth and reference at "face value."

16. Ben-Menahem is referring to Hilary Putnam, "Strawson and Skepticism," in *The Philosophy of P. F. Strawson,* Lewis E. Hahn, ed., (La Salle, Ill.: Open Court, The Library of Living Philosophers, 1998), 273–287; reprinted in this volume as Chapter 31.

17. Ben-Menahem, "Putnam on Skepticism," 134.

Here I want to say that the no-miracles argument was *never* intended as an argument against skepticism about the existence of the external world. Ben-Menahem is right that I am very concerned in recent writing to show that certain forms of antirealism are *incoherent*. (This is especially true of my recent writing about Dummett's position.)[18] But I do not agree that refuting antirealism makes it unnecessary to defend anti-instrumentalism.

Arthur Fine and Axel Mueller

In a paper in the same volume provocatively titled "Realism, beyond Miracles," Fine and Mueller make a strange error concerning the terms I used over and over for many years. They write, "Are empirical descriptions really capable of objective truth or falsity? In reaction to this challenge, Putnam experimented with a defense of these core commitments by re-presenting them as a substantive view alternatively called 'scientific realism' or 'metaphysical realism.'"[19] In fact, I *never* used these two terms as synonyms, and in fact, as I just pointed out, I defended both "internal realism" (the very antithesis of "metaphysical realism") and "scientific realism" in "Realism and Reason." What is more important is that the idea that the no-miracles argument is supposed to establish the existence of "independent reality" in a sense that ipso facto excludes instrumentalism gets combined with the further mistake of supposing that I *no longer believe* that it makes sense to say that our truth claims are answerable to a reality not of our making. They write:

> According to [the no-miracles argument], the correspondence of our true claims and referential terms with a unique mind-independent reality explains the success and the communicability of scientific claims. . . . We shall see, as Putnam came to realize, that such a defense fails badly; that it neither lends support to realism nor excludes anti-realism; in particular, instrumentalism.

18. See Hilary Putnam, "Between Scylla and Charybdis: Does Dummett Have a Way Through?" in R. E. Auxier and L. E. Hahn, eds., *The Philosophy of Michael Dummett* (Chicago: Open Court, 2007), 155–167; Putnam, "Between Dolev and Dummett: Some Comments on 'Antirealism, Presentism and Bivalence'," *International Journal of Philosophical Studies* 18, no. 1 (February 2010): 91–96; and Putnam, "Reply to Michael Dummett," in R. E. Auxier, ed., *The Philosophy of Hilary Putnam* (Chicago: Open Court, forthcoming).

19. Arthur Fine and Axel Mueller, "Realism, beyond Miracles," in Yemima Ben-Menahem, ed., *Hilary Putnam* (Cambridge: Cambridge University Press, 2005), 83–124. The quotation is from p. 85.

Putnam's reaction to the failure is not to give up the referential princi-ples, but to revert to the participant's perspective and to deepen his reflec-tion on their use in evaluating empirical claims. In the course of this reflection he gradually dismantles the metaphysical realist picture of a mind-independent reality. . . . The key element here is Putnam's insistence on the fact that, from the participant's perspective, there is no access to any reality but by describing it in a certain way—that is, by using certain conceptual systems. The idea of an absolutely mind-independent, totally unconceptualized reality, since indescribable, is also not usable for any purposes. The upshot is that our notion of a statement's objective correct-ness does not entail commitment to any theory-neutral domain, but only commitment to the public revisability of our claims.[20]

On this, my comments are as follows:

1. "The correspondence of our true claims and referential terms with a unique mind-independent reality" was exactly the thesis of the position I called "metaphysical realism." I repeat, it was never the thesis of the position I called "scientific realism."

2. "The idea of an absolutely mind-independent, totally unconceptualized reality, since indescribable, is also not usable for any purposes." That there is water on the side of the moon we could not see or photograph until space travel became a reality is certainly causally independent of the existence of minds, and no one supposes otherwise. So "mind-independent" cannot mean "causally independent." That proposition is also (obviously) logically independent of the statement that organisms with minds have evolved, and I do not suppose that Fine and Mueller think otherwise. So "logically independent" cannot be what "mind-independent" means. *So what does "mind-independent" mean?* When I misguidedly used that expression in describing "metaphysical realism," what I meant was that the metaphysical realist believes that there are truths that *do not depend on whether or not human beings or other sentient beings could or could not verify them.* Now, I do believe that there are such truths. Many cosmologi-cal truths must be of that sort, partly for logical reasons—the impossibility, for example, of verifying a negative existential statement like *P*: "There are no in-telligent extraterrestrials," in case it is true[21]—and partly for empirical reasons

20. Ibid., 84–85.
21. More precisely, the impossibility of verifying the conjunction $P \wedge N \wedge S$ if it is true, where S is the statement that an observer cannot verify whether there is intelligent life in a

(the inaccessibility of information from beyond the "event horizon," or from the interiors of black holes).[22] But are there facts that human beings could not even *conceive?* Doubtless yes. It certainly seems possible that there should be organisms with minds much better than ours, and why should they not be able to conceive of states of affairs we cannot even understand, and, with luck, even verify their existence? "But such states of affairs would not be usable for any purpose." Not by us, of course, but for them they would be.

So have I become a metaphysical realist in my old age? Yes and no. As I used to explain "metaphysical realism,"[23] what it came to was the conjunction of (1) the rejection of verificationism and (2) the denial of conceptual relativity. I believe that the rejection of verificationism was right, but not the denial of conceptual relativity. My "metaphysical realist" believed that a given thing or system of things can be described in exactly one way if the description is complete and correct, and that that way fixes exactly one "ontology" and one "ideology" in Quine's sense of those words, that is, exactly one domain of individuals and one set of predicates of those individuals. Thus, according to my "metaphysical realist," it cannot be a matter of convention (as I have often argued that it is) whether there are such individuals as mereological sums; either the "true" ontology includes mereological sums or it does not. And it cannot be a matter of convention (as I have often argued that it is) whether space-time points are individuals or mere limits.

To be sure, this is *one form* that metaphysical realism can take. But if we understand "metaphysical realist" more broadly, as applying to all philosophers who reject verificationism and all talk of our "making" the world, then it is perfectly possible to be a metaphysical realist in *that* sense and to accept

---

region of space-time that that observer is unable to receive causal signals from and *P* is the possible empirical theory that tells us that (1) causal signals do not travel faster than light, (2) it is physically possible (and highly probable) that there are intelligent extraterrestials, but (3) it is also physically possible that there are not, and (4) there are large regions of space-time that any particular physically possible observer is unable to receive causal signals from. $P \wedge N \wedge S$ is a statement that it is *logically* impossible to verify if it is true, and yet it is a statement that P itself tells us could be true (and even assigns a probability to).

22. See Hilary Putnam, "Pragmatism," *Proceedings of the Aristotelian Society* 95, no. 3 (March 1995): 291–306; and Putnam, "When 'Evidence Transcendence' Is Not Malign: A Reply to Crispin Wright," *Journal of Philosophy* 98, no. 11 (November 2001): 594–600.

23. See particularly Hilary Putnam, "A Defense of Internal Realism," in Putnam, *Realism with a Human Face* (Cambridge, Mass.: Harvard University Press, 1990), 30–42.

the existence of cases of such "conceptual relativity." And that is the sort of realist I am.

(In fact, as I pointed out in Chapter 2, in mathematical physics conceptual relativity is a ubiquitous phenomenon.)

3. "The upshot is that our notion of a statement's objective correctness does not entail commitment to any theory-neutral domain, but only commitment to the public revisability of our claims." I do not agree, and I am troubled that this is presented as a description of *my* position. The system described in "Bosonization as Duality"[24] is certainly theory neutral (what do bosons know about theories?), and the objective correctness of statements about it does depend on whether they describe it as it is. But I do not say this because I accept the no-miracles argument, or anyway not *just* because I accept the no-miracles argument. I reject the idea that there is no objective correspondence between true descriptive statements and mind-independent states of affairs because all the alternatives to that sort of "representationalism" seem to me to be either demonstrably incoherent (for example, how can we talk about *public* revisability of *claims* without referring to *people* and acts of claiming?) or postmodernist gobbledygook.

It may be, however, that what drives Fine and Mueller to suppose that I do not regard realism as having any factual content is the fact that (after giving up "internal realism") I argued that various forms of antirealism are incoherent. They may well think that this means that I must regard realism as at most a tautology because it is the denial of an incoherent set of views. Most "Wittgensteinians" think, for example, that if we cannot make sense of a philosophical claim, then the negation of that claim must be a "grammatical" truth, if it is intelligible at all. On this reading of the later Wittgenstein, if Wittgenstein thinks that such "hinge propositions" as "the world has existed for a long time" do not have intelligible negations, then he must also think that the fact that the world has existed for more than five minutes is something like an *analytic* truth. But (apart from confusing Wittgenstein's "hinge propositions" with Carnapian "framework propositions") this depends on just the sort of notion of a philosophically contentful "analytic sentence" that Quine demolished. In my view, even when the principles of Euclidean geometry were still "conceptual truths," they were also meaningful statements about the world. To say that

24. C. P. Burgess and Fernando Quevedo, "Bosonization as Duality," *Nuclear Physics* B 421 (1994): 373–390; e-Print Archive: hep-th/9401105; http://arxiv.org/abs/hep-th/9401105.

scientific claims are answerable to reality is not, in my view, the same as saying that they are answerable to our practices (which would make them "grammatical truths" in a certain "language game").

However, to return to my claim that finding metaphysical *antirealism* incoherent is not *sufficient* reason for accepting realism with respect to theoretical entities, the point here is that belief in objective representation does not ipso facto exclude the possibility that theoretical terms in science are not genuine representations, but just useful devices. Instrumentalism does not, to repeat (I hope not ad nauseam), stand or fall with realism about reference (nor does realism about reference exclude recognizing cases of conceptual relativity, as I pointed out earlier). That the combination of realism about observables plus antirealism about *unobservables* makes the success of science a miracle is a separate argument.

Fred Tauber

Last but very much not least, I come to Fred Tauber. Readers who are acquainted with his *Science and the Quest for Meaning*[25] and have looked at the back cover will know that it is a book that I am enthusiastic about. In my blurb I say, among other things, that "it suggests a program for what the author calls a 'moral epistemology' as a way to do justice to the claims of both sides in that struggle [between humanists and scientists]. The effort is vitally important, even if one has, as I am sure many will, disagreements with this or that part of Tauber's argument, and the synoptic vision of 'how we got here' is amazing, as are Tauber's moral sensitivity and breadth of knowledge. This is a book that anyone who cares about repairing the fractures in our culture should read and ponder." Here I will consider the chapter with which I *disagreed* in that book, but I hope that my readers will understand that ever since Socrates' time the way one honors a philosopher one admires is by quarreling with him in public.

The chapter with which I disagree is chapter 4, "The Science Wars." In that chapter Tauber defends a moderate "social constructivism" that comes too close, to my ears, to Nancy Cartwright's position in *How the Laws of Nature Lie*[26] or

25. Alfred Tauber, *Science and the Quest for Meaning* (Waco, Tex.: Baylor University Press, 2009).

26. Nancy Cartwright, *How the Laws of Nature Lie* (Cambridge: Cambridge University Press, 1983). Tauber cites her approvingly in *Science and the Quest for Meaning*, 128.

even to Richard Rorty's view that science is valuable only as a paradigm of free discussion and mutual criticism.[27] What goes missing in this chapter is *my* side, the scientific realist side, in this debate. In one way, this is to the good. By not mentioning my no-miracles argument, also known as "the argument from the success of science," Tauber completely avoids the charge that he misunderstood/misrepresented that argument, the charge I have leveled against Ben-Menahem, Fine, and Mueller. Here, it seems to me, is a group of sentences that lies at the very heart of his argument against scientific realism:

> "Truth" then is a moving target, a category that helps organize the investigative endeavor, but no criteria avail themselves for approximating how close current theories are to some final vision. If this is relativism, then it is only interesting as a sociological depiction, namely, reality looks different now as opposed to then (as well as different from here as opposed to there). Accepting such a relativist position does not allow the dismissal of scientific knowledge as somehow an arbitrary construction or that reality is only a construct. If the relativist is confident to go in an airplane or subject himself to open heart surgery, then debunking science as a construct (and thus somehow arbitrary) cannot be his game. He cannot be saying that science provides as "truthful" a picture of nature in 2011 as it did in 1611. Rather, he holds to a picture of truth that is *the best consensual approximation to some unknown ideal and the most pragmatically useful instrument for various ventures, again as determined by the group's own notions of human flourishing.*[28] *Accordingly, each era holds reality within its own cultural moment, and that scheme changes with time.*[29]

Although Tauber identifies agreement with reality with "some unknown ideal," unlike Fine and Mueller, he clearly thinks that reality does somehow constrain science (it is not just because we believe in the "public revisability of claims" that airplanes fly and open-heart surgery prolongs many lives). However, he virtually takes back this concession to realism by saying that whether such things are the case is determined by "the group's" (*what* group's?) "own notions of human flourishing." "Fred," I would say, "while it *is* a matter of our conception of human flourishing that traveling long distances quickly

---

27. Cited approvingly by Tauber, *Science and the Quest for Meaning,* 122.

28. Here Tauber has "Latour 1999." The reference is to Bruno Latour, *Pandora's Hope: Essays on the Reality of Science Studies* (Cambridge, Mass.: Harvard University Press, 1999).

29. Tauber, *Science and the Quest for Meaning,* 121 (emphasis added).

is a good thing and that living longer is a good thing, it is a fact that airplanes enable one to do the first and that open-heart surgery sometimes enables one to do the second *whether or not one thinks that these things are conducive to human flourishing.* Having gone part way to realism, you should not have retreated to Rortianism." Nevertheless, in this passage he does raise what seem to me far more difficult issues than the ones canvassed by Fine and Mueller.

To explain those issues, let me remind the reader of one of my favorite Woody Allen movies, *Sleeper,* in which Miles Monroe, the owner of a health-food store, is unfrozen (after several centuries of cryogenic suspended animation), only to discover that medical science has "established" that smoking is good for you. Medical science has not, of course, changed its mind about the badness of smoking, but it has changed its mind about many things. And if science constantly changes its mind, then why should one be a "realist" of any kind about its findings? And if one says that most scientific "findings" are at least *approximately true*, then where are the criteria by which we can determine how good the approximation is?

Of course, the problem becomes even worse if one agrees with Eddington[30] that the layman's table does not really exist (because physics has shown that tables consist mainly of empty space) and with Sellars[31] that the manifest image is false (another way of saying the same thing). Even if "the scientific table" really exists, as Eddington thought, the person on the street knows from beans about "the scientific table." So *nothing* the person on the street actually talks about really exists, if Eddington and Sellars are right! And if each century's physics has a different "image" of the table, then the previous centuries' "scientific table" does not really exist either!

Against this, I have long argued[32] that the principle of charity in interpretation[33] requires that we interpret the term "solid" in such a way that the

30. Sir A. S. Eddington, *The Nature of the Physical World* (Cambridge: Cambridge University Press, 1928). The introduction begins: "I have settled down to the task of writing these lectures and have drawn up my chairs to my two tables. Two tables! Yes; there are duplicates of every object about me—two tables, two chairs, two pens." (Eddington's "two tables" are the commonsense table and "the scientific table.")

31. Wilfrid Sellars, "Philosophy and the Scientific Image of Man," in Robert Colodny, ed., *Frontiers of Science and Philosophy* (Pittsburgh: University of Pittsburgh Press, 1962), 35–78; reprinted in Sellars, *Science, Perception and Reality* (New York: Humanities Press, 1963).

32. See Putnam, *Meaning and the Moral Sciences,* 22–25.

33. Cf. my "principle of benefit of the doubt," ibid., 24.

discovery that tables consist mostly of empty space is *not* incompatible with saying that most tables are solid; indeed, although Newton already knew that tables consist mostly of empty space (ignoring, of course, *fields!*), there is a *branch* of physics called "solid-state physics"; so physicists themselves have for the most part automatically applied the principle of charity and have not adopted the Eddington-Sellars view. It is true that the "images" associated with one century's physics are often incompatible with the images associated with the next century's, but I have long argued that we do not have to accept what I once called "the pessimistic meta-induction"[34] to the conclusion that all scientific theories sooner or later turn out to be completely false. We do not accept the picture of space as Euclidean or of time as absolute, but it is well known that we usually compute trajectories, even the trajectories of satellites, by employing Newtonian physics. We can do this because Newtonian explanations can easily be reformulated in relativistic terms (when the velocities are not too great and other conditions are met). In this sense Newtonian physics is *robust;* it is "approximately true," and later physics tells us just how good the approximation is.

But can we say that a theory is approximately true *before* we have a later theory that tells us how good the approximation is? Well, Newton certainly thought so. Not only did he *not* claim that his theory of gravitation was perfectly accurate; he speculated that, for example, the power "2" in the famous inverse square law might itself be only an approximate figure, and he also emphasized that his "action-at-a-distance theory" might itself be replaceable by a "local-action" theory when we had more knowledge. To say that the concept "approximately true" must itself be made mathematically precise or else discarded is, I think, a form of scientism.

In sum, Newtonian physics worked *because* it was approximately true (whether or not there was a "consensus" that it was, pace Rorty, Brandom, Habermas, and all the public-consensus philosophers), regardless of anyone's notion of "human flourishing," and regardless of the fact that Newton did not have a "criterion" for the goodness of the approximation. Nevertheless, I repeat, Tauber has obviously touched on a deep issue here.

Have we learned nothing, then, from the "science studies" that Tauber describes? Speaking for myself, I have derived entertainment and not much more from reading Bruno Latour. But there is another area from which we have learned a good deal, and that is the area of *socially constructed social facts.*

34. Ibid., 25.

Good examples of the successive construction of such "facts" can be found in the history of sexuality, a field pioneered by Michel Foucault and continued in a much more fine-grained way by Arnold Davidson.[35]

Although he was inspired to study the subject by Foucault's unfinished multivolume *History of Sexuality*,[36] Davidson's focus was not on defending sweeping theses concerning the relation of knowledge to power, but on the study of what he called "styles of reasoning," for example, styles of reasoning about what used to be called "perversion." His papers on this subject describe how "perversion" was first treated simply as a "sin" and not as a subject calling for medical theories at all. Later, perversion came to be "explained" by genital abnormalities. When the anatomical data failed to support this explanation, doctors postulated anatomical deformations in the *brain;* finally, toward the end of the nineteenth century, such explanations were first "supplemented" and then supplanted by theories that treated homosexuality as a neurotic symptom, a style of explanation that allowed wholly novel uses of the concept of homosexuality; and more recently it has been treated as indeed a psychological phenomenon, but not an illness. Before the psychiatric stage, people could perform "sinful" acts (such as sodomy); after the psychiatric stage such sentences as "he is a homosexual who has never performed a homosexual act" became intelligible.

Davidson's work obviously supports the idea that putative "facts" about human beings can be socially constructed, and it also connects with Ian Hacking's modest reminder that people will sometimes behave differently as we "construct" different social categories, and we relate to them (and they come to understand themselves) in terms of those categories.[37] But there is nothing in any of this that blocks a "realist" story about how various errors have been gradually overcome, and about how we have gradually developed more nuanced understanding of various sexual phenomena.

35. Arnold Davidson, "Sex and the Emergence of Sexuality," *Critical Inquiry* 14, no. 1 (Autumn 1987): 16–48; Davidson, "How to Do the History of Psychoanalysis: A Reading of Freud's Three Essays on the Theory of Sexuality," *Critical Inquiry* 13 (Winter 1987): 252–277; and Davidson, "Closing Up the Corpses: Diseases of Sexuality and the Emergence of the Psychiatric Style of Reasoning," in Tim Dean and Christopher Lane, eds., *Homosexuality and Psychoanalysis* (Chicago: University of Chicago Press, 2001), 59–90.

36. Michel Foucault, *The History of Sexuality*, 3 vol. (London: Allen Lane, 1979–1988).

37. Ian Hacking, *The Social Construction of What?* (Cambridge, Mass.: Harvard University Press, 1999).

The Rortian alternative to the realist account was to take over from Dewey the valorization of discussion that allows novel ideas to be raised and objections to be voiced while *scrapping*—that is what makes it the "Rortian" alternative—the pragmatist insistence that that kind of discussion helps us arrive at what Dewey tellingly called *objective* resolutions of problematic situations.[38] For Rorty, if there is progress here, it is, indeed, progress only from the point of view of *our* (Western liberal) notion of human flourishing. It sometimes seems (e.g., in *Contingency, Irony, and Solidarity*[39]) that for Rorty the kind of discussion pragmatists advocated is good simply for *aesthetic* reasons—reasons that appeal to *our* tastes. Tauber, happily, does not want to go that far; but then I urge him to pay more sympathetic attention to the other side in "the science wars."

38. John Dewey, *Logic; The Theory of Inquiry*, in Dewey, *The Later Works, 1925–1953*, ed. Jo Ann Boydston, vol. 12 (Carbondale: Southern Illinois University Press, 1984), 287.

39. Richard Rorty, *Contingency, Irony, and Solidarity* (Cambridge: Cambridge University Press, 1989).

5

The Content and Appeal
of "Naturalism"

As my title suggests, there are just two questions about "naturalism" that I mean to address: what it means to say that one is a "naturalist" (or, more precisely, what it means if a certain popular definition of the term is accepted); and why, in spite of (what we shall find to be) the extreme unclarity of the position and the host of problems it faces, this sort of "naturalism" seems to be so appealing.

The Content of "Naturalism"

Today the most common use[1] of the term "naturalism" might be described as follows: philosophers—perhaps even a majority of all the philosophers writing about issues in metaphysics, epistemology, philosophy of mind, and philosophy

---

1. There is a different use, due to John McDowell, characterized by the ideas—which McDowell draws from Aristotle—that although we are indeed animals, and thus part of nature, we acquire a "second nature" as we become sharers of a culture (*Mind and World* [Cambridge, Mass.: Harvard University Press, 1994]). My subject in the present essay is not this sort of naturalism, which I agree with, but which is either ignored or actually scorned by the "naturalists" I shall be talking about.

of language—announce in one or another conspicuous place in their essays and books that they are "naturalists" or that the view or account being defended is a "naturalist" one. This announcement, in its placing and emphasis, resembles the placing of the announcement in articles written in Stalin's Soviet Union that a view was in agreement with Comrade Stalin's; as in the case of the latter announcement, it is supposed to be clear that any view that is not "naturalist" (not in agreement with Comrade Stalin's view) is anathema and could not possibly be correct. A further very common feature is that, as a rule, "naturalism" is not *defined*.

One happy exception to this rule is that in the glossary to Boyd, Gasper, and Trout's *The Philosophy of Science,* naturalism is actually defined, namely, as "the view that all phenomena are subject to natural laws, and/or that the methods of the natural sciences are applicable in every area of inquiry."[2] However, the definition is a *disjunctive* one, and the two disjuncts are actually very different. In effect, we are being offered *two* definitions of "naturalism" rather than one. Let us consider them in turn.

According to the first definition (or the first disjunct), a naturalist is a philosopher who believes that "all phenomena are subject to natural laws." But what exactly is this supposed to mean? Consider the following "phenomenon": someone whose prose is usually very clear writes a paragraph that is quite difficult to interpret. A naturalist (in the sense of this first definition) must believe that this phenomenon is subject to natural laws. Is it clear what this means? Certainly the writer did not *violate* any natural laws when he wrote the difficult paragraph. If all that is involved in being a naturalist is thinking that there are not phenomena that actually *violate* natural laws, then who *is not* a "naturalist"? Or is it required to be a naturalist that one believe that there are "natural laws" containing the concept "difficult to interpret" (*and* every other concept used to describe a "phenomenon")? Or would it be enough (to count as a "naturalist") to think that the token event of writing the unclear paragraph is identical with some token physical event (à la Davidson)? (Would thinking that "token identity" has no clear definition at all make one a "nonnaturalist"?)

According to the second definition (the second disjunct), on the other hand, a naturalist is a philosopher who believes "that the methods of the natural sciences are applicable in every area of inquiry." Well, what would it

---

2. Richard Boyd, Philip Gasper, and J. D. Trout, eds., *The Philosophy of Science* (Cambridge, Mass.: MIT Press, 1991), 778.

mean to say that the methods of the natural sciences apply to *the interpretation of texts?* Or is the interpretation of texts not an "area of inquiry"? If what is involved is believing that there is a science that resembles physics (complete with laws, theories, experiments, and so on) of every single thing (for example, that history is a science, as the logical positivists used to claim), then why would it not be perfectly respectable *not* to hold such an implausible view?

At first blush, the fact that so many philosophers are proud of calling themselves "naturalists," without spelling out what the term means, might suggest that "naturalism" has no definite content at all.[3] But this would be a mistake; there *is* a content to "naturalism" (in the scientistic understanding of the term that I am criticizing), but the unfortunate term "naturalism" conceals it instead of making it clear. To find what that content is, we have to consider what the *opponents* of (scientistic) naturalism really defend. What they defend is, of course, not "supernatural" or "occult" explanations (although the term "naturalism" is intended to suggest that that is what they defend). What they defend is, rather, *conceptual pluralism.* But what is "conceptual pluralism"?

Pluralism

Conceptual pluralism might be briefly defined as the denial that any one language game is adequate for all our cognitive purposes, but this description is not yet very helpful. "Naturalists" too can easily concede that this is the case—but with qualifications. The most common qualification is represented by Quine's distinction between a first-grade conceptual system (science, or rather science properly formalized) and what he called "our second-grade conceptual system."[4] Quine, as everyone knows, simply ruled that only our

3. The history of the term is no help. Dewey gave it currency, and he certainly insisted that "the methods of the natural sciences" are applicable in every area of inquiry, but for Dewey, what that meant was that one should be *fallibilistic* and *experimental* in every area of inquiry (including when one is painting or criticizing a picture—see his *Art as Experience,* in John Dewey, *The Later Works, 1925–1953,* ed. J. A. Boydston, vol. 10 [Carbondale: Southern Illinois University Press, 1987]). It also meant that one should see the "scientific method," in his sense, as presupposing the interpenetration of fact and value (including aesthetic value)—the interpenetration of problems of use and enjoyment and abstract scientific problems. But this is not what philosophers who talk about "the methods of the natural sciences" mean by that expression today.

4. W. V. Quine, *Ontological Relativity and Other Essays* (New York, Columbia University Press, 1969), 24.

first-grade conceptual system represents an account of what the world contains that we have to take seriously. Because nothing in the conceptual scheme of physics, for example, corresponds to a *meaning fact,* the closest we can come to such facts is bare behaviorist psychology in the style of Skinner. (Today many "naturalists" would say "neurophysiology.") And if Skinnerian psychology cannot provide an account of meaning or reference, so much the worse for meaning and reference! Although other "naturalists" would draw the line between the first-grade and the second-grade elsewhere (no two "naturalists" seem to draw it in the *same* place), what is common to most versions of "naturalism" is that those conceptual resources and conceptual activities that do not fit into the narrowly scientific first-grade system are regarded as something less than bona fide rational discourse.

The heart of my own conceptual pluralism is the insistence that the various sorts of statements that are regarded as less than fully rational discourse, as somehow of merely "heuristic" significance, by one or another of the "naturalists" (whether these statements be ethical statements or statements about meaning and reference, or counterfactuals and statements about causality,[5] or mathematical statements, or whatever) are bona fide statements, "as fully governed by norms of truth and validity as any other statements," as James Conant has put it.[6]

5. In *The Cement of the Universe* (Oxford: Clarendon Press, 1974), John Mackie described the ordinary notion of causality as a New Stone Age ("neolithic") notion. He tried to provide a substitute, but I have argued that he did not succeed in Hilary Putnam, "Is the Causal Structure of the Physical Itself Something Physical?" in P. A. French, T. E. Uehling, and H. K. Wettstein, eds., *Causation and Causal Theories,* Midwest Studies in Philosophy, vol. 9 (Minneapolis: University of Minnesota Press, 1984), 3–16; reprinted in Putnam, *Realism with a Human Face* (Cambridge, Mass.: Harvard University Press, 1990), 80–95.

6. James Conant, "Wittgenstein's Philosophy of Mathematics," *Proceedings of the Aristotelian Society* 97, pt. 2 (1997): 195–222; quotation from p. 202. The passage reads: "Putnam wants . . . to hang on to . . . the idea that ethical and mathematical propositions are *bona fide* instances of assertoric discourse: ethical and mathematical thought represents forms of reflection that are as fully governed by norms of truth and validity as any other form of cognitive activity. But he is not friendly to the idea that, in order to safeguard the cognitive credentials of ethics and mathematics, one must therefore suppose that ethical and mathematical talk bear on reality *in the same way* as ordinary empirical thought, so that in order to safeguard the truth of such propositions as 'it is wrong to break a promise' or '2 + 2 = 4,' one must suppose that, like ordinary empirical propositions, such propositions, in each sort of case, 'describe' their own peculiar states of affairs. There is an assumption at work here that Putnam wants to reject—one which underlies Blackburn's way of distinguishing realism and antirealism—the

The Instability of "Naturalism"

While Quine (and today, Simon Blackburn) and—as it seems at times—
Bernard Williams represent what might be called "minimalism" with respect
to what can be included in the first-grade conceptual system,[7] the system that
alone can be taken seriously when our interest is (in a famous phrase of
Quine's) "limning of the most general traits of reality,"[8] many naturalists are
understandably uncomfortable with the idea of dismissing *so much* of our
discourse to the murky reaches of our "second-grade conceptual system."
This discomfort pushes some "naturalists" to try to show that large parts of
what Quine, Blackburn, Williams, and others would push out of the realm
of the first-grade system (what Williams calls "the absolute conception of
the world")[9] can actually be *reduced to* first-grade properties and thus shown
to be first-grade after all. In short, one wing of the "naturalist" camp gets
pushed in the direction of (physicalist) *ontological reductionism*. At their most
ambitious, ontological reductionists seek to rehabilitate even the property of

---

assumption that there are just two ways to go: either (i) we accept a general philosophical account of the relation between language and reality according to which all indicative sentences
are to be classified equally as 'descriptions of reality'; or, (ii) we accept an alternative account
of the relation between language and reality which rests on a metaphysically-grounded distinction between those sentences which do genuinely describe reality (and whose cognitive
credentials are therefore to be taken at face value) and those which merely purport to describe
reality (and whose claims to truth are therefore to be taken as chimerical)." Of course (as
Conant is aware), there are other statements that Blackburn and some others would reject as
not fully worthy of being taken cognitively seriously—statements about what *refers* to what
and statements about what *caused* what, for example—that I would regard as bona fide statements *and* as (when true) "descriptions of reality." The point is that the claim that the supposedly "second-grade" statements are bona fide in the sense of being governed by norms of truth
and validity must not be confused with the claim that *all* of them are "descriptions of reality."
This is the error that Conant accuses Sabina Lovibond of making. I will return to this point at
the end of the present essay.

7. Simon Blackburn, *Spreading the Word* (Oxford: Clarendon Press, 1984); Bernard Williams, *Descartes: The Project of Pure Enquiry* (Harmondsworth, Middlesex: Penguin, 1978);
and Williams, *Ethics and the Limits of Philosophy* (Cambridge, Mass.: Harvard University
Press, 1985).

8. Quine wrote, "The quest for a simplest clearest overall pattern of canonical notation is
not to be distinguished from a quest of ultimate categories, a limning of the most general
traits of reality," in *Word and Object* (Cambridge, Mass.: MIT Press, 1960), 161.

9. Bernard Williams, *Descartes*, 236–303, and *Ethics and the Limits of Philosophy*, 135–139.

being ethically *good* (Richard Boyd and, perhaps, Peter Railton),[10] but philosophers who are willing to go this far are comparatively rare. Others—Jerry Fodor is a famous example—try to give physicalist accounts of *meaning*.[11] Still others have tried to give physicalist accounts of the content of mathematics (Hartry Field and Penelope Maddy, albeit in very different ways). The trouble is that none of these ontological reductions gets *believed* by anyone except the proponent of the account and one or two of his friends and/or students. So—as always in philosophy—a "recoil" sets in. But whither is a "naturalist" to recoil?

The area of discourse that has been recognized by "naturalists" themselves as most resistant to reductive explanation, and hence as problematic from their own point of view, is "intentional" discourse (talk of reference, meaning, belief, desire, and the like), although mathematics is also frequently regarded as a problem because, as it is often put, "we don't causally interact with mathematical entities." Already in the nineteenth century Brentano had taken the irreducibility of intentionality to refute naturalist and physicalist accounts of mind.[12] Quine was willing to agree with Brentano that intentionality is irreducible, but he concluded that it was not a genuine phenomenon at all: "One may accept the Brentano thesis either as showing the indispensability of intentional idioms and the importance of an autonomous science of intention, or as showing the baselessness of intentional idioms and the emptiness of a science of intention. My attitude, unlike Brentano's, is the second."[13]

10. Richard Boyd, "How to Be a Moral Realist," in Geoffrey Sayre-McCord, ed., *Essays in Moral Realism* (Ithaca, N.Y.: Cornell University Press, 1988), 181–228; Peter Railton, "Moral Realism," *Philosophical Review* 95 (1986): 163–207.

11. Jerry Fodor, "Meaning and the World Order," in Fodor, *Psychosemantics* (Cambridge, Mass.: MIT Press, 1987), 97–127; and Fodor, *A Theory of Content* (Cambridge, Mass.: MIT Press, 1990). For some criticisms, see the chapter on Fodor (chap. 3) in Putnam, *Renewing Philosophy* (Cambridge, Mass.: Harvard University Press, 1992), 35–59.

12. The classic formulation of the problem is Paul Benacerraf's "Mathematical Truth," *Journal of Philosophy* 70 (1973): 661–679. For a recent attempt to deny that the problem is genuine, while defending naturalism, see Jody Azzouni, *Metaphysical Myths, Mathematical Practice* (Cambridge: Cambridge University Press, 1994). For my own view, see Hilary Putnam, "Was Wittgenstein *Really* an Antirealist about Mathematics?" in Timothy McCarthy and Sean Stidd, eds., *Wittgenstein in America* (Oxford: Oxford University Press, 2001), 140–194; reprinted in this volume as Chapter 23

13. Quine, *Word and Object*, 221.

Moreover, it is not only the "nonextensional" intentional idioms that Quine viewed as "baseless." The fundamental notion of extensional semantics is *reference;* and reference, Quine famously held, is "indeterminate." One of his routes to this conclusion is via his doctrine of "ontological relativity." Here is an example. Suppose that we define *"X complement-drinks Y"* to mean that whatever object (counting regions of space as "objects") is left when we take away *X* from the entire physical universe (the "cosmos") drinks whatever is left over when we take away *Y* from the entire cosmos. Quine once had a cat named Tabitha. A moment's reflection suffices to convince oneself that Tabitha drinks milk when and only when the cosmos minus Tabitha complement-drinks the cosmos minus the milk. Quine contended that it is objectively indeterminate *whether, in the sentence "Tabitha is drinking the milk," the name "Tabitha" refers to Tabitha, "is drinking" to the act of drinking, and "the milk" to the milk, or "Tabitha" refers to the cosmos minus Tabitha, "is drinking" to complement-drinking, and "the milk" to the cosmos minus the milk.*[14]

Here Quine was characteristically up front about the counterintuitive consequences of his philosophical theories. A number of present-day "naturalists" seem either to be unaware of them or to believe that they can accept much the same account of reference without "going as far" as Quine. An example is Stephen Leeds's account in a fascinating paper in the *Pacific Philosophical Quarterly* some years ago.[15] The most common view of reference—and one that I agree with, as long as it is not confused with the quite different position that reference can be *defined* in terms of causal connection—is that reference to empirical particulars and properties *presupposes* information-carrying causal interaction with those particulars and properties, or at least with particulars and properties in terms of which identifying descriptions of those particulars and properties can be constructed. Leeds rejects this idea entirely. Instead, he simply posits that the familiar "disquotational" formulas such as " 'Caesar' refers to Caesar" define the relation of reference. But Quine's

---

14. W. V. Quine, *The Pursuit of Truth* (Cambridge, Mass.: Harvard University Press, 1990), 33.

15. Stephen Leeds, "Brains in Vats Revisited," *Pacific Philosophical Quarterly* 77 (June 1996): 108–131. Leeds's earlier paper, "Theories of Reference and Truth," *Erkenntnis* 13 (1978): 111–129, is one of the most widely cited statements of the "deflationist" position on truth and reference. Hartry Field defended a similar view in a paper about my book, *Reason, Truth and History* (Cambridge: Cambridge University Press, 1981), titled "Realism and Relativism," *Journal of Philosophy* 79, no. 10 (1982): 553–567. I replied to Field in "A Defense of Internal Realism," in Putnam, *Realism with a Human Face*, 30–42.

Tabitha example already showed the error in this claim. If there is no determinate correspondence between any of our terms and nonlinguistic objects—if there is neither a context-sensitive correspondence nor a context-insensitive one—then I can utter (or inscribe in the mythical "belief box" in my brain that some "naturalists" like to posit) the sentence

"'Tabitha' refers to Tabitha"

and as many other instances of the "disquotation" scheme

"'T' refers to Ts"

as I want, and no relation between the term "Tabitha" and any particular object external to my words is thereby defined. Even saying "I specify that the extension of the two-place predicate 'refers' is the set of ordered pairs <'Tabitha,' Tabitha>, <'Taj Mahal,' the Taj Mahal>, <'cat,' the set of all cats>, and so on" still leaves totally undetermined what I am talking about when I say "Tabitha," "Taj Mahal," "cat," or whatever.

To put the same point in model-theoretic terms: a model[16] according to which the word "Tabitha" names the cosmos minus Tabitha immediately extends to a model for the metalanguage of the relevant portion of my language in which "'Tabitha' refers to Tabitha" is *true* (not false, as one might expect)—true because the second occurrence of the proper name "Tabitha" in this sentence (the occurrence without quotation marks around it, the one that is "used" and not "mentioned") corresponds to *the cosmos minus Tabitha* in that model of the metalanguage, as well as in the submodel that is the model of the object language. (As Quine might have put it, in all the different models corresponding to the various "proxy functions" it is true that "Tabitha" refers to Tabitha—whatever *Tabitha* is.) Leeds's account simply abandons the idea that there is even a partly determinate correspondence between terms and objects without acknowledging that that is what it does.

Sometimes philosophers who hold views like Quine's or Leeds's have replied as follows when I have raised the foregoing objection: "Yes, we give up the idea that reference is a words-world correspondence, but that does not mean that language *is not* connected to objects. It is connected to objects *causally*, not referentially." It is remarkable that this response happens to be

16. Quine speaks in terms of "proxy functions" rather than models, but the point is unaffected.

the central thesis of a philosopher whom the "naturalists" I am discussing regard as an opponent: Richard Rorty. (To be sure, Rorty might also call himself a "naturalist," but not in the scientistic sense that the definition in Boyd, Gasper, and Trout's glossary attempted to capture.) The "naturalists" whom Boyd and Gasper were thinking of, and whose views I have been criticizing, typically say that they are "scientific realists" (Quine called himself a "robust realist"). For them, science is first philosophy. For Rorty, however, "realist" is a dirty word.[17] And rejecting the idea that we possess "discrete component capacities to get in touch with discrete hunks of reality"[18] (in other words, that there is a semantic connection between a given utterance of the word "Tabitha" and a given discrete furry hunk of reality) seems to Rorty (but evidently not to Quine, Leeds, Field, and others) virtually to *require* that we abandon realism. And he is not being unreasonable, for once one loses the thought that there is a determinate relation between one's words and scientific or other objects, what does calling oneself a "realist" actually come to?

Well, it is clear what it came to for Quine: a positivist insistence that all our language (cognitively speaking) is a machine for anticipating and controlling the stimulations of one's nerve endings. And it is true that to choose this as one's criterion for what one will take "seriously" is very different from accepting Rorty's more elastic criterion of whatever helps us "cope." But if neither criterion has any pretension to providing a sense in which our propositions are capable of mapping the behavior of specific hunks of reality (and how can there be *mapping* without any *mapping relation?*), then valorizing prediction of nerve stimulations over "coping" broadly construed is (as Rorty tirelessly points out) utterly arbitrary. Quine, it seems to me, gave up realism without noticing that he did so, because he thought that as long as he valorized scientific discourse above all other discourse, that *made* him a "realist."

17. At a discussion between us in Cerisy (June 25, 1995), Rorty chided me for calling myself a "commonsense realist," saying, "Commonsense realism is as bad as Metaphysical Realism: one leads to the other." Nor is this the only occasion on which he has virtually equated being a "realist" and being a "metaphysician"—the latter term being clearly pejorative.

18. "The fallacy comes in thinking that the relationship between vocable and reality has to be piecemeal (like the relation between individual kicks and individual rocks), a matter of discrete component capacities to get in touch with discrete hunks of reality." Richard Rorty, "Pragmatism, Davidson and Truth," in Ernest Lepore, ed., *Truth and Interpretation* (Oxford: Blackwell, 1986), 333–355; reprinted in Rorty, *Philosophical Papers*, vol. 1, *Objectivity, Relativism, and Truth* (Cambridge: Cambridge University Press 1986), 126–150; quotation from pp. 145–146 of the reprint.

And I suspect that Leeds, Hartry Field, Paul Horwich, and the other "deflation-ists" follow him in this. "Naturalism" is unstable indeed if it slides so easily into Rortian antirealism.

(My aim in this essay is simply to examine the content and appeal of "natu-ralism," not to offer an alternative picture. But if I were to offer an alternative account, I would begin by pointing out that although Frege was right to say that only in the context of a statement [or a question, or a command, and so on] does a word have reference or meaning at all, it is only in the context of such a prac-tice as *talking about cats* that such a sentence as "Tabitha won't drink milk" constitutes a *statement*. And not only is there no reason to think that the practice of talking about cats (or about cabbages or kings or sealing wax, and so on) can be defined in physicalistic terms, but there is not even a way to begin to talk about that practice without talking about the various things we do with cats (with cabbages, with kings, with sealing wax), starting with *perceiving* them— that is, *talking about things* is a vast and ever-expanding motley of *world-involving practices*. Just uttering marks and noises that are "causally connected" to things is not yet *engaging in a practice*—not under the description "uttering marks and noises that are causally connected to things," anyway.)[19]

However, not all "naturalists" recoil from the problem of "naturalizing semantics" in a deflationist direction. David Lewis famously insisted that certain properties (or rather, certain classes) are "objective similarities," where the property of being an objective similarity is not to be understood as in any way relative to our interests. Reference, for Lewis, is largely fixed by the re-quirement that typical predicates in the language denote objects that are ob-jectively similar.[20] But for Lewis's idea to work, one similarity that has to be recognized as "objective" is being designed for the function of being sat on. (Think of the predicate "chair.") But to be designed for a function presup-poses being *thought of* as capable of fulfilling that function, and this is an intentional condition. Only a being that can *think about things as fulfilling a function* can design things to fulfill it. So the primitive notion (for that is what it was in Lewis's metaphysics) "objective similarity" must not just refer to similarities in, say, color or shape; it must refer to similarities in "form" in

---

19. I develop an account along these lines in Hilary Putnam, *The Threefold Cord: Mind, Body, and World* (New York: Columbia University Press, 1999).

20. David Lewis, "Radical Interpretation," *Synthese* 23 (1974): 331–344; reprinted in Lewis, *Philosophical Papers,* vol. 1 (Oxford: Oxford University Press, 1983), 108–118.

the old Aristotelian sense. If one sort of contemporary naturalist is a Rortian antirealist in disguise, another sort—Lewis being the most brilliant recent example—is a high medieval Aristotelian metaphysician in disguise. The laudable aim of a modest nonmetaphysical realism squarely in touch with the results of science is lost in both cases.

The Appeal of "Naturalism"

Considering the failures and instabilities of the various forms of "naturalism" we have briefly canvassed, one can be sure that there would not be so many philosophers still proudly announcing "I am a naturalist" or "This is a naturalist view" if the view did not have strong appeal. In some cases the appeal may be genuinely "appealing"; that is, the vision of a state of knowledge in which everything—everything that we have to take seriously when we are "limning the most general traits of reality"—is made clear in the manner in which physics is imagined to make things clear[21] may correspond to a deep scientistic or positivistic outlook on the part of a few philosophers. Positivists,[22] in this loose sense, are always with us. In the great majority of cases, I believe, the "appeal" is more likely the "appeal" of a medicine that tastes awful (and even has some unpleasant side effects) but that one takes to avert or cure a serious illness. "Naturalism," I believe, is often driven by fear, fear that accepting conceptual pluralism will let in the "occult," the "supernatural." But before I say a word or two more about the nature of this fear, let us first look at the arguments for naturalism that seem to convince more philosophers (more students, at any rate) than any others. In my experience there are two:

1. *The argument from evolution/composition.* There was a time when the world contained nothing but fields and particles, and everything we

21. At the most fundamental level, present-day physics does *not* say that the entities of which the world consists are either particles (not if "particles" are supposed to be objects to which one can assign number and determinate identity through time) *or* (pace Quine) fields with real number values or ordinary vector values at points in a space-time that is supposed to be "there" independently of the fields. It is curious how many philosophers who praise modern physics as first philosophy do not trouble to learn it.

22. Here I use "positivist" in the sense it had before logical positivism appeared on the scene, the sense introduced by Auguste Comte. In this older sense "naturalism" seems to be precisely the late twentieth- and early twenty-first-century form of "positivism."

find in the world now came into being as a result of physical processes (including Darwinian evolution, at a relatively recent stage in the history of the cosmos). Hence things *are* nothing but mereological sums of physical objects (counting space-time regions as physical objects). Hence all explanations in terms of intentional notions and propositional attitudes, all value judgments, all counterfactuals (some, but decidedly not all, "naturalists" would add "mathematical state-ments") must either be shown to be reducible to the nonintentional, noncounterfactual level of something like Quine's "canonical nota-tion" or else shown to be cognitively spurious, part of the heuristic "second-grade conceptual system" at best.

2. *The "you are leaving something unexplained" argument.* It is contended that in some cases (the case of semantics, but usually not the case of the so-called special sciences like geology),[23] failure to give a reductive explanation of something—say, of what it is to *refer* to something, or how there can be such things as *true counterfactual conditionals* or *true mathematical sentences*—is to leave it "unexplained" and hence to admit a "mystery" (in the sense of something occult) into the (funda-mentally unmysterious) "physical world."

Let us look at these arguments in turn.

Re the first argument: it is simply not the case that "things are nothing but mereological sums of physical objects."

To be sure, the Quinean conception of philosophy that influenced so much subsequent "naturalism," the conception of philosophy as the *theory that tells us what we can and what we cannot quantify over in that part of our language that alone we can take with metaphysical seriousness, namely, the properly formal-ized scientific part of our language,* fits very well with the idea of taking a concrete (nonabstract) thing to be simply a mereological sum of particles or, in

---

23. That the failure of reductionism in cases like that of geology is perfectly acceptable to a "naturalist" is made clear in Jerry Fodor, "Special Sciences (or the Disunity of Science as a Working Hypothesis)," *Synthese* 28 (1974): 97–115; at the same time, Fodor tries very hard to explain the notion of reference reductively with the aid of the notion of "causation" precisely because he supposes—wrongly, in my opinion—that there are no intentional (or "semantic") elements in the latter notion. Hence Fodor's revealing statement in "Meaning and the World Order," 126–127, that he has "wasted a lot of time that I could have put in sailing" if "the cause" is indeed intentional/semantic.

more recent physics, a mereological sum of fields and particles, even if this was not Quine's own preference.[24] Here, in the mereological sums of particles (or fields and particles) we are supposed to have a nice set of physical objects, describable in what is supposedly a precise language (if we pass over the fictitious physics involved),[25] the language of symbolic logic, counting mereology as a part of symbolic logic. The remaining tasks of philosophy then become, first, to decide how many abstract entities (if any) we need to quantify over in order to do first-class science and which ones are avoidable (for example, Quine thought that *intentional* entities, meanings and attributes, are avoidable, but sets are not), and, second, to show how other things that we apparently have to quantify over can be reduced to the privileged scientific objects, or dismissed to our "second-grade conceptual system."

But there is a deep problem with this whole line of thought, one pointed out by Saul Kripke when he introduced the idea of "rigid designation."[26] Although I owe the point to Kripke, I will develop it in a somewhat different way. I suggest that one way of seeing the inadequacy of the whole idea of seeing a (concrete) object as a mereological sum of molecules or atoms or elementary particles (or anything of that kind) is to note a certain ambiguity in the very claim that something "is" a mereological sum. Am I (or is my body) the mereological sum of the atoms in my body? In one sense, yes. Those atoms are parts of my body, and there is no part of my body that is wholly disjoint from every one of those atoms. In that sense, likewise, Massachusetts is the mereological sum of the counties that make it up. Those counties are parts of Massachusetts, and no part of Massachusetts is wholly disjoint from (that is, fails to touch) at least one of those counties. But typically the statement that I am the mereological sum of the atoms in my body, or that

24. Quine preferred to take the basic entities to be space-time points, which he identified with *quadruples of real numbers,* and regions, which he identified with *sets of quadruples of real numbers,* but I do not know any philosopher who followed him in this (self-styled) "Pythagoreanism."

25. The physics is fictitious because in our most fundamental physical theory, quantum field theory, "what there is" is not fields and particles in the classical sense at all. What is associated with each space-time point (or rather, with each point in a postulated background space-time, which is not the same as physical space-time) is not a field vector, but a "_-function," a vector in an infinite-dimensional space (a Hilbert space or some other abstract space), and there are no definite particles and no definite field intensities at all, in the general case.

26. Saul Kripke, *Naming and Necessity* (Cambridge, Mass.: Harvard University Press, 1980).

Massachusetts is the mereological sum of the various counties, is understood to mean more than this; it is understood as a statement of "logical identity," the identity (symbolized by the sign "=" in symbolic logic) that satisfies Leibniz's law, the principle that if $x = y$ (x is identical with y), then every property of x is a property of y and vice versa. The statement that I am (or my body is) the mereological sum of the atoms in my body is easily seen to be false if it is understood as a statement of identity in this logical sense.

Let us suppose (this is certainly a physical possibility) that the atoms in my body all existed a thousand years ago. Of course, they were not in my body at that time; some were in the earth, some in the air, and so forth. The mereological sum of the atoms in my body exists whenever those atoms exist, and that means that the mereological sum of the atoms in my body existed a thousand years ago. By Leibniz's law, if I am that mereological sum, it follows that I existed a thousand years ago, which is clearly false. Moreover, if—and here we come closer to Kripke's own version of the argument—I had simply eaten different food for dinner last night, I would now consist of different atoms, and hence I would not be identical with the particular mereological sum with which I *am* identical, on the supposition that the sense in which "I am" the mereological sum of those atoms is logical identity.[27] But why should I care about what would happen if a *different person* had a different diet than I do? As Kripke emphasizes, our use of counterfactuals depends on the supposition that the proper names in them designate *rigidly,* that is, that they designate *identical* individuals in the various possible worlds we contemplate. And it does not help (to meet this) to say that I am identical with a mereological sum of *time-slices* of atoms (where a "time-slice" is supposed to be a temporal "part" of an atom, say, the atom between time t_1 and time t_2), for if I had had a different dinner I would now consist not just of different atoms but of different time-slices of atoms.

David Lewis, whose views Kripke discusses in *Naming and Necessity,* would bite the bullet here and say that if I had had something else for dinner last night, then I would not exist now, but a *counterpart* of me would exist

---

27. Kripke does not discuss mereological sums as such in *Naming and Necessity,* but his rejection of "counterpart theory" (ibid., lecture I, esp. p. 45) turns on the claim that when I say that *I* would have done such and such if I had eaten different food, I mean precisely that; I am not saying that an *X* that is not identical with me would have done such and such. Moreover, my statement may well be *true.*

now; but very few philosophers are willing to follow Lewis here. In short, if we accept an ontology in which the world consists only of atoms and mereological sums of atoms (let us pretend for simplicity that all objects consist of atoms, and that atoms are indivisible), it will not be the case that that ontology contains all the *familiar* physical objects, and the only price one has had to pay for adopting the basic postulate of mereology (to the effect that for any *x* and *y*, there is a mereological sum of *x* and *y*) is a certain counterintuitive extravagance; that is, it is not the case that one now has all the familiar objects (without having had to worry about principles of unity and persistence), and what one has done is to add certain unfamiliar objects to them, such as the mereological sum of my nose and the Eiffel Tower (an addition that might be thought justified on the grounds that it does *at least* provide us with a "clean," "simple" ontology that *includes* everything [physical] that we want to include). *Just* the familiar objects, the "ships and shoes and sealing wax and cabbages and kings," are completely *missing* in this ontology. This ontology is, in fact, the ontology of *physics* (and only of an oversimplified physics at that).

What alternatives are there, however, to an ontology of time-slices of atoms (or of particles, or of fields and particles) and mereological sums thereof? Kripke would say that I am not *identical* with the mereological sum of the atoms that are parts of me (or with the mereological sum of time-slices of atoms). I am a different thing from my matter, although I *consist* of my matter. I am a thing with different persistence conditions from my matter, and different identity conditions across possible worlds. I consist of certain matter, as things actually are, but had things been different, had I had pot roast for dinner last night, I would consist of different matter, but I would be the very same person.[28]

---

28. I agree with this much, but in addition, I am attracted to an idea that I know Kripke does not like, the idea of "sortal" identity: that is, the idea that things can be *identical in* one *respect but not in another respect.* For example, I am inclined to say (still idealizing the physics, by the way) that a certain mereological sum of time-slices of atoms *is,* as things actually stand (I did not have pot roast for dinner last night), *identical with me qua physical system,* but *not identical with me qua person.* This does not rescue the ontology of basic physical entities and mereological sums thereof, at least as usually understood, because as usually understood, the identity of formalized ontology is not sortal identity. And if we do allow sortal identity, then the ontology tells us only what there is *in a respect* (say, what there is *qua* physical objects); sortal identity is essentially *pluralistic,* and unless we postulate that the number of *sorts* can be limited in

Accept Kripke's point, however, and the idea that the ontology of basic physical entities and mereological sums includes all the familiar objects that we want to include has to be given up. Mereology is an elegant mathematical theory, but no mathematical theory can overcome the fact that there is no closed set of scientific objects that includes all the things that we quantify over—and find it indispensable to quantify over—in our actual life with our language.

Re the second argument: to say—as I, for one, do say—that no clear meaning at all has been given to the idea of "reducing" our ability to refer to something (for example, our ability to talk about cats and dogs) to the "nonsemantic," and further, that "deflationism" provides no account at all of this ability, is not at all to say that something "occult" is going on when we talk about cats and dogs. There are all sorts of nonreductive studies of various aspects of language acquisition and use in existence, none of them a *definition* of "talking about" in "nonsemantic" terms, but none of them an account of something "occult" either.

The very fact that no "naturalist" philosopher thinks that *geology* is "occult," even though the predicates used in geology cannot be reductively defined in the language of fundamental physics, gives the show away. What is really behind the invidious distinction between the "semantic" and the, so to speak, merely "geological" is that describing language use involves *normative* concepts: things we say are variously *true* and *false,* are *justified* and *unjustified,* succeed in *referring* and fail to *refer,* are appropriate and inappropriate, and so on. The fact is that naturalists regularly assume that if the normative cannot be eliminated or reduced to the nonnormative, then some "occult" realm of values must be postulated. The possibility that goes missing here, as I remarked at the outset, is that an indicative sentence can be a bona fide statement without being a "description of reality." Developing this theme has to be the subject of a different set of essays.[29] But I cannot close without mentioning that the same missing possibility accounts for (1) the widespread failure of most philosophers to notice the inadequacy of Quinean definitions

---

advance—which I would deny—sortal identity *subverts* the question "What is there?" by countering: "What is there in *which respect?*"

29. My Hermes Lectures (given in Perugia in October 2001), which appear as part 1 of Hilary Putnam, *Ethics without Ontology* (Cambridge, Mass.: Harvard University Press, 2004).

of logical validity[30] and (2) the weird oscillation between nominalism and "Platonism" that we find in the philosophy of mathematics.[31]

In sum, to the extent that the appeal of "naturalism" is based on fear, the fear in question seems to be a horror of the normative. In the case of logical positivism, there was a not-dissimilar horror, the horror that the slightest trace of realism about scientific objects was tantamount to the acceptance of "metaphysics." We got over that horror when we realized that talk of unobserved entities did not need either metaphysical interpretation or positivist reinterpretation. We need to learn that the same is true of normative language.

30. In *Methods of Logic* Quine defines validity—for first-order logic only (he rules that second-order logic is not logic but "mathematics")—thus: a sentence is valid if and only if it is a substitution instance of at least one valid schema, and a schema is valid if and only if all its substitution instances are *true*. The inadequacies of these definitions that I refer to in the text include the following: (1) In the Tarskian semantics that Quine accepts, "true" is defined for only one formalized language at a time. In other words, given the conceptual resources Quine is willing to accept, there is, strictly speaking, *no such predicate as "true,"* but only a potentially infinite number of particular predicates "True-in-L_0," "True-in-L_1," "True-in-L_2," . . . , corresponding to a potential infinity of formalized (and interpreted) languages L_0, L_1, L_2, and so on. (Note that the names "L_0," "L_1," "L_2," . . . do not, from a logical point of view, occur as *meaningful* parts in the expressions "True-in-L_0," "True-in-L_1," "True-in-L_2," . . . , any more than the English word "cat" occurs as a *meaningful* part in the English word "cattle," and similarly the word "True" does not occur as a *meaningful* part in the names "True-in-L_0," "True-in-L_1," "True-in-L_2," . . . ; "True-in-L_0" and its mates are all, from a logical point of view, arbitrary and *undecomposable* names. [That Truth is not *relative* to a language but rather is *defined one language at a time* if one follows Tarski is something that Donald Davidson seems to me to be confused about, and perhaps other people share this confusion.] If we follow Quine, then, *validity* will likewise not be *relative* to a language but rather defined one language at a time.) (2) Pending success in the process of completely formalizing a natural language, Quinean validity is defined only for *formalized* languages. In no natural language do the foregoing definitions "work." Moreover, according to Quine himself, formalizing a natural language is not discovering a preexisting content (if it were, a notion of "correct translation" would be presupposed, contradicting Quine's indeterminacy thesis). This means that for a strict Quinean, the failure of the definitions of validity in the case of natural languages could not be repaired even by formalizing them—formalizing a language is just deciding to use a different language altogether. (3) The *modal* character of the notion of validity—the fact that substitution instances of a valid schema are not just true but *necessarily* true—is ignored. (Quine famously rejects modal concepts altogether, but those who do not follow him in this should be bothered by this fact.)

31. See Chapters 9 and 12 in this volume.

6

A Philosopher Looks at Quantum Mechanics (Again)

Background

In 1965 I published a paper titled "A Philosopher Looks at Quantum Mechanics"[1] that many people found useful as an introductory explanation of the controversies about the interpretation of quantum mechanics. I have been aware for many years of the need to return to the subject because much relevant theorizing was not known when I wrote that paper (for example, "A Philosopher Looks at Quantum Mechanics" was written in 1963–1964, and I had not seen Bell's famous paper on the Einstein-Podolsky-Rosen paradox).[2] Bell's central claim was that if quantum mechanics were right, the measured values of spin on certain pairs of separated particles (electrons in an "entangled"

---

1. Hilary Putnam, "A Philosopher Looks at Quantum Mechanics," in Robert G. Colodny, ed., *Beyond the Edge of Certainty: Essays in Contemporary Science and Philosophy* (Englewood Cliffs, N.J.: Prentice-Hall, 1965), 75–101; reprinted in Putnam, *Philosophical Papers*, vol. 1, *Mathematics, Matter and Method* (Cambridge: Cambridge University Press, 1975), 130–158.

2. J. S. Bell, "On the Einstein-Podolsky-Rosen Paradox," *Physics* 1 (1964): 195–200; reprinted in Bell, *Speakable and Unspeakable in Quantum Mechanics* (Cambridge: Cambridge University Press, 1987, revised edition 2004), 14–21.

state) would be incompatible with the classical postulate of "locality."[3] Simply put, what "locality" means is that these experiments could be set up in such a way that the measurement of the spin of particle 1 produces no physical disturbance in particle 2. As David Albert explains in an excellent introduction to the topic, locality, in this sense, seemed almost self-evident. There seemed to be any number of ways you could do it: you could, for example, separate the two particles by some immense distance so large that there is no time for a "signal" from one of the particles to reach the other without traveling faster than light, or build impenetrable shields between them, or "set up any array of detectors you like in order to verify that no measurable signals ever pass from one of the electrons to the other in the course of the experiment (since quantum mechanics predicts that no such array, in such circumstances, whatever sorts of signals it may be designed to detect, will ever register anything)."[4] The experiment described by Aspect, Dalibard, and Roger[5] showed that the "nonlocality" that Bell had derived from quantum mechanics in 1964 really exists, and ever since the question of how to understand nonlocality has been at the very center of discussions of the interpretation of quantum mechanics.[6]

3. However, nonlocality can also be demonstrated with experiments involving other magnitudes than spin and other particles than electrons, and also with experiments involving fields rather than particles. See David Finkelstein, "The Quantum Paradox," in *Encyclopedia Britannica Yearbook of Science and the Future* (Chicago: Encyclopedia Britannica, 1987), 186–207, for a very clear explanation of nonlocality using photons as the particles and direction of polarization as the relevant observable.

4. D. Z. Albert, *Quantum Mechanics and Experience* (Cambridge, Mass.: Harvard University Press, 1992), 64.

5. Alain Aspect, Jean Dalibard, and Gérard Roger, "Experimental Test of Bell's Inequalities Using Time-Varying Analyzers," *Physical Review Letters* 49 (1982): 1804–1807.

6. Most laypeople, when they hear that quantum mechanics predicts nonlocal correlations, immediately conclude that this shows that "causal signals can travel faster than light," thus contradicting Einstein's theories of special and general relativity. But things are not so simple! What the "disturbance of particle 2" turns out to be is a statistical matter: the probabilities of outcomes of measurements on particle 2 are altered. And this kind of disturbance, it turns out, is not necessarily incompatible with relativity. Whether the particular form of nonlocality we have in quantum mechanics is or is not compatible with relativity is searchingly examined in T. W. Maudlin, *Quantum Non-locality and Relativity: Metaphysical Intimations of Modern Physics* (Oxford: Basil Blackwell, 1994). One reason that "nonlocal correlations" cannot be used to synchronize watches (which would immediately contradict relativity) is that there does not seem to be any experimentally determinable fact of the matter about

In addition, at least one major interpretation of quantum mechanics that I will discuss in this essay had not yet been proposed. In brief, we now know many more facts and also many more possibilities of interpretation than I did when I wrote "A Philosopher Looks at Quantum Mechanics." Moreover, in that paper I failed to discuss the "many-worlds" interpretation proposed by Everett[7] (although I subsequently discussed it elsewhere).[8] So there are many reasons for my returning to the subject and taking a second look at quantum mechanics.

As I did in the earlier paper, I will try to write in a way that is intelligible to readers who do not know quantum mechanics. I tried very hard to make the earlier paper self-contained and to explain *the logical structure* of the problem rather than to go into mathematical details, and that is what I hope to do again.

Scientific Realism Is the Premise of My Discussion

I will begin by quoting "A Philosopher Looks at Quantum Mechanics,"[9] omitting footnotes and references. (I still agree with those pages, and they will set the stage for what I want to say now.)

> Before we say anything about quantum mechanics, let us take a look at the Newtonian (or "classical") view of the physical universe. According to that view, nature consists of an enormous number of particles. When Newtonian physics is combined with the theory of the electromagnetic field, it becomes convenient to think of these particles as dimensionless (even if there is a kind of conceptual strain involved in trying to think of something as having a *mass* but not any *size*), and as possessing electrical properties— negative charge, or positive charge, or neutrality. This leads to the well-known "solar system" atom—with the electrons whirling around the nucleus

---

when the "signal" from particle 1 (if there is a "signal") reaches particle 2—it could even reach particle 2 before the measurement is made on particle 1!

7. Hugh Everett III, "Relative State Formulation of Quantum Mechanics," *Reviews of Modern Physics* 29 (1957): 454–462; reprinted in J. A. Wheeler and W. H. Zurek, eds., *Quantum Theory and Measurement* (Princeton, N.J.: Princeton University Press, 1983), 315–323.

8. See Hilary Putnam, "The Uncertainty Principle and Scientific Progress," *Iride* 7 (1991): 9–27; and David Albert and Hilary Putnam, "Further Adventures of Wigner's Friend," *Topoi* 14, no. 1 (1995): 17–22.

9. Putnam, "A Philosopher Looks at Quantum Mechanics," 130–132.

(neutrons and protons) just as the planets whirl around the sun. Out of atoms are built molecules; out of molecules, macroscopic objects scaling in size from dust motes to whole planets and stars. These latter also fall into larger groupings—solar systems and galaxies—but these larger structures differ from the ones previously mentioned in being held together exclusively by gravitational forces. At every level, however, one has trajectories (ultimately that means the possibility of continuously tracing the movements of the elementary particles) and one has causality (ultimately that means the possibility of extrapolating from the history of the universe up to a given time to its whole sequence of future states).

When we describe the world using the techniques of Newtonian physics, it goes without saying that we employ *laws*—and these laws are stated in terms of certain *magnitudes,* e.g. distance, charge, mass. According to one philosophy of physics—the operationalist view so popular in the 1930s—statements about these magnitudes are mere shorthand for statements about the results of measuring operations. Statements about distance, for example, are mere shorthand for statements about the results of manipulating foot rulers. I shall assume that this philosophy of physics is *false.* Since this is not a paper about operationalism, I shall not defend or discuss my "assumption." I shall simply state what I take the correct view to be.

According to me, the correct view is that when the physicist talks about electrical charge, he is talking quite simply about a certain magnitude that we can distinguish from others partly by its "formal" properties (e.g. it has both positive and negative values, whereas mass has only positive values), partly by the system of laws this magnitude obeys (as far as we can presently tell), and partly by its *effects.* All attempts to *literally* "translate" statements about, say, electrical charge into statements about so-called observables (meter readings) have been dismal failures, and from Berkeley on, all *a priori* arguments designed to show that all statements about unobservables must ultimately reduce to statements about observables have contained gaping holes and outrageously false assumptions. It is quite true that we "verify" statements about unobservable things by making suitable *observations,* but I maintain that without imposing a wholly untenable theory of meaning, one cannot even *begin* to go from this fact to the wildly erroneous conclusion that talk about unobservable things and theoretical magnitudes *means the same* as talk about observations and observables.

Now then, it often happens in science that we make inferences from measurements to certain conclusions couched in the language of a physical theory. What is the nature of these inferences? The operationalist answer is that these inferences are *analytic*—that is, since, say "electrical charge" *means by definition* what we get when we measure electrical charge, the step from the meter readings to the theoretical statement ("the electrical charge is such-and-such") is a purely conventional matter. According to the non-operationalist view, this is a radical distortion. We know that this object (the meter) measures electrical charge not because we have adopted a "convention," or a "definition of electrical charge in terms of meter readings," but because we have accepted a body of theory that includes *a description of the meter itself in the language of the scientific theory.* And *it follows from the theory,* including this description, that the meter measures electrical charge (approximately, and under suitable circumstances). The operationalist view disagrees with the actual procedures of science by replacing a probabilistic inference within a theory by a non-probabilistic inference based on an unexplained linguistic stipulation.

If the non-operationalist view is generally right (that is to say, correct for physical theory in general—not just for Newtonian mechanics), then *the term "measurement" plays no fundamental role in physical theory as such.* Measurements are a subclass of physical interactions—no more or less than that. They are an important subclass, to be sure, and it is important to study them, to prove theorems about them, etc.; but "measurement" can never be an *undefined* term in a satisfactory physical theory, and measurements can never obey any "ultimate" laws other than the laws "ultimately" obeyed by *all* physical interactions.

For myself, and for any other "scientific realist,"[10] the whole so-called interpretation problem in connection with quantum mechanics is just this: *whether* we can understand quantum mechanics—no, let me be optimistic—*how* to understand quantum mechanics in a way that is compatible with the antioperationalist philosophy to which I subscribed in the pages I just quoted, and to which I have always subscribed. But it took a long time for physicists

---

10. At the suggestion of Yemima Ben-Menahem, let me say that by "scientific realism" I mean what I called "convergence realism" in Hilary Putnam, "Three Kinds of Scientific Realism," in Putnam, *Words and Life* (Cambridge, Mass.: Harvard University Press, 1994), 492–498. As I argued there, and in fact as far back as Putnam, "Realism and Reason," 1976 Presidential Address to the Eastern Division of the American Philosophical Association, in *Meaning and the Moral Sciences* (London: Routledge and Kegan Paul, 1978), 123–138, scientific realism in this sense does not presuppose either metaphysical realism or metaphysical antirealism, although it is incompatible with all forms of operationalism and phenomenalism.

to admit that there is such a problem. I can tell you a story about that. In 1962 I had a series of conversations with a world-famous physicist (whom I will not identify by name). At the beginning, he insisted, "You philosophers just *think* there is a problem with understanding quantum mechanics. We physicists have known better from Bohr on."[11] After I forget how many discussions, we were sitting in a bar in Cambridge, and he said to me, "You're right. You've convinced me there is a problem here; it's a shame I can't take three months off and solve it."

What "Quantum Mechanics" Says—and Some Problems

If I am to explain why there is a problem with understanding quantum mechanics in a way that is compatible with scientific realism, I first have to say what "quantum mechanics" says. I will simplify considerably,[12] of course, but the logical structure of the problem will not be affected by these simplifications.

"Pure states" of a physical system—states about which we have as much information as quantum mechanics allows us to have—are represented by vectors in an abstract space called "Hilbert space." I will call these vectors "state vectors" (and I will freely identify states and their state vectors to simplify exposition). Like vectors in a real vector space, they can be multiplied by a scalar,[13] and they are subject to appropriately defined operations of vector addition and inner product.

These states can also be represented as functions over what is called a "basis." For example, if we have a system of just two particles[14] A and $B,$ and we let the position coordinates of A be x_1, x_2, and x_3, and the position coordinates

11. The physicist in question was, of course, referring to Niels Bohr and the so-called Copenhagen interpretation of quantum mechanics.

12. For example, I will not distinguish between a vector and a "ray" (a one-dimensional subspace of a Hilbert space), and I will not employ "bra-ket" notation. Most important, I will confine attention to pure states, ignoring the problem of understanding mixed states. On that problem, I still stand by what I wrote in "A Philosopher Looks at Quantum Mechanics," 155–156.

13. The scalars can be complex numbers, and not only real numbers, a fact that has a mathematical connection with the phenomenon of "interference."

14. For the time being I confine attention to elementary nonrelativistic quantum mechanics, with a finite number of particles.

of B be x_4, x_5, and x_6, then any pure state can be represented as a complex-valued function of x_1, x_2, x_3, x_4, x_5, x_6, and t, where t represents *time*. This is called a "representation in a position basis."[15] (In general, if the system consists of N particles, the basis will have $3N$ coordinates.)

Naturally, we want to know how the states of a system change with time, and the answer is that, represented as functions in the way just described, they obey a famous differential equation, the Schrödinger equation. This equation is the heart of the "dynamics" of quantum mechanics. (Whether this is the *only* way in which states can change is, as we will see, at the heart of the controversies about quantum mechanics.)

With just this minimum of information about what quantum mechanics says, I can already state the first problem that I want to discuss, the famous problem of "Schrödinger's cat." I think that everyone has heard of Schrödinger's cat by now. Schrödinger imagined that one has a cat in a well-isolated system, say, a satellite in space, and one has a device that shoots exactly one photon in the direction of a half-silvered mirror, beyond which is a detector. The half-silvered mirror has the property that the quantum-mechanical probability that the photon will go through is ½. If the photon goes through and hits the detector, in Schrödinger's bloody experiment the cat gets electrocuted;[16] if the photon is reflected, the cat lives. According to the Schrödinger equation, the cat should end up in a "state" that would be (represented by) the vector sum or "superposition" of a vector that represents the cat surviving the experiment and a vector that represents the cat being electrocuted, a state of the form $1/\sqrt{2}$(live cat) $+ 1/\sqrt{2}$(dead cat). Problem one is, what are we to make of a state that is a superposition of two states like this, two states in which a *macro-observable* has different values? (Problem two is simply that the existence of such a state is *unbelievable,* and this is connected with a joke of Einstein's that I will mention later.)

15. Instead of the position coordinates, x_1, x_2, x_3, x_4, x_5, x_6, we can also use the components of momentum in the corresponding directions as a basis, and there are still other possible choices. The function will be different, but the instructions for "decoding" the function—interpreting it as the description of a state vector—will ensure that it represents the state we want it to.

16. The reader may be happy to know that in Hilary Putnam, "Three Kinds of Scientific Realism," in Putnam, *Words and Life* (Cambridge, Mass.: Harvard University Press, 1994), and in Albert and Putnam, "Further Adventures of Wigner's Friend," the cat is tickled rather than electrocuted.

If we never observe such a state, why do we not? All interpretations of quantum mechanics are required to give some answer to that question.

One version of the Copenhagen interpretation of quantum mechanics—a version that really *is* an interpretation, unlike Bohr's own remarks, which I interpret as a *rejection* of the possibility of a scientific realist interpretation—was given by von Neumann.[17] According to von Neumann's axiomatization, quantum mechanics says that physical states change in *two* ways—and not always in accordance with the Schrödinger equation. When no measurement is made, changes do obey that equation. But when a measurement is made, states change in a different way: they "collapse," according to the von Neumann axioms. This is so important that I will display it explicitly, and with more detail.

Von Neumann's Interpretation

1. When no measurement is made—and "measurement" is a primitive notion; one is simply supposed to know what a "measurement" is—states change in time in accordance with the Schrödinger equation.

2. When a measurement is made—when one measures the position of some system, or one measures its kinetic energy, or whatever—the state "collapses" (changes discontinuously) into one in which the "observable" (the physical quantity) that was measured has a definite value. Mathematically the collapse is represented not by a solution to a differential equation but by wiping out, erasing, the state that was there before and "putting in by hand" the state that was found by the measurement. The probability that any given possible result of measurement (any given "eigenstate" of the observable measured) will actually be the one found by the measurement is postulated to be proportional to the square of the absolute value of the vector that is the inner product of the unit vector corresponding to that eigenstate and the vector that we "erased"—the vector that *would have been* the state of the system *if it had evolved according to the Schrödinger equation.*[18]

17. John von Neumann, *Mathematische Grundlagen der Quantenmechanik* (Berlin: Springer 1932); English translation, *The Mathematical Foundations of Quantum Mechanics* (Princeton, N.J.: Princeton University Press, 1955).

18. An important terminological note: the word "observable" in the foregoing has little or nothing to do with the epistemological use of "observable" in my explanation of scientific realism

The answer to our problem one—the problem of Schrödinger's cat—on the von Neumann interpretation is straightforward but hardly satisfactory. We never see a cat in a state of the form $1/\sqrt{2}$(live cat) $+ 1/\sqrt{2}$(dead cat) because our looking to see whether the cat survived is a "measurement," and so the cat obligingly collapses into the state (live cat) or into the state (dead cat) when we look!

Other Interpretations of Quantum Mechanics

I mentioned at the start that today we know of many different interpretations of quantum mechanics. Table 1 presents two unconventional ones. The Bohm interpretation, first proposed by Bohm in 1952,[19] was rejected by me in the 1965 article for reasons I now want to retract, and the Ghirardi-Rimini-Weber (GRW) interpretation was proposed long after my paper was published.[20]

Bohm's theory, a successor to the earlier "pilot-wave" interpretation described by De Broglie,[21] has become the classical example of a hidden-variable theory. According to it, particles do have definite positions and momenta at all times. They even have continuous trajectories. These trajectories are determined by two things: (1) a "velocity field" and (2) the initial positions and momenta of the particles, which are assumed to be distributed (at whatever is chosen to be the initial time t_0) randomly, but in accordance with the probability distribution given by the state vector. The velocity field is determined by what is called "the probability current" (which depends on the state vector of the system in accordance with the classic rule for determining probabilities in quantum mechanics first suggested by Max Born and employed by von Neumann).

---

at the outset of this essay. Unfortunately, "observable" has become the standard term in quantum mechanics for any physical magnitude, whether we are able to measure it or not. It is not a philosophical term.

19. David Bohm, "A Suggested Interpretation of the Quantum Theory in Terms of 'Hidden' Variables: I and II," *Physical Review* 85 (1952): 166–193.

20. Giancarlo Ghirardi, Alberto Rimini, and Tullio Weber, "An Attempt at a Unified Description of Microscopic and Macroscopic Systems," in Vittorio Gorini and Alberto Frigerio, eds., *Fundamental Aspects of Quantum Theory: Proceedings of the NATO Advanced Research Workshop*, Como, September 2–7, 1985, NATO ASI Series B 144 (New York: Plenum Press, 1986), 57–64.

21. Louis-Victor De Broglie, *Non-linear Wave Mechanics: A Causal Interpretation* (Amsterdam: Elsevier, 1960). The idea of interpreting the state functions of quantum mechanics as "pilot waves" was also entertained early on by Schrödinger, who seems to have quickly given it up.

Table 1   Two unconventional interpretations of quantum mechanics

Bohm	GRW (Ghirardi-Rimini-Weber)
The classical "hidden-variable" theory. Particles have definite positions and momenta all the time. The trajectories are determined by a velocity field that is itself determined by the quantum-mechanical "state."	The spontaneous-localization model. Each particle has a tiny probability of jumping into a close to definite position state. Objects that contain millions of atoms will, as a result, always have definite positions by macroscopic standards, thus solving the problem of "Einstein's bed."

It is a consequence of the Bohm theory that these initial positions and momenta of the particles are in principle impossible to determine; all that we know about these initial positions is that they are distributed in such a way that the *probability* that any one of these particles is at any given place at t_0 is the standard quantum-mechanical probability. That property, that the probability distribution is quantum mechanical, is then preserved through all time. In this way the Bohm interpretation explains why the probability of finding any given particle in any given place, or, indeed, of finding the whole system of particles in any given position configuration at any given time, is in accordance with standard quantum-mechanical calculations (e.g., with those given by von Neumann).

In "A Philosopher Looks at Quantum Mechanics" I rejected Bohm's interpretation for several reasons that no longer seem good to me. Even today, if you look at Wikipedia on the Web, you will find it said that Bohm's theory is mathematically inelegant. Happily, I did not give *that* reason in the 1965 article, but in any case it is not true. The formula for the velocity field is extremely simple: you have the probability current in the theory anyway, and you take the velocity vector to be proportional to the current. There is nothing particularly inelegant about that; if anything, it is remarkably elegant! However, I rejected Bohm's theory for reasons I got from my teacher, Hans Reichenbach, who rejected an earlier version of the "pilot-wave" theory due to de Broglie (an incompletely worked-out hidden-variable theory) as leading to unacceptable "causal anomalies."[22] I was also misled by Bohm's unfortunate choice of a term for the velocity field; he called it a "potential." I took it to be

22. Hans Reichenbach, *Philosophic Foundations of Quantum Mechanics* (Berkeley: University of California Press, 1944), part I, paragraphs 7–8.

literally a potential energy distribution and argued that that is an interpretation that it will just not bear.

It is certainly true that the Bohm theory implies certain "causal anomalies." But the most obvious of these are no longer reasons for rejecting the theory, because we know that they really occur in nature. Bohm's theory implies nonlocality. Both Reichenbach and I worried about the question "How come this Bohm 'potential' (which we thought of as a force) does not get weaker with distance?" The answer, as we now know, is "Because nonlocal correlations can appear over any distance." If, as Maudlin has suggested,[23] we think of the "Bohm field" as *a mathematical representative of nonlocality,* then we need no longer be bothered by the "causal anomaly." In sum, I do not think that the Bohm theory can simply be dismissed, removed from consideration, in the way I did in 1965.

In the interpretation on the right of Table 1, the interpretation proposed by Ghirardi, Rimini, and Weber, which is called "the spontaneous-localization model," each particle has a *tiny* probability—*truly* tiny—of spontaneously jumping into a definite position state. The probability is so tiny that, for example, if the system consists of just one isolated particle, one would have to wait, on average, many thousands of years for it to jump into a definite position state. In other words, don't hold your breath. *But,* as we all know, an object such as the table in front of me as I type these words consists of millions upon millions of particles. If there are many millions of particles, the object will, according to GRW theory, always have a definite position.[24] As I say in Table 1, that solves "the problem of Einstein's bed," and I will tell you what that problem is.

The reason it solves the problem is that, according to quantum mechanics, if even one of the particles, even one of the electrons, for example, of which the table consists, "jumps" into a definite position state, that state *multiplies* the state of the whole system by a Gaussian that forces that state to become definite with respect to position (to come close to being an eigenstate of position). These "spontaneous collapses" may be happening at times that are very far apart, as far as any one particle is concerned, but they are happening *all the time* as far as this very large collection of particles is concerned. For that reason, the probability that the table will fail to have a definite position (by

23. Maudlin, *Quantum Non-locality and Relativity.*

24. More precisely, it will have a definite position by macroscopic standards of definiteness with a probability that is so close to one that it can be treated as certainty.

macroscopic standards of definiteness) becomes virtually zero. It is not *impossible,* but it will never happen.

The Problem of Einstein's bed

Why do I call this "the problem of Einstein's bed"? I said early on that the existence of superpositions of states in which macro-observables have different values, states such as $1/\sqrt{2}$(live cat) $+ 1/\sqrt{2}$(dead cat), is *unbelievable,* and that the problem this poses is connected with a joke of Einstein's. Here is the story of Einstein's joke.

I met Einstein just once, in 1953. He was a friend of my teacher, Hans Reichenbach. When I joined the Princeton faculty in 1953, Reichenbach arranged for me to meet Einstein. I had tea with him in the little house on Mercer Street. Not unexpectedly, he talked about his dissatisfaction with quantum mechanics. He did *not* say, "God doesn't play with dice." He did not say or imply anything to the effect that he could not accept a theory that is indeterministic (and we now know from Don Howard[25] and other historians who have studied Einstein's unpublished correspondence that the failure of determinism was not Einstein's most significant problem with quantum mechanics). What he said on that occasion was something like the following: "Look, I don't believe that when I am not in my bedroom my bed spreads out all over the room, and whenever I open the door and come in it jumps into the corner."

In other words, Einstein could not believe von Neumann's "collapse" assumption.

To tell you the truth, I do not like the term "measurement problem," which is often used in connection with quantum mechanics. I think that "collapse problem" would be a better term. In other words, the real problem is, *do we or do we not need to postulate a "collapse," and if we do assume a "collapse," what should we say about it?* Ever since GRW has been in the field, there have been interpretations of quantum mechanics in which collapse has nothing to do with *measurement.* It happens whether a human makes a measurement or not. For that reason, the term "measurement problem" is out-of-date.

In "A Philosopher Looks at Quantum Mechanics" I said that the view that I regarded as the most plausible one to take was a variant of the Copenhagen

---

25. Don Howard, "Einstein on Locality and Separability," *Studies in History and Philosophy of Science* 16 (1985): 171–201.

interpretation.[26] In order to avoid taking "measurement" as a primitive notion, I said that von Neumann's assumption that the state "collapses" into an
eigenstate of the observable measured upon measurement should be replaced
by a different assumption.

Instead of saying henceforth that observables do not exist unless they are
measured, we will have to say that *micro*-observables do not exist unless they
are measured. We will take it as an assumption of quantum mechanics that
macro-observables retain sharp values (by macroscopic standards of sharp-
ness) at all times. The formulation of the Copenhagen interpretation that
I am now considering, then, comes down to this: that macro-observables
have sharp values at all times in the sense just explained, while micro-observables
have sharp values only when they are measured, where measurement is to be
defined as a certain kind of interaction between a micro-observable and a
macro-observable.

And I said that the remaining open problem for quantum mechanics was
to say what is so special about macro-observables: "The result we wish is that
although micro-observables do not necessarily have definite numerical values
at all times, macro-observables do. And we want this result to come out of
quantum mechanics in a natural way. We do not want simply to add it to
quantum mechanics as an *ad hoc* principle. So far, however, attempts to derive
this result have been entirely unsuccessful."[27]

Classification of the Possible Kinds of Interpretation

For many years after that, I tried to come up with an interpretation of quantum mechanics that would solve the problems I have described. In 1968
I even proposed an interpretation that involved a nonstandard logic,[28] but, as
I explained in 1994,[29] that interpretation collapsed (although not in the
quantum-mechanical sense of "collapse"). More recently, however, it has oc-

---

26. Putnam, "A Philosopher Looks at Quantum Mechanics," 149–155. Here I understood
"Copenhagen interpretation" to mean von Neumann's axiomatization, not Bohr's rejection of
scientific realism.

27. Ibid., 157.

28. Hilary Putnam, "The Logic of Quantum Mechanics," in Putnam, *Philosophical Papers*,
vol. 1, *Mathematics, Matter and Method*, 174–197.

29. Hilary Putnam, "Michael Redhead on Quantum Logic," in Peter Clark and Bob Hale,
eds., *Reading Putnam* (Oxford: Blackwell, 1994), 265–280.

Table 2 Kinds of interpretations of quantum mechanics

Collapse	No collapse
Produced by something external to the system and not subject to superposition (e.g., von Neumann)	No hidden variables (many worlds)
versus	versus
Spontaneous (e.g., GRW)	Hidden variables (e.g., Bohm)

curred to me that I should instead attempt to classify *all possible interpretations* (or all possible interpretations that do not involve giving up classical logic, because I had satisfied myself that the approach via a nonstandard logic would not work). In this way, I arrived at the four classes shown in Table 2, which obviously rests on little more than the law of the excluded middle: either an interpretation says that there is a collapse or it does not; if it says that there is a collapse, then either that collapse is spontaneous or it is explained by something external to the system (something that does not itself undergo collapse). If it says that there is no collapse, then either there are hidden variables or there are not. After the possible interpretations are laid out, each of us can say which of these possibilities should be ruled out as "nonstarters"—as I will now attempt to do—but it seems to me that considerable clarification will result if we can agree that these *are* the possibilities before us if we want a scientific realist interpretation.

In Table 2 of the four kinds of possible (scientific realist) interpretations, I give an example of the currently most plausible version of each to stand for the kind, but really I am interested in these as *classes* of possible interpretations.

On the left we have collapse interpretations, and on the right we have the no-collapse interpretations. (Even Brouwer would have to agree that this is a legitimate use of the law of the excluded middle!) Either an interpretation of quantum mechanics says that there is a "collapse of the state function" or it says that there is no collapse, that it never happens.

In the left column, the "collapse" column, the first possibility is that the collapse is produced by something external to the system and not subject to superposition, which is von Neumann's proposal. For example, one might propose that macro-observables are not subject to superposition, as I did in my 1965 paper, or one might say—von Neumann hints at this in his book, and

Eugene Wigner famously advocated it [30]—"No, the collapse occurs when the result of a measurement is registered by a consciousness." I do not know of anyone who currently advocates this "psychic" view, but those are the two "classical" versions of the view that collapse is produced by a "measurement," where "measurement" means that the system interacts with something that is intrinsically not subject to superposition.

However—as GRW has taught us—there is another possibility to be included in the "collapse" column: the collapse could be spontaneous. For example, it could be an ultimate statistical law of nature that a particle has a certain fixed probability of "jumping into a position eigenstate" (or, as in GRW, a state very close to but not exactly the same as a position eigenstate). Alternatively, the spontaneous collapse could be provoked by a "trigger"—it could, for example, as suggested by Penrose,[31] occur whenever the global gravitational field would otherwise be superimposed to a certain unacceptable extent, although this "gravitational-trigger" suggestion has not, to my knowledge, been worked out in detail.

What about the interpretations in the other column—the "no-collapse" column?

The most famous of these is the so-called many-worlds interpretation. This interpretation was proposed by the late Hugh Everett III, who suggested in 1957,[32] in a paper that, as Albert has put it, is "both extraordinarily suggestive and hard to understand,"[33] that he had found a way of coherently entertaining the possibility that there is no collapse and there are no hidden variables: the world (the state function of the whole universe) just rolls on and on as predicted by the Schrödinger equation. Again quoting Albert, "The idea of what has become the canonical *interpretation* of Everett's paper[34] is that the means of coherently entertaining the possibility that Everett must have had in mind (or perhaps the one that he ought to have had in mind) is to take the two components of a state [such as $1/\sqrt{2}$(live cat) $+ 1/\sqrt{2}$(dead cat)][35] . . . to represent

30. Eugene Wigner, *Symmetries and Reflections* (Bloomington: Indiana University Press, 1967), 200–208.

31. Roger Penrose, *Shadows of the Mind* (Oxford: Oxford University Press, 1994), part 2.

32. Everett, "Relative State Formulation of Quantum Mechanics."

33. Albert, *Quantum Mechanics and Experience,* 112–113.

34. Albert has in mind B. S. Dewitt, "Quantum Mechanics and Reality," *Physics Today* 23 (1970): 30–35.

35. Albert uses a different example.

(literally!) two physical *worlds*."[36] The idea, applied to the thought experiment of Schrödinger's cat, is that when the photon reaches the half-silvered mirror, the number of physical worlds there are literally increases from one to two: in one of the two worlds the photon goes through the half-silvered mirror, the cat is electrocuted, and the observer has the determinate belief that the cat is dead; in the other the photon is reflected, the cat lives, and the observer (and presumably the cat as well) has the determinate belief that the cat is alive.

Finally, there are no-collapse interpretations in which there *are* hidden variables, and the most famous of these is the Bohm interpretation that I described earlier. In that interpretation, positions are the hidden variables: they can sometimes be localized, it is true, but there is no way of determining the precise trajectories that the particles follow.

But why did I say there is no collapse in the Bohm interpretation? What happens to the state function when I make a position measurement on a particle, and I say, "Now it is here"? The answer is that the state continues to evolve according to the Schrödinger equation on the Bohm interpretation, exactly as in the many-worlds interpretation! Bell, who showed that quantum mechanics implies nonlocality, referred to the particle locations in the Bohm interpretation—which he took very seriously—as the "beables," rather than as "hidden variables." These beables are physical quantities that have definite values at all times. The quantum-mechanical state "guides" the particles, but it never collapses, on the Bohm interpretation.

So these are the four families of interpretations. To repeat: either there is a collapse or there is no collapse. If there is a collapse, either it is produced by something not subject to superposition, as in von Neumann's formulation of quantum mechanics, or it is spontaneous. If there is no collapse, either there is no collapse and no hidden variable, no beables, or there is no collapse plus beables. That is how I propose that we draw the map. Even if we subsequently rule out some of the interpretations, we should not begin by ignoring them. Now I will say which interpretations in Table 2 I think we should discard.

Which Interpretations I Think We Can Rule Out

As I have said, in "A Philosopher Looks at Quantum Mechanics" I chose the left column ("collapse"), I chose the first interpretation in that column, taking

36. Albert, *Quantum Mechanics and Experience,* 113; the emphases are in the original.

macro-observables to be the "something external to the system and not subject to superposition" that brings about the collapse, and I took it to be an unsolved problem why this is the case. I took it to be a problem because it cannot be an ultimate postulate of physics that macro-observables do not undergo superposition. "Macro-observable" is not the sort of term that can be an irreducible primitive in an ultimate physical theory, so I called for some future extension of quantum mechanics that would explain why macro-observables do not go into such states as $1/\sqrt{2}$(live cat) $+ 1/\sqrt{2}$ (dead cat).

That proposal seems less and less attractive to me. That there is something special about macro-observables seems tremendously unlikely, unless what is special is, for example, that they involve large numbers of particles and the particles spontaneously collapse, or something of that sort. So I am now inclined to give up this most classic interpretation—it was, in a way, *the* interpretation that was actually used in the first decades of quantum mechanics. And this leaves the second interpretation—spontaneous collapse—as the open possibility in the left-hand column.

I do not necessarily mean GRW theory. There are certain problems with the GRW account of collapse. For example, if a system collapses into an eigenstate of position at a certain time—if it collapses into a really *absolutely sharp* position state (which Ghirardi, Rimini, and Weber do not say, for just the reason I am going to describe)—then, according to quantum mechanics, the "complementary" observable, momentum, would be infinitely indeterminate. It could be as huge as you like, and then one would have violations of the conservation of energy. So, what GRW postulates is that the collapse is not infinitely sharp, but also not too "spread out" (otherwise Einstein's bed would lose its nice definite position). The upshot is that GRW predicts violations of the conservation of energy that are *too small to be observed*. This is a little distressing (but then life is hard, you cannot please everyone!). Anyway, if GRW turns out to be true, I hereby predict that somebody will come up with a redefinition of "energy" that saves the law of the conservation of energy. (Mathematicians are very clever, and the history of physics suggests that they would somehow manage.)

On the other side of my table, the "no-collapse" side, the many-worlds interpretation now has distinguished advocates.[37] I myself think that it is untenable, and my reason is very simple.

37. Murray Gell-Mann and J. B. Hartle, "Quantum Mechanics in the Light of Quantum Cosmology," in W. H. Zurek, ed., *Complexity, Entropy and the Physics of Information,* Santa Fe Institute Studies in the Sciences of Complexity, vol. 8 (Boston: Addison-Wesley, 1990).

Suppose that I perform an experiment. It can have just two outcomes, as in the Schrödinger's cat thought experiment, except that instead of having a complicated system such as a cat, let us just have two light bulbs, a red one and a green one, and let the experimental setup be arranged so that either the red light goes on or the green light goes on when the interaction takes place (but not both). The setup need not be such that the two outcomes are equi-probable, as was the case with Schrödinger's cat. The two outcomes could have probabilities of 0.9 versus 0.1, or 0.00001 versus 0.99999, or whatever you wish. The probabilities can be very unequal, and it does not matter. I per-form this experiment. Then I perform it again. Then I perform it a third time. Let us suppose that I repeat it thirty times.

According to the many-worlds interpretation, because the world obeys the Schrödinger equation at all times, in the course of the first trial of the experi-ment it really does go into a superposition of the form p(green light lights) $+ (1-p)$(red light lights), where p is the probability that the outcome is that the green light goes on and $(1-p)$ is the probability that the red light goes on, and this means, on Everett's interpretation as interpreted by Dewitt, that there are now (after the first trial) *two* physical universes, in one of which I observe the green light go on (and only that light go on) and in the second of which I observe only the red light go on.

So far, this is just a simplified version of "Einstein's bed." If Einstein had chosen to talk about this experiment rather than (unfortunately for his own case[38]) about his bed, he would have asked, "Why don't I see a *superposition* of red light on and green light on?" The answer that Everett-Dewitt would have given him would have been, "Because the state vector p(green light lights) $+ (1-p)$(red light lights) is *coupled* with Einstein's state vector, the Ein-stein in the first of the two physical universes that exist after the interaction *sees* the green light go on, and the Einstein in the *second* of the two physical universes *sees* the red light go on." In the total reality—we might call it the Everett "multiverse"—after the experiment is performed the first time, there will be two Einsteins, one seeing a green light shining and the other seeing a red light shining. After the experiment is performed a second time, there will be *four* Einsteins and four Einstein histories: (1) a history that goes Einstein sees green followed by green again, (2) a history that goes Einstein sees green

38. I say "unfortunately for his own case" because the Schrödinger equation does not imply that the bed would go into a superposition of states in which it is in different places in the room between times when Einstein enters the bedroom.

followed by red, (3) a history that goes Einstein sees red followed by green, and (4) a history that goes Einstein sees red followed by red. After the experiment has been performed thirty times, there will be 2^{30} Einsteins with 2^{30} histories. And, unlike Schrödinger's cat, which is only a thought experiment, performing a simple two-outcome experiment thirty times is something any of us can do.

I repeat, on the many-worlds interpretation, there will be 2^{30} Einstein histories—"parallel worlds"; science fiction is literally right!

Now suppose that instead of saying "Einstein," I say "I" (and I ask you, dear reader, to think of yourself as the "I" in question). Suppose that I do this. And I ask myself, "*What is the probability in the naïve sense*—not the 'probability' in the quantum-mechanical sense, this real number that I calculate by finding the square of the absolute value of a certain vector—but the probability in the sense of *the number of my future histories in which I will observe that, say, the green light went on half of the time plus or minus 5 percent of the time divided by the total number of my future histories?*" You can figure this out by simple combinatorics. (All it takes is the binomial theorem.) And the answer is independent of the quantum-mechanical "probability." So why should I use quantum-mechanical "probability" to predict what I am likely to observe?

Not surprisingly, the largest number of my future histories are ones in which the green light will have gone on approximately 50 percent of the time, and this is so regardless of what the quantum-mechanical "probability" p happens to be.

Let me put it another way. If histories in which what one of the 2^{30} Hilary Putnams observes confirms quantum mechanics are so rare—as they will be if p is very small, as we can easily arrange for it to be—how come we are so lucky? How come my observations have so far been in accordance with quantum-mechanical probability? On the many-worlds interpretation, quantum mechanics is the first physical theory to predict that the observations of most observers will disconfirm the theory.

The philosophical moral behind my question is this: once you give up the distinction between actuality and possibility—as the many-worlds interpretation in effect does by postulating that all the quantum-mechanical possibilities are actualized, each in its own physical universe—and once you say that all possible outcomes are, ontologically speaking, equally actual, the notion of "probability" loses all meaning. "No collapse and no hidden variables" is incoherent.[39]

---

39. Defenders of the many-worlds interpretation often appeal to a set of "decoherence" theorems to show why detectable interference between the different "histories" does not occur.

The "Moral" of This Discussion

What we are left with, if what I have said so far is right, is a conclusion that I initially found very distressing: *either GRW or some successor, or else Bohm or some successor, is the correct interpretation*—or, to include a third possibility to please Itamar Pitowski,[40] we will just fail to find a scientific realist interpretation that is acceptable. (And the ghost of Bohr will laugh and say, "I told you all along that the human mind cannot produce a realist interpretation of quantum mechanics!") But if we are optimists and think that there is somehow a realistic interpretation to be found, then—as argued in Maudlin—we are left with GRW and Bohm.

Why did this conclusion make me unhappy? Not just because of the paradoxical consequences of each of these theories (after all, the phenomena they are called on to account for are "paradoxical"). In the case of the Bohm theory, there is the fact that it has not yet been extended to a quantum field theory, let alone to a quantum cosmology, although Callender and Huggett have already made proposals in this direction.[41] It is not just the technical problems—which are a good thing, a stimulus to productive research. It is that neither of these theories is Lorentz invariant, and it seems likely to me that there is no rigorous proof, but, as Maudlin argues, it is pretty clear that no theory in either of the classes that they represent (the "no-collapse and hidden-variables class" and the "spontaneous-collapse" class) can do without an "absolute time" parameter. An absolute time will come back into the picture if either *sort* of theory is destined to be the future physics.[42]

(My commentator at the conference at which this paper was originally read, Itamar Pitowski, said, "You are saying that before we can interpret

---

These theorems answer objections to the interpretation from within the many-worlds picture. But if my argument above is correct, the entire picture is incoherent, and the decoherence theorems in no way speak to that problem.

40. Itamar Pitowski was my very helpful and challenging commentator at the meeting of the Israel Association for History and Philosophy of Science at which this article was delivered.

41. Craig Callender and Nick Huggett, eds., *Physics Meets Philosophy at the Planck Scale: Contemporary Theories in Quantum Gravity* (Cambridge: Cambridge University Press, 2001).

42. Amazingly, this has turned out not to be correct. Roderich Tumulka has produced an example of a GRW-type "collapse" theory that is Lorentz invariant and accounts for the electron paramagnetic resonance phenomenon. An excellent account of the details will appear in the forthcoming fourth edition of Maudlin's *Quantum Non-locality and Relativity*. [Added April 5, 2011.] On this see Chapter 7.

quantum mechanics we have to *change* it." My reply was that von Neumann already "changed" quantum mechanics, certainly from Bohr's point of view. *All* interpretations of quantum mechanics are in a sense "changes" of quantum mechanics, because it is an *incomplete* theory—one cannot "regiment" it, formalize it in standard logical notation [e.g., quantificational logical notation], *unless* one adds an "interpretation." And up to now, Putnam's law, the law that an interpretation of quantum mechanics is believed by its inventor and up to six other physicists, has held good.)

Can we soften the "bad news" that we may need to return to a notion of "absolute time"? My final suggestion is this: when it comes to quantum cosmology—and, as yet, neither GRW nor the Bohm theory has been extended to quantum cosmology—in my view, *present-day* quantum cosmology *does already* involve a "background" time parameter. It is sometimes concealed, as when cosmologists say that they are not really taking an absolute time as the parameter in the Schrödinger equation but are taking something such as the "radius" of the universe as the time parameter (and hoping that this is a well-behaved quantity). But this parameter plays exactly the role of an absolute time in which the cosmos is supposed to evolve.

The reason for its presence is that in present-day quantum cosmology one does not talk about *one single space-time.* Quantum mechanics depends on the idea that all physical "states of the world" (live cat, dead cat, red light lights, green light lights, and so on) are represented by mathematical objects, vectors, that can be multiplied by scalars and can be added, so that one gets such states as p(green light lights) + $(1 - p)$(red light lights). In quantum cosmology the state vectors can represent *different geometries of space-time.*[43] In effect, one superimposes whole space-times. And this superposition of space-times evolves in the background time.

So what relieves my initial distress at the idea of an absolute time coming back into the picture is the following thought: it might not be quite as bad a contradiction of Einstein's vision as it first seems. It might be that before we "superimpose," *each space-time is perfectly Einsteinian*—each space-time is a Minkowski space-time that knows nothing about any "simultaneity." And it may be that the time parameter that both GRW and Bohm need is just the absolute time parameter that quantum cosmology seems to need. Of course,

---

43. A classic presentation is in C. W. Misner, K. S. Thorne, and J. A. Wheeler, *Gravitation* (San Francisco: W. H. Freeman, 1973).

this is just a speculation. But it would mean that although Einstein would have to admit that there is such a thing as simultaneity, it comes from "outside" any one well-defined space-time; it comes from the quantum-mechanical "interference" between whole space-times. And with this speculative suggestion I will close.[44]

44. The views I defend in this paper owe a great deal to Maudlin's *Quantum Non-locality and Relativity* and to David Albert, with whom I have discussed these issues for more than twenty years.

7

Quantum Mechanics and Ontology

Ontology

Quine is generally supposed to have taught us two important lessons about ontology: (1) The proper task for a fallibilistic and naturalistic philosopher (both of which I count myself as being) is not to say what there is but only to tell us what our best scientific theories claim there is. (I chafe at the restriction to "scientific" theories, but that is not my topic at present.) (2) That task—telling us what our best scientific theories claim there is—can be performed by looking at those theories and seeing what entities their existential quantifiers purport to range over.

Although I have criticized this idea and the notion of an "entity" that it presupposes in a recent book,[1] it is not my intention to present that criticism once again here. Instead, I shall reflect on why Quine's apparently straightforward advice is so little help in the case of quantum mechanics.

To do this, it will be useful to explain a term introduced into the discussion by John S. Bell, the term "local beable." A "beable" is an entity that a

---

1. Hilary Putnam, *Ethics without Ontology* (Cambridge, Mass.: Harvard University Press, 2004).

theory postulates as being physically real. In "The Theory of Local Beables," Bell wrote,

> The word "beable" will . . . be used here to carry [the distinction] familiar already in classical theory between "physical" and "non-physical" quantities. In Maxwell's electromagnetic theory, for example, the fields E and H are "physical" (beables, we will say) but the potentials A and φ are "non-physical." Because of gauge invariance the same physical situation can be described by very different potentials. It does not matter that in Coulomb gauge the scalar potential propagates with infinite velocity. It is not really supposed to be there. It is just a mathematical convenience.[2]

Further:

> We will be particularly concerned with the *local* beables, those which (unlike, for example, the total energy) can be assigned to some bounded spacetime region. For example, in Maxwell's theory, the beables local to a given region are just the fields E and H and all the functionals thereof. It is in terms of local beables that we can hope to formulate some notion of local causality.[3]

Employing this terminology, Tim Maudlin has argued that measuring apparatuses (dials and pointers, for example)—not to mention tables and chairs!—had better be local beables if talk of "measurements" is to have any clear sense. As he says,

> We take the world to contain localized objects of unknown composition in a certain disposition that changes through time. These are the sorts of beliefs we *begin with*. A physics that cannot somehow account for these beliefs is a physics we would not have any use for. This is not to say that a physics with no local beables at all could not in principle account for those beliefs, but it is to say that understanding such a theory, and its relations to our pre-theoretical beliefs, is going to be a much, much more complicated business than understanding a theory with observable local beables.[4]

---

2. John S. Bell, "The Theory of Local Beables," TH-2053-CERN, 1975 July 28; reprinted in Bell, *Speakable and Unspeakable in Quantum Mechanics* (Cambridge: Cambridge University Press, 1987; cited from the revised edition 2004), 53.

3. Ibid.; emphasis in the original.

4. T. W. Maudlin, "Completeness, Supervenience and Ontology," in "The Quantum Universe," special issue, *Journal of Physics A: Mathematical and Theoretical* 40, no. 12 (March 23, 2007): 3151–3172. The quotation is from p. 3160; emphasis in the original.

He goes on to make an important remark:

> It is worth noting here that a persistent abuse of terminology has helped to obscure these basic points. In discussions of quantum theory, the postulation of anything in the ontology beside the wavefunction is commonly called the postulation of *hidden variables,* such as the particles in Bohmian mechanics. Since the wavefunction itself is not a local beable, any version of quantum mechanics that has local beables at all will risk having those local beables denominated "hidden." But if they really *were* hidden, i.e., if we could not easily tell just by looking what they are, then the postulation would not help solve the problem of contact with evidence at all. It is exactly because the local beables are *not* hidden, because (according to the theory) it is easy to physically produce correlations between the disposition of those beables and the state of a "measuring apparatus" (or the state of our brain), that they can play the right role in our epistemology. The local beables—at least some of them—had better be manifest rather than hidden. In Bohm's theory, they are.[5]

Local Beables and Scientific Realism

Maudlin's point deserves careful discussion. The following is only the beginning of such a discussion.

The reason that the behavior of measuring apparatuses is not a deep conceptual problem for physics before quantum mechanics is that all macroobjects, including scientific instruments, *consist of* the particles and fields whose motion physics describes.

Physics predicts the trajectories of the measuring instruments in the same way in which it predicts the trajectories of any other physical system. In that sense there is no "cut" between the measuring instrument, whose indications are the epistemological foundation of physics, and the systems described by the physical theories.

The Copenhagen interpretation of Bohr and Heisenberg postulated such a "cut," however, and thus generated a deep epistemological problem. In this paper I shall be considering interpretations that reject the idea of such a "cut," as well as the idea of a "special role of consciousness," etc., etc. (Bell called these "unromantic" interpretations.) But there is still a problem. The problem,

5. Ibid., 3160.

as we shall see, is that it is not clear that particles or classical fields (fields in 3 + 1 space-time) exist as individual things in many of those theories (or many interpretations of those theories). If, for example, "there is only the wave function" (an issue that will come up several times in what follows), then how can there also be tables and chairs and measuring instruments? Indeed, because the wave function does not "live" in 3 + 1 space-time (four-dimensional space-time, or even eleven-dimensional space-time with seven compactified dimensions), but in a higher-dimensional configuration space, how can there be such a thing as space-time, let alone individuals in space-time?

One possible reply is that the time evolution of the wave function described by the theory can be connected by well-known rules to observable phenomena. But this is just the answer of logical positivism! In effect, the wave function is given an "empirical interpretation" by "coordinating definitions," just as Carnap would have said. But this is not the sort of account that a scientific realist—someone who want to understand quantum mechanics as describing reality, and not just as a device for making predictions—is seeking. And I am a scientific realist.

Quinean "Ontological Commitment" and Physics

Armed with these observations, let us now see what happens when we try to follow Quine's advice and look at quantum-mechanical theories to see what entities their existential quantifiers purport to range over. The advice already runs into problems with Maxwell's theory. A formalization of that theory might well quantify over potentials, but those potentials are, for the reason Bell gives, not in fact part of the physical ontology of the theory in any reasonable sense. To be sure, the theory could be formalized so as to avoid such quantification, but there is no end to the different ways in which *any* theory can be formalized, and Quine's advice completely overlooks the *real* issue: how to choose between formalizations with different "ontologies" in his sense of "ontology" (and when to say that the "choice" is simply a matter of convention). But pursuing this criticism would involve issues I cannot consider here.[6]

In the case of "standard" quantum mechanics (not Bohm's theory, or GRW theory, or the Everett "many-histories" theory, but the theory as it is usually taught in a college course, with or without a "collapse" postulate), there

6. See Putnam, *Ethics without Ontology,* especially the last two chapters, for a discussion.

is clearly an entity the theory assumes to exist, namely, the "system"—which, if we apply the theory to the physical universe as a whole, is simply that universe, call it U. But what of the wave function, of which Maudlin speaks in the quotation above? Here there is a small inaccuracy in what he says. The wave function is *not* a part of the physical ontology of quantum mechanics because, as its name implies, it is a *mathematical* object. But here is the way I like to think of what it is doing in the theory: My view of the wave function is that it is a mathematical object that we use to represent a property of U, much as we use a real number x to represent a physical property when we say that an object has a mass of x grams. The property represented by the wave function is a "holistic" property in the literal sense: a property of the whole system. It is not, of course, a bivalent property like "married" (although maybe that one is becoming less bivalent lately?), or even a property capable of degrees like "mass," but a property with a more complex structure, whose possible "values" are representable as wave functions, which I think of as "generalized numbers" (like numbers and vectors, they can be added and "multiplied"). Call this holistic property the "state" of U. The "state" belongs to the second-level ontology, not to the level of individuals. So when Maudlin says that "the postulation of anything in the ontology beside the wave-function is commonly called the postulation of *hidden variables*" (which are, of course, usually thought to be a no-no), what he is saying could suggest the following application of Quine's idea of ontological commitment to standard quantum mechanics: there is exactly one physical object, namely, U, and that physical object has a complex property called its "state", period! "State" can be formalized by introducing a relation S between U and vectors in a higher-dimensional space (a Hilbert space), just as mass can be formalized as a relation M between physical objects (such as planets, asteroids, rocks, and tables) and a real number. But, of course, there are other ways to formalize "standard" quantum mechanics. One could, for example, also quantify over "measuring apparatuses"—this is clearly the way the fathers of quantum mechanics and its Copenhagen interpretation thought. In one way of thinking, when I see that there is a bowl of cherries on the table, what I "really" see is that $S(U)$ lies in a certain subspace of a Hilbert space H. In that way of thinking, that "ontology" talk of bowls and cherries is only a *façon de parler*. In the other "ontology," there really are bowls and cherries, and not only the single individual "object" U. Which is the right way of thinking (if we accept the theory)?

That is not a straightforward empirical question, but it may well determine the choice of one research program over another (as Maudlin argues, by the way, and I think that he is right). The answer may also determine whether we find the relation between what the theory says is going on and what we observe relatively transparent or deeply mysterious (which is the point of the paragraphs from Maudlin's paper that I quoted). (For example, if seeing that a pointer lies between the "5" and the "6" on the dial *is just* seeing that $S(U)$ lies in a certain subspace of a Hilbert space H, how can that "measurement" also *cause* $S(U)$ to "collapse" to a vector in that subspace?) Obviously, quantum mechanics in its standard form does not wear its beables, local or otherwise, on its sleeve! The currently most fundamental physical theories are all quantum mechanical; and quantum mechanics in its mathematical form does not divide the world into "entities" and "properties."

The Instrumentalist Reaction

The reaction to the foregoing remarks of most physicists would, I fear, be somewhat as follows: "Why bother imposing an 'ontology' on quantum mechanics at all? For that matter, why bother formalizing it in the notation of mathematical logic that you philosophers love so much, anyway? Formalization might have some value in the case of discourses in ordinary language, as a way of clearing up ambiguities or revealing lapses in the reasoning, but standard quantum mechanics has a precise mathematical language of its own. If there are problems with that language, they are problems for mathematical physicists, not for philosophers. And in any case, we know how to use that language to make predictions accurate to a great many decimal places. If that language does not come with a criterion of 'ontological commitment,' so much the worse for 'ontology.'"

Although I am sympathetic to the idea that first-order formalisms (and associated models to determine what the variables "purport to range over") should not be made into a fetish, the questions raised by Bell and Maudlin do not presuppose any such fetishism. What they are asking is what quantum mechanics is *about,* what it takes to be physically real. To say "We physicists are just technicians making predictions; don't bother us with that 'physically real' stuff" is effectively to return to the instrumentalism of the 1920s. But physical theories are not just pieces of prediction technology. Even those who claim that that is all they are do so only to avoid having to think seriously about the content of their theories; in other contexts they are, I have observed, quite happy to talk about the same theories as descriptions of reality—as, indeed, they aspire to be.

Perhaps for that reason, various halfway houses to realism have appeared on the scene. One such halfway house is the idea that the wave function describes "information," and not a physical state. But information about *what?* If the answer is, "The purpose of the information approach is precisely to reject that question," then this is simply a more sophisticated route back to instrumentalism. Instrumentalism with postmodern sauce?

The Everett Many-Histories Interpretation

I said that quantum mechanics in its standard form does not wear its beables, local or otherwise, on its sleeve. As we shall now see, the same thing is true of quantum mechanics in various nonstandard forms. Let me begin with the Everett many-histories interpretation. The first thing to note is that this interpretation itself *bears more than one interpretation.* In its earliest form, a paper published by Hugh Everett III in 1957, this was usually known as the "many-worlds" interpretation, or after a publication by B. S. DeWitt in 1970 that attracted a great deal of attention,[7] the "Everett-DeWitt" interpretation. To explain the term "many worlds," I shall employ Bell's humane version of the famous Schrödinger's cat thought experiment. In Bell's version, a cat in an isolated laboratory (Sally's cat) is automatically fed if and only if a given atom in the laboratory has decayed. (Let the probability of this be ½.) On Everett's interpretation, this means that when Sally looks to see whether her cat has been fed or is hungry, the part of the wave function corresponding to the outcome "atom decayed/Sally's cat fed" corresponds to a physically real environment in which Sally observes a fed cat, and the part of the wave function corresponding to the outcome "atom did not decay/Sally's cat hungry" corresponds to a physically real environment in which Sally observes a hungry cat.

The "many-worlds" interpretation was usually thought to have a simple and clear ontology, namely, what Maudlin calls the ontology of "only the wavefunction" (although, as we saw, this is a little misleading). More precisely, the

---

7. B. S. DeWitt, "Quantum Mechanics and Reality," *Physics Today* 23 (1970): 30–35.

real world in this ontology is the system U, and U does not "live" in $3+1$ space-time, but in the Hilbert Space H. As the wave function $S(U)$ evolves according to the Schrödinger equation, it can happen that it takes the form of a superposition, say, $1/\sqrt{2}$(fed cat and Sally seeing fed cat) $+ 1/\sqrt{2}$(hungry cat and Sally seeing hungry cat); and this is the ontological reality corresponding to the fact that one Sally sees a fed cat and one Sally sees a hungry cat. As Maudlin describes this interpretation in the paper from which I quoted at the beginning of this chapter,

> What is the many-worlds theory but the claim that what we call "observation" of a Schrödinger cat is not a process by which many people can come to agreement about the state of the cat, but rather a process by which many people all subdivide into many many people, largely unaware of each other's presence, with the illusion that everyone who looked saw the same thing?[8]

An objection to the many-worlds theory that I shall not review here—I develop it in my 2005 "A Philosopher Looks at Quantum Mechanics (Again)," reprinted in this volume as Chapter 6—is that *if all outcomes of all experiments are equally real, then assigning different probabilities to these outcomes is meaningless.* People who hear this objection either agree at once, in which case they find the Everett-Dewitt interpretation incoherent (as I do), or they somehow think that it does not matter that all outcomes are equally real as long as we can assign real numbers to them that obey the axioms of probability theory.

I have described the Everett-DeWitt interpretation as it was originally understood as a "many-worlds" interpretation, and there are those who still understand it in this way today (Simon Saunders, for example). But a very different understanding is available today, and that understanding goes with a preference for a different name: the "decoherent-histories" approach. That approach is connected with a set of theorems that certainly have bearing on our understanding of the measurement process, no matter how we interpret quantum mechanics. What those theorems say is that under the conditions that prevail when humans actually make measurements, "histories" corresponding to different possible outcomes of an experiment (think of these as represented by wave functions, such as $1/\sqrt{2}$[fed cat and Sally seeing fed cat]

---

8. Maudlin, "Completeness, Supervenience and Ontology," 3161.

and 1/√2[hungry cat and Sally seeing hungry cat]) quickly cease to cohere with one another in the higher-dimensional space H. (I ignore technical details concerning the "grain" of these histories.) This means that the phenomena represented by the alternative histories do not exhibit the sort of interference that we have in, say, a two-slit experiment. An essential innovation of the decoherent-histories approach is that it provides us with a probability formula for the histories. (This probability formula is just the Wigner formula for the probability distribution for the results of a sequence of ideal measurements.)[9]

This does not, of course, mean that one cannot think of all the histories as "equally real," as Everett and DeWitt did. But it does mean, to some interpreters at least, that we do not *have* to adopt the radical "many-worlds" ontology of Everett and DeWitt. (At this point we already see that the Everett interpretation also does not wear its ontology on its sleeve.) For example, as Gell-Mann and Hartle understand the decoherent-histories account, only *one* of the histories is objectively real. But if that is so, a host of problems arise. For example, because a given history can belong to more than one decoherent family (and the union of all the decoherent families is not a decoherent family), we need to be told *what singles out one decoherent family as the "right" one*.[10] There are even decoherent families that consist of "histories" that contain superpositions like 1/√2(fed cat and Sally seeing fed cat) + 1/√2(hungry cat and Sally seeing hungry cat). Why are such decoherent families "wrong"? At best, what we have is an *approach* to an interpretation of quantum mechanics—one that may prove successful in the future, but that is certainly not without difficulties as it stands.

In his 1989 Trieste Lecture, Bell distinguished between first-class difficulties with quantum mechanics, that is, profound conceptual problems such as the measurement problem, and second-class difficulties, for example, the problem of the infinities that appear in certain quantum-mechanical calculations, and he closed by remarking that "I know very well that you can get along very well without paying any attention to these first class difficulties and

---

9. This is something I learned from Sheldon Goldstein.

10. For a list of related difficulties with the decoherent-histories approach, see Giancarlo Ghirardi, "Some Reflections Inspired by My Research Activity in Quantum Mechanics," in "The Quantum Universe," special issue, *Journal of Physics A: Mathematical and Theoretical* 40, no. 12 (March 23, 2007): 2891–2918.

many of these people ['practical physicists'] are so confident in their practice that they can even dismiss the existence of the first class difficulties. I am always wanting to argue with such people, but I have learned from experience that there is no hope of converting them."[11] The difficulties I just alluded to with the decoherent-histories approach are, perhaps, second-class difficulties. But there is also a profound conceptual difficulty, a first-class difficulty that, as a philosopher, I find more important.

The first-class difficulty I have in mind is this: what qualifies the Gell-Mann and Hartle "many-histories" approach as an interpretation of the "many-worlds" theory at all? The whole idea of Everett and DeWitt was that the ontology of quantum mechanics is "only the wave function," that is, only U and its changing "state" evolving according to the Schrödinger equation, and that measurement is a sort of illusion produced by the fact that the state naturally evolves into a mixture of histories (more precisely, into a superposition of histories that can be treated as if it were a statistical mixture, because the different histories do not cohere). To each of the Sallys in the mixture it *seems* as if there were a measurement with a unique result, but what there really is, as Maudlin said, is a process by which many people all subdivide into many many people, largely unaware of each other's presence, with the illusion that everyone who looked "saw the same thing." But Gell-Mann and Hartle, as I understand them, think that only *one* of the histories is objectively real. This seems precisely to *give up* what was most characteristic of Everett's many-worlds view.

Moreover, it raises just the question Everett thought he had succeeded in dismissing (as based on a false assumption): what happens when one history becomes objectively real? Is the many-histories interpretation not a *collapse* interpretation after all? Partisans of many histories cannot have it both ways.

The Bohm Theory

The Bohm theory is the great exception to my claim that quantum-mechanical theories do not wear their ontology on their sleeve. This is the case because it was constructed with precisely the intention of being a theory that repro-

11. John Stewart Bell, "The Trieste Lecture of John Stewart Bell," in "The Quantum Universe," special issue, *Journal of Physics A: Mathematical and Theoretical* 40, no. 12 (March 23, 2007): 2919–2934.

duces the predictions of standard quantum mechanics while having an intelligible and mathematically precise ontology.[12]

The attitudes of physicists to Bohmian mechanics are discussed in a published (online) correspondence between Steven Weinberg and Sheldon Goldstein.[13] There Goldstein writes that "unfortunately Bohmian mechanics is a nonrelativistic theory, and so it is of value primarily for the lessons it conveys about finding a sensible interpretation of quantum mechanics that is relativistic, rather than for the specific details of the theory itself. Now the question [Weinberg] raised about pair creation is, of course, very important. Bohmian mechanics itself is not a theory with particle creation or annihilation. However, I see no reason why some Bohm-type theory should not permit these things."

In fact, in an article in the same issue of *Journal of Physics A* that published Maudlin's paper and Bell's Trieste Lecture, Roderich Tumulka describes progress in constructing such a Bohm-type theory.[14] The theory uses, however, what Tumulka calls a "time-foliation," a preferred set of simultaneities, and is thus not fully compatible with special relativity. An interesting fact is that in the theory Tumulka describes, the Bohmian motion of the particles is interrupted by "jumps" of particle creation and annihilation whose "jump rate" is dependent on the wave function; these jumps constitute a Markov process, and thus Tumulka's Bohmiam quantum field theory is *indeterministic,* unlike the original Bohmian elementary particle theory.

GRW Quantum Mechanics

In the Ghirardi-Rimini-Weber version of quantum mechanics (suggested in 1976), each particle has a *tiny* probability of spontaneously jumping into a definite position state. For example, if the system consists of just one isolated atom, one would have to wait, on average, many thousands of years for it to

12. An excellent account is the article by Sheldon Goldstein on "Bohmian Mechanics" in E. N. Zalta, ed., *The Stanford Encyclopaedia of Philosophy,* http://plato.stanford.edu/entries /qm-bohm/.

13. "The Debate on Bohmian Mechanics." This is part of the correspondence between Sheldon Goldstein and Steven Weinberg on Bohmian Mechanics, http://www.mathematik .uni-muenchen.de/~bohmmech/BohmHome/weingold.htm.

14. Roderich Tumulka, "The 'Unromantic Pictures' of Quantum Theory," in "The Quantum Universe," special issue, *Journal of Physics A: Mathematical and Theoretical* 40, no. 12 (March 23, 2007): 3245–3274.

jump into a definite position state. *But,* as we all know, something like the table in front of me as I type these words consists of millions upon millions of particles. If there are many millions of particles, the object will, according to GRW theory, always have a definite position.

The reason it will have a definite position is that, according to quantum mechanics, if even one of the particles, even one of the electrons, for example, of which the table consists, jumps into a definite position state, that state *multiplies* the state of the whole system by a factor that forces that state to become definite with respect to position (to come close to being an eigenstate of position). These "spontaneous collapses" may be happening at times that are very far apart as far as any one particle is concerned, but they are happening *all the time* as far as this very large collection of particles is concerned. For that reason, the probability that the table will fail to have a definite position (by macroscopic standards of definiteness) becomes virtually zero. It is not *impossible,* but it will never happen.

In Chapter 6 I concluded my survey of the different interpretations of quantum mechanics by saying that "*either GRW or some successor, or else Bohm or some successor, is the correct interpretation* [because, as mentioned above, I found "many-worlds" incoherent]—or . . . we will just fail to find a scientific realist interpretation that is acceptable." This conclusion made me unhappy because, I thought, neither theory can do without an "absolute time."

But when the issue of *Journal of Physics A: Mathematical and Theoretical* whose special issue in honor of Ghirardi contained most of the papers I have cited appeared in 2007, those words were already out of date! The paper by Tumulka I cited earlier shows that progress is being made in constructing a Bohmian quantum field theory (although that theory still requires what I called "an absolute time"), and recent work by Tumulka[15] shows that a GRW-type theory can be Lorentz invariant, although so far this has been achieved only in the case of systems of noninteracting particles (including, notably, particles whose states are entangled). On the other hand, quantum cosmology (which is itself a speculative but highly active field) also seems to need an absolute "background time" to make sense of its talk of "superimposing" whole space-times. Before we "superimpose," *each space-time is perfectly*

15. Roderich Tumulka, "A Relativistic Version of the Ghirardi-Rimini-Weber Model," *Journal of Statistical Physics* 125, no. 4 (2006): 821–840.

Einsteinian—each space-time is a Minkowski space-time that knows nothing about any "simultaneity." Thus it could be that although Einstein would have to admit that there is such a thing as simultaneity, it comes from "outside" any one well-defined space-time; it comes from the quantum-mechanical "interference" between whole space-times.

But what is the *ontology* of the GRW theory? The *mathematics* of the theory consists of two laws: the Schrödinger equation, which is obeyed between collapses, and the law determining the frequency of collapses. As it stands, this does not rule out the "only the wave function" ontology, but, like Maudlin, those sympathetic to (or at least willing to consider) GRW believe that some account of the "local beables" needs to be added if GRW is to be a scientifically realist alternative to both instrumentalism and various forms of subjectivism (e.g., the idea, favored by Wigner, that consciousness reduces wave packets). Two main candidates have emerged: the idea that the local beables are mass densities, favored by Ghirardi himself, and what has come to be called the "flash ontology," suggested by Bell. The paper by Maudlin I recommended at the beginning of this paper is largely a discussion of these two proposals. As he points out, they lead to very different pictures of "what is going on"; for example, the mass densities are typically nonzero over a large area, and if there is a collapse, they may suddenly disappear from a large part of that area. As Maudlin points out, this makes the mass-density picture seriously nonrelativistic. He imagines a situation in which an electron has an equal chance of being in either side of a box. If we make a position measurement, there will almost immediately be a collapse, with the result that the mass density on the left side of the box will suddenly either double or be reduced to essentially zero. "So if we could see that sudden jump, we could determine the exact moment that the distant measurement was made. We could determine that two distant events (the measurement and the jump in mass density) took place at the same moment of absolute time just by keeping track of the local beables."[16]

Bell's proposal turns on the fact that although the quantum jumps postulated by GRW are events in the career of the wave function (which "lives" in configuration space), they are associated with points in space-time; the Gaussian that represents a particular jump is centered on a particular space-time point (x, t). Bell writes, "So we can propose these events as the basis of the

16. Maudlin, "Completeness, Supervenience and Ontology," 3168.

'local beables' of the theory. These are the mathematical counterparts in the theory of real events at definite places and times in the real world."[17]

Here is my way of spelling out Bell's idea for interpreting GRW: the local beables are spatial objects (points and regions) plus the particles at such times as these exist. Here is my suggestion about when they exist: a particle exists precisely when there is a "collapse" of the wave function with respect to that particle and is located at the space-time point (*x, t*) mentioned by Bell. At all other times, each particle has only potential existence. Hence the name "flashes" for such particles. (So the holistic property represented by the wave function—the "state" property—is simply a mathematical object that represents the probability of a "beable" being spontaneously created at a certain point in space-time.)

Moral

The moral of this paper is one you have heard several times now, but perhaps it bears repeating: mathematically presented quantum-mechanical theories do not wear their ontologies on their sleeve. The moral *is not* "so much the worse for ontology"—although there are cases in which apparently incompatible ontologies turn out to be cases of what some physicists call "duality," and what philosophers call "equivalent descriptions," that is not what is going on in the cases I have discussed. The moral is rather that the mathematics does not transparently tell us what the theory is *about*. Not always, anyhow.

17. Bell, *Speakable and Unspeakable in Quantum Mechanics,* 205. See the discussion in Maudlin, "Completeness, Supervenience and Ontology," 3166.

8

The Curious Story of Quantum Logic

In 1940 the *New Yorker* carried a famous cartoon by Charles Addams show-ing single ski tracks on each side of a tree, as though the skier seen farther down the hill has passed right through it. (The cartoon can be found online by searching Google for "Charles Addams skier" and clicking "Images.") I first saw this cartoon when my teacher, Hans Reichenbach, used it to explain why he believed that quantum mechanics should be interpreted with the aid of three-valued logic rather than classical logic.[1] Early in my own career as a philosopher of science I published a paper defending Reichenbach's view,[2] but in 1960 the physicist David Finkelstein acquainted me with a different logic for quantum mechanics (due to John von Neumann[3]) that seemed to me decidedly preferable, and Finkelstein and I proposed that this logic was

1. Hans Reichenbach, *Philosophic Foundations of Quantum Mechanics* (Berkeley: University of California Press, 1944; reprint, Mineola, N.Y.: Dover Books, 1998).

2. Hilary Putnam, "Three-Valued Logic," *Philosophical Studies* 8 (1957): 73–80; reprinted in Putnam, *Philosophical Papers*, vol. 1, *Mathematics, Matter and Method* (Cambridge: Cambridge University Press, 1975), 166–173.

3. John von Neumann, *The Mathematical Foundations of Quantum Mechanics* (Princeton, N.J.: Princeton University Press, 1955). (The German original was published in 1932.)

the "real" logic of the physical world, a view that I defended until 1990.[4] On both my view and Reichenbach's, classical logic was merely a limiting case of a new logic. My most strident declaration of this was published as a seven-page article in the journal *Synthese*.[5] Today this seem totally wrong to me, but I want to tell you how Reichenbach, Finkelstein (inspired by what he took to be von Neumann's view, although von Neumann himself would not have interpreted his own proposal as we did), and I (inspired by Finkelstein) were led to such a radical view.

Let me backtrack to Reichenbach's use of the Charles Addams cartoon. Although in the cartoon the startled viewer on the right side of the picture (call him "the observer") looks as if he may have actually *seen* the skier on the lower left (call him "the system") "pass through" the tree without being hurt or disturbed, for Reichenbach's purposes we have to suppose that the observer looked just a moment *after* the hypothetical passage through the tree (or whatever it was that happened—that is exactly the problem). Moreover, let us suppose that "normal people" like the observer often observe the system when (judging from his/its ski tracks) it has passed magically through a tree, but they never observe the system actually "in the process" of passing through a tree. As Reichenbach put it (I am quoting from memory, not his own words), "we" know by experiment that skiers never pass through large and solid trees (without injury or any sign whatsoever of an impact) *while observed,* but we are supposed to imagine that we frequently observe skiers who *must have done so.*

Here is a seemingly analogous situation in quantum mechanics:[6] If we observe a system consisting of a star and the planets orbiting around it, and we calculate an upper bound on the total energy of the system, then we can easily calculate the diameter of a "sphere" around the star such that none of its planets can possibly be outside that sphere (because if it were, the potential energy of that planet would be greater than the total energy of the system). The imaginary sphere in question is a "potential barrier" that none of the planets can pass through. But if we have an atom, say, a hydrogen atom, the

4. I first advanced this view in "Is Logic Empirical?" in Robert S. Cohen and Marx W. Wartofsky, eds., *Boston Studies in the Philosophy of Science,* V (Dordrecht: Reidel, 1968), 216–241; reprinted as "The Logic of Quantum Mechanics," in Putnam, *Philosophical Papers,* vol. 1, *Mathematics, Matter and Method,* 174–197.

5. Hilary Putnam, "How to Think Quantum-Logically," *Synthese* 29 (December 1974): 55–61.

6. This is mentioned in Reichenbach, *Philosophic Foundations of Quantum Mechanics,* 165.

situation is different. We can measure the total energy of any number of similar atoms and determine that none of them has a total energy greater than, say, γ. So, if nature is not playing tricks on us, we should be able to conclude that the next atom we choose must also have a total energy less than or equal to γ. Assuming that this is right, we can determine a distance d such that the "satellite" (in this case, the electron) in any such atom cannot be farther than d from the nucleus (the proton), because otherwise its potential energy would be greater than the total energy of the system. But if instead of measuring the total energy, we chose to measure the *distance* between the electron and the nucleus (and, according to quantum mechanics, if we do this, we cannot *also* measure the total energy at the same time), the distance will sometimes be found to be greater than d. In short, like skiers whose trajectories are observed, atoms whose total energy is observed behave as we expect (all have the total energy we expect), but, like Charles Addams's skier who is unobserved at the crucial moment, atoms whose energies are *not* observed sometimes seem to have "done the impossible" by possessing electrons that are "on the wrong side of the potential barrier."

To see how phenomena such as this suggested three-valued logic to Reichenbach,[7] we have to mention the "split between the observer and the system" that was a part of orthodox quantum mechanics for many years. J. S. Bell refers to this as the "shifty split" in a book that is must reading for anyone who wants to think about the foundations of quantum mechanics.[8] (Bell calls the split "shifty" because, according to, for example, von Neumann's 1932 *Mathematische Grundlagen der Quantenmechanik,* the theorist is allowed to "put" the split in different places.)

More about Reichenbach

In sum, according to the received view—and it was still the received view in 1944 when the first edition of Reichenbach's *Philosophic Foundations of Quantum Mechanics* was published—quantum mechanics presupposes (1) a split

---

7. I write "suggested" and not "showed the necessity for" because Reichenbach did not think that causal anomalies *violate* logic. Even Charles Addams's cartoon does not show a violation of logic. (What would that mean?) What Reichenbach claimed is that changing the logic enables us to avoid causal anomalies.

8. J. S. Bell, *Speakable and Unspeakable in Quantum Mechanics* (Cambridge: Cambridge University Press, 1987; quoted from the revised edition 2004), 228.

between the apparatus (or, alternatively, the "observer") and the system to which quantum mechanics is applied and (2) that for the purpose of saying how the apparatus succeeds in measuring this or that particular magnitude (e.g., the position or momentum of a particle), it is *necessary* to use *classical* physics.[9] These features of the received view are preserved by Reichenbach's interpretation: magnitudes that are measured by "classical" means—a paradigmatic example was measuring the position of a particle by allowing it to collide with a screen or a photographic emulsion—are described by Reichenbach as "phenomena," and the hypothetical values of the magnitudes that are not measured (if one is allowed to speak of them at all) are called by Reichenbach "interphenomena." Statements about phenomena are true or false, according to Reichenbach's scheme; statements about interphenomena receive a "third truth-value," "indeterminate." (For example, the statement that a particle whose position was determined at time t had, at that same time t, a precise momentum p has the truth-value indeterminate.) A simple set of truth tables goes with this. For example, a conjunction with one false conjunct is false (even if the other conjunct is indeterminate), but a conjunction with two indeterminate conjuncts or with one true conjunct and one indeterminate conjunct is indeterminate. What is interesting about Reichenbach's book, however, is not the formal features of the logic but the *motivation* for this interpretation.

Reichenbach recognizes two sorts of alternatives to his proposal. One sort, which he identifies with the view of Bohr and Heisenberg, is the same as his except that the "indeterminate" statements are regarded as *meaningless* and hence not properly part of the scientific language at all, rather than as having a third truth-value. He calls such an interpretation a "restrictive" interpretation because it restricts or prohibits us from forming statements about interphenomena. The other sort of alternative is what has come to be called a "hidden-variable" theory, an interpretation that does in one way or another assign bivalent truth and falsity to statements about interphenomena, or equivalently, that assigns values to physical magnitudes at times when they are not "measured." Reichenbach calls such interpretations "exhaustive" interpretations.

9. E.g., Niels Bohr writes in "Discussion with Einstein on Epistemological Problems in Atomic Physics," in P. A. Schilpp, ed., *Albert Einstein: Philosopher-Scientist* (LaSalle, Ill.: Open Court, 1949), 199–242, "It is decisive to recognize that however far the phenomena transcend the scope of classical physical explanation, the account of all evidence must be expressed in classical terms." The quotation is from p. 209.

What Reichenbach thinks about the program of constructing "exhaustive" interpretations is best understood by comparison with his brilliant discussion of the use of non-Euclidean geometries in cosmology in his early (1920) *The Theory of Relativity and A Priori Knowledge.*[10] There Reichenbach pointed out that experiments with light rays, etc., do not by themselves show that space is non-Euclidean, because one might postulate that some strange forces are altering the paths of the light rays, changing the lengths of measuring rods, or having other effects. However, if the topology of space is non-Euclidean (as it may be, according to general relativity), then postulating such universal distorting forces will require us to countenance a variety of "causal anomalies," including violations of what he was later to call the common-cause principle, that is, the principle that correlations between distinct phenomena are to be explained by showing *either* that one phenomenon causally affects the other *or* that they are both affected by a third phenomenon, a "common cause." In sum, an interpretation of cosmological phenomena that preserves Euclidean geometry is not ruled out by experiment alone (by the "phenomena" per se), but the non-Euclidean structure of physical space is revealed by the fact that all descriptions that preserve Euclidean geometry contain "causal anomalies."

As Reichenbach saw it (he died in 1953), the situation in quantum mechanics is also describable in terms of interpretations with and without causal anomalies. He summed this up in a reply to a critic written in 1946:

> I have shown . . . that in a physical world of a certain kind the use of Euclidean geometry leads to causal anomalies; the determination of that geometry which describes the world free from causal anomalies, therefore, reveals structures of the physical world. Similarly, it must be regarded as a property of the world of the quanta that a language free from causal anomalies, if it is not to employ the metalanguage for the statement of physical laws [e.g., by saying that certain well-formed sentences of the object language are "meaningless"], must be constructed in terms of a three-valued logic. It is for this reason that I regard a three-valued logic as the adequate linguistic form of quantum mechanics.[11]

10. Hans Reichenbach, *Relativitätstheorie und Erkenntnis apriori* (Berlin: Springer, 1920); translation, *The Theory of Relativity and A Priori Knowledge* (Berkeley: University of California Press, 1965).

11. Hans Reichenbach, "Reply to Ernest Nagel's Criticism of My Views on Quantum Mechanics," *Journal of Philosophy* 43, no. 9 (1946): 239–247. The quotation is from p. 247.

Putting all this together, we see that Reichenbach was making two claims:

1. He claimed that "every exhaustive interpretation leads to casual anomalies"[12] (i.e., that any "hidden-variable theory" that recovers the predictions of quantum mechanics must lead to them).
2. He further claimed that if we restrict ourselves to statements about results of measurement ("phenomena"), then we do not see any causal anomalies.[13] Causal anomalies only occur when and where we are not "looking."

Together these two claims can be summed up by saying that if Reichenbach is right, then *we live in a Charles Addams universe*. Indeed, using my earlier example of the quantum-mechanical phenomenon of an electron "tunneling through a potential barrier," one could draw a comic strip showing in the first panel an experimenter turning off his apparatus; in the second panel, an electron sneaking through the potential barrier with a balloon showing it (the electron) thinking, "I fooled him again!"; and in the third panel, the apparatus locating the electron "beyond" the potential barrier, with a balloon that shows the experimenter thinking, "How on earth does it do that?!!" That might be thought to show exactly how our universe is "Charles Addams–ey." But things are not so simple.

12. Reichenbach, *Philosophic Foundations of Quantum Mechanics,* 136; his proof (pp. 111–135) requires an exhaustive interpretation to include all the standard parameters of quantum mechanics. (In Bohm's hidden-variable theory, for example, only *position* is a genuine physical quantity; the eigenvalues of the momentum operator are mathematical conveniences and do not represent values of a physical magnitude in the classical sense.) To be nonanomalous, by Reichenbach's definition, action-by-contact must be obeyed, joint probability distributions have to exist for all these parameters, and the laws must be (locally) deterministic or stochastic. Bell's theorem shows that the assumption that all the standard parameters are "beables" (a term introduced by J. S. Bell for "real" physical quantities) is not needed to demonstrate nonlocality and hence is not needed to show that an exhaustive interpretation must have causal anomalies. For an excellent account of what Reichenbach proved, see Roger Jones, "Causal Anomalies and the Completeness of Quantum Theory," *Synthese* 35, no. 1 (May 1977): 41–74.

13. "We see that a three-valued logic is the adequate form of a system of quantum mechanics in which no causal anomalies can be derived." Reichenbach, *Philosophic Foundations of Quantum Mechanics,* 166. The "system of quantum mechanics" referred to is Reichenbach's three-valued interpretation, and because results of measurement receive classical truth-values in that interpretation, Reichenbach is claiming that those results do not themselves ever exhibit "causal anomalies."

The Verdict of History on Reichenbach's Claims

What we know today (although it certainly does not follow merely from the phenomenon of "tunneling" through potential barriers, which it is known that a hidden-variable theory can explain[14]) is that Reichenbach's first claim is right. J. S. Bell showed that the statistical predictions of quantum mechanics cannot be recovered by any local hidden-variable theory, and violations of locality are paradigmatic causal anomalies in Reichenbach's sense.[15] Reichenbach makes it clear that a nonanomalous theory has to exhibit action by contact. But that one cannot produce an "exhaustive interpretation" that does not have what he would clearly have regarded as causal anomalies is true, whereas von Neumann's claim to have excluded hidden-variable theories altogether was not true. And Reichenbach's second claim was certainly assumed to be true when he wrote; indeed, it was assumed to be true by Einstein in his debate with Bohr about the Einstein-Podolsky-Rosen thought experiment.[16] But Bell's theorem and experiments by Alain Aspect and others have shown that observable phenomena do show clearly nonlocal effects.

Here is what this means for Reichenbach's view: Reichenbach had shown (in *The Theory of Relativity and A Priori Knowledge*) that we can construct an account that agrees with relativistic physics and nevertheless interprets all physical phenomena in terms of Euclidean geometry, but any such account will postulate causal anomalies. Those anomalies will vary from one such account to another, and they can all be eliminated by giving up Euclidean geometry and adopting the geometry (of variable curvature and possibly Riemannian topology as well) that goes with Einstein's general relativity. The claim of his *Philosophic Foundations of Quantum Mechanics* was that, similarly, we can construct an account that agrees with quantum physics and nevertheless counts statements about unmeasured physical parameters ("interphenomena") as meaningful and subject to classical two-valued logic, but

14. This phenomenon (or, rather, the "measurements" that seem to show that the electron has "tunneled") is predicted by both of the hidden-variable theories that I discuss in "A Philosopher Looks at Quantum Mechanics (Again)," *British Journal for the Philosophy of Science* 56, 4 (December 2005): 615–634; reprinted in this volume as Chapter 6.

15. J. S. Bell, "On the Einstein-Podolsky-Rosen Paradox," *Physics* 1 (1964): 195–200; reprinted as the second chapter of *Speakable and Unspeakable in Quantum Mechanics*.

16. Bohr, "Discussion with Einstein on Epistemological Problems in Atomic Physics," and Albert Einstein, "Reply," in Schilpp, ed., *Albert Einstein: Philosopher-Scientist*, 199–242, 663–688.

any such account will postulate causal anomalies. As was the case with Euclidean geometry and relativity theory, Reichenbach no doubt expected that the anomalies would vary from one such account (one such "exhaustive interpretation") to another, but he claimed (this is what I called his "second claim") that they can all be eliminated by giving up the assumption that statements about interphenomena are true or false. However, Bell's theoretical predictions, predictions that unquestionably follow from quantum mechanics and have, moreover, been experimentally confirmed, show that there are causal anomalies that are *observable* without postulating any exhaustive interpretation whatsoever. The phenomena *themselves* are "causally anomalous," in classical terms. The particular anomaly ("nonlocality") is observable at the level of the phenomena. Unlike the "anomalies" Reichenbach described in *The Theory of Relativity and A Priori Knowledge,* they cannot be transformed away by adopting a different theory, even one that uses a nonstandard logic. They appear unmistakably in the experimental results themselves.

Bell's Inequalities and Their Violation

Bell's theorem states that no "local" hidden-variable theory can recover the statistical predictions of quantum mechanics.[17] I give here a simplified version of his argument.[18]

The idea is to show that certain inequalities ("Bell inequalities") hold for all correlations, and to show in addition that quantum-mechanical measurements exhibit what appear to be violations of those inequalities, violations that are impossible if "locality" is assumed, that is, impossible if we rule out both superluminal signaling and the sort of "holistic" causation we see in such famous hidden-variable theories as David Bohm's and the "spontaneous-collapse" theory of Ghirardi, Rimini, and Weber. I speak of "holistic causation" because in those theories a physical interaction in one place affects the wave function—which is not "in" three-dimensional space at all—in such a way as to affect the possible course of events in every part of the system. It is important that the wave function is a function of the entire system and is not factorable into separate wave functions for its parts in the case of the spin-singlet

17. Bell, "On the Einstein-Podolsky-Rosen Paradox."

18. This version of Bell's argument is due to N. D. Mermin, "Is the Moon There When Nobody Looks? Reality and the Quantum Theory," *Physics Today* 38 (April 1985): 38–47.

system that figures in Bell's argument. (Strange as it may seem, nonlocality is not necessarily incompatible with relativity, although the theories I mentioned in their present forms are nonrelativistic.)

Here is a simple example of a "Bell inequality": Suppose that you have three series of events (e.g., coin tosses), A, B, C. If the events in the A series and corresponding events in the B series are 99 percent correlated (both events are heads or both events are tails 99 percent of the time), and the events in the B series and corresponding events in the C series are likewise 99 percent correlated, then the events in the A series and corresponding events in the C series must be at least 98 percent correlated.

And here is the (apparent) violation predicted by quantum mechanics: Suppose that we have two particles whose state is "entangled" in such a way that a measurement of the spin of the first particle in any given direction always gives the opposite result to a measurement of the spin of the second particle in the same direction, no matter how widely separated the particles may be (this is called a "spin-singlet system," and for this argument the anti-correlation is equivalent to a correlation). If the apparatus that interacts with the second particle measures the spin at an angle that differs (say, by an angle of θ) from the direction of the spin measured on the first particle, then the results will not be perfectly correlated, as they are when $\theta = 0$. But for the angle $\theta = 0.1$, the correlation will be 99 percent. If, instead, the spin measurement of the first particle is made in the direction $\theta = 0.1$ and the spin measurement of the second particle is made in the direction 2θ, the correlation between the results will again be 99 percent. If no "nonlocality" is at work, the situation must be exactly analogous to our three series of coin tosses, and the correlation between the results when the angle of the spin measurement of the first particle is 0 and the results when the angle is 2θ must be no less than 98 percent. But the result predicted by quantum mechanics (and confirmed by experiments) is a correlation of 96 percent. And this is a "causal anomaly" that does not disappear when we use Reichenbach's three-valued logic.

These results do not actually contradict logic or mathematics because if the second particle "knows," via superluminal signaling, for example, what direction has been chosen by the operator of the apparatus measuring the spin of the first particle, then it can easily "arrange" to have the predicted perfect correlation appear when the direction is the same as the direction of the apparatus that is measuring its own spin. But if locality is assumed, the only way the results of the two measurements can always correlate when the angle between the relevant directions is 0 (even though the choice of those directions may be made by the independent free wills of the two operators or may depend on local random processes) is if (1) the hidden variables in the vicinity of each apparatus *determine* what the result of a spin measurement in an arbitrary direction must be; and hence (2) what we are observing when we measure any spin is simply what values those hidden variables already determine for each choice of an angle. But in that case the two correlations of 99 percent we observe must be genuine correlations between states of the hidden variables, and the third correlation we observe cannot possibly be less than 98 percent—but it is! Hence locality must be false.

Reichenbach's "Language Free from Causal Anomalies"

It may seem odd that Reichenbach could claim that adopting his three-valued logic yields "a language free from causal anomalies." Does not the Charles Addams cartoon that he himself used to illustrate his position show observables ("phenomena"), namely, the ski tracks, that are decidedly "anomalous"? Reichenbach's answer would be that it does not actually show an event that violates classical physics *happening*. Reichenbach inherited a received view that held that one *must* use classical physics when one describes what the measuring apparatus does, and his interpretation permits statements about this classically described measuring apparatus and about the classically described measurements that they are used to make to receive definite truth-values in such a way that classical physics is not violated. I do not know what Reichenbach would have said if he had learned that these classically described events show *correlations* that make no classical sense. Probably he would not have considered this a "causal anomaly,"[19] but that would not have been very reasonable.

It is instructive to compare Reichenbach's and von Neumann's views at this point. Von Neumann mistakenly thought that he had proved that there could not be a hidden-variable theory that recovered the predictions of quantum

19. Reichenbach's discussion of the debate between Einstein and Bohr concerning spin-singlet systems (Bohr, "Discussion with Einstein on Epistemological Problems in Atomic Physics," and Einstein, "Reply to Niels Bohr") in *Philosophic Foundations of Quantum Mechanics,* 170–176, suggests that he would have said that the answer to the question whether the correlations are produced by "interphenomena" locally or nonlocally is that both statements have an indeterminate truth-value.

mechanics. If this were right, then there would be no "exhaustive interpreta-tions" of quantum mechanics in Reichenbach's sense at all, anomalous or not. But von Neumann was not prepared to go as far as Bohr and say, "There is no quantum world. There is only an abstract quantum mechanical description."[20] If there is no possibility of assigning truth-values to physical parameters at times at which those parameters are not measured, we can still associate each possible result of a measurement of such a parameter with a subspace of Hilbert space, and the various relations among the results of such possible measure-ments are represented by the inclusion (and exclusion) relations among those subspaces. That is what von Neumann's "logic" represents, on *his* interpretation— relations among possible results of measurement. And that is the best von Neumann thought we could do. But, as J. S. Bell showed, von Neumann's "no-go" theorem supposedly excluding hidden-variable theories rested on physically unrealistic assumptions.[21] In fact, a number of hidden-variable theories have been constructed that do reproduce the predictions of quantum mechanics, the earliest being David Bohm's.[22] Bohm constructed his theory in 1952, eight years after the publication of Reichenbach's *Philosophic Founda-tions of Quantum Mechanics,* but Reichenbach knew of an incompletely worked-out hidden-variable theory (the direct ancestor of von Neumann's theory, in fact), the "pilot-wave" theory of De Broglie (which did not include an account of measurement, which is what Bohm supplied). In any case, Reichenbach obviously did not believe von Neumann's claim, for his whole discussion presupposes that hidden-variable theories *could* be constructed, although they would contain "anomalies." And Reichenbach was right, which is what makes his book still of interest.

My Interest in Von Neumann's Quantum Logic

Von Neumann's logic was based on the idea that the lattice of propositions about a quantum-mechanical system corresponds precisely to the lattice of

20. Max Jammer, *The Philosophy of Quantum Mechanics* (New York: John Wiley, 1974), 204, quoting Aage Petersen, "The Philosophy of Niels Bohr," *Bulletin of the Atomic Scientists* 19, no. 7 (1963): 8–14.

21. J. S. Bell, "On the Problem of Hidden Variables in Quantum Mechanics," in Bell, *Speakable and Unspeakable in Quantum Mechanics,* 1–13.

22. David Bohm, "A Suggested Interpretation of the Quantum Theory in Terms of 'Hidden' Variables: I and II," *Physical Review* 85 (1952): 166–193.

projection operators on the Hilbert space of the system. My interest in von Neumann's logic was neither in using it to display relations among possible results of measurement (von Neumann's purpose) nor in showing that it avoided causal anomalies. In fact, Finkelstein and I positively *reveled* in the fact that the world was (according to our interpretation) full of anomalies, all of which we interpreted as manifestations of one giant fact one just had to get used to: *the world has a nonclassical logic!* My interest came, indirectly, from the influence of another of my teachers, W. V. Quine. In what is still his most influential essay, the famous "Two Dogmas of Empiricism," Quine had written:

> Any statement can be held true come what may, if we make drastic enough adjustments elsewhere in the system. Even a statement very close to the periphery can be held true in the face of recalcitrant experience by pleading hallucination or by amending certain statements of the kind called logical laws. Conversely, by the same token, no statement is immune to revision. Revision even of the logical law of the excluded middle has been proposed as a means of simplifying quantum mechanics; and what difference is there in principle between such a shift and the shift whereby Kepler superseded Ptolemy, or Einstein Newton, or Darwin Aristotle?[23]

I believed and still believe that Quine is right in this. That does not mean that I think that we can describe circumstances under which the law of the excluded middle should be given up (Quine was, of course, referring to Reichenbach's view, which we have been examining), or circumstances under which the distributive law of propositional calculus $p \wedge (q \vee r) \Leftrightarrow (p \wedge q) \vee (p \wedge r)$ (which fails in von Neumann's logic) should be given up. Perhaps there are no such circumstances, but that is not something we can ever be certain of. I agree with Quine that the idea of a metaphysical *guarantee* that any part of our supposed "knowledge" will never turn out to be wrong, perhaps wrong in a way we cannot now even conceive, is a fallacious idea, even if philosophers have sought for such a guarantee, for a fool-proof line between the truly certain and that which might turn out to need revision, for thousands of years. The "proof of the objective validity of our categories" that Kant thought he had succeeded in finding was supposed to be such a guarantee,

---

23. W. V. Quine, "Two Dogmas of Empiricism," *Philosophical Review* 60, no. 1 (January 1951): 20–43; reprinted in Quine, *From a Logical Point of View* (Cambridge, Mass.: Harvard University Press, 1961), 20–46.

but there is no proof of that sort to be had—no proof of the security of any part of our knowledge whose premises might not themselves turn out to need revision.

I met David Finkelstein at an American Mathematical Society Summer Seminar on "Modern Physical Theories and Associated Mathematical Developments" in Boulder, Colorado, in 1960. David was an enthusiastic and brilliant proponent of von Neumann's logic, and it was from him that I first heard the idea that, far from being just a way of formalizing the operational meaning of quantum mechanics, von Neumann's logic was *the real logic of the world.* Here is an example to illustrate what that was supposed to mean: Imagine a two-slit experiment in which only one photon is emitted every second, so there is no interaction whatsoever between the individual photons. Still, an interference pattern will build up—one caused by the interference of each photon *with itself.* Let B be the statement that such a photon goes through the slit on the north, let C be the statement that it goes through the slit on the south, and let A be the statement that it strikes a plate at a position p (which is where we observe it to strike). We have $A \wedge (B \vee C)$. But $(A \wedge B)$ and $(A \wedge C)$ both correspond to the null space in the von Neumann lattice and are thus contradictory in his logic. In either case, I said, $A \wedge (B \vee C) \Leftrightarrow (A \wedge B) \vee (A \wedge C)$ fails! The reason we, knowing A to be true, cannot "find out" whether B or C is true is that if we were to find out that, for example, B is true, then we would know that A and B are both true, and $(A \wedge B)$ is a contradiction! Nevertheless, I maintained (under Finkelstein's influence), the particle *does* "go through one slit or the other"—that is to say, $B \vee C$ is true, and in fact, because A is known, $A \wedge (B \vee C)$ is true! $A \wedge (B \vee C)$ is true, but "you cannot distribute!"

I was won over, partly by Finkelstein's enthusiasm, but even more by the way in which this claim, if right, would perfectly support what Quine had claimed. Quine had given Reichenbach's three-valued logic as his example of how logic might need to be revised, but I had already concluded that what Reichenbach had done was simply to formalize the Copenhagen view that one could not really *say* anything about "interphenomena"; now, I thought, we had the real thing! After all, relativity theory had forced us to give up a geometry that had long been thought to be a priori; why should it be surprising if quantum mechanics were to force us to give up a *logic* that had also been thought since the days of the ancient Greeks to be a priori?

I gave up this view in 1990 for reasons that are somewhat complicated to explain (they have to do with the famous case of "Schrödinger's cat");[24] but there is a much simpler reason that I should have given it up sooner, and that one should also not rest content with Reichenbach's proposal that we assign an "indeterminate" truth-value to statements about interphenomena.

But Can We Just Rule Out Realist Interpretations?

The reason is that before we do anything as radical as giving up one of the distributive laws of propositional calculus, or as counterintuitive as restricting what we regard as true or false to statements about phenomena that can be described using classical physics (to what is on the classical side of the split between measuring apparatus and quantum system), as Reichenbach proposed, we ought to be sure that there is no coherent "exhaustive interpretation," no coherent realistic interpretation, to use a terminology I prefer, that conforms to classical logic.[25] And there are such at least three such interpretations: the Bohmian pilot-wave theory and its successors, the Ghirardi-Rimini-Weber (GRW) spontaneous-collapse theory and its successors,[26] and the many-worlds interpretation and its various versions.[27] All three have vigorous supporters and are the subject of serious research—research that would never have occurred if everyone had accepted either Reichenbach's view or Finkelstein and Putnam's view. Surely, before we accept views that require us to revise our logic, we need to be sure that it is *necessary* to go *that* far to make sense of quantum phenomena. And we now know that it is not. (Of course, none of the "interpretations" of quantum mechanics I just mentioned were known to Reichenbach in 1944. I had not really studied the Bohm interpretation when I wrote "Is Logic Empirical?" although, of course, I should have.)

---

24. I explain my reasons for giving it up in my reply to Michael Redhead in "Comments and Replies" in Peter Clark and Robert Hale, eds., *Reading Putnam* (Oxford: Blackwell, 1994), 242–295.

25. See Putnam, "Philosopher Looks at Quantum Mechanics (Again)."

26. Giancarlo Ghirardi, Alberto Rimini, and Tullio Weber, "An Attempt at a Unified Description of Microscopic and Macroscopic Systems," in Vittorio Gorini and Alberto Frigerio, eds., *Fundamental Aspects of Quantum Theory: Proceedings of the NATO Advanced Research Workshop*, Como, September 2–7, 1985, NATO ASI Series B 144 (New York: Plenum Press, 1986), 57–64.

27. I discuss all these in "Philosopher Looks at Quantum Mechanics (Again)."

Although the many-worlds interpretation has attracted support from Murray Gell-Mann[28] and other famous physicists, Bohmian mechanics and GRW are usually regarded as questionable because, as originally put forward, they require an objective time foliation (an objective "now"), which contradicts the spirit of relativity theory, although in none of these theories is superluminal signaling actually possible. But the situation is more complicated than this quick statement suggests, and certainly much more research is needed.

It is more complicated because (1) contrary to what is often assumed, many worlds is *not* actually a Lorentz-invariant interpretation. This is something that follows from results by David Albert and Yakir Aharonov that show that the wave function in one frame cannot in general be obtained from the wave function in a different frame without additional information; a fortiori, it cannot be obtained by a Lorentz transformation. Albert and I are now writing a paper showing that, in fact, no known interpretation of quantum mechanics is fully Lorentz invariant![29] So the fact that Bohm and GRW are not is not as damning as it first looks. (2) A Lorentz-invariant version of GRW for noninteracting particles that fully recovers predictions about nonlocality, two-slit experiments, and other problematic cases now exists, thanks to Roderich Tumulka.[30] Tumulka is optimistic that this will be extended to the case of interacting particles. So we cannot take it as known that a realistic interpretation must violate relativity. (3) In any case, the existence of a preferred time foliation does not necessarily mean that ours is not a relativistic space-time. Such a foliation could be an additional structure on top of Minkowski's geometry. In practice, after all, cosmologists do use a preferred frame (the background radiation).

To sum up: Reichenbach was right to say that an "exhaustive interpretation" will exhibit a "causal anomaly" (namely, *nonlocality*). His was *the*

---

28. Murray Gell-Mann, *The Quark and the Jaguar: Adventures in the Simple and the Complex* (New York: Henry Holt, 1994).

29. For example, the Copenhagen interpretation postulated that physical magnitudes take on definite values "when they are measured." But those values are spacelike separated in the case of the Einstein-Podolski-Rosen experiment, and Bohr, while denying that the experiment reveals values that were there before the measurement, made no attempt at all to explain how this taking on of a value could be described as Lorentz invariantly.

30. Tim Maudlin, *Quantum Non-locality and Relativity: Metaphysical Intimations of Modern Physics,* 3rd ed. (Oxford: Wiley-Blackwell, 2011), contains a fine explanation of Tumulka's results.

first book on the interpretation of quantum mechanics to put the issue of locality at the center of attention (note that "contact-action" was a central feature of what he called freedom of anomaly). But he did not consider that the inevitability of nonlocality might be because *there is a correct exhaustive interpretation, and it is nonlocal.* By assigning all "exhaustive interpretations" the truth-value "indeterminate," Reichenbach, to all intents and purposes, did what the Copenhagen interpretation did by labeling them "meaningless." "Is there a true exhaustive interpretation?" is the question Reichenbach *should have asked*, and not "Is there an exhaustive interpretation that obeys the principle of contact-action?" But, of course, he did not know what we know now. And I was wrong to think that a tenable realistic interpretation must give up classical logic. It was David Bohm, Hugh Everett, John Stewart Bell, Giancarlo Ghirardi, Alberto Rimini, Tullio Weber, Roderich Tumulka, and those who have continued and extended their investigations who have put us back on a sensible path.

PART TWO

Mathematics and Logic

9

Indispensability Arguments
in the Philosophy of Mathematics

If one consults *The Stanford Encyclopedia of Philosophy* on the topic "Indispensability Arguments in the Philosophy of Mathematics,"[1] one finds (as part of a moderately lengthy entry written by Mark Colyvan) the following paragraph:

> From the rather remarkable but seemingly uncontroversial fact that mathematics is indispensable to science, some philosophers have drawn serious metaphysical conclusions. In particular, Quine . . . [2] and

1. Mark Colyvan, "Indispensability Arguments in the Philosophy of Mathematics," in E. N. Zalta, ed., *The Stanford Encyclopedia of Philosophy,* Fall 2004 ed., http://Plato.stanford .edu/archives/fall2004/entries/mathphil-indis/. Colyvan is also the author of *The Indispensability of Mathematics* (Oxford: Oxford University Press, 2001).

2. The author of this entry, Mark Colyvan, is referring to W. V. Quine, "Carnap and Logical Truth," reprinted in Quine, *The Ways of Paradox and Other Essays* (New York: Random House, 1966), 107–132, and in Paul Benacerraf and Hilary Putnam, eds., *Philosophy of Mathematics: Selected Readings,* 2nd ed. (Cambridge: Cambridge University Press, 1983), 355–376; Quine, "On What There Is," *Review of Metaphysics* 2 (1948): 21–38, reprinted in Quine, *From a Logical Point of View,* 2nd ed (Cambridge, Mass.: Harvard University Press, 1980), 1–19; Quine, "Two Dogmas of Empiricism," *Philosophical Review* 60, no. 1 (January 1951): 20–43, reprinted in Quine, *From a Logical Point of View,* 20–46; Quine, "Things and Their Place in Theories," in

Putnam . . .[3] have argued that the indispensability of mathematics to empirical science gives us good reason to believe in the existence of mathematical entities. According to this line of argument, reference to (or quantification over) mathematical entities such as sets, numbers, functions and such is indispensable to our best scientific theories, and so we ought to be committed to the existence of these mathematical entities. To do otherwise is to be guilty of what Putnam has called "intellectual dishonesty."[4] Moreover, mathematical entities are seen to be on an epistemic par with the other theoretical entities of science, since belief in the existence of the former is justified by the same evidence that confirms the theory as a whole (and hence belief in the latter). This argument is known as the Quine-Putnam indispensability argument for mathematical realism. There are other indispensability arguments, but this one is by far the most influential, and so in what follows I'll concentrate on it.

From my point of view, Colyvan's description of my argument(s) is far from right. The fact is that in "What Is Mathematical Truth?" I argued that *the internal success and coherence of mathematics* is evidence that it is true under *some* interpretation, and that its *indispensability for physics* is evidence that it is true under a realist interpretation—the antirealist interpretation I considered there was intuitionism. This is a distinction that Quine nowhere draws. It is true that in *Philosophy of Logic* I argued that at least some set theory is indispensable in physics *as well as logic* (Quine had a very different view on the relations of set theory and logic, by the way), but both "What Is Mathematical Truth?" and "Mathematics without Foundations" were published in *Mathematics, Matter and Method* together with "Philosophy of Logic," and in both of *those* papers I said that set theory did not have to be interpreted Platonistically. I said that modal-logical mathematics (i.e., mathematics that takes mathematical possibility as primitive and not abstract entities of any kind[5]) and

---

Quine, *Theories and Things* (Cambridge, Mass.: Harvard University Press, 1981), 1–23; and Quine, "Success and Limits of Mathematization," in Quine, *Theories and Things,* 148–155.

3. Colyvan is referring to Hilary Putnam, "What Is Mathematical Truth?" *Historia Mathematica* 2 (1975): 529–543, reprinted in Putnam, *Philosophical Papers,* vol. 1, *Mathematics, Matter and Method* (Cambridge: Cambridge University Press, 1975), 60–78; and Hilary Putnam, *Philosophy of Logic* (New York: Harper and Row, 1971); reprinted as an essay with the same title in *Philosophical Papers,* vol. 1, *Mathematics, Matter and Method,* 323–357.

4. Ibid., 347.

5. I proposed such an interpretation of mathematics in "Mathematics without Foundations," *Journal of Philosophy* 64, 1 (1967): 5–22; repr. in Putnam, *Philosophical Papers,* vol. 1,

mathematics that takes sets as primitive are "equivalent descriptions." In fact, in "What Is Mathematical Truth?" I said, "The main burden of this paper is that one does not have to 'buy' Platonist epistemology to be a realist in the philosophy of mathematics. The modal logical picture shows that one doesn't have to 'buy' Platonist ontology either."[6] Obviously, a careful reader of *Mathematics, Matter and Method* would have had to know that I was in no way giving an argument for realism about sets as opposed to realism about modalities.

In sum, my "indispensability" argument was an argument for the objectivity of mathematics in a realist sense—that is, for the idea that mathematical truth must not be identified with provability. Quine's indispensability argument was an argument for "reluctant Platonism," which he himself characterized as accepting the existence of "intangible objects" (numbers and sets).[7] The difference in our attitudes goes back at least to 1975.

In addition, there is a premise in Colyvan's formalization of the supposed "Quine-Putnam indispensability argument" the "and only" part of which I have never subscribed to in my life, namely:

(P1) We ought to have ontological commitment to all *and only* the entities that are indispensable to our best scientific theories.[8]

Nevertheless, there was a common *premise* in my argument and Quine's, even if the *conclusions* of those arguments were not the same. That premise was "scientific realism," by which I meant the rejection of operationalism and kindred forms of "instrumentalism." I believed (and *in a sense*[9] Quine also believed) that fundamental physical theories are intended to tell the truth about physical reality, and not merely to imply true observation sentences.

Mathematics, Matter and Method (1975), 43–59. For a detailed development, see Geoffrey Hellman, *Mathematics without Numbers: Towards a Modal-Structural Interpretation* (Oxford: Oxford University Press, 1989).

6. Putnam, "What Is Mathematical Truth?" 72.

7. E.g., Quine wrote, "[The words 'five' and 'twelve'] name two intangible objects, numbers, which are *sizes of* sets of apples and the like." Quine, *Theories and Things,* 149.

8. My italics.

9. For the reasons that I say "in a sense," see Hilary Putnam, "The Greatest Logical Positivist," in Putnam, *Realism with a Human Face,* ed. James Conant (Cambridge, Mass.: Harvard University Press, 1990), 268–277.

Objections to Indispensability Arguments

A common objection to arguments from indispensability for physics to realism with respect to mathematics (I shall consider more sophisticated objections shortly) is that we do not yet have, and may indeed never have, the "true" physical theory; my response is that, at least when it comes to the theories that scientists regard as most fundamental (today that would certainly include quantum field theories), we should regard all the rival theories as candidates for truth or approximate truth, and that *any philosophy of mathematics that would be inconsistent with so regarding them should be rejected.* This "indispensability argument" has continued to appear in my work right up to the present day.[10]

I believe that the most serious objections to *my* "indispensability arguments" depend, albeit in very different ways, on considering *nominalist* alternatives to present-day theoretical physics. The two most important such objections, I believe, are due to Hartry Field and to Gideon Rosen, and I will consider them in the next two sections. (In the Appendix I will argue briefly that if my arguments are accepted, not only intuitionism and nominalism need to be rejected as philosophies of mathematics, but also—and I know that this will be controversial—the idea that "predicative" mathematics is all that science needs.) But because I began by quoting Mark Colyvan, in the rest of the present section I shall discuss the objections that he mentions in his entry in *The Stanford Encyclopedia of Philosophy.* I shall discuss them, of course, on the assumption that they are supposed to be objections to *my* arguments. It may be that Colyvan intended them to apply only to *Quine's* argument. (I suspect that although Quine certainly accepted something like [P1], the idea that he accepted or used the idea of "confirmation" is wildly off the mark, and several of the objections depend on that idea. But interpretation of Quine is not the purpose of this essay.)

The objections in question are the following:

1. "The debate continues in terms of indispensability, so we would be well served to clarify this term. The first thing to note is that 'dispensability' is not the same as 'eliminability.' If this were not so, *every* entity would be dispensable (due to a theorem of Craig)."

---

10. See Chapters 25 and 26 in this volume. Both of these chapters are critical of some of Wittgenstein's unpublished remarks on mathematics.

2. "These issues naturally prompt the question of *how much* mathematics is indispensable (and hence how much mathematics carries ontological commitment)."

3. "Confirmational holism is the view that theories are confirmed or disconfirmed as wholes.[11] So, if a theory is *confirmed* by empirical findings, the *whole* theory is confirmed. In particular, whatever mathematics is made use of in the theory is also confirmed.[12] Furthermore, as Putnam stressed in "What Is Mathematical Truth?" it is the same evidence that is appealed to in justifying belief in the mathematical components of the theory that is appealed to in justifying the empirical portion of the theory[13] (if indeed the empirical can be separated from the mathematical at all). Naturalism and holism taken together then justify (P1). Roughly, naturalism gives us the 'only' and holism gives us the 'all' in (P1)."

4. "There have been many objections to the indispensability argument, for example Charles Parsons's ('Mathematical Intuition'[14]) concern that the obviousness of basic mathematical statements is left unaccounted for by the Quinean picture and Philip Kitcher's (*The Nature of Mathematical Knowledge*[15]) worry that the indispensability argument doesn't explain *why* mathematics is indispensable to science. The objections that have received the most attention, however, are those due to Hartry Field, Penelope Maddy and Elliott Sober. In particular, Field's nominalisation program has dominated recent discussions of the ontology of mathematics."

5. "Maddy's first objection to the indispensability argument is that the actual attitudes of working scientists towards the components of

11. Quine, "Two Dogmas of Empiricism," 41 (of the reprinted version). What Quine actually wrote is that "our statements about the external world face the tribunal of sense experience not individually but only as a corporate body." This does not mention (1) "theories" or (2) "confirmation." When asked how exactly we go about judging statements about the external world in the light of experience, Quine's famous advice was to "settle for psychology." Quine was endorsing a Duhemian thesis in "Two Dogmas," not propounding a claim about the logic of "confirmation."

12. In support of this, Colyvan cites Quine, "Carnap and Logical Truth," 120–122.

13. No such claim appears in Putnam, "What Is Mathematical Truth?"

14. Charles Parsons, "Mathematical Intuition," *Proceedings of the Aristotelian Society* 80 (1979–1980): 145–168.

15. Philip Kitcher, *The Nature of Mathematical Knowledge* (New York: Oxford University Press, 1984).

well-confirmed theories vary from belief, through tolerance, to outright rejection.[16] The point is that naturalism counsels us to respect the methods of working scientists, and yet holism is apparently telling us that working scientists ought not to have such differential support to the entities in their theories. Maddy suggests that we should side with naturalism and not holism here. Thus we should endorse the attitudes of working scientists who apparently do not believe in *all* the entities posited by our best theories. We should thus reject (P1)."

6. "Thus if mathematics is confirmed along with our best empirical hypotheses (as indispensability theory claims), there must be mathematics-free competitors. But Sober points out that *all* scientific theories employ a common mathematical core. Thus, since there are no competing hypotheses, it is a mistake to think that mathematics receives confirmational support from empirical evidence in the way other scientific hypotheses do."

7. "The two most important arguments *against* mathematical realism are the epistemological problem for Platonism—how do we come by knowledge of causally inert mathematical entities?[17]—and the indeterminacy problem for the reduction of numbers to sets—if numbers are sets, which sets are they?"[18]

Although this is a large budget of objections, I can reply to them fairly quickly because they all assume that my "indispensability argument" (as opposed to the fictitious "Quine-Putnam indispensability argument") depends on P1 and on "confirmational holism," and neither supposition is correct. In addition, they all assume that my indispensability argument was an argument for "Platonism" in Quine's sense, and, as I have already pointed out, this was not the case.[19] Here follows a brief word about each of them:

16. Penelope Maddy, "Indispensability and Practice," *Journal of Philosophy* 89, no. 6 (1992): 275–289; quotation from p. 280.

17. Paul Benacerraf, "Mathematical Truth," *Journal of Philosophy* 70 (1973): 661–679; reprinted in Benacerraf and Putnam, *Philosophy of Mathematics,* 403–420, and in W. D. Hart, ed., *The Philosophy of Mathematics* (Oxford: Oxford University Press, 1996), 14–30.

18. Paul Benacerraf, "What Numbers Could Not Be," *Philosophical Review* 74 (1965): 47–73; reprinted in Benacerraf and Putnam, *Philosophy of Mathematics,* 272–294.

19. For a detailed discussion of Quine's ontological views, see Hilary Putnam, *Ethics without Ontology* (Cambridge, Mass.: Harvard University Press, 2004), especially chaps. 2, 3, and 4.

Re (1): ["The first thing to note is that 'dispensability' is not the same as 'eliminability.'] If this were not so, *every* entity would be dispensable (due to a theorem of Craig)."] Craig's theorem shows only that it is logically possible to replace a theory T that contains terms for unobservable entities (mathematical or nonmathematical) with a theory—in effect, a Turing machine—that recursively enumerates exactly those theorems of the first theory that contain only "observation terms."[20] Such an "elimination" is acceptable only to an instrumentalist who thinks of theories as mere prediction machines. I repeat that my "indispensability arguments" were intended to show that it is incoherent to attempt to be a scientific realist with respect to physics (that is, a realist with respect to the unobservable entities postulated by physics) and a verificationist, intuitionist, nominalist, or what-have-you with respect to mathematics. They were not arguments against instrumentalism, although I have published a number of such arguments in my time.[21]

As for clarifying the notion of "indispensability": it is enough for my purposes that W the different versions of, say, present-day quantum field theory so far proposed presuppose enough mathematics to develop the theory of unitary transformations of the state vectors in certain abstract spaces (Hilbert spaces and Fock spaces).

Re (2): ["These issues naturally prompt the question of *how much* mathematics is indispensable (and hence how much mathematics carries ontological commitment)."] Well, I reject "and hence" root and branch. I have never said that *only* what is indispensable to natural science carries "ontological commitment" (if you insist on speaking that way). Nor do I think that only as much mathematics as is used in physics "carries ontological commitment" in the sense of being true under a realist interpretation. In my view, once we have agreed that number theory and the real and complex analysis, differential geometry, and the other branches of mathematics that are used in physics are neither to be rejected as "fiction" nor reinterpreted in a verificationist way, then I think that it would be absurd to say, "Well, I accept 'ontological

---

20. See Hilary Putnam, "Craig's Theorem," in Putnam, *Philosophical Papers*, vol. 1, *Mathematics, Matter and Method*, 228–236.

21. One such argument—the one in "Explanation and Reference," in Glenn Pearce and Patrick Maynard, eds., *Conceptual Change* (Dordrecht: Reidel, 1973), 199–221; reprinted in *Philosophical Papers*, vol. 2, *Mind, Language and Reality* (Cambridge: Cambridge University Press, 1975) 196–214 —is quoted later in this essay.

commitment,' in that sense, to sets of integers and even sets of reals, but not to sets of sets of reals." I do not think that "set" is a notion that it makes sense to interpret in a realist way up to a certain rank α and a nonrealist way at rank $\alpha+1$. But I agree with Colyvan that the question how much mathematics is indispensable in physics *is* an interesting and difficult one.

Re (3) [My alleged "confirmational holism"]: I have never claimed that mathematics is "confirmed" by its applications in physics (although I argued in "What Is Mathematical Truth?" that there is a sort of quasi-empirical confirmation of mathematical conjectures *within* mathematics itself). What I claimed in "What Is Mathematical Truth?" (and subsequently) is, I repeat, that a prima facie attractive position—realism with respect to the theoretical entities postulated by physics, combined with *antirealism* with respect to mathematical entities and/or modalities—does not work.

Re (4): ["If a theory is *confirmed* by empirical findings, the *whole* theory is confirmed. In particular, whatever mathematics is made use of in the theory is also confirmed."] I agree with Parsons's "concern that the obviousness of basic mathematical statements is left unaccounted for by the Quinean picture." That is why I do not think it at all plausible to think that numbers are "intangible objects" whose existence we "confirm" in the same way in which we confirm the existence of, say, mesons. My argument was never intended to be an "epistemology of mathematics." If anything, it is a *constraint* on epistemologies of mathematics from a scientific realist standpoint. (Recall that in "What Is Mathematical Truth?" I said that there are reasons internal to mathematics itself for concluding that mathematics is "true under *some* interpretation.")

Re (5): ["Maddy suggests that we should side with naturalism and not holism here. Thus we should endorse the attitudes of working scientists who apparently do not believe in *all* the entities posited by our best theories."] I agree with Maddy that the actual attitudes of working scientists toward the components of well-confirmed theories vary from belief through tolerance to outright rejection. But this does not invalidate my "indispensability arguments," because they were never offered as an account of which "ontological commitments" are "confirmed."[22] For my purposes, it suffices to point out that there is *no* serious quantum field theorist who does not believe something like the following: (1) There are such things as quantum-mechanical "states" of physical systems. Those states determine the probabilities with

22. I criticize Quine's "ontological commitment" talk in chapter 4 of *Ethics without Ontology*.

which various physical interactions have various observable effects. (2) The "time evolution" of physical systems (that is, the way in which those "states" change over time) is governed (except possibly at times of a quantum-mechanical "collapse"[23]) by the Dirac equation, or by an equation (possibly a nonlinear one) to which the Dirac equation is typically an extremely good approximation. (3) In virtue of (1) and (2), present-day quantum field theories (with their various "representations") are legitimately believed to be *approximately correct* descriptions of the relevant physical systems. If one understands this *approximate correctness* in a realist way, then an argument against "verificationism" in the philosophy of mathematics, which I published[24] in criticism of an unpublished remark of Wittgenstein's,[25] still seems pretty good to me. In brief, the argument is that when a mathematical physicist accepts a theory according to which the time evolution of a physical system obeys a differential equation or system of equations, she is committed to the claim that those equations have solutions (in real numbers or, in certain cases, in complex numbers) for each real value of the time parameter t. Suppose, now, that the statement "The solution to such and such equations for the given value of t is in the interval between r_1 and r_2" is *true*. If the physicist makes the mistake of thinking that (mathematical) truth is the same as provability by human beings, then she must construe this statement as implying that it is *possible to calculate the solution to the equations in question for the specified value of t, and the solution will be found to lie in the interval between r_1 and r_2.* But, given present knowledge,[26] this is something to which it is unsafe to commit oneself, even if the value of the magnitude in question does lie in that interval.

I repeat that my indispensability arguments address the question whether antirealism with respect to mathematics is compatible with realism with respect

23. See Hilary Putnam, "A Philosopher Looks at Quantum Mechanics (Again)," *British Journal for the Philosophy of Science* 56, no. 4 (December 2005): 615–634; reprinted in this volume as Chapter 6.

24. See Chapter 25 in this volume. That argument can also be found in full in Chapter 11 of this volume, in the section titled "Where Wittgenstein Went Wrong."

25. Ludwig Wittgenstein, *Remarks on the Foundations of Mathematics,* ed. by G. H. von Wright, Rush Rhees, and G. E. M. Anscombe (Oxford: Blackwell, 1978), V, §34.

26. For the reasons that, given present knowledge, the truth of such a statement probably does not in all cases imply its provability by a calculation human beings could actually carry out, see Chapter 11 in this volume, in the section titled "Where Wittgenstein Went Wrong."

to physics. That scientists do not take seriously all the "components of well-confirmed theories" that they accept does not refute my argument, unless what they "do not take seriously" includes the whole mathematical apparatus of fundamental physics. And obviously it does not.

Re (6): ["Sober points out that *all* scientific theories employ a common mathematical core. Thus, since there are no competing hypotheses, it is a mistake to think that mathematics receives confirmational support from empirical evidence in the way other scientific hypotheses do."] I completely agree with Elliott Sober that it is a mistake to think that mathematics receives confirmational support from empirical evidence in the way other scientific hypotheses do.

Re (7): ["The two most important arguments *against* mathematical realism are the epistemological problem for Platonism—how do we come by knowledge of causally inert mathematical entities? and the indeterminacy problem for the reduction of numbers to sets—if numbers are sets, which sets are they?"] This objection illustrates both the way in which it is assumed that any defense of realism in the philosophy of mathematics must be a defense of "Platonism" and the way in which the idea that there are "equivalent descriptions" in mathematics—an idea that, as mentioned earlier, was already defended in "What Is Mathematical Truth?"—has been ignored for over thirty years. For my most recent defense of that idea, see my *Ethics without Ontology,* especially chapters 2 and 3. There I argue that the decision to identify the natural numbers with one ω-sequence of sets rather than with another is simply a matter of convention—yet another instance of "equivalent descriptions."

Hartry Field's Position

Hartry Field understands very well what my arguments were, and he attempts to meet them on their own terms. What he tries to do is to show that mathematical physics can be *restated* in a way that avoids *all* quantification over abstract entities. For this purpose, as is well known, he assumes that *points* (in space or space-time) are concrete particulars. Although this assumption has been attacked, it does not seem illegitimate to me (whether the assumption is *correct* as a matter of contingent physical fact is another question). Then, rather than construe such a statement as

(S_1) The charge of O is r

as asserting that a substitution instance of the open sentence "the charge of 1 is 2" is true—namely, a substitution instance in which "1" is replaced by the name of a physical object and "2" is replaced by the name of an abstract entity, a real number—Field will *reformulate* (S$_1$) thus:

$$(S_2) \text{ Charge } (O, G),$$

where "Charge (x, y)" is a two-place predicate that replaces the real-number-valued function *charge*(x) of classical physics, and G is a suitable geometric object (e.g., a pair of line segments the ratio of whose lengths is r). Line segments, are, of course, construed as mereological sums (not sets) of points, and a pair of line segments is just the mereological sum of the two line segments. (I have not followed Field's own construction exactly, but this will give the philosophical "flavor" of that construction.) In this way, all quantifications over sets and numbers become quantifications over mereological sums of points, which are taken to be nominalistically acceptable particulars. So sets and numbers are *not* indispensable in physics after all, as both Quine and I assumed. *Voila!*

Well, I agree that, assuming the nominalistic acceptability of such geometrical objects as line segments and pairs of line segments, and such properties/relations of line segments as congruence and equal ratios of lengths (in the case of two pairs of line segments), Field has shown that much or perhaps even all of classical physics can avoid any use of set theory at all. (A somewhat improved and clarified version of Field's mathematical claims is presented by John Burgess and Gideon Rosen in their book *A Subject with No Object*.)[27] My objection to Field's proposed solution is that although it may work for classical physics, it is not sufficiently *robust;* it probably cannot be adapted to the postclassical physical theories that are today the live candidates for quantum theories of gravity. Here I will have to be moderately technical.

Very roughly, my objection is that *all* theories of quantum gravity assume that *space-times* can be superimposed. That means that space-time does not have a determinate metric at all, and hence one cannot speak of determinate "straight lines" (geodesics) at all, or even of curve segments with equal or unequal lengths (or of two pairs of curve segments whose lengths have the same ratio) unless future physics abandons the idea that space-time itself is quantum mechanical, a point implicitly made by Wheeler, Thorne, and Misner in

---

27. John Burgess and Gideon Rosen, *A Subject with No Object: Strategies for Nominalistic Interpretation of Mathematics* (Oxford: Oxford University Press, 1997).

their classic book *Gravitation.* (They famously wrote that space-time is a "foam of worm-holes" at the quantum scale.)[28]

In sum, if the theories of quantum gravity are right that our world exists in a superposition of different space-times, then if someone says, "Consider a line on a spacelike hypersurface," what she says is literally meaningless: there is no such thing as "one of the lines on the surface" (in fact, there is no such thing as "the surface"). *There is only a superposition of space-times.*

If we go back to the sentence (S₂) above:

$$(S_2) \ Charge \ (O, \ G),$$

the problem is that the geometric object (the pair of line segments) *G* is not an object that makes any sense in quantum gravitation.

As everyone knows, we certainly do not know what a satisfactory theory of quantum gravity will look like in any detail. But for the reasons just given, I do not see how, for example, string theory or quantum loop gravity could be "nominalized" à la Field, nor how any other theory that lacks a classical space-time could be. For this reason, I am skeptical that Field's proposal can work. To reject theories that need more set theory than Field allows would require us to ban precisely the theories or theory sketches that cutting-edge physicists are working on.

A historical remark: that the whole mathematics of physics, including the calculus, was just a part of geometry is not a new idea. This is something that Leibniz and Kant regarded as obvious (in Kant's case, because that is how Newton understood the calculus). In a sense, Field's proposal is an attempted revival of the seventeenth- and eighteenth-century view (in a more sophisticated form, of course). When I say that the proposal is not "robust," what I mean is that it depends essentially on features of that classical view that are no longer accepted, and that are not likely to be features of future fundamental physics.

Gideon Rosen's Position

Gideon Rosen is not a nominalist, but he believes that neither the arguments for nor the arguments against nominalism are rationally compelling. In a deep and beautiful paper titled "Nominalism, Naturalism, Epistemic

---

28. C. W. Misner, K. S. Thorne, and J. A. Wheeler, *Gravitation* (San Francisco: W. H. Freeman, 1972).

Relativism,"[29] he argues that both nominalism and antinominalism are rationally permissible. But like both myself and Hartry Field, he believes that "it would be patently unreasonable for an informed citizen of the modern world simply to reject modern science," and so he goes on to say,[30]

> This means that any case for the permissibility of the suspension of judgment about the abstract must show how nominalism of this sort is compatible with taking science seriously as a source of information about concrete nature. This is of course the point of the familiar "nominalization" programs. When these programs are not pitched as dubious hermeneutic proposals, they are best understood as attempts to reconcile doubts about the literal truth of extant science with a policy of *accepting* Platonistic theories as instruments for description and explanation. I shall not discuss these programs in detail here. Instead I shall describe what I regard as a simple trick for nominalizing at a stroke any theory whatsoever. Given any theory T formulated in the usual mathematical terms, the trick returns a theory that does without abstract objects, but which is nonetheless in a certain sense equivalent to the original.[31]

Rosen believes that "we" (and his "we" certainly includes scientific realists of my ilk) "indulge in Platonistic discourse;[32] we often believe what we say; we harbor no secret reservations; and all this is rationally permissible by our lights. Indeed let us make the further assumption that so far as we can see, the cumulative scientific case for certain abstract entities is so utterly compelling that it would be unreasonably cautious to demur."[33]

Rosen illustrates (what he argues to be) the possibility of an *equally rational* nominalistic version of science—one that might have been our version

---

29. Gideon Rosen, "Nominalism, Naturalism, Epistemic Relativism," *Philosophical Perspectives* 15, *Metaphysics* (2001): 69–91.

30. Ibid., 73.

31. However, Rosen is not himself a nominalist. He adds (ibid., 73–74), "Before I describe the trick, I should emphasize that I do not endorse the novel theories it delivers. The nominalistic version of, say, quantum electrodynamics is in some sense a different theory from standard QED. But it is not a better, more acceptable, theory. The nominalistic version is of interest, not as an account of what science actually says, and certainly not as an account of what it *ought* to be saying, but rather as an account of what science might have said, and might have been justified in saying had cultural history gone somewhat differently."

32. "Platonistic discourse" is Rosen's term for discourse that quantifies over numbers and sets.

33. Rosen, "Nominalism, Naturalism, Epistemic Relativism," 74.

had "cultural history had gone somewhat differently"[34]—by imagining a world called "Bedrock":[35]

> That is how we are, but in Bedrock things are different. . . . Bedrocker science and mathematics are indiscernible from ours. Their textbooks, their advanced teaching, the transcripts of their laboratory conversations, and the rest are all thoroughly Platonistic in the sense that the sentences they contain imply the existence of abstract entities.
>
> The difference is that in Bedrock they have reservations about this aspect of what they say—carefully considered and fully articulate reservations. Bedrockers are encouraged from early childhood on to suspect that only concrete things exist, and that discourse about the abstract is at best a useful fiction. If you ask them how they square this with what they say when they're doing science, they smile as if they've heard the question a thousand times and deliver themselves of a speech along the following lines.

Because the speech is very long, I shall "set the stage" in my own words. Consider the following sentence:

(1)   The number of Martian moons $= 2$.

Now, define the *concrete core* of a world W to be the largest wholly concrete part of W: the mereological sum of all the concrete objects that exist in W.
    Let me now quote the Bedrockers' words:[36]

> Now suppose ours is a numberless world and that (1) is therefore false. If we were concerned to speak the truth, we would never countenance its assertion. But the fact is, we are rarely concerned to speak the truth. Our unhedged assertoric utterances normally aspire to a weaker condition we call *nominalistic adequacy*. S is nominalistically adequate iff the concrete core of the actual world is an exact intrinsic duplicate of the concrete core of some world at which S is true—that is, just in case things are in all concrete respects *as if* S were true.

The "trick" to which Rosen referred, the one that, given a theory that quantifies over numbers and sets, is supposed to return a theory that does without abstract entities, but that is nonetheless in a certain sense equivalent to the

---

34. Ibid., 73.
35. Ibid., 74.
36. Ibid., 75.

original, is to do what the Bedrockers do: replace the claim that, to use Rosen's own example, quantum electrodynamics is *true* with the claim that quantum electrodynamics is *nominalistically adequate.*

Have the Bedrockers really found a way to exhibit a "rationally permissible" substitute for any physical theory that quantifies over numbers and sets, including quantum electrodynamics, as Rosen claims? I shall suggest three reasons in favor of a negative answer to this question.

First, a reason for doubting that the Bedrocker's "trick" has really been described in a way that makes sense.

Both the notions on which the "trick" depends—namely, "concrete object" and "exact intrinsic duplicate"—are difficult and perhaps impossible to make sense of in what we take to be our best theoretical physics. Already in relativistic electromagnetic theory a problem arises: are *fields* "concrete objects" or not? They have mass and carry energy and momentum, so the answer would seem to be yes. But what nominalistically acceptable predicates can we use to describe them? Normally one describes them by specifying the components of the field vectors at points in space. If one undertakes to do this nominalistically, then one undertakes to carry out a program like Hartry Field's—but the Bedrockers' "trick" was supposed to bypass the need to do that.

As for "intrinsic duplicate": if an intrinsic duplicate of, say, my brain (or an electromagnetic field) is required to have *all* the properties of my brain—including properties described in the language of mathematical physics—why should we agree that a nominalist is entitled to any such notion? And if an intrinsic duplicate is only required to "match" the possible entity of which it is a duplicate with respect to *nominalistically acceptable predicates* ("concrete respects"), then which predicates are they?

If the problem was bad in classical physics, it may well be hopeless in quantum field theories, whether or not they include gravitation. Here is a dilemma I propose for Gideon Rosen to consider: are *electrons* "concrete" or not?

Assume that Rosen answers yes. Then the following problem arises: Quantum physicists have long rejected the idea that particles in states that are "indeterminate" in one or another respect (particles in a superposition of states in which a property has different values) "really" possess a definite value of the indeterminate property, only we cannot *know* what it is; that way of looking at quantum-mechanical indeterminacy literally makes no sense in contemporary quantum mechanics, and in field theories in particular. But among the properties that enter into superpositions are (1) the number of electrons in the system and (2) the number of protons in the system.

So suppose that we have a state that is a superposition of a state in which the system consists of N electrons and one in which it consists of M protons. (Such states are perfectly possible.) Then if someone says, "Consider one of the electrons in the system," what she says is literally meaningless: there is no such thing as "one of the electrons in the system"; *there is only a system in a superposed state.* There are definite expectation values for different measurements, including measurements designed to find electrons and measurements designed to find protons, and we can say what these expectation values are, but before the measurement interaction there is not a definite number of "electrons" and "protons" there. This means that strictly speaking, there are not such "objects" as electrons in the "ontology" of the theory; what there are are quantized fields, and talk of "electrons" is a useful way of talking of, for example, certain components of vectors in the Fock space that we use to describe those fields. (Note that I am not saying, "Electrons do not exist"; I am saying that electron talk is not object talk).[37] If, in spite of this, Rosen (or better, the Bedrocker) claims that electrons are "concrete," what nominalistically acceptable predicates does he use to describe them? And how does he formalize such a sentence as "the number of electrons in the box is indeterminate, but the state is $1/\sqrt{2}$(two electrons in the box)$+1/\sqrt{2}$(three electrons in the box)"? If the answer is "We Bedrockers don't recognize any such statement as truth-apt, but only as nominalistically adequate," then can the Bedrocker also explain what it means to say that a "concrete" system can be an "intrinsic duplicate" of a system in a possible world in which there are numbers, and in which that sentence is true? Again we face the question, does not the notion

---

37. This argument now [April 5, 2011] seems bad to me because it totally ignores the Bohm and Ghirardi-Rimini-Weber (GRW) interpretations that I myself argue we should take seriously in Chapters 6 and 7. In those interpretations particles do have positions, or (in the case of GRW) at certain moments (when there is a "collapse") they do. A better argument would have been to point out that to "nominalize" quantum mechanics, one would have to nominalize the mathematical notion of probability, because an "intrinsic duplicate" of a quantum-mechanical world obviously has to have the same probabilities of nominalistically describable occurrences (trajectories in the case of a Bohmian world, if they can be described nominalistically, and "flashes" in the case of a GRW world with a "flash" ontology); but how can the Bedrocker say that without talking mathematics? I should also have pointed out that Rosen's interpretation is not nominalistic in my sense; it is, rather, a somewhat vague modal-logical interpretation.

of an "intrinsic duplicate" presuppose some notion of a nominalistically acceptable description of a "concrete object" and its "concrete respects"?

Now assume that Rosen answers no, electrons are not "concrete." Short of reverting to instrumentalism, what then are the "concrete objects" that quantum field theory describes? Fields? But fields too can be superimposed.

I suspect that the "trick" Rosen describes depends, in fact, on assuming that all possible things are good old-fashioned "concrete objects," which we know how to describe in nominalistic language, or good old-fashioned "abstract entities," which are not in space-time, and which are causally inert. But the entities of modern physics are neither of the above.

Second, a reason for doubting that the Bedrockers' "trick" produces inferences that the Bedrockers themselves have any reason to see as justified.

In "Explanation and Reference,"[38] a paper I wrote back when I was still battling logical positivism, I gave an argument that I think can be easily adapted against Bedrockism. Here is the original argument:

> When a realistically minded scientist . . . accepts a theory, he accepts it as true (or probably true, or approximately-true, or probably approximately-true). Since he also accepts *logic,* he knows that certain moves *preserve truth.* For example, if he accepts a theory T_1 as true and he accepts a theory T_2 as true, then he knows that $T_1 \wedge T_2$ is also true, by logic, and so he accepts $T_1 \wedge T_2$. If we talk about probability, we have to say that if T_1 is very highly probably true and T_2 is very highly probably true, then the conjunction $T_1 \wedge T_2$ is also highly probable (though not as highly as the conjuncts separately), provided that T_1 is not negatively relevant to T_2— i.e. provided that T_2 is not only highly probable on the evidence, but also no less probable on the assumption of T_1 (this is a judgment that must be made on the basis of what T_1 *says* and background knowledge, of course). If we talk about approximate-truth, then we have to say that the approximations probably involved in T_1 and T_2 need to be compatible for us to pass from the approximate-truth of T_1 and T_2 to the approximate-truth of their conjunction. None of these matters is at all deep, from a realist point of view. But even if we confine ourselves to the simplest case, the case in which we can neglect the chances of error and the presence of approximations, and treat the acceptance of T_1 and T_2 as simply the acceptance of them as true, I want to suggest that the move from this acceptance to the

38. Hilary Putnam, "Explanation and Reference," 210 in the reprinted version.

acceptance of their conjunction is one to which one is not entitled on positivist philosophy of science. One of the simplest moves that scientists daily make, a move they make as a matter of propositional logic, a move which is central if scientific inquiry is to have any *cumulative* character at all, is *arbitrary* if positivist philosophy of science is right.

As I went on to explain:[39]

> The difficulty is very simple. Acceptance of T_1, for a positivist, means acceptance of T_1 as leading to successful predictions (i.e. all *observation sentences* which are theorems of T_1 are true . . .). Similarly, the acceptance of T_2 means acceptance of T_2 as leading to successful predictions. But from the fact that T_1 leads to successful predictions and the fact that T_2 leads to successful predictions it does not follow at all that the conjunction $T_1 \land T_2$ leads to successful predictions. The difficulty, in a nutshell, is that the predicate which plays the role of truth—the predicate "leads to successful predictions"—does not have the *properties* of truth. The positivist may teach in his philosophy class that acceptance of a scientific theory is acceptance of it as "simple and leading to true predictions," and then go out and do science . . . by verifying theories T_1 and T_2, conjoining theories which have previously been verified, etc.—but then there is just as great a discrepancy between what he teaches in his philosophy seminar and his *practice* as there was between Berkeley's teaching that the world consisted of spirits and their ideas and continuing in practice to daily rely on the material object conceptual system.

The relevance to Bedrock is this: the Bedrockers say that "our unhedged assertoric utterances normally aspire to a weaker condition [than truth] we call *nominalistic* adequacy." But the predicate "is nominalistically adequate" no more has the properties of truth than the positivists' "leads to true predictions" (or "is simple and leads to true predictions") does. Whenever their scientists employ one of the simplest truth-functional inferences there is, conjunction-introduction, they make a move to which they are not entitled. If they really "aspire to" nominalistic adequacy and not truth, then why do they constantly make inferences that lead from true premises to true conclusions, but not, in general, from nominalistically adequate premises to nominalistically adequate conclusions (assuming that the difficulties I have raised about even defining "nominalistic adequacy" can be met)?

39. Ibid., 210–211.

Third, a more traditional sort of objection.

Suppose that we define a "correct" explanation to be one that, in addition to being acceptable as an explanation by the standards both we and the Bedrockers are assumed to accept, possesses *true premises* (and, of course, a true conclusion). A question I should like to put to the Bedrockers is this: "You tell me that in science you do not aspire to what I just called correct explanations; you say that you have only the 'weaker' aspiration of finding nominalistically adequate explanations. Moreover, all the major explanations you and we accept are *fictions* by your lights. Very well, then, do you have any idea what the *correct* explanation of, say, the orbits of the planets is, given that the one taught in your textbooks has false premises?"

What are the Bedrockers supposed to say? The notion of "correct explanation" does not involve any concepts they do not *have*. So why are they not interested in correct explanations? Unlike Rosen, I think that disinterest in correct explanations is prima facie a rational shortcoming, and nothing in his brilliant philosophical fiction about Bedrock exonerates the Bedrockers from criticism here. To quote one of my former selves, in the light of the last two objections, I am inclined to say that the Bedrocker's epistemology makes the success of science a miracle.[40]

APPENDIX

As I promised in the main text of this essay, in this Appendix I will argue briefly that if my arguments are accepted, not only intuitionism and nominalism need to be rejected as philosophies of mathematics, but also the idea that "predicative" mathematics is all that science needs.

Recall that I began by saying that my answer to the objection that we do not yet have the "true" physical theory is that we should regard each of the

---

40. In "What Is Mathematical Truth?" I wrote, "The positive argument for realism is that it is the only philosophy that doesn't make the success of science a miracle" (73). As an argument against a sophisticated antirealism such as Dummett's, or Crispin Wright's, or my own former "internal realism," this was not a good argument—indeed, as I pointed out in *Reason, Truth and History* (Cambridge, Mass.: Cambridge University Press, 1981), chapter 5, a sophisticated antirealist need not deny that "terms in a mature science typically refer and theories accepted in a mature science are typically approximately true"—this was my characterization of "scientific realism." But as an argument against positions such as logical positivism, van Fraassen's "constructive empiricism," and Bedrocker nominalism, I still believe it.

rival theories as a candidate for truth or approximate truth, and that *any philosophy of mathematics that would be inconsistent with so regarding them should be rejected.* (This is what I mean by "scientific realism.")

An example of such a view is Wittgenstein's view (which he did not publish) that the notion of a real number is vague ("there is no set of irrational numbers").[41] Classical physics assumes that there is a point corresponding to every triple of real numbers (alternatively, if you prefer an ontology of regions, a sphere the coordinates of whose center are any given triple of real numbers). If we accepted that there is no determinate totality of real numbers, as Wittgenstein urged, then this important claim—the claim that the geometric continuum is isomorphic to the analytic continuum—would become inexpressible.

Of course, it is often said that the continuum is only an "idealization." This may indeed turn out to be true for *physical* reasons—indeed, in Chapter 6 I argued that it has, in a sense, turned out to be true not because space-time has turned out to be discrete, but because we live in a world that features a *superposition* of space-times—but a scientific realist should insist that the issue not be prejudged by adopting a finitist philosophy of mathematics. Philosophers should not be in the business of telling physicists what space-time "really" is, or what it is "meaningless" to say it is.

Let us turn now to the claim that predicative set theory is adequate to the needs of physics. I would not dare to challenge Solomon Feferman's claim (in, e.g., the papers collected in *In the Light of Logic*[42]) to have shown that the theorems that are needed in the "applications" of physical science can all be derived within predicative mathematics. What I am skeptical about is whether the predicative mathematician can answer the difficulty that I just raised as an objection to Wittgenstein's finitism. The statement "There is a point corresponding to every triple of real numbers" (alternatively, "There is a sphere the

---

41. Wittgenstein wrote, "It might be said, besides the rational points there are *diverse systems* of irrational points to be found in the number line. There is no system of irrational numbers but also no super-system, no 'set of irrational numbers' of higher-order infinity." Ludwig Wittgenstein, *Remarks on the Foundations of Mathematics,* ed. G. H. von Wright, R. Rhees, and G. E. M. Anscombe (Cambridge, Mass.: MIT Press, 1978), II, §33. In the 1956 edition, this material was in what was then appendix II of part I. The present part II includes a few remarks that were left out of the 1956 edition, and, in addition, "the arrangement of sentences and paragraphs into numbered Remarks corresponds to the original text (which was not wholly the case in the 1956 edition)." See the "Editors' Preface to the Revised Edition," 31. The numbers of the remarks are, therefore, different from those in the 1956 edition, as is the order of some of the remarks.

42. Solomon Feferman, *In the Light of Logic* (Oxford: Oxford University Press, 1998).

coordinates of whose center are any given triple of real numbers") is not, as far as I can see, *expressible* without quantifying over *all* triples of real numbers— which is just what predicative analysis forbids!

If I understand Feferman correctly, he would recommend that the physicist replace the statement I just mentioned with a *schema* to the effect that "there is a point corresponding to every triple of real numbers of order α," where α is a "dummy letter" for any predicatively attainable ordinal. In other words, the statement "There is a point corresponding to every triple of real numbers of order α" might be asserted with what in the context of *Principia Mathematica* has long been called "typical ambiguity"—except that the ambiguity here would be one of "orders" of predicativity, not types. Indeed, in the context of set theories that have either types or orders of predicativity or both, the device of "typical ambiguity" is familiar from the time of *Principia Mathematica*. But it is a highly unnatural device in the context of an empirical theory. What could the justification for such a schema be, except that one believes (but thinks for philosophical reasons that one is not allowed to say) that *every* triple of real numbers is the coordinate triple of a point in space (or the center of a unit sphere in space)?[43]

Of course, one could (and many do) wax skeptical, not on physical but on supposedly "conceptual" grounds, about whether physics "really needs" to talk about every point in space or every region in space; but I have difficulty in seeing the difference between such skepticism and outright "instrumentalism." Indeed, if adequacy for "applications" just means adequacy for testable predictions, then the view that *that* is all that physics is after is precisely the denial of what I called "scientific realism."

But what of the quantum theories of gravitation that I spoke about in Chapter 6? It is true that here we lose the notion of points and spheres as determinate objects in physical space-time, but if anything, it is even easier to show that the worldview of quantum mechanics needs *all* the real numbers— and, in fact, we do not have to talk about quantum gravity at all to do this. The worldview of quantum mechanics is that there is a definite totality of possible "states" of a physical system—these are what the "state vectors" represent. And if it is even physically possible that future time is infinite, then it is easy to show that there are as many distinct states of a very simple system as there are real numbers (I will spare you the details).

---

43. One might also ask how a physicist could be expected to discover such a schema. By first forming the classical hypothesis and then weakening it to fit the predicative straitjacket?

10

Revisiting the Liar Paradox

The Liar Paradox

The best presentation I know of the liar paradox is due to Charles Parsons,[1] and in the end the view I shall defend is, I believe, an elaboration of his. In his paper "The Liar Paradox," a paper I have thought about for almost twenty years, the paradox is stated in different ways. One of these ways is in terms of three alternatives: either a sentence expresses a true proposition, or it expresses a false proposition, or it does not express a proposition at all. A second way is the one I followed in my presentation of the liar paradox in "Realism with a Human Face" (the title paper of the book with that title), in which talk of propositions is avoided, and I mostly employ that way here in order to facilitate comparison with Tarski's work.[2] Later I will, however, give some quotes from Parsons's paper that employ the first way of presenting the paradox.

1. Charles Parsons, "The Liar Paradox," *Journal of Philosophical Logic* 3 (1974): 381–412; reprinted in Robert Martin, ed., *Recent Essays on Truth and the Liar Paradox* (Oxford: Oxford University Press, 1984), 9–45.
2. Alfred Tarski, "Der Wahrheitsbegriff in den formalizierten Sprachen," *Studia Philosophica*, 1 (1935): 261–405; translation, "The Concept of Truth in Formalized Lan-

It is an empirical fact that the one and only sentence numbered "(I)" on page 11 of my *Realism with a Human Face* is the following:

(I) The sentence (I) is false.

Is the sentence numbered (I) on page 11 of my *Realism with a Human Face* true? Tarski famously used "Snow is white" as his example of a typical sentence, and his "Convention T" requires that a satisfactory treatment of truth must enable us to show that

"Snow is white" is true if and only if snow is white.

If we suppose that the sentence (I) has a truth-value at all, it follows by Convention T that

(i) "The sentence (I) is false" is true if and only if the sentence (I) is false.

But, as just mentioned, the sentence (I) = "The sentence (I) is false," and hence

(ii) The sentence (I) is true if and only if the sentence (I) is false,

which is a contradiction!

So far we do not have an actual inconsistency. We assumed that the sentence (I) has a truth-value, and that assumption has now been refuted. We cannot consistently assert either that (I) is true or that (I) is false. But now we come to the "strong liar." The form I considered in *Realism with a Human Face*[3] is:

(II) The sentence (II) is either false or lacks a truth-value.

The sentence (II) is paradoxical because, if we try to avoid the previous argument by denying that (II) has a truth-value, that is, by asserting that

(II) lacks a truth-value,

then it obviously follows that

(II) is either false or lacks a truth-value,

---

guages," in Tarski, *Logic, Semantics, Metamathematics* (Oxford: Clarendon Press, 1956), 152–278.

3. Hilary Putnam, *Realism with a Human Face* (Cambridge, Mass.: Harvard University Press, 1990), 12.

and the sentence (II) is one that we discover ourselves to have just asserted. So we must agree that (II) is true, which means that we have contradicted ourselves.

Tarski showed us how to avoid such paradoxes by relativizing the predicate "is true" to whichever language we are speaking of, and by introducing a hierarchy of languages. If I say of a sentence in a language L that it is true or false, my assertion belongs to a language of a higher level—a metalanguage. No language is allowed to contain its own truth-predicate. The closest I can come to such sentences as (I) or (II) is to form a sentence (III) with a relativized truth-predicate:

(III) The sentence (III) is not true-in-L,

but this sentence does not belong to L itself, but only to meta-L. Because it does not belong to L, it is true that it is not true-in-L. And because this is exactly what it says in meta-L, it *is* true in meta-L. (III) is not even well formed in the "object language" L, and is true in the metalanguage, meta-L, and this dissolves the paradox.

In *Realism with a Human Face* I asked "if Tarski succeeded, or if he has only pushed the antinomy out of the formal language and into the informal language which he himself employs when he explains the significance of his formal work." If each language has its own truth-predicate, and the notion "true-in-L," where L is a language, is expressible in meta-L, but not in L itself, the semantic paradoxes can all be avoided, I agreed. "But in what language is Tarski himself supposed to be saying all this?" I asked.[4]

"Tarski's theory introduces a 'hierarchy of languages,'" I continued.

There is the "object language" . . . there is the meta-language, the meta-meta-language, and so on. For every finite number n, there is a meta-language of level n. Using the so-called transfinite numbers, one can even extend the hierarchy into the transfinite—there are meta-languages of higher and higher *infinite* orders. The paradoxical aspect of Tarski's theory, indeed of any hierarchical theory, is that one has to stand outside the whole hierarchy even to formulate the statement that the hierarchy exists. But what is this "outside place"—"informal language"—supposed to be? It cannot be "ordinary language," because ordinary language, according to Tarski, is semantically closed and hence inconsistent. But neither can it be

4. Ibid., 13.

a regimented language, for no regimented language can make semantic generalizations about itself or about languages on a higher level than itself.[5]

I also considered Parsons's way out; as I explained it then, this way involves the claim that the informal discourse in which we say such things as "every language has a metalanguage, and the truth-predicate for the language belongs to the metalanguage and not to the language itself" *is not* part of any language but a kind of speech that is sui generis (call it, "systematic ambiguity"). And I found difficulty in seeing what this comes to. After all, one can formally escape the paradox by insisting that all "languages" properly so called are to be written with ink other than red, I pointed out, and reserving red ink for discourse that generalizes about "languages properly so called." Because generalizations about "all languages" would not include the red-ink language in which they are written (the red-ink language is sui generis), we cannot derive the liar paradox. But is this not just a formalistic trick? How does Parsons's "systematic ambiguity" differ from red-ink language? I asked. I hope to answer my own previous objection here, and thereby to deepen our understanding of what systematic ambiguity is and why it is necessary

Black-Hole Sentences?

When I wrote the title paper of *Realism with a Human Face,* I regarded Tarski's hierarchical solution as just a technical solution, a way of constructing restricted languages in which no paradox arises. It seemed to me that as a *general* solution, the hierarchical solution can only be "shown but not said"; it is literally inexpressible. But I proposed no solution of my own. In an earlier paper, a memorial lecture for James Thompson, I did propose a solution, but I was dissatisfied and did not publish that lecture. In that unpublished lecture I set up a language that is not hierarchical, and in which the truth-predicate can be applied to any and all of the sentences of the language. Semantic paradoxes were avoided by assuming Convention T only for a subset of the sentences of the language, the "Tarskian" sentences. I did not define the set of Tarskian sentences once and for all; instead, there were axioms

5. Ibid., 13–14.

enabling us to prove that certain sentences (sentences that we are sure are paradox free and that we are likely to need) are all Tarskian, and the idea was that just as we add stronger axioms of set existence to set theory when we discover we need them, we could add axioms specifying that additional sentences are "Tarskian" as these become necessary. With respect to the obviously paradoxical sentences such as the liar paradox, the position I recommended was a sort of logical quietism, that is, "Don't say anything (semantic) about them at all." (This idea was, perhaps, an anticipation of Haim Gaifman's idea[6] that there are "black-hole" sentences, that is, sentences that are paradoxical and, moreover, such that the application of any semantic predicate to one of them simply generates a further paradox.) But this seemed to me desperately unsatisfactory, because if we are content not to say anything at all about the paradoxical sentences, why do we not just stick to Tarski's solution? The problem with that solution, after all, arises only if we try to state it as a *general* solution; and one could just as well be a quietist about the principle *underlying* the Tarskian route to avoidance of the paradoxes in particular cases as about the black-hole sentences. Both forms of "quietism" are *so* unsatisfactory that I want to make another attempt to see whether we can find something more satisfactory to say about the paradoxes. But I must warn you in advance that what I will end up with will not be a "solution" to the paradoxes, in the sense of a point of view that simply makes all appearance of paradox go away. Indeed, I still agree with the main moral of "Realism with a Human Face," which is that such a solution does not seem to be possible.

We Begin the Reconsideration of Parsons's Solution

As I indicated, I now accept Parsons's solution. As I explained above, my objection to that solution was that it rests on the notion of "systematic ambiguity," but it was not clear to me why systematic ambiguity was not just another "language," a language that was simply *stipulated* to be outside the hierarchy and thus available to serve as a kind of Archimedean point. Another problem (one I did not mention in "Realism with a Human Face") was that I had dif-

6. Haim Gaifman, "Operational pointer semantics: Solution to self-referential puzzles I," in *Proceedings of the Second Conference on Theoretical Aspects of Reasoning about Knowledge* (Los Angeles: Morgan Kauffman, 1988), 43–59, and "Pointers to truth," *Journal of Philosophy*, 89 (1992): 223–261.

ficulty in understanding the following: when Parsons applies his solution to natural language, he asserts that

> however vaguely defined the schemes of interpretation of the ordinary (and also not so ordinary) use of language may be, they arrange themselves naturally into a hierarchy, though clearly not a linearly ordered one. A scheme of interpretation that is "more comprehensive" than another or involves "reflection" on another will involve either a larger universe of discourse, or assignments of extensions or intensions to a broader body of discourse, or commitments as to the translations of more possible utterances. A less comprehensive interpretation can be appealed to in a discourse using the more comprehensive interpretation as a metadiscourse.
>
> To many the hierarchical approach to the semantical paradoxes has seemed implausible in application to natural languages because there seemed to be no division of a natural language into a hierarchy of "languages" such that the higher ones contain the "semantics" of the lower ones. Indeed there is no such neat division of any language as a whole. What the objection fails to appreciate is just how far the variation in the truth-conditions of sentences of a natural language with the occasion of utterance can go, and in particular how this can arise for expressions that are crucially relevant to the semantic paradoxes: perhaps not "true," but at all events quantifiers, "say," "mean," and other expressions that involve indirect speech.[7]

The fact is that I found it difficult to understand what Parsons meant by his claim that talk of "meaning" and "saying" and "expressing" in natural language (if not the use of the predicate "true" itself) presupposes interpretations that can be arranged into hierarchies. But what I want to do now is to apply Parsons's idea to a context in which it will be clear what the "interpretations" are and how they form hierarchies.

Parsons considered the following sentence:[8]

(2) The sentence written in the upper right-hand corner of the blackboard in Room 913-D South Laboratory, the Rockefeller University, at 3:15 P.M. on December 16, 1971, does not express a true proposition.

(Note for later use that I will sometimes abbreviate the sentence description in (2) as "A.")

---

7. Parsons, "Liar Paradox," 406.
8. Ibid., 386.

We are given that (2) was, in fact, written in the upper right-hand corner of the blackboard in the room mentioned on December 16, 1971, and was the only sentence so written. The guiding idea behind Parsons's solution to the liar paradox is contained in a page of his paper,[9] in which he says, in effect (I have slightly simplified the exposition): (2) says of itself that it does not express a true proposition. Because it does not express *any* proposition, in particular it does not express a true one. Hence it seems to say something true. Must we then say that (2) expresses a true proposition? In either case we will be landed in a contradiction. A simple observation that would avoid this is as follows: the quantifiers in one object language could be interpreted as ranging over a certain universe of discourse U. Then a sentence such as

$$(\exists x)(x \text{ is a proposition} \wedge A \text{ expresses } x)$$

is true just in case U contains a proposition expressed by A, that is, by (2). But what reason do we have to conclude from the fact that we have made sense of (2), and even determined its truth-value, that it expresses a proposition that lies *in the universe U*?

It is this rhetorical question that leads Parsons to speak of a hierarchy of interpretations of paradoxical sentences such as (2). In order to generate a hierarchy of interpretations that can serve as a kind of formal model for what Parsons is suggesting here, I shall begin by using Saul Kripke's idea to generate an initial interpretation.

A Hierarchy Beginning with a Kripkean Interpretation

Saul Kripke is, of course, *the* contemporary logician who put the idea that there is an alternative to Tarski's method, that is, a way to construct a consistently interpreted formalized language so that the truth-predicate for the sentences of the language belongs to the language itself, "on the map."[10] Since his famous paper, research into that alternative has never ceased. It should not be surprising that I drag in Kripke here, because sentences like (II) cannot generate a paradox if we follow Tarski's method; the very fact that they are not in the language about which they speak aborts the paradox. If we wish to formalize the paradoxical reasoning at all, we need a language in which

9. Ibid., 230.
10. Saul Kripke, "Outline of a Theory of Truth," *Journal of Philosophy* 72 (1975): 690–716.

the truth-predicate belongs to the language L and not only to a metalanguage, and this is the kind of language Kripke showed us how to interpret.

Kripke himself admits that his solution does not *wholly* avoid hierarchy, for a reason that I shall mention shortly, and, of course, the whole moral of Parsons's paper was that even if the language itself (or natural language itself) is not stratified into object language, metalanguage, and so on—that is, even if its *syntax* is not hierarchical—still, the best way to think about what is going on with the liar paradox is to think of a hierarchy: not a hierarchy of formalisms, but a hierarchy of *interpretations* of the syntactically unstratified formalism.

What Kripke achieved was to find a natural way (or, actually, a whole class of natural ways) to do the following: to assign to the predicate "true" (using a device from recursion theory called "monotone inductive definition") not simply an extension, but a *triple* of sets of sentences, say, <Trues, Undecideds, Falses>. The first set in the triple—let us call it the Trues—consists of sentences in the object language (henceforth simply "L") that are assigned the truth-value "true"; the third set, the Falses, consists of sentences in L that are assigned the truth-value "false"; and the middle set, the Undecideds, consists of sentences whose truth-value is *undefined*. All this is done in such a way that, of course, the liar sentence itself turns out to be one of the Undecideds.

I mentioned monotone inductive definition. In fact, the triple <Trues, Undecideds, Falses> is itself the limit of a monotone increasing sequence of (triples of) sets <Trues$_\alpha$, Undecideds$_\alpha$, Falses$_\alpha$>, indexed by ordinals; and this whole sequence is a precise mathematical object, as is its limit. Thus each of the sets "Trues," "Undecideds," "Falses" is itself definable (explicitly and precisely) in a strong-enough language, although not in L itself. This is why Kripke himself speaks of "the ghost of Tarski's hierarchy" as still being present in his construction.

The way in which I am going to formally model Parsons's remarks about the liar paradox is the following. To simplify matters, I will not speak of sentences as expressing propositions, as Parsons does. Instead, I will simply think of sentences as true, false, or lacking in truth-value. To say of a sentence S that it is not true, that is, to write "$\sim T(S)$," will be to say something that is (intuitively, as opposed to what happens on Kripke's scheme) *true* if S is either *definitely* false or *definitely* lacking in truth-value. In short, "$\sim T(S)$" says "S is either false or lacking in truth-value," or, in Parsons's formulation, "S does not express a true proposition."

I will assume that, as is usual in formal work, sentences are identified with their Gödel numbers, and that the language L is rich enough to do elementary number theory. Then, by a familiar diagonal lemma, given any open sentence $P(x)$ of the language, we can effectively construct an arithmetical term σ such that the numerical value of the term σ is the Gödel number of the very sentence $P(\sigma)$; that is, we have a uniform technique for constructing self-referring sentences. Because the language contains the predicate "T" for truth—that is, for truth in an interpretation (although the interpretation will be allowed to vary in the course of our discussion), and hence contains its negation "~T," we can effectively find a numerical term τ such that the numerical value of τ is the Gödel number of the very sentence

$$\sim T(\tau).$$

In loose terms, the sentence $\sim T(\tau)$ says of itself that it is not true, or in Parsons's language, that it does not express a true proposition, and this is precisely the liar paradox.

I will model Parsons's discussion as follows: Suppose that when a student, call her Alice, first thinks about the liar sentence, that is, about the sentence $\sim T(\tau)$, her first reaction is to say that this is a "meaningless" sentence, that is, not true or false; and I will also follow Parsons by supposing that she (implicitly) has an interpretation in mind. Parsons, reasonably enough, supposes that the schemes of interpretation that actual speakers have in mind are only "vaguely defined,"[11] but because I am idealizing, I will assume that Alice, implausibly of course, has in mind precisely one scheme of interpretation, and that it is given by a Kripkean construction.[12] Thus when Alice says of the liar sentence that it is neither true nor false, she means that it is one of Kripke's Undecideds. But when we ask her, following Parsons's scenario, whether it does not follow from the fact that the liar sentence is neither true nor false that *in particular it is not true,* and we bring her to say yes, then what has happened (according to Parsons's analysis—and this seems reasonable) is that she has subtly shifted her understanding (her "interpretation") of the predicate "true" ("T"). The sense in which the liar sentence ("$\sim T(\tau)$") is not

11. Parsons, "Liar Paradox," 388.

12. This may not be as implausible as it sounds. A speaker may well intend that all "paradoxical" sentences should be left un-truth-valued, and the above might be regarded as a way of rendering that inexact intention precise.

true is that it does not belong to what we might call the "positive extension" of "T" in Alice's initial (Kripkean) interpretation, that is, to the set of Trues. She has now shifted to a *bivalent* interpretation of "T" under which "$T(\sigma)$" is true (where σ is any numerical term of the language L) just in case the statement that the numerical value of σ lies in the set of Trues is a true sentence of meta-L, the language in which the Kripkean interpretation of L is explicitly defined. Parsons's own discussion ends at this point; he is content to point out that Alice need not be contradicting herself when she says that the liar sentence is not true, because the interpretation presupposed by this second remark is not the interpretation of L presupposed by her initial statement that the liar sentence is neither true nor false. But what interests me is something else—something pointed out some years ago by Professor Ulrich Blau of Munich University, who has recently published the long work on the paradoxes on which he has been working for many years.[13] What interests me is that the situation is now *unstable*.

Note first that the second interpretation—let us henceforth refer to the initial interpretation as Interpretation$_0$ and the second as Interpretation$_1$—has a paradoxical feature. For on Interpretation$_1$, $T(\tau)$ is true just in case the numerical value of the term τ lies in the set of Trues (generated by Interpretation$_0$), and it does not. Hence $T(\tau)$ is not true, and hence $\sim T(\tau)$ is true (because Interpretation$_1$ is *bivalent*). But Convention T requires that if the numerical value of any term σ is (the Gödel number of) a sentence S, then

$$T(\sigma) \Leftrightarrow S$$

is true; and σ *is* (the Gödel number of) the sentence $\sim T(\tau)$. Hence

$$T(\tau) \Leftrightarrow \sim T(\tau)$$

should be true. Of course, this failure of Convention T is not surprising, because under Interpretation$_1$ "T" does not refer to truth *under Interpretation$_1$ itself* (so Convention T does not really apply) but to truth under the initial interpretation.

The instability, of course, arises because reflection on this new interpretation will generate still another interpretation and, by iteration, an infinite series of interpretations. To spell this out: under the next interpretation,

---

13. Ulrich Blau, *Die Logik der Unbestimmtheiten und Paradoxien* (Heidelberg: Synchron, 2009).

Interpretation$_{n+1}$, $T(\tau)$ is true just in case the numerical value of the term τ lies in the set of sentences (identified, as we stipulated, with their Gödel numbers) that are true under Interpretation$_n$. Because Interpretation$_1$ is simply the bivalent interpretation of L generated by letting "T" stand for the set of Truths of Interpretation$_0$, and that set is definable in meta-L, the set of sentences that are true under Interpretation$_1$ is itself definable in meta-meta-L (or meta^2-L). As we have just seen, under Interpretation$_1$, the liar sentence is true; hence $T(\tau)$ is true under Interpretation$_2$, and hence the liar sentence is *false* under Interpretation$_2$. In short, the truth value of the liar sentence *flips* when we go from Interpretation$_n$ to Interpretation$_{n+1}$, $n > 0$. (Interpretation$_n$ is, of course, definable in meta^{n+1}-L.)

The series of interpretations can be extended into the transfinite. I will define a sentence S to be true at a limit ordinal λ if it has *become stably true* at some ordinal $< \lambda$, that is, if there is an ordinal $\kappa < \lambda$ such that S is true under Interpretation$_\gamma$ for every γ such that $\kappa < \gamma < \lambda$. Sentences that have become stably true at a stage before λ are true at λ. Similarly, sentences that have become stably false at some stage less than λ are false at λ, and sentences that have not become stably true or stably false (e.g., the liar sentence) are undecided at λ. (Limit interpretations are not bivalent.)

What I want to come to now is the point hinted at in the closing sentences (before the postscript) of Parsons's paper:[14]

> In a simple case, such as that of the word "I," we can describe a function that gives it a reference, depending on some feature of the context of utterance (the speaker). We could treat the "scheme of interpretation" in this way as argument to a function, but that, of course, is to treat it as an object, for example a set. But a discourse quantifying over *all* schemes of interpretation, if not interpreted so that it did not really capture *all*, like talk of sets interpreted over a set, would have to have its quantifiers taken more absolutely, in which case it would not be covered by any scheme of interpretation in the sense in question. We could produce a "superliar" paradox: a sentence that says of itself that it is not true under any scheme of interpretation. We would either have to prohibit semantic reflection on this discourse or extend the notion of a scheme of interpretation to cover it. The most that can be claimed for the self-applicability of our discussion is that if it is given a precise sense by one scheme of interpretation, then

---

14. Parsons, "Liar Paradox," 406.

there is *another* scheme of interpretation of our discourse which applies the discourse to itself under the *first* interpretation. But of course this remark applies to the concept "scheme of interpretation" itself. Of it one must say what Herzberger says about truth: in it "there is something schematic . . . which requires filling in."

The sequence of "schemes of interpretation" of the semantic paradoxes that I just described is a well-defined set-theoretic construction. So far, one has simply associated a scheme of interpretation with each ordinal. (Of course, if one continues it through all the ordinals, then, by cardinality considerations, at some point one will get only interpretations that are extensionally identical with ones already constructed.) But—this is the point at which Parsons, citing Herzerger, hints, and the point that Ulrich Blau emphasizes—there is still another source of paradox here. To see this source of paradox, we need to imagine a different scenario than the one Parsons imagined earlier in his paper (our scenario with Alice). There[15] Parsons imagines someone who looks at the liar sentence, decides that it is not true or false (that it is "meaningless," or, in Parsons's terminology, that it does not express a proposition), and then concludes from that very fact that it is *true* that it does not express a true proposition; and he is concerned to argue that that judgment may be totally in order provided we recognize that the scheme of interpretation has changed in the course of the reflection itself. But it seems to me unlikely that this could be the terminus of Alice's reflections. If she is sophisticated, Alice will naturally be led to investigate just the hierarchy of interpretations we constructed, the hierarchy that would result if her act of reflection were iterated through the transfinite.

At this point a new temptation may arise for Alice, the temptation to land herself in what Parsons refers to as the "superliar paradox." This need not be a temptation to suppose that one can *stand outside* the hierarchy (although one can do that, because the whole inductive definition is carried out, so far at least, within set theory); it is the temptation to suppose that even *standing within the hierarchy* (and "gazing up," as it were), one can define an *ultimate* sense of "stably true," namely, stably true with respect to the whole hierarchy, and see now that in an *ultimate* sense the liar sentence is not true (does not express a true proposition), namely, that it does not *ever* become stably true. But this, of course, will simply generate a new hierarchy.

15. Ibid., 387 and passim.

Can we go still further? It seems to me that we can. To do so in an interesting way (there are some obvious but uninteresting ways of going further), we will need to use some such phrase as "all the hierarchies one might ever arrive at by continuing reflection," and that means that we will no longer be dealing with precisely defined set-theoretic constructions. This is important because it may indicate what the answer to my question how "systematic ambiguity" is supposed to differ from just another "language" might be. When we imagine continuing reflection without limit, creating new hierarchies and then summing them up—going to the "ultimate interpretation" with respect to a hierarchy, and then taking that ultimate interpretation as the zero stage of a new hierarchy, and so on—we are no longer speaking of what is mathematically well defined. Because we are no longer in the realm of what is mathematically well defined, we cannot assume bivalence (and hence classical logic). Nevertheless, it does seem that there are things that can be seen to be true in the sense of *provable from the very description of the procedure.* (Compare, in the Tarskian hierarchy—which also can be imagined as extended without limit in similar fashion—the way in which we can see the truth of "For every language L, there is a metalanguage ML that contains a truth-predicate for L.") For example, we see from the very description of the procedure by which any hierarchy is constructed from a given initial interpretation that the liar sentence never becomes stably true. We cannot imagine an Archimedean point here. We cannot regard the vague "hierarchy of all hierarchies" as something that we can describe *from outside,* as it were. But *one can see from below* how things must go. That is, we can see that *no matter what we "get to"* in the way of reflection on the liar sentence, no matter what scheme of interpretation we arrive at, we can always use that scheme as the beginning of a new hierarchy, and we can see that, vague as the notion of a hierarchy is, at least *this* much is true of it: the liar sentence will never become stably true. In fact, using Parsons's device of systematic ambiguity, I can say things like "If the liar sentence has no truth-value at a stage, it gets one at the next stage, and if it has a truth-value at a stage, that truth-value 'flips' at the next stage." But now it seems to me that Alice may well become the victim of a supersuperliar paradox. The temptation now will be to think something like this: "When we talk of all the hierarchies of interpretation we could produce, I know that we are not talking about something precise and well defined, but nevertheless, as you have just shown, there is a sense—a *last sense*—in which the liar sentence is not true: namely, it does not become stably true in *any* hierarchy, not even in, so to speak, the hierarchy of all hierarchies. But surely being eventually stably true in the hierarchy of all hierarchies is the last sense of being stably true, and so there is an absolute sense, namely, the *last sense*, in which the liar sentence is not true." At this point, of course, she will generate yet another interpretation—an ill-defined one, of course, but nonetheless an interpretation that can also be used as the basis of a hierarchy (even if we have to use intuitionist logic rather than classical logic to talk about it, in view of the fact that the only notion of "truth" we appear to have in connection with it is some species of provability).

In short, the final temptation is the temptation to suppose that the notion of a *last* scheme of interpretation makes sense. What Parsons says, using a term of Herzberger's, in a sentence I quoted a little while ago seems to be exactly right, namely, that when we talk of hierarchies in general, rather than of a specific hierarchy constructed in a specific set-theoretic way, we are necessarily talking "schematically"; and the schematic character of such talk is, it seems to me, just the difference between talking with "systematic ambiguity" and merely using "red-ink language."

But there is a further point I want to make. It is indeed possible to think of an interpretation of the language L according to which (1) sentences that are provably unstable with respect to *every* hierarchy are "undecided," (2) sentences that will eventually receive the value "true" in any hierarchy that goes far enough are true, and (3) sentences such that we can prove that they will eventually become false in any hierarchy that goes far enough are false. (Because provability is not the same as classical truth, there will, however, be sentences such that we cannot say that they are true, false, *or* undecided according to such an interpretation. As I already mentioned, there are logics that do not assume bivalence—for example, intuitionist logic—that one might employ in this connection; but it is not my purpose here to attempt a formal treatment.) But if it is all right or possibly all right to treat the hierarchy of all hierarchies as something we can reason about, at least in an intuitionist setting, then the mistake that Alice would be making if she gave in to the temptation that leads to what I called the supersuperliar paradox would not be that what I imagined her calling the last interpretation does not *exist*. The mistake is more subtle, it seems to me. (This is a point that I first learned from Ulrich Blau when I attended his seminar in Munich in the 1980s.) The mistake, rather, lies in thinking of it as the "last." The phrase "last interpretation" assumes that limits have some kind of finality. But if we allow talk of the last interpretation, we must also allow that there is a *successor* to the last interpretation. That is, it is quite true that the liar sentence is undecided in the so-called last interpretation, but it is equally true that it becomes true again just after the "last" interpretation. In short, the phrase "last interpretation" is a misnomer. The illusion is that by this very act of "looking up from below" at what happens in our hierarchies, we can somehow generate an absolute sense of a "last interpretation," and the paradox itself shows this to be an illusion. Our desire to have a *final thing we can say about the liar sentence*, or an absolutely best thing to say about the liar sentence, is what always causes the liar sentence to spring back to life from the ashes of our previous reflections.

I am led back, in a way, to my own rejected solution in the unpublished James Thompson Memorial Lecture. If you want to say something about the liar sentence, in the sense of being able to give final answers to the questions "Is it meaningful or not? And if it is meaningful, is it true or false? Does it express a proposition or not? Does it have a truth-value or not? And which one?" then you will always fail. And the paradox itself shows why this desire to be able to say one of these things must always fail. In closing, let me say that even if Tarski was wrong (as I believe he was) in supposing that ordinary language is a theory and hence can be described as "consistent" or "inconsistent," and even if Kripke and others have shown that it is possible to construct languages that contain their own truth-predicates, still, the fact remains that the totality of our desires with respect to how a truth-predicate *should* behave in a semantically closed language, in particular, our desire to be able to say without paradox of an arbitrary sentence in such a language that it is true, or that it is false, or that it is neither true nor false, *cannot* be adequately satisfied. The very act of interpreting a language that contains a liar sentence creates a hierarchy of interpretations, and the reflection that this generates does not terminate in an answer to the questions "Is the liar sentence meaningful or meaningless, and if it is meaningful, is it true or false?" On the other hand, Tarski's own suggestion of giving up on unrestricted truth-predicates and contenting ourselves with hierarchies of stronger and stronger languages, each with its own truth-predicate, creates much the same situation as does Parsons's hierarchy of interpretations of a single language. In the end we are led to see that the things we say about formal languages must be (to use Herzberger's term) "schematic."

11

Set Theory: Realism, Replacement, and Modality

In 1995 I delivered a pair of Alfred Tarski Lectures at Berkeley, under the titles "Paradox Revisited I: Truth" and "Paradox Revisited II: Sets—A Case of All or None?"[1] The first of those lectures (Chapter 10 in the present volume, retitled "Visiting the Liar Paradox"), on the semantic paradoxes, I am still satisfied with. However, the second, on the set-theoretic paradoxes, I am now dissatisfied with, and the present chapter will tell you what I *wish* I had said in that lecture.

Gödel's Views

What concerned me in that lecture, and what will concern me now, is the attitude toward the paradoxes expressed by Gödel in two famous papers. Gödel felt that we do not yet have a completely satisfactory view with respect to the "intensional" paradoxes (i.e., the liar paradox and the other "semantic"

1. Hilary Putnam, "Paradox Revisited I: Truth" and "Paradox Revisited II: Sets—A Case of All or None?" in Gila Sher and Richard Tieszen, eds., *Between Logic and Intuition: Essays in Honor of Charles Parsons* (Cambridge: Cambridge University Press, 2000), 3–15 and 16–26.

paradoxes), but that we do now have a satisfactory view of the nature of sets, and not just a way of pushing such paradoxes as Russell's out of the formal language.

What bothered me about Gödel's position was that, as I understand it, it presupposes quantification over absolutely *all* sets, and in my second Alfred Tarski Lecture I regarded such quantification as problematic in much the way in which quantification over absolutely all "levels of language" is problematic. If I have changed my mind about that, it is not, I hasten to add, because I now believe that there is a sort of Platonic universe in which all the sets are already laid up (this is the view with which, rightly or wrongly—this is not a lecture about Gödel interpretation!—I saddled Gödel in my Tarski Lecture). It is because I forgot a way to make sense of that kind of quantification that I myself proposed a quarter century earlier that does not presuppose that Platonic view.

On the other hand, I strongly agree with Gödel that we do not have a completely satisfactory view with respect to the liar paradox (indeed, it may be intrinsic to the nature of the case that there is no such thing as "complete satisfaction" to be had with respect to the liar paradox). I argued in my first Alfred Tarski Lecture that the hierarchy of metalanguages and the hierarchy of interpretations of the uses of "true" in natural language are intrinsically puzzling hierarchies—they are "schematic" (to use a term of Hans Herzberger's) and not perspicuously surveyable either from above or below. In the end they are, as we say, "paradoxical" even if they are not formally contradictory. All this I still believe. Thus the present rethinking of my second Alfred Tarski Lecture will, in the end, express agreement with Gödel on the very point with which that lecture expressed disagreement, even if I come to that agreement via a different set of philosophical arguments.

Realism

In the Tarksi Lecture that this chapter is meant to replace, I also pooh-poohed the philosophical question of *realism*. There I wrote:

> Philosophers typically take the hard philosophical questions to be whether sentences about "abstract objects" (a notion which lumps together numbers, lines and circles in Geometry, functions in Analysis, and Cantorian sets under the supposedly univocal label "object") are true or false even

when undecidable, and if so, what "makes them" one or the other. My view with respect to that question is that it is not really a question we understand. But I mention that issue only to set it aside for now. As far as I am concerned, *of course* statements of number theory are true or false— what is the third possibility? Speaking mathematically, there is none, and the supposed philosophical sense of the question is, I think, chimerical.[2]

What is unfortunate about that paragraph is that the view I have held at least since I wrote "What Is Mathematical Truth?" (published in 1975), and that I have not given up, is that *the internal success and coherence of mathematics* are evidence that it is true under some interpretation, and that its *indispensability for physics* is evidence that it is true under a *realist* interpretation (the antirealist interpretation I considered there was intuitionism).

For the record, let me emphasize that (1) I certainly take the question whether particular scientific theories should or should not be interpreted realistically to be a meaningful and moreover an extremely important question, both for philosophy and for the sciences in question, and (2) as I just said, I regard the indispensability of mathematics for physics as a powerful argument for interpreting mathematics "realistically," in the sense of rejecting all attempts to account for the meaning and the objectivity of mathematical claims in a verificationist way.

I have to admit that I simply cannot account for my failure even to mention my own scientific realist view in the Alfred Tarski Lecture. Did I, perhaps, momentarily succumb to a "Wittgensteinian" tendency to dismiss such metaphysical questions as nonsense? But that would be peculiar, because in a footnote to the very paragraph in which I described the question as "not really one we *understand*," I cite a paper of my own in which I criticize a remark of Wittgenstein's that I took as showing that he sometimes felt the temptation to give a verificationist account of mathematical truth, and in which I appeal to precisely the way in which that verificationist account is incompatible with scientific realism as grounds for rejecting it. This is the remark in question (written in 1944):

Suppose that people go on and on calculating the expansion of π. So God, who knows everything, knows whether they will have reached "777" by the end of the world. But can his *omniscience* decide whether they *would*

2. Putnam, "Paradox Revisited II," 18–19.

have reached it after the end of the world? It cannot. I want to say: Even God can determine something mathematical only by mathematics. Even for him the rule of expansion cannot decide anything that it does not decide for us.[3]

Obviously, human beings cannot possibly calculate at all "after the end of the world"—especially if, as is reasonable to suppose, the "end of the world" means the end of space and time.[4] The only reasonable interpretation is that Wittgenstein meant us to suppose that human beings go on calculating to the very limit of what is possible for such beings. That is, contrary to Michael Dummett's well-known interpretation, Wittgenstein is not here saying that not even God could know what human beings would count as a correct calculation before their actually accepting it as correct (which would involve attributing a strongly "antirealist" attitude toward counterfactuals to Wittgenstein, on no evidence that I can see); but he is explicitly saying that not even God can decide whether the pattern does or does not occur in the expansion of π except by a calculation that is actually—not just "mathematically"—possible for human beings.

Where Wittgenstein Went Wrong

When we make the statement that a physical system obeys a certain equation (this is an example of a "mixed" statement, an empirical statement that contains mathematical terms), for example, when we say that the state vector of a physical system obeys the Dirac equation, or, in Newtonian physics, when we say that gravitational forces obey the Newtonian law of gravitation, or even

---

3. Wittgenstein, *Remarks on the Foundations of Mathematics,* ed. G. H. von Wright, Rush Rhees, and G. E. M. Anscombe (Oxford: Blackwell, 1978), V, §34. It is important to contrast what Wittgenstein published on this question, in §516 of *Philosophical Investigations* (Oxford: Blackwell, 1953), with this *unpublished* paragraph in *Remarks on the Foundations of Mathematics.* The passage in *Philosophical Investigations.* §516, reads, "It seems clear that we understand the meaning of the question: 'does the sequence 7777 occur in the development of p?' It is an English sentence; it can be shown what it means for 415 to occur in the development of p; and similar things. Well. Our understanding of that question reaches just so far, one may say, as such explanations reach." There is nothing here about what God can or cannot determine!

4. The passage could, of course, be read in another way: there is a Last Judgment in time, and that is "the end of the world," and there is still more time available for calculation after the Last Judgment—but then why should not human beings go on calculating in eternity?

when we say that a certain phenomenon obeys the wave equation, what is the situation? As long as we accept the correctness of Newton's law of gravitation, we are committed to the statement that the evolution of an N-body system will be in accordance with the solutions to the appropriate system of differential equations. In general, any application of mathematical physics to a physical system involves treating the system as behaving in accordance with some equations—usually differential equations—or others. If a physicist believes that the equations are correct, and furthermore that they describe the behavior of the system at each time point, then she is committed to the claim that those equations have solutions (in real numbers or, in certain cases, in complex numbers) for each real value of the time parameter t.

Now suppose that the solution to the equations, in a particular physically given case, for a particular rational value of t is in a certain rational interval, say, between r_1 and r_2, where r_1 and r_2 are rational numbers. That is, suppose that the statement "the solution to such and such equations for the given value of t is in the interval between r_1 and r_2" is *true*.[5] If truth is the same as provability (in the empirical sense of "provability"), then what the physicist is committed to, if she believes the view that Wittgenstein flirted with in *Remarks on the Foundations of Mathematics*, V, §34,[6] is that it is *possible to calculate the solution to the equations in question for the specified value of t, and that the solution will be found to lie in the interval between r_1 and r_2*. But, given present knowledge, this is something a sensible physicist *better not* commit herself to!

Because this point is at the heart of my argument, let me state it again for clarity: if truth is coextensive (in the mathematical case) with the *empirical* possibility of being proved by a calculation or a proof from axioms, then no mathematical statement can be true but not demonstrable by calculation or proof. Not even "God's omniscience" can know the truth-value of such a humanly undecidable statement, which is to say that it does not have a truth value. If the statement is that the equations of motion of a system S have a solution (say, "$P(t)$ is in the interval between 3.2598 and 3.2599," where P is some physical parameter), then *if it is not physically possible for human beings to compute $P(t)$, or to prove by some deduction from acceptable axioms that $P(t)$*

---

5. I have confined attention to rational values of t, r_1, and r_2 so as to avoid the objection that not all real numbers have names in the language.

6. See *Remarks*, chapter 25.

is or is not in the interval between 3.2598 and 3.2599, then there is no fact of the matter about whether $P(t)$ is in that interval or not. But to accept this is precisely to be a verificationist in one's physics. It is to give up a claim that is part of our best physical theory of the world, the claim that the equations of that theory describe the behavior of certain systems, in the sense that those equations have *solutions* for each real value of the time parameter t, and those solutions give the value of the physical parameter P in question (or do so to a high degree of accuracy) regardless of whether it is feasible to verify that they do by a calculation it is physically possible for human beings to make. Systems of equations are, on a verificationist view, just prediction devices, and when it is not feasible to derive a prediction from them (even if we are allowed to go on calculating until "the end of the world"), then there is nothing that they *say* about the case in question.[7]

My Indispensability Argument

What I have just rehearsed is, of course, my "indispensability argument" for realism with respect to mathematical truth. And because Gödel was also a realist with respect to mathematical truth, this is another point of convergence between my views and Gödel's, although the arguments are, as I mentioned, very different. But there are a number of misunderstandings of my indispensability argument floating around in the literature that I would caution you against.[8]

First, I do not intend the argument as an epistemology of mathematics. My argument does not address the question, "How do we know that the axi-

---

7. But why do I say that the physicist "better not commit herself to" the claim that it is physically possible to calculate the solution to the differential equations of physics for arbitrary specified values of t? One important reason has to do with chaos (a phenomenon represented mathematically by certain kinds of differential equations): when a phenomenon is sufficiently "chaotic," it may be "empirically" (though not "mathematically") impossible to actually calculate the time evolution of certain parameters even though we know the equations they obey. (A possible additional reason has to do with nonrecursiveness; for example, it is known that there are cases in which it has been proved that the solutions to the wave equation are not recursively calculable, even given recursive initial data.) In sum, short of being a verificationist about physics, one cannot consistently sustain the identification of mathematical correctness with provability and with calculation that seems to be asserted in *Remarks on the Foundations of Mathematics*, V, §34.

8. See Chapter 9 in this volume.

oms of mathematics are true?" It addresses the quite different question, "Is it consistent to be a verificationist with respect to mathematical discourse and a realist with respect to the discourse of physics?" And it argues that the answer is in the negative. Second, and this is something that will play a role later in this chapter, the argument does not assume that mathematics *must* be interpreted as about "immaterial objects"; it holds just as well if mathematics is interpreted "modally," as concerning what *could* and what *could not* be the case (in a sense of "could" peculiar to mathematics), rather than as concerning what is actually the case. Thus both so-called structuralists and so-called Platonists are free to make use of my argument, if they wish, against verificationists of all stripes.

As I said in Chapter 9, my "indispensability" argument is an argument for the objectivity of mathematics in a realist sense (i.e., for the idea that mathematical truth must not be identified with provability), while Quine's indispensability argument (although also based on scientific realism with respect to natural science) was an argument for "reluctant Platonism," which Quine characterized as accepting the existence of "intangible objects,"[9] that is, numbers and sets.

A common objection to arguments from indispensability for physics to realism with respect to mathematics is, of course, that we do not yet have, and may indeed never have, the "true" physical theory; my response (which I gave in Chapter 9) is that, at least when it comes to the theories that scientists regard as most fundamental (today that would certainly include quantum field theories), we should regard all the rival theories as candidates for truth or approximate truth, and that *any philosophy of mathematics that would be inconsistent with so regarding them should be rejected.*

The "Modal-Logical Interpretation"

In "Mathematics without Foundations," a paper I published in the *Journal of Philosophy* in 1967, I proposed an interpretation of mathematics (and particularly of set theory) that I called "mathematics as modal logic." (In recent decades the modal interpretation has been developed in detail by Geoffrey

---

9. E.g., Quine wrote "[The words 'five' and 'twelve'] name two intangible objects, numbers, which are *sizes of* sets of apples and the like." W. V. Quine, *Theories and Things* (Cambridge, Mass.: Harvard University Press, 1990), 149.

Hellman.)[10] In "Mathematics without Foundations" I explained the idea thus:[11]

> My purpose is not to start a *new* school in the foundations of mathematics (say, "modalism"). Even if in some contexts the modal-logical picture is more helpful than the mathematical-objects picture, in other contexts the reverse is the case. Looking at things from the standpoint of many different "equivalent descriptions," considering what is suggested by *all* the pictures, is both a healthy antidote to foundationalism and of real heuristic value in the study of scientific questions.
>
> Now, the natural way to interpret set-theoretic statements in the modal-logical language is to interpret them as statements of what would be the case if there were standard models for the set theories in question. Because the models for Zermelo-Fraenkel set theory and its strengthenings are also models for Zermelo set theory, let me concentrate on Zermelo set theory. In order to "concretize" the notion of a model, let us think of the model as a graph. The "sets" of the model will then be pencil points (or some higher-dimensional analogue of pencil points, in the case of models of large cardinality), and the relation of membership will be indicated by "arrows." (I assume that there is nothing inconceivable in the idea of a space of arbitrarily high cardinality,[12] so models of this kind need not be denumerable and may even be standard.) The model will be called a "standard concrete model" if (1) there are no infinite-descending "arrow" paths; and (2) it is not possible to extend the model by adding more "sets" without adding some "ranks" in the model.[13] If we regard the "points" and "arrows" of the

10. Geoffrey Hellman, *Mathematics without Numbers: Towards a Modal-Structural Interpretation* (Oxford: Oxford University Press, 1989).

11. Hilary Putnam, "Mathematics without Foundations," *Journal of Philosophy* 64, no. 1 (1967): 5–22; reprinted in Philosophical Papers, vol. 1, *Mathematics, Matter and Method* (Cambridge: Cambridge University Press, 1975), 43–59; quotation from p. 57. I have abbreviated the passage and also slightly revised the wording.

12. In the original I had "nothing inconceivable in the idea of a physical space of arbitrarily high cardinality." "Physical" was a blunder. Nothing hangs on whether such spaces could or could not be "physical"; what matters is only their mathematical possibility.

13. In Zermelo-Fraenkel set theory and its successors, the ranks are defined thus: the collection of all hereditarily finite sets (the finite sets all of whose members, and members of members. and members of members of members, and so on are also finite) are rank 0, and the set of all subsets of a given rank κ (its "power set") is rank $\kappa+1$. When λ is a limit ordinal, rank λ is defined to be the union of all the ranks below λ. Thus ranks, unlike Russellian "types," are *cumulative*; the sets of a given rank also belong to all higher ranks.

graph as possible individuals, then the statement that a graph G is a standard concrete model for Zermelo set theory can be expressed using no "nonnominalistic" notions except the "$\square$" (the modal-logical symbol for "it is necessarily the case that").

If S is a statement of bounded rank, and if we can characterize the "given rank" in question in some invariant way (invariant with respect to standard models of Zermelo set theory), then the statement S can easily be translated into modal-logical language. The translation is just the statement that if G is any standard model for Zermelo set theory and G contains the invariantly characterized rank in question, then necessarily S holds in G. (It is trivial to express "S holds in G" for any *particular* S without employing the set-theoretic notion of "holding.") Our problem, then, is how to translate statements of *un*bounded rank into modal-logical language.

The method is best indicated by means of an example. If the statement has the form $(x)(\exists y)(z)\mathrm{M}xyz$, where M is quantifier free, then the translation is this:

Necessarily: If G is any standard concrete model for Zermelo set theory and if x is any point in G, then it is possible[14] that there is a graph G' that extends G and is a standard concrete model for Zermelo set theory and a point y in G' such that $\square$ (if G'' is any standard concrete model for Zermelo set theory that extends G' and z is any point in G'', then Mxyz holds in G'').

I do not propose the modal-logical interpretation as a step to arguing that numbers do not really exist, as the title of Hellman's book *Mathematics without Numbers* unfortunately suggests. The idea that we are saying something false when we say things like "There is always a prime between n and $2n$" seems obviously wrong to me. And I have argued elsewhere that the idea that using quantifiers always "commits us" to the existence of *objects* is at best unclear and at worst hopelessly misleading. Nor, by the way, do I think that the modal-logical interpretation yields an "epistemology" for mathematics. On that (the epistemology of mathematics), I agree with a view that John Burgess has expressed, that the best way to find out how mathematical knowledge is obtained is to look at what mathematicians do. I suspect that Burgess is also right in thinking that what we will find will not fit any epistemological picture so far proposed. So why do I like the modal-logical interpretation?

14. The modal-logical notion "it is possible that p" (symbolized "$\lozenge p$") is defined as "$\sim\square\sim p$."

One thing I like about it is that it yields a natural resolution of Benacerraf's famous problem about "what numbers could not be"[15] and its generalizations. Benacerraf's problem is that although the natural numbers can, as is well known, be identified with sets—for example, with the von Neumann ordinals, $\varnothing$, $\{\varnothing\}$, $\{\varnothing, \{\varnothing\}\}$, $\{\varnothing, \{\varnothing\}, \{\varnothing, \{\varnothing\}\}\}$, . . . —they can be identified with sets in *infinitely many ways*. (For example, they could also be identified with the Hao Wang ordinals, $\varnothing$, $\{\varnothing\}$, $\{\{\varnothing\}\}$, $\{\{\{\varnothing\}\}\}$, and so on.) And to stamp one's feet and insist that "the natural numbers are not identical with sets at all" seems a bit arbitrary—in set theory we do often identify them with progressions of sets, after all, from the first days of modern mathematical logic.

Shall we then say that quantification over natural numbers is a sort of deliberately *ambiguous* quantification over the successive elements of any infinite series of sets (any "ω-sequence") you like? The problem, as I am sure Benacerraf well knows, is that the same problem arises with quantification over sets. Sets can, after all, be identified with "characteristic functions" (functions whose range is $\{0, 1\}$), as is standardly done in a good deal of recursion theory and hierarchy theory, for example. On the other hand, functions can be identified with ordered n-tuples. And ordered n-tuples can be identified with sets in infinitely many different ways. Can there really be a "fact of the matter" about whether sets are a kind of function or functions are a kind of set? Can there really be a "fact of the matter" about what the "correct" definition of "ordered pair" is? The mind boggles.

If, however, quantification over sets, functions, and so on is simply quantification over possibilia and not over actually existing entities, then the problem disappears in the sense that all the different "translations" of number theory into set theory, all the different translations of set theory into function theory, and all the different translations of function theory into set theory are just different ways of showing what structures have to *possibly exist* in order for our mathematical assertions to be true. In my view, then, what the modal-logical "translation" of a mathematical statement gives us is a statement with the same mathematical content that does not have even the appearance of being about the actual existence of "immaterial objects." With all this as background, I now turn to set theory.

15. Paul Benacerraf, "What Numbers Could Not Be," *The Philosophical Review* 74 (1965): 47–73.

The Concept of a "Set"

According to Gödel, there are two principal notions of a "set," namely, "set as the extension of a predicate" and "set as an arbitrary collection of things,"[16] where the "things" may be sets of some rank. Standard Zermelo-Fraenkel set theory is the theory of sets conceived according to the second notion, "arbitrary" sets, provided the operation of taking the arbitrary collections of things in all the ranks so far obtained to be sets in a new rank is iterated through all the ordinals.

Gödel's view, that there is no problem with thinking of all the sets (on this "iterative" conception) as forming a definite totality faces a number of problems, however, which is why (in the Alfred Tarski Lecture that is the direct ancestor of the present lecture) I was led to the view that quantification over all sets is problematic in the same way in which quantification over absolutely all "levels of language" is problematic.

Here are two of the problems: (1) "All the ordinals" means "all the ordinals that are order-types of well-orderings that are *sets*" (the ordering of all the ordinals is an example of a well-ordering that is definable in set theory but whose order-type is not an "ordinal"). Thus there seems to be an obvious *circularity* in explaining "set" in terms of "ordinal." (2) We still have to make clear what "arbitrary" means.

In my second Alfred Tarski Lecture I suggested that this second problem might be solved by employing a notion due to the late George Boolos, the notion of "plural quantification."[17]

Boolos was concerned with a well-known problem concerning the notion of validity in first-order logic.[18] On the one hand, the quantifiers in first-order logic (for simplicity, I shall confine attention to monadic quantifiers) are supposed to range over arbitrary bunches of individuals (in an arbitrary universe), and that has to mean arbitrary subsets of the universe, does it not? But first-order logic can be used to reason about totalities that are not sets, for

16. For a fuller account of these notions, see Putnam, "Paradox Revisited II."

17. George Boolos, "To Be Is to Be the Value of a Variable (or to Be Some Values of Some Variables)," *Journal of Philosophy* 81 (1984): 430–449.

18. George Boolos, *Logic, Logic, and Logic,* ed. Richard Jeffrey and John Burgess (Cambridge, Mass.: Harvard University Press, 1998), 10, reports "an old observation of Georg Kreisel" that "the standard definition of validity considers only interpretations in which the quantifiers range over some *set* of things."

example, the universe of set theory itself. And the sub*sets* of that universe are not *all* the "bunches of individuals" in *that* universe. They are not even all the *extensions of predicates*. (For example, the von Neumann ordinals are the extension of a predicate of set theory, but not a set.) So the standard model-theoretic definition of validity as validity in all models (in all *sets* with a certain structure) badly fails to capture the intended notion of validity, which is truth in *all* interpretations, including interpretations in which the extensions of some of the predicates are not sets.

Boolos's view was that there is a natural way of understanding second-order logic that does not involve the notion of a set at all. He employs, in fact, a linguistic device found in all natural languages—one as intuitive and perhaps even more primitive than talk of the numbers—namely, "plural quantification."

Rather than try to define plural quantification, I will give an example:

(1) Some boys at the party talked only to one another.

Instead of hearing this as a disguised statement to the effect that

(2) there is an object *x* (a "set") whose "members" are all boys at the party, and such that no "member" of *x* spoke to anyone at the party who was not a "member" of *x*.

(Note that this reading of (1) makes it part of the content of (1) that the boys form a set), Boolos suggests that we hear (1) as related to the quantifier in

(3) There was a boy at the party who talked only to himself (which is expressible in first-order logic) as plural to singular.

Given a formula of first-order logic, we can say that it is valid without restricting the admissible universes of discourse to sets by using an appropriate formula of second-order logic and interpreting the quantifiers as plural quantifiers over the individuals in the universe in question. For example, to say that

$$(4) \qquad\qquad\qquad (x)Fx \Rightarrow Fy$$

is valid is to say (in second-order language) that

$$(5) \qquad\qquad (U)(y)_U\,(F)[(x)(Ux \Rightarrow Fx) \Rightarrow Fy],^{[19]}$$

and, interpreted à la Boolos, what that means is that for any universe *U,* for any *y* in *U,* and for any *F*s, if *x* is in *U,* and if everything in *U* is one of the *F*s, then *y* is an *F,* or in natural language, "For any things you please, and for any one of those things you please, and for any other things you please, if all of the

---

19. Actually, validity is better expressed in second-order logic with *modality,* thus: "$\Box\,(U)(y)_U\,(F)[(x)(Ux \Rightarrow Fx) \Rightarrow Fy]$" expresses the validity of "$(x)Fx \Rightarrow Fy.$"

first things are among the other things, then that one is one of the other things as well."

This says that (4) is valid over all universes, including the universe of "all sets" (if there is/were such a universe), provided we do not construe the plural quantifiers as set quantifiers. Of course, this is only an illustration of Boolos's idea. His famous claim was that all second-order quantification can and should be interpreted in this way, and that if it is so interpreted, there is no reason not to view second-order logic as "logic." But I made a somewhat different application.

What I said was that Boolos's observations show that by virtue of our possession of the notion "some *As*," we do have a way of speaking of arbitrary bunches of objects (at least we have a way of doing this whenever a notion of an "object" is already in place). However, that notion does not, of itself, automatically generate a new notion of an "object." I claimed that we may usefully think of the formation of the notion of an "arbitrary set" of objects in *U* (where *U* is some already-given universe of discourse) as split into two stages: (1) extending plural quantification to the universe *U*; and (2) "reifying" bunches of *U*s by taking them to be "objects." If, as is often the case, forming *U* already involved an extension of the notion of an "object," then, I said, this second stage will constitute a further extension of the notion, and it is this extension that constitutes the notion of an arbitrary set, just as the reification of extensions of predicates generated the Fregean and Russellian notions of a class. And like those extensions, this extension can be iterated indefinitely, and the result will again be a hierarchical structure. Indeed, if the iteration is just through the natural numbers, it will be the hierarchy of the simple theory of types (with the members of the original universe *U* as the individuals). Of course, one can also iterate through any infinite well-ordering that becomes available at any stage of the construction, and in this way (taking the types as cumulative) one can get models for Zermelo set theory.

What of Zermelo-Fraenkel set theory? Here I "boggled." What I wrote was:

What of Zermelo-Fraenkel set theory? *Quite frankly, I see no intuitive basis at all for the characteristic axiom of that theory, the axiom of Replacement.* Better put: I don't see that a *notion* of a set on which that axiom is clearly true has ever been explained. Instead, it seems to me, we have a formal maneuver which Gödel tried to justify as news from Mount Olympus.[20]

20. Putnam, "Paradox Revisited II," 24.

In addition to wanting to apologize to Gödel (even if he is no longer alive, his memory is certainly very much alive) for the appallingly bad taste of that last remark, I want now to explain *why* I was skeptical about the axiom of replacement, and how I came to overcome that skepticism.

The Axiom of Replacement

I said that the notion of an "arbitrary" set might be explained in terms of plural quantification. If that is right, and it still seems to me to be right, then there is a problem with the idea of a totality of actually existing objects that is "the universe of iterative set theory," one also noted by Boolos in another famous paper.[21] The problem is that any bunch of actually existing objects, including sets, if *they* are actually existing objects, is a "value" of a plural quantifier: we can talk about that bunch by saying "some objects" ("some sets"). But then it is arbitrary (in the pejorative sense of "arbitrary") to say that we cannot speak of *all* the arbitrary collections of sets. But if we can, then we can *extend* the universe of sets still further. (To say that you can extend the universe of iterative set theory just once, but we call the resulting objects "classes" and not "sets," is an idea that Boolos rightly dismissed as a desperate artifice.)

The moral is that any "universe of sets"—or, to speak in terms of the modal-logical interpretation, any graph G that is a standard concrete model for Zermelo set theory in any "possible world"—must be *extendable*. And this I still believe.

Why did this lead me to doubt the axiom of replacement? First, let us recall what it is. Replacement, in its first-order form, is actually a schema. Specifically, suppose that P is any formula with two free variables x and y.[22] Then an instance of replacement states that *if* for every set x there is a unique set y such that $P(x, y)$, *then* given any set A, there is a set B whose members are precisely those sets y such that $P(x, y)$ for some x in A. Note that there is one such axiom for every predicate P.

The intuition behind replacement is often described as "limitation of size." The idea is that if a collection of sets B is "small enough to be a set"—that is,

21. George Boolos, "Iteration Again," *Philosophical Topics* 17 (1989): 5–21; reprinted in Boolos, *Logic, Logic, and Logic* (Cambridge, Mass.: Harvard University Press, 1998): 88–104.

22. If P contains additional free variables, then the instance of the axiom schema begins with a string of universal quantifiers binding these variables.

of the same cardinality as some set A—then B is also a set, or, by contraposition, a collection B can fail to be a "set" only by being "too big." To see how a model can satisfy the axioms of Zermelo's theory without satisfying the replacement schema (added by Fraenkel in 1921), consider the smallest standard model for Zermelo set theory that has a countable number of infinite ranks. If we assume the generalized continuum hypothesis, the cardinality of this model is $\aleph_\omega$, and every set in the model has a cardinality smaller than $\aleph_\omega$. But the model does not satisfy replacement, because the two-place predicate "x is a natural number, and $y = \aleph_x$" is definable in the language of Zermelo set theory, and the positive integers form a set in the model, but the collection $\{\aleph_0, \aleph_1, \aleph_2, \ldots\}$ does not. What suffices for replacement to hold in a standard model for Zermelo set theory is that the model be such that every "small" set of ordinals (every set of ordinals of the same cardinality as a set in the model) has a least upper bound that is a set in the model. What occasioned my doubts was the question, why should we suppose that there is such a standard model, or that such a standard model is *possible,* to speak in terms of the modal-logical interpretation?

The example just given, in which the predicate $P(x, y)$ is taken to be "x is a natural number, and $y = \aleph_x$" (or the formula that "translates" that English phrase into the notation of set theory), does not bring out the difficulty, because that predicate can be expressed in a way that is absolute with respect to models of Zermelo set theory that contain all the ranks less than rank $\aleph_\omega$. Because of this absoluteness (or invariance of the extension of the predicate), the above instance of replacement, which failed if the model contains only the ranks less than rank $\aleph_\omega$, can be guaranteed to hold by making sure that the model contains rank $\aleph_\omega$ itself. But full replacement requires that all instances of replacement hold, including instances in which the formula $P(x, y)$ may change its extension when the standard model being used to interpret it is extended, no matter how "big" that standard model may be.

The usual way to get around this is to postulate that there are "large cardinals," cardinals Θ with properties that guarantee that rank Θ will satisfy replacement. (The assumption that Θ is strongly inaccessible is the most familiar such postulate, although when set theorists today speak of "large cardinals," they usually have in mind axioms that postulate the existence of cardinals much "bigger" than that.) But why should one believe that there are standard models of Zermelo set theory that contain such cardinals (or, in modal-logical terms, that it is possible ["$\Diamond$"] that there is a graph G that is a standard concrete model of Zermelo set theory containing a strongly inaccessible cardinal)? The problem is not that we cannot *prove* this, of course; I did not even see how to make replacement intuitive.

What I would now say is that I was looking in the wrong place for an intuitive argument. Although there may be such intuitive arguments (in fact, I now think that there are), it is not necessary that we convince ourselves that there is (or possibly ["◊"] is) such a graph *G*. For, as we saw earlier—as I had already seen by 1967—*one can make sense of quantification over absolutely all sets even if one thinks (as I argued we should) that it is not possible for a world to contain a G* that is a standard concrete model for Zermelo set theory that cannot be extended to still higher ranks. I repeat the passage:

> Our problem, then, is how to translate statements of *un*bounded rank into modal-logical language.
>
> The method is best indicated by means of an example. If the statement has the form $(x)(\exists y)(z)\mathrm{M}xyz$, where M is quantifier-free, then the translation is this:
>
> Necessarily: If *G* is any standard concrete model for Zermelo set theory and if *x* is any point in *G*, then it is possible that there is a graph *G'* that extends G and is a standard concrete model for Zermelo set theory and a point *y* in *G'* such that □ (if *G''* is any standard concrete model for Zermelo set theory that extends *G'* and *z* is any point in *G''*, then M*xyz* holds in *G''*).

This means that the intuition behind replacement need only be that no "small" set of ordinals—no set of ordinals the same size as any collection of things that could exist in one and the same "possible world"—can be cofinal with *all possible ordinals*. I do not claim that this is self-evident, of course (self-evidence is not to be hoped for in set theory, it seems), but it does seem to me a compelling intuition in much the way in which the other axioms of set theory seem compelling. Thus it seems to me that an advantage of the modal-logical interpretation is that in addition to providing a nice way of resolving Benacerraf's problem and its generalizations, it provides a more intuitive way of thinking about replacement.

Intuitiveness is not everything, of course. In set theory we also ask that axioms pay their way, and replacement has richly paid its way. Indeed, the formalization of modern mathematics within set theory—one of the most important aims of set theory itself, from the time of Zermelo—would have been hopelessly difficult, if not impossible, without it. But one does ask for *some* intuitiveness, after all.

What of the "large cardinal axioms" that set theorists are beginning to investigate? Here too, I think that the attitude I expressed in the ancestor to this lecture was way too dismissive. Here I was in the grip of the picture that axioms of set theory ought to be as "self-evident" as the axioms of number theory. Now it seems obvious to me that in set theory the best we can hope for is a mixture of *some* intuitiveness and a great deal of mathematical fruitfulness. I now find Hugh Woodin's program in this area extremely exciting,[23] but I cannot discuss it further in the present context.

Concluding Remark

That last sentence would be a nice place to close, but I fear that there is one objection to the realism that I am defending about which I must say a word. That is the objection that even if the "indispensability" argument that I deployed shows that we should be realists about *some* mathematics, it does not show that we should be realists about *all* mathematics, and particularly not about replacement, large cardinals, and the like. I will make only three brief remarks about that objection.

First, I would like to know what part of mathematics one should *not* be realistic about. If someone proposes that we confine ourselves to predicative mathematics, for example, then I have to just say that I have argued elsewhere[24] that the world picture of modern physics is not one that I know how to express in that language, and for a realist the question is not only how much mathematics the *calculations* of physics require, but how much mathematics we need to explain what physics is saying. But I cannot review the details of that argument now.

Second, if someone proposes that we confine ourselves to analysis (integers and sets of integers), then, as I explained in Chapter 9 I do not believe that "set" is a notion that can be interpreted in a realist way up to a certain rank α and in a nonrealist way at rank $\alpha+1$.

Third, if one thinks, as most people do, that one should be a realist (in the sense I explained) about number theory, then one should note that set theory is indispensable to the proving of some of the most exciting results we have about the natural numbers. I am thinking of Wiles's proof of Fermat's so-called

23. W. H. Woodin, *The Axiom of Determinacy, Forcing Axioms, and the Nonstationary Ideal* (New York: Walter de Gruyter, 2010).

24. See Chapter 9 in this volume.

last theorem, which employed techniques due to Grothendieck that themselves require set theory to formalize in a straightforward way in standard mathematical logic. (Even if it proves possible to find work-arounds that enable us to code Wiles's proof into second-order or even first-order number theory, the result would not be the proof that mathematicians accept and that convinced them that Fermat's last theorem is *true*.) So I would like to suggest a "bootstrapping" argument here: if the indispensability of mathematics (number theory and analysis) for physics is a good reason for interpreting at least number theory and analysis "realistically," *then why is not the indispensability of set theory to number theory a reason for interpreting set theory realistically?*

12

On Axioms of Set Existence

Among the axioms of logic and mathematics, the axioms asserting the existence of sets and classes have long been of special interest to philosophers. I proved the following theorems in a lecture I delivered in 1957. Because I published only a summary of that lecture, and that summary is now hard to find, it seems appropriate to state those theorems and give their proofs in the present volume.

Axioms of Set Existence

The axioms of set existence we are concerned with have the form:

(1) $(y_1)(y_2) \ldots (y_n)(\exists w)(x)(x \in w \Leftrightarrow \Delta)$, where Δ does not contain w.

(When the initial string of universal quantifiers is empty, such axioms are traditionally referred to as "comprehension axioms.")

The following theorem says that for finitely axiomatized set theories, the number of such axioms can always be reduced to one:

Theorem 1. The conjunction of two or more sentences of the form (1) is logically equivalent to a single sentence of the form (1).

Proof: It obviously suffices to show that the conjunction of two sentences of the form (1) is equivalent to a single sentence of form (1). Suppose that

$$(2) \qquad\qquad (y_1)(y_2) \ldots (y_n)(\exists w)(x)(x \in w \Leftrightarrow \Delta)$$

and

$$(3) \qquad\qquad (z_1)(z_2) \ldots (z_m)(\exists w)(x)(x \in w \Leftrightarrow \Phi)$$

are two such sentences, and choose $y_1, y_2, \ldots y_n, z_1, z_2, \ldots z_m$ to be distinct variables. Then the conjunction of (2) and (3) is equivalent to

$$(4) \qquad (y_1)(y_2) \ldots (y_n)(z_1)(z_2) \ldots (z_m)(\exists w)(x)(x \in w \Leftrightarrow P\Phi \vee \sim P\Delta),$$

where P is used to abbreviate the whole of sentence (2).

We verify this as follows:

To verify that the conjunction of (2) and (3) implies (4):
Using the fact that P truth-functionally implies $[(P\Phi \vee \sim P\Delta) \Leftrightarrow \Phi]$, we have that (2) (that is, P) implies that

$$(4) \Leftrightarrow (y_1)(y_2) \ldots (y_n)(z_1)(z_2) \ldots (z_m)(\exists w)(x)(x \in w \Leftrightarrow \Phi).$$

Because the quantifiers "$(y_1)(y_2) \ldots (y_n)$" are vacuous in the right side of this biconditional, it (the right side of the biconditional) is equivalent to (3). Thus we have that (2) implies (4) $\Leftrightarrow$ (3). Hence the conjunction of (2) and (3) implies (4).

To verify that (4) implies the conjunction of (2) and (3):
Assume (4). Using the fact that $\sim$P truth-functionally implies $[(P\Phi \vee \sim P\Delta) \Leftrightarrow \Delta]$, we see that

$$\sim P \Rightarrow [(4) \Leftrightarrow (y_1)(y_2) \ldots (y_n)(z_1)(z_2) \ldots (z_m)(\exists w)(x)(x \in w \Leftrightarrow \Delta)].$$

Because the quantifiers "$(y_1) (y_2) \ldots (y_n)$" are vacuous in the right side of this biconditional, it (the right side of the biconditional) is equivalent to (2). Thus we have that $\sim P \Rightarrow [(4) \Leftrightarrow (2)$ (that is, P)], but we have assumed (4), and hence $[\sim P \Rightarrow P]$. Therefore P, which is (2). But we already have that (2) implies (4) $\Leftrightarrow$ (3), and we have now shown (2) (on the assumption that (4)), so (3). Thus (4) implies the conjunction of (2) and (3). This completes the proof.

Application to Von Neumann–Bernays-Gödel Set Theory

Von Neumann–Bernays-Gödel set theory is two sorted (the two sorts being sets and classes); however, the device just described can be used to reduce the number of axioms of set existence to one and the number of class existence to one. It should be noted that the initial string of universal quantifiers in these axioms need not be homogenous. Thus Zermelo's separation axiom (as formalized in this set theory) reads:

$$(5) \qquad (A)(y)(\exists w)(x)(x \in w \Leftrightarrow (x \in y \wedge x \in A)),$$

and Gödel's replacement axiom, although an axiom of set existence, contains an initial universal quantifier ranging over such classes as happen to be one-one functions, but this does not obstruct the argument.

Application to Quine's System New Foundations

The system whose axioms are the axiom of extensionality:

$$(6) \qquad (x)(y)[(w)(w \in x \Leftrightarrow w \in y) \Rightarrow x=y]$$

and

$$(7) \qquad \text{all instances of (1) with } \Delta \text{ stratified,}[1]$$

is Quine's system New Foundations.[2]

Hailperin showed that the schema (7) can be replaced by eleven of its instances. By applying theorem 1 and noting that the formula (4) will be stratified if (2) and (3) are, we have the following:

Theorem 2. In New Foundations the schema (7) can be replaced by a single one of its instances.

The next theorem refers to New Foundations minus the axiom of extensionality:

---

1. A stratified formula is one that can be reconciled with the theory of types by some assignment of type-indices to its variables.

2. W. V. Quine, "New Foundations for Mathematical Logic," *American Mathematical Monthly* 44 (1937): 70–80; reprinted in Quine 1953), 80–101.

Theorem 3. Let NF* be the system obtained from New Foundations by deleting (6). Then Con(NF*) is equivalent to the following statement: (8) For all *S,* if *S* is of the form (1) with Δ stratified, then ~*S* is not a theorem of quantification theory.

Although the single axiom that we obtain to replace (7) by starting with Hailperin's axioms and using the device illustrated by (4) is extremely complicated, (8) is very simple, because it says, not that some particular statement of the form (1) with Δ stratified is consistent, but that every such instance is consistent. This is strikingly simple for the consistency statement of a powerful set theory.

For my final theorem, I turn from finitely axiomatized set theories to set theories with infinitely many axioms of the form (1).

Ideal Set Theories Cannot Be Axiomatized

"Naïve set theory" assumes all instances of the form (1) as axioms. The Russell paradox shows that this is inconsistent. An ideal substitute would be to assume some maximal consistent subset of these sentences as axioms. This raises the question whether any axiomatizable and nontrivial set theory can be ideal in this sense. Our next theorem answers this question in the negative.

Theorem 4. If a maximal consistent set of sentences of the form (1) is axiomatizable (recursively enumerable), then the set theory generated by those axioms is complete, and hence decidable. The proof follows easily from the following lemma:

Lemma: If a theory has even a single theorem of the form (1), then every sentence of the theory is equivalent in the theory to some sentence of the form (1).

Proof: Let $(y_1)(y_2) \ldots (y_n)(\exists w)(x)(x \in w \Leftrightarrow \Delta)$ be a theorem of a theory *T,* and let *S* be any sentence of *T.* Then S is equivalent in *T* to

$$(9) \qquad (y_1)(y_2) \ldots (y_n)(\exists w)(x)(x \in w \Leftrightarrow S\Delta \lor (\sim S(\sim x \in x))).$$

To verify, simply note that *S* implies (9) $\Leftrightarrow (y_1)(y_2) \ldots (y_n)(\exists w)(x)(x \varepsilon w \Leftrightarrow \Delta)$, and the right side of this biconditonal was assumed to be a theorem of *T;* but ~*S* implies [(9) $\Leftrightarrow ((\exists w)(x)(x \in w \Leftrightarrow \sim x \in x))$], which is the Russell contradiction.

13

The Gödel Theorem and
Human Nature

In the *Encyclopedia of Philosophy* edited by Paul Edwards, the article by Jean van Heijenoort titled "Gödel's Theorem" begins with the following terse paragraphs:[1]

> By Gödel's theorem the following statement is generally meant:
>
> In any formal system adequate for number theory there exists an undecidable formula—that is, a formula that is not provable and whose negation is not provable. (This statement is occasionally referred to as Gödel's first theorem.)
>
> A corollary to the theorem is that the consistency of a formal system adequate for number theory cannot be proved within the system. (Sometimes it is this corollary that is referred to as Gödel's theorem; it is also referred to as Gödel's second theorem.)
>
> These statements are somewhat vaguely formulated generalizations of results published in 1931 by Kurt Gödel, then in Vienna.

1. Jean van Heijenoort, "Gödel's Theorem," in Paul Edwards, ed., *Encyclopedia of Philosophy*, vol. 3 (New York and London: Macmillan, Free Press, and Collier Macmillan, 1967), 348–349.

In spite of the forbidding technicality of the 1931 paper,[2] the Gödel theorem has never stopped generating enormous interest. Much of that interest is aroused by the fact that with the proof of the Gödel theorem, the "human mind" succeeded in proving that, at least in any fixed consistent system with a fixed finite set of axioms that are at least minimally adequate for number theory and with the usual logic as the instrument with which deductions are to be made from those axioms, there has to be a mathematical statement it cannot prove. (In fact, the theorem is much stronger than this: even if we allow the system in question to contain an infinite list of axioms and to have an infinite list of additional rules of inference, the theorem still applies, provided the two lists can be generated by a computer that is allowed to go on running forever.) Moreover, instead of speaking of formal systems adequate for (at least) number theory, one can speak of computers. In such a version the theorem says that if a computer is allowed to write down sentences of number theory forever, subject to the constraints that (1) the list contains a certain subset of the axioms of Peano arithmetic, and (2) the list is not inconsistent (i.e., the list does not contain both some sentence and the negation of that sentence), then there is a true sentence of number theory—in fact, the sentence that says that the output of the machine is consistent is an example—that is not included in the list of sentences listed by the computer. If we speak of the sentences listed by the computer as "proved" by the computer, and of a computer of the kind just mentioned as an "acceptable" computer, we can say this succinctly as follows:

There is a procedure by which, given any acceptable computer, a human mathematician can write down a sentence of number theory that the computer cannot prove or disprove.

Does this or does it not "tell us something about the human mind"? That is what I shall discuss.

What provoked the research described in this essay is a conversation I had with Noam Chomsky many years ago. I asked Chomsky whether he thought

---

2. Kurt Gödel, "Über formal unentscheidbare Sätze der *Principia Mathematica* und verwandter Systeme, I," *Monatshefte für Mathematik und Physik* 38 (1931): 173–198; reprinted in Kurt Gödel, *Collected Works,* vol. 1, *Publications, 1929–1936,* ed. Salomon Feferman, J. W. Dawson, S. C. Kleene, G. H. Moore, R. M. Solovay, and John van Heijenoort, with the German original and the English translation on facing pages (New York: Oxford University Press, 1986), 144–195.

that our *total* competence—not just the competence of the "language organ" that he postulates, but also the competence of the "scientific faculty" that he also postulates—can be represented by a Turing machine (where the notion of "competence" is supposed to distinguish between our true ability and our "performance errors"). He said yes. I at once thought that a Gödelian argument could be constructed to show that if that were right, then we could never *know*—not just never prove mathematically, which is obvious after the Gödel incompleteness theorems, but never know even with the help of empirical investigation—*which* Turing machine it is that simulates our "scientific competence." That Gödelian argument was given in a paper that I published in 1985,[3] but I have always been dissatisfied with one of the assumptions that I used in that proof (I called it a "criterion of adequacy" for formalizations of the notion of justification), namely, the assumption that no empirical evidence can justify believing *p* if *p* is *mathematically false*. (Although many would regard this assumption as self-evident, those who would allow "quasi-empirical" methods in mathematics have good reason to reject it.) Now I am ready to show you a proof whose assumptions seem to me unproblematic. (Or at least they *would be* unproblematic if the notion of a formalizable "scientific competence" made sense. Why I do not think that it makes sense is the subject of the final part of this essay.)

Gödel's Incompleteness Theorems

Gödel's second incompleteness theorem shows that if the set of mathematical truths that it is within human "competence" to prove were the output of an acceptable computer (as defined above), then the consistency of the computer's output could not be one of those mathematical truths. For reasons I will go into in the final part of this lecture, this *does not* show that the human mind (or the competence of a human mathematician) cannot be perfectly simulated by a Turing machine. (This remark does *not* mean, by the way, that I want to defend the view that it *can* be simulated by a Turing machine; in the final part of this lecture I will argue that thinking of us as *having* a "mathematical competence" that is so well defined that that question makes sense is a mistake.) But—postponing that point for the time being—the Gödel

3. Hilary Putnam, "Reflexive Reflections," *Erkenntnis* 22 (January 1985): 143–154; reprinted in Putnam, *Words and Life* (Cambridge, Mass.: Harvard University Press, 1994), 416–417.

incompleteness theorems do not even prove that we cannot *know* which Turing machine simulates our mathematical competence, assuming that there is such a machine, for to know *that* would require empirical, not mathematical, research.

To see why, note that the proposition that we are "consistent" is ambiguous. For example, this proposition may mean that we know the English sentence "Our mathematical competence is consistent" is true; this is very different from knowing an *arithmetized version* of the claim that this is consistent. This latter requires knowing which Turing machine we are. So even if we in *some* sense "know" that we are consistent, to go from that fact to the statement "Turing machine T_k has a consistent output," where T_k is a machine that simulates our mathematical competence, would require the *empirical premise* that T_k is such a machine. But then "T_k is consistent" would be a quasi-empirical mathematical statement, that is, a statement of mathematics that we believed on the basis of an argument some of whose premises were empirical, and there is nothing in the Gödel theorems to rule out our knowing the truth of such a quasi-empirical statement, in addition to knowing the mathematical theorems that we are able to prove in the strict sense. However, I do want to say that reflecting on Gödel's methods and their implications can lead us to a deeper philosophical understanding of the issues raised by Lucas, Penrose, and (on the other side) Chomsky.

Thinking about Chomsky's Conjecture in a Rigorous Gödelian Way

In order to think about Chomsky's conjecture in a Gödelian way, let "COMPETENCE" abbreviate the empirical hypothesis that a particular Turing machine T_k (the kth machine in the standard Kleene enumeration) perfectly simulates the competence of our "scientific faculty," in Chomsky's sense. To make this concrete, we may take this to assert that for any proposition p, and for any evidence u expressible in the language that T_k uses for expressing propositions and expressing evidence (and if Chomsky is right, that language is innate), T_k sooner or later prints out *Justified$_u$*(p) if and only if it is justified to accept p when u is our total relevant evidence. (I use the convention of underlining a sentence to indicate the notation for the Gödel number of the sentence. Thus "T_k sooner or later prints out '*Justified$_u$*($\underline{p}$)'" means that T_k sooner or later prints out the word "*Justified*" followed by the symbols for the

Gödel number of the sentence u [in subscript font] followed by a left parenthesis followed by the symbols for the Gödel number of the sentence p followed by a right parenthesis.) In addition, COMPETENCE asserts that whatever hypotheses and evidence we can describe in an arbitrary natural language can also be represented in T_k's language—that is, that that language has the full expressive power of the "Mentalese" that Chomskians talk about. I am going to sketch a proof (given in full in the Appendix) that if this is true, then COMPETENCE cannot be both true and justified. In other words, if some particular Turing machine T_k correctly enumerates the sentences that our "competence" would take to be justified on arbitrary evidence, then that statement is one that cannot be empirically justified on any evidence.

The proof proceeds, of course, via reductio ad absurdum. The idea is to assume that COMPETENCE is true and that some sentence e states evidence that justifies accepting COMPETENCE. (Note that e and k are constants throughout the following argument.) Using Gödel's technique, we can construct a sentence GÖDEL in the language of Peano arithmetic, and hence in the language of T_k (because that language is capable of expressing all humanly accessible science), that "says" that T_k never prints out $Justified_e(\underline{\text{GÖDEL}})$. So GÖDEL has the following *property:*

It is justified to believe GÖDEL on evidence e *if and only if it is justified to believe (on that same evidence) that the machine* T$_k$ *never says (even if allowed to run forever) that it is justified to believe GÖDEL on evidence* e.

We are now ready to describe the idea of the proof of the following theorem (the actual proof is given in the Appendix).

An Anti-Chomskian Incompleteness Theorem: "COMPETENCE" Cannot Be Both True and Justified

It is well known that Gödel's proof "mimics" the liar paradox. The proof of the "anti-Chomskian incompleteness theorem" given in the Appendix similarly mimics the following paradox, which we might call the justification paradox:

By the Gödel technique for constructing self-referring sentences (or the technique in Quine's *The Ways of Paradox and Other Essays*[4]), it is easy to construct a sentence that "says" of itself that belief in it is not justified, for example (using Quine's technique):

4. W. V. Quine, *The Ways of Paradox and Other Essays* (New York: Random House, 1966), 9.

(B) *"When appended to its own quotation yields a sentence such that believing it is not justified" when appended to its own quotation yields a sentence such that believing it is not justified.*

Now reflect: Is believing (B) justified? Well, if belief in (B) is justified, then I am justified in believing that belief in (B) is not justified, because I see that that is what (B) says. But if belief in (B) is justified, then believing that belief in (B) is justified is justified. So believing that belief in (B) is justified is both justified and not justified, which is impossible. So I now have justified believing that belief in (B) is not justified. But what (B) "says" is precisely that belief in (B) is not justified. So I have justified believing (B). So believing (B) *is* justified, which we just proved leads to a *contradiction*.

Just as Gödel's proof turned the liar paradox from a contradiction into a proof that the arithmetized version of the sentence "I am not provable" is not provable (unless the system in question is inconsistent), so the proof I give in the Appendix turns the justification paradox into a proof that the arithmetized version of the sentence "Believing me is not justified" is not printed out by the machine T_k. But if COMPETENCE were true and also verified (by evidence e) to be true, we could know this proof and so justify "Believing me is not justified)"—which would lead to a real contradiction.

Lucas and Penrose

In 1961 the Oxford philosopher John Lucas claimed that the Gödel theorem shows that "noncomputational processes" (processes that cannot in principle be carried out by a digital computer, even if its memory space is unlimited) go on in our minds.[5] He concluded that our minds cannot be identical with our brains or with any material system (because, if standard physics is assumed, the latter cannot carry out noncomputational processes), and hence that our minds are immaterial. More recently, a thinker whose scientific work I very much admire, Roger Penrose, used a similar but more elaborate argument to claim that we need to make fundamental changes in the way we view both the mind and the physical world. Penrose too argued that the Gödel theorem shows that noncomputational processes must go on in the human mind, but instead of positing an immaterial soul for them to go on in, he concluded that these noncomputational processes must be physical

5. John Lucas, "Minds, Machines and Gödel," *Philosophy* 36 (1961): 112–127.

processes in the brain, and that our physics needs to change to account for them.[6]

Although neither Lucas nor Penrose is a "Chomskian," there is an important similarity between their notion of our mathematical ability and Chomsky's notion of "competence." From *Syntactic Structures* on,[7] Chomsky has insisted that our competence in his sense includes the ability to generate *arbitrarily long* grammatical sentences, together with the proofs of their grammaticality (which had the form of "trees" in Chomsky's *Syntactic Structures*). It is well known[8] that this ability (in the case of unrestricted languages in the Chomsky-Schützenberger hierarchy) is equivalent to the ability to simulate the computations of an arbitrary Turing machine. Thus our "competence" is supposed to include the ability to verify all truths of the form "Machine T_i eventually prints out sentence p." But Lucas and Penrose must also be making this assumption, because they assume that our competence is strictly *greater* than the ability to prove all the theorems of Peano arithmetic, and all truths of the form "Machine T_i eventually prints out sentence p" are (upon suitable arithmetization) theorems of Peano arithmetic.

In 1994 I reviewed Penrose's argument, and like many others who studied it, I found many "gaps."[9] But relax! I am not going to ask you to follow the details of Penrose's argument, nor am I going to give a listing of the places at which I believe it goes wrong.

It may be that Penrose himself recognized that there are gaps in his argument (even though at the beginning he promised a "clear and simple proof"), because he offered a number of arguments that are not mathematical at all, but philosophical.

The heart of his reasoning goes as follows: Let us suppose that the brain of an ideally competent mathematician can be represented by the program of an

6. Roger Penrose, *Shadows of the Mind* (Oxford: Oxford University Press, 1994).

7. Noam Chomsky, *Syntactic Structures* (The Hague/Paris: Mouton, 1957; 2nd ed., Berlin: Mouton de Gruyter, 2002).

8. See, e.g., J. E. Hopcroft and J. D. Ullman, *Introduction to Automata Theory, Languages, and Computation* (Reading, Mass.: Addison-Wesley, 1979),

9. Hilary Putnam, review of Roger Penrose, *Shadows of the Mind, New York Times Book Review,* November 20, 1994, 7; reprinted in *Bulletin of the American Mathematical Society* 32, no. 3 (1995): 370–373. A number of criticisms of Penrose's argument can be found online at http://psyche.cs.monash.edu.au/symposia/penrose/; Penrose's replies can be found at http://psyche.cs.monash.edu.au/v2/psyche-2-23-penrose.html.

"acceptable computer" (as defined above), a program that generates all and only correct and convincing proofs. The statement that the program is consistent is not one that the program generates, by Gödel's second incompleteness theorem. So if an ideal human mathematician could prove that statement, we would have a contradiction. This part of Penrose's reasoning is uncontroversially correct.

It is possible, however, that the program is too long to be consciously apprehended at all—as long as the London telephone book, or even orders of magnitude longer. In that case, Penrose asks, *how could evolution have produced such a program?* It would have to have evolved in parts, and there would be no evolutionary advantage to the parts.

As I shall explain shortly, there are things I find wrong with the notion of "encapsulating all humanly accessible methods of proof,"[10] and hence with the question, "Could a computer do such a thing?" But even if we accept that notion, it is important to recall that just this objection has been made to just about every evolutionary explanation ("How could *wings* have evolved in parts?" "How could eyes have evolved in parts?"), and in case after case an answer has been found.[11] But the answer or answers (animal wings evolved at least four separate times) typically involve generations of researchers, and we are hardly at a stage at which study of the aspects of the brain that subserve our mathematical abilities are well enough understood to even begin evolutionary study. Still, if you will permit me to offer what are obviously speculations, here are two ways in which our mathematical abilities have "parts" that could have utilities apart from being combined in their present "package":

1. As Wittgenstein suggested, the basic laws of arithmetic and geometry could, for example, have first arisen as empirical generalizations that were later "hardened" into conceptual truths.[12] Most of this process would belong to the history of our cultural, rather than our biological, evolution, although it would presuppose the evolution of language and of the skills that underlie, for example, counting.

10. Penrose employs this notion in his replies to his critics, http://psyche.cs.monash.edu.au/v2/psyche-2-23-penrose.html.

11. On the evolution of the eye, see, for example, Richard Dawkins, "Where D'you Get Those Peepers?" *New Statesman & Society* 8 (June 16, 1995): 29. On the evolution of wings, see Michalis Averof and S. M. Cohen, "Evolutionary Origin of Insect Wings from Ancestral Gills," *Nature* 385 (1997): 627–630.

12. For an account of this view, see Mark Steiner, "Mathematical Intuition and Physical Intuition in Wittgenstein's Later Philosophy," *Synthese* 125, no. 3 (2000): 333–340.

2. Reflection on one's practice (in *any* area) is historically an important way in which humans arrive at new practices. But the biological capacity for such reflection could certainly have evolved independently of mathematics itself. *Meta*mathematical reflection, which is what fascinates Penrose, is, in fact, a very late comer on the mathematical scene. For several millennia mathematics was exhausted by the two subjects of arithmetic (or number theory) and (Euclidean) geometry. And the explosion in what we today consider mathematics, including set theory and its less well-known (by non-mathematicians) rival, category theory, was not driven by either Gödelian reflection or by new a priori insights; instead, new axioms, such as the axiom of choice and the axiom of replacement, were added on strongly pragmatic grounds (again, a matter of cultural rather than biological evolution). Only with Gödel's incompleteness theorems does reflection on mathematical practice become a way of proving *number-theoretic* propositions that could not have been proved previously. (The most elementary form of metamathematical reflection is semantic reflection, illustrated by the following: argument: [1] all the axioms of Peano arithmetic are true; [2] the rules of first-order logic preserve truth; [3] therefore, all the theorems of Peano arithmetic are true; but [4] "1 = 0" is not true; and therefore [Conclusion], "1 = 0" is not a theorem of Peano arithmetic; i.e., Peano arithmetic is consistent. Tarski's work shows how this argument can be formalized in a weak second-order extension of Peano arithmetic, and the Gödel theorem shows that the result cannot be proved in Peano arithmetic itself.) But nothing about this so far suggests any problem for standard evolutionary theory.

What about *iterated* reflection? Let L_0 be Peano arithmetic. Because we can see, by the kind of metamathematical reflection just described, that the soundness of L_0 (which we accept) entails the truth of $Con(L_0)$, we can accept the soundness of L_1, where L_1 is the result of adding $Con(L_0)$ as an additional axiom to L_0. Continuing this procedure, we generate a hierarchy of stronger and stronger intuitively acceptable systems, L_0, L_1, L_2, L_3, and so on ad infinitum.

Alan Turing proposed to extend this series into the transfinite, using notations for computable ordinals.[13] At limit ordinals λ, the associated L_λ is the union of all the earlier Ls. (The notations for the ordinals are indispensable in defining the sentence $Con(L_\lambda)$ precisely.) What if our "idealized" mathematical

---

13. Alan Turing, "Systems of Logic Defined by Ordinals," *Proceedings of the London Mathematical Society,* ser. 2, 45 (1939): 161–228.

competencies were represented by the totality of such L_λ with $\lambda < \phi$, for some computable ordinal ϕ? (The ordinal ϕ would have to be a computable ordinal such that no recursive well-ordering of that order-type could be "seen" by a human mathematician to be a well-ordering or proved to be a well-ordering "from below.") Such a totality might well be "longer than the London telephone book." But it would be constructed from L_0, a system that is intuitively simple, using a method of extension that is also intuitively simple. (And, as we just remarked, the ability to reflect on our own practices, which is what enabled us to arrive at that method of extension and also enables us to see that various computable orderings are well-orderings [normally, by seeing that all their initial segments are well-orderings] could well have developed independently of L_0.)

But what if the ordinal logics procedure Turing described did not break off at a computable ordinal? Feferman's article "Transfinite Recursive Progressions of Axiomatic Theories"[14] shows that with stronger reflection principles one can obtain all true number-theoretic propositions on a suitable path through the computable ordinal notations. In that case Penrose would be right—our mathematical competence would exceed the powers of any Turing machine. But my point is that Penrose cannot exclude the possibility that our powers do break off at some computable ordinal.

Ah, but Penrose will certainly ask, "What about us determines the *limits* of metamathematical reflection? What determines where we break off?" Here I have to say that I do not think that there is a precise limit to what human mathematicians can do via reflection. The very notion of an "ideal mathematician" is too problematic for us to speak of a determinate limit here—a point I shall return to shortly. There are different limits on what different individuals can do via metamathematical reflection (and the possibilities of metamathematical reflection have barely been scratched to date). But I would say that what the limits are in the case of any given individual is obviously an empirical question, and not necessarily a very interesting or important one.

One thing I will grant Penrose, however: I do find it mysterious that we have the mathematical abilities we have. But the simple fact that we can intuitively understand the notion that the natural number sequence has no end amazes me more than the fact that we can "Gödelize" any given consistent

---

14. Solomon Feferman, "Transfinite Recursive Progressions of Axiomatic Theories," *Journal of Symbolic Logic* 27 (1962): 259–316.

formal system. I also find it mysterious that we have the natural-scientific abilities, the linguistic abilities, the aesthetic abilities, and so on that we have. If a change in our physical, biological, and neurological theories makes it possible to explain them, then of course I would welcome it. We all would. But I do not believe that Penrose has given compelling reasons to think that our brains *must* have noncomputable powers, and that we must seek a scientific theory to explain them.

To see what we *can* learn about the human mind, or at least about how we should (or, rather, should not) think about the human mind, from the Gödel theorem, I want now to raise two objections against the very question that Penrose asks, the question whether the set of theorems that an ideal mathematician can prove could be generated by a computer.

My first objection—and this is a point that I want to emphasize—is that *the notion of simulating the performance of a mathematician is highly unclear.* Perhaps the question whether it is possible to build a machine that behaves as a *typical* human mathematician behaves is a meaningful empirical question, but a typical human mathematician *makes mistakes.* The output of an *actual* mathematician contains inconsistencies (especially if we are to imagine that she goes on proving theorems forever, as the application of the Gödel theorem requires); so the question of proving that the whole of this output is consistent does not even arise.

To this, Penrose's reply seems to be that the mathematician may make errors, but she corrects them upon reflection. As I pointed out before, even if (via this or some other philosophical argument) we know that *the English sentence* "Our mathematical competence is consistent" is true, this is very different from knowing an *arithmetized version* of the claim that this is the case. This latter would require knowing which Turing machine simulates our mathematical competence. (That we cannot know a similar fact about our natural-scientific competence by empirical investigation was what our "anti-Chomskian incompleteness theorem" showed.)

My second objection is that to confuse these questions is to miss the *normativity* of the notion of *ideal* mathematics. The notion of "ideal" mathematics is the notion of a human mathematician who always behaves (in his mathematical work) as a mathematician *should* (or "ideally should"). To the extent that we can succeed in describing human beings as physical systems, physics can say (at best) how they *will* behave, "competence errors" and all. But physics is not in the business of telling us how human beings "should" behave.

A Caution against a Widespread but Naïve Argument
for "Algorithms" in the Brain

It is not my aim merely to criticize those who make the mistake of thinking that the Gödel theorem proves that the human mind or brain can carry out noncomputational processes. That Lucas and Penrose have failed to prove their claims about what the Gödel theorem "shows about the human mind" is widely recognized. At the same time, however, a view that may seem just at the opposite end of the philosophical spectrum from the Lucas-Penrose view seems to me to be vulnerable to the two objections I just made. I refer to the widespread use of the argument, in philosophy as well as in cognitive science, that whenever human beings are able to recognize that a property applies to a sentence in a "potentially infinite" (i.e., a fantastically large) set of cases, there must be an "algorithm" in their brains that accounts for this, and the task of cognitive science must be to describe that algorithm. I think that the case of mathematics shows that that cannot be a *universal* truth, and my criticism of Penrose may enable us to see why we should not draw any mystical conclusions about either the mind or the brain from the fact that it is not a universal truth.

But first a terribly important qualification. I have no doubt that human beings have the ability to employ recursions (another word for "algorithms") both consciously and unconsciously, and that there are recognition abilities that cognitive science should explain (and in many cases *does* successfully explain) by appealing to the notion of an algorithm. The work of Chomsky and his school, in particular, shows how our ability to recognize the property of *grammaticality* might be explainable in this way. Why, then, do I say that the case of mathematics shows that it cannot be a *universal* truth that whenever human beings are able to recognize that a property applies to a sentence in a "potentially infinite" (i.e., a fantastically large) set of cases, there must be an "algorithm" in their brains that accounts for it?

Well, just as it is true that we can recognize grammaticality in a fantastically large set of cases, it is true that we can recognize *being proved* in a fantastically large set of cases. And as we saw (via the anti-Chomskian incompleteness theorem), Gödel's techniques can be used to show—via a *very* different argument from Penrose's—that *if* there is an "algorithm" that accounts for this ability, we are not going to be able to verify that there is. But I think that it would be wrong to stop with this conclusion.

It would be wrong because talk about "a potentially infinite set of cases" makes sense only *when* there is an algorithm that we *can* verify. Why do I say this?

Think for a minute about the mathematician that Penrose imagines he is talking about. If the mathematician could *really* recognize a potential infinity of theorems, then she would be able to recognize theorems of *arbitrary finite length*—for example, theorems too long to be written down before all the stars are cooled down, or sucked into black holes, or whatever. But any adequate physics of the human brain will certainly entail that the brain will disintegrate long before it gets through trying to prove any such theorem. In short, *in real actual fact,* the set of theorems any *physically* possible human mathematician can prove is finite.

Moreover, it is not really a "set," because "sets," by definition, are two valued: an item is either "in" or "out." But the predicate "prove" is vague. We can, of course, make it precise by specifying a fixed list of axioms and rules to be used. But then the Gödel theorem does show that there is a perfectly natural sense in which we can "prove" a statement that is not derivable from those axioms by those rules. This may drive us to say that "by 'prove' we mean '*prove by any means we can* "see" *to be correct.*'" But this uses the essentially vague notion "see to be correct." In short, there is not a "potentially infinite set of theorems a mathematician can prove" concerning which we can ask: is it recursive or nonrecursive? What there is is a *vague finite collection of theorems a (real flesh-and-blood) mathematician can prove.* And vague finite collections are neither "recursive" nor "nonrecursive"; those concepts apply only to well-defined sets.

But, it may be objected, did Chomsky not teach us that we can—and *should*—idealize language users by imagining that (like ideal computers) they *can* go on generating grammatical sentences *forever*—sentences of *arbitrary* length? Yes, he did, but the idealization he showed us how to make is precise only because it corresponds to a well-defined algorithm. If we wish the question "What can an ideal mathematician prove?" to be a precise one, we must specify just how the mathematician's all-too-finite brain is to be idealized, just as Chomsky did. And if we do this by specifying an algorithm, then, of course, the result will be "computational."[15]

---

15. I thank the referees of this paper for calling to my attention two important papers that make points connected with the fact that Penrose's and Lucas's arguments depend on *idealizing* the human mathematician, namely, Stewart Shapiro, "Mechanism, Truth, and Penrose's

If this is right, then we should not allow ourselves to say that the set of statements that a mathematician can "prove" is a "potentially infinite set." We should say only that there are "fantastically many" statements a mathematician—an *actual* mathematician—can prove and let it go at that. It does *not* follow that there is an algorithm such that everything an "ideally competent" mathematician would (should?) count as proving a mathematical statement is generated by that algorithm, just as it does not follow from the fact that there are fantastically many jokes that a person with a good sense of humor can see to be "funny" that there is an algorithm such that everything a person with an "ideal" sense of humor would see to be funny is generated by that algorithm.

APPENDIX

As above, we assume that T_k is a machine that perfectly represents our scientific competence in the sense that for any proposition p and for any evidence u (expressible in the language that T_k uses for expressing propositions and expressing evidence), T_k sooner or later prints out *Justified$_u$(p)* if and only if it is justified to accept p when u is our total relevant evidence. As before, COMPETENCE also asserts that whatever hypotheses and evidence we can describe in an arbitrary natural language can also be represented in T_k's language—that is, that that language has the full expressive power of the "Mentalese" that Chomskians talk about. Using Gödel's technique, we can construct a sentence GÖDEL in the language of Peano arithmetic, and hence in the language of T_k (because that language is capable of expressing all humanly accessible science), that "says" that T_k never prints out *Justified$_e$(GÖDEL)*. So GÖDEL has the following property:

It is justified to believe GÖDEL on evidence e *if and only if it is justified to believe (on that same evidence) that the machine* T$_k$ *never says (even if allowed to run forever) that it is justified to believe GÖDEL on evidence* e.

We assume two axioms concerning the notion of justification:

Axiom 1: It is never both the case that it is justified to believe p on evidence e and that it is justified to believe ~p on evidence e.

New Argument," *Journal of Philosophical Logic* 32 (2003): 19–42; and Solomon Feferman, "Are There Absolutely Undecidable Problems? Gödel's Dichotomy," *Philosophia Mathematica,* ser. 3, 14 (2006): 134–152.

(Remark: we know various statements of science that are at a certain stage justified on all the available relevant evidence but are subsequently rejected on the basis of new relevant evidence, possibly within the framework of new theories. Thus a Turing machine that simulates our scientific competence as it proceeds chronologically may list inconsistent statements because our evidence is not the same at different times. But T_k is not supposed to mimic the diachronic behavior of our brains, but simply to represent a recursive enumeration of the sentences that it is within the competence of "the scientific faculty"—which Chomsky claims to be computational—to accept as justified on any evidence expressible in "Mentalese." Thus axiom 1 represents the requirement that the "scientific faculty," when acting in accordance with its "competence," not regard both p and $\sim p$ as justified *on the same evidence.* Even this has been questioned by a friend, who writes: "I suppose that if an agent had purported justifications of a sentence and its negation, he would realize that at least one of these justifications is bogus, and should be withdrawn. But the subject may not realize that he has justified a given sentence and its negation. To give our agents the ability to tell whether their (justified) beliefs are consistent is to give them a nonrecursive ability." My response is that it is part of the notion of justification that justified beliefs are not flatly contradictory [i.e., of the form $p \wedge \sim p$]. I believe that this is something all epistemologists would agree on. If there is some other notion of "justification" on which both p and $\sim p$ can sometimes be justified on the very same evidence, I am sure that it is not the notion of justification either Chomsky or Penrose or Lucas or I have in mind. [The ability to tell whether two beliefs are flatly contradictory is a primitive recursive ability.])

Because COMPETENCE is the statement that T_k perfectly represents our competence, we also have the following:

Axiom 2: If *Justified$_e$*(the machine T_k eventually prints out *Justified$_e$*(p) and *Justified$_e$*(COMPETENCE), then *Justified$_e$*(p).

("*Justified$_e$*(p)" is, of course, simply the sentence in the language of T_k—which was assumed to be a formalized version of Chomsky's "Mentalese"—that expresses the statement "It is justified to believe the sentence p on evidence e." Note that by well-known results of Gödel and Turing, "the machine T_k eventually prints out *Justified$_e$*(p)" is expressible by a sentence of Peano arithmetic.) As before, I underline a propositional expression to express the formal numeral for the Gödel number of that expression.

Proof of the Anti-Chomskian Incompleteness Theorem

To guide the reader, here is an outline of the proof I will give. (In both parts of the proof we will assume that COMPETENCE is both true and justified by the evidence e.)

Part 1: I will prove (by assuming the opposite and deriving a contradiction) that it is not the case that the machine T_k eventually prints out $Justified_e(\text{GÖDEL})$) (i.e., the machine T_k, which we assumed to correctly enumerate the sentences that our epistemic competence would tell us are justified on arbitrary evidence, does not generate the Gödelian sentence, on the evidence that justifies our acceptance of COMPETENCE].

Outline of Part 2:

The Conclusion of Part 1 was as follows:

It is not the case that the machine T_k eventually prints out $Justified_e(\text{GÖDEL})$. Because we can know this proof if we have empirically justified COMPETENCE, and a proof is a justification, it follows immediately that

(a) $Justified_e(\underline{\text{The machine } T_k \text{ never prints out } Justified_e(\text{GÖDEL})})$.

Then the rest of Part 2 will derive a contradiction from (a), thus showing that the assumption that COMPETENCE is both true and justified must be false.

The proof in full:

Part 1 of the proof:

Assume that the machine T_k prints out $Justified_e(\text{GÖDEL})$

(1) $Justified_e(\text{GÖDEL})$. Reason: For the reductio ad absurdum proof of my incompleteness theorem I am assuming COMPETENCE. By assumption, the machine T_k prints out $Justified_e(\text{GÖDEL})$. So by COMPETENCE, $Justified_e(\text{GÖDEL})$.

(2) It is justified$_e$ to believe that the machine T_k eventually prints out $Justified_e(\text{GÖDEL})$. Reason: I just showed that $Justified_e(\text{GÖDEL})$. Then by COMPETENCE, $Justified_e(\underline{\text{the machine } T_k \text{ eventually prints out } Justified_e(\text{GÖDEL})})$—that is, we are justified$_e$ in believing that the machine T_k eventually prints out $Justified_e(\text{GÖDEL})$.

(3) We are justified in believing "The machine T_k does not eventually print out $Justified_e(\text{GÖDEL})$." Reason: because, I pointed out above, on Chomsky's idealized notion of "competence" (and also on the notion of our mathematical competence assumed by Lucas and Penrose), it is within our competence to recognize all truths of the form "Machine T_i eventually prints out

sentence p," and, by assumption, the machine T_k prints out *Justified$_e$* (GÖDEL), this is a truth that our scientific faculty is able to prove. Because a proof is a justification, we have *Justified$_e$* (the machine T_k eventually prints out *Justified$_e$* (GÖDEL)). But:

It is justified to believe GÖDEL on evidence e if and only if it is justified to believe (on that same evidence) that the machine T_k never says (even if allowed to run forever) that it is justified to believe GÖDEL on evidence e.

Because this equivalence is known to us, and by (1) we are justified$_e$ in believing the left side, we are justified$_e$ in believing the right side—that is, we are justified$_e$ in believing "The machine T_k does not eventually print out *Justified$_e$* (GÖDEL))."

But (2) and (3) violate our consistency axiom, axiom 1. Thus (still assuming that COMPETENCE is both true and justified on evidence e) we conclude that the assumption of our subproof is false: it is not the case that the machine T_k prints out *Justified$_e$* (GÖDEL). But this is reasoning that we can easily go through if we have discovered COMPETENCE to be true! So, without any additional empirical evidence, we have justified$_e$:

"The machine T_k does not eventually print out *Justified$_e$* (GÖDEL)."

To complete the proof of my incompleteness theorem, I therefore now need a proof that this too leads to a contradiction. Here it is:

Part 2 of the proof:

I have proved in part I that

(1) T_k does not eventually print *Justified$_e$* (GÖDEL).

From this, by COMPETENCE we have

(2) ~*Justified$_e$* (GÖDEL).

But we have from before

(B) *Justified$_e$* (GÖDEL) if and only if (*Justified$_e$* (T_k does not eventually print *Justified$_e$* (GÖDEL)).

So

(3) ~*Justified$_e$* (T_k does not eventually print *Justified$_e$* (GÖDEL)).

But from (1) and the obvious rule of inference that proving X allows writing *Justified$_e$* (X) we have

(4) *Justified$_e$* (T_k does not eventually print *Justified$_e$* (GÖDEL)).

Now (3) and (4) are a contradiction.

This completes the proof of the anti-Chomskian incompleteness theorem.

14

After Gödel

Some years ago I heard a famous logician say that "Alfred Tarski was the greatest logician after Aristotle." Tarski was indeed a great logician, but for reason that I soon will explain, it seems clear to me that the title "greatest logician after Aristotle" belongs to Kurt Gödel and not Alfred Tarski (or, as one might be allowed to refer to him in Israel, "Alfred Teitelbaum"—Tarski changed his name to a more "Polish" one because of the anti-Semitism in the Polish universities). On another occasion the famous mathematician David Mumford (who later went into computer science) said to me, "As far as I am concerned, Gödel wasn't a mathematician. He was just a philosopher." I do not know whether Mumford still thinks this, but if he does, he is wrong too.

Gödel is, of course, known best for the famous incompleteness theorems, but they are only part of his contribution to logic. If they were the whole, or the only part of such fundamental importance, then the claim that Tarski's formalization of the notion of "satisfaction" and his use of that notion to show us how to define truth of a formula in a formalized language over a model were at least as great a contribution might be tenable.[1] However, the very field

1. However, I have argued that the philosophical, as opposed to the mathematical, significance of Tarski's achievement has been both overestimated and misrepresented. See Hilary

for which Tarski is most famous, model theory, was launched by two theorems, one of which bears Gödel's name: the Skolem-Löwenheim theorem and the Gödel completeness (or completeness and compactness) theorem. Moreover, Alonzo Church's theorem was clearly known to Gödel before Church, as a careful reading of the footnotes to the famous paper on undecidable sentences makes clear. And without question, what I may call the "Gödel-Cohen theorems"—that the axiom of choice and the continuum hypothesis are both independent of the other axioms of set theory (unless those axioms are inconsistent) are by far the most stunning results ever obtained in set theory. (Gödel proved that the axiom of choice and the continuum hypothesis are consistent with the axioms of Zermelo–Fraenkel in the 1930s, and in 1962 Paul Cohen showed that their negations are likewise consistent. These results are extremely robust in the sense that it is unlikely that any further axioms that one could add to set theory will both resolve the continuum problem and be found sufficiently "intuitively evident" to command acceptance by the mathematical community, although Gödel hoped the contrary.)

One could also mention many other contributions by Gödel, including important contributions to recursion theory. For example, in the course of proving the incompleteness theorems, Gödel showed that for any predicate $F(x)$ in a formal system that contains expressions for all the primitive recursive functions, there is an expression of the form $F(N)$, where f is a primitive recursive function and N is an integer,[2] such that the Gödel number of the formula $F(f(N))$ is equal to precisely $F(N)$ (i.e., the formula is true if and only if its own Gödel number has the property F). (I shall refer to this as "Gödel's diagonal lemma," although he did not give it a name.) This means that, for example, there is a formula that is true if and only if its own Gödel number is prime (just take $F(n)$ to be the formula Prime(n)), a formula that is true if and only if its own Gödel number is even, and so on. This is not just similar to the fixed-point theorem of recursion theory; it introduced precisely the technique of proof we need for that fixed-point theorem (Kleene's recursion theorem).

What I want to do now is to describe some of the ways logicians after Gödel have built on and extended his results.

Putnam, "A Comparison of Something with Something Else," *New Literary History* 17, no. 1 (Autumn 1985): 61–79; reprinted in Putnam, *Words and Life* (Cambridge, Mass.: Harvard University Press, 1994), 330–350.

2. I am deliberately ignoring the "use-mention distinction" to simplify exposition.

Diophantine Equations

Gödel's own undecidable sentence can easily be put into the form "There does not exist a natural number n such that $f(n) = 0$," where $f(n)$ is primitive recursive. But primitive recursive functions are not a topic the average mathematician is particularly interested in. (Probably David Mumford was not when he made the remark I quoted earlier.)

However, in the late 1950s Martin Davis, Julia Robinson, and I proved that the decision problem for exponential Diophantine equations is recursively unsolvable.[3] Julia Robinson had already shown that the decision problem for ordinary Diophantine equations is equivalent to the decision problem for exponential Diophantine equations if and only if there exists a single ordinary Diophantine equation whose solutions (considered as functions of any one of the unknowns) have a roughly exponential rate of growth, and a few years later Yuri Matiyasevich proved that such an equation exists. This Davis-Matiyasevich-Putnam-Robinson theorem showed that the decision problem for ordinary Diophantine equations is recursively unsolvable, thus providing a negative solution to Hilbert's tenth problem. In fact, we showed that for every recursively enumerable set S, there is a polynomial P with integral coefficients such that $P(n, x_1, x_2, \ldots x_k) = 0$ has a solution in natural numbers exactly when n belongs to S. Applied to Gödel's original paper on undecidable sentences, this yields the fact that the undecidable sentence can have the mathematically very familiar form "The Diophantine equation $P = 0$ has no solution." To quote I do not know who, "Who woulda thunk it?"

Model Theory

In 1930 Gödel also showed that the standard axioms and rules of quantification theory (which was already known by the name the American philosopher Charles Peirce, the founder of pragmatism, gave it—"first-order logic")

---

3. Equations of the form $P = 0$ to be solved in natural numbers, where P is a polynomial with integral coefficients, are called Diophantine equations. "Exponential Diophantine equations" are equations of the form $P = 0$ to be solved in natural numbers, where P is an expression that is like a polynomial with integral coefficients except for having some variable exponents. For example, the Fermat equation can be written as $x^n + y^n - z^n = 0$, which is an exponential Diophantine equation.

are *complete* in the sense that every valid formula, every formula that is true in all possible models, is a theorem. His proof also establishes that if every finite subset of an infinite set of formulas of quantification theory has a model, then the whole infinite set has a model. This is the compactness theorem, and it is still of fundamental importance in model theory.

To illustrate the importance of the compactness theorem, let P be any set of axioms for Peano arithmetic, or, for that matter, for any consistent system that extends Peano arithmetic. Let a be a new individual constant, that is, one not used in P. Consider the theory T with the following recursively enumerable set of axioms: the axioms of P plus "a is a natural number," "$a \neq 0$," "$a \neq 1$," "$a \neq 2$," "$a \neq 3$," and so on ad infinitum. Let s be any finite subset of these axioms, and let N be the largest integer such that $a \neq N$ is a member of S. Then s obviously has a model—just take any model for P and interpret a as denoting $N+1$. By the compactness theorem T has a model. In that model the object denoted by a is an infinite integer, and so are $a+1$, $a+2$, $a+3$, and so on. Moreover, a has a predecessor $a-1$ in the model (otherwise, by a theorem of Peano arithmetic, and hence a theorem of P, it would be 0, violating the axiom "$a \neq 0$"), and that predecessor has a predecessor, and so on, and all these "natural numbers" $a-1$, $a-2$, $a-3$, and so on are likewise infinite integers. Thus we have the existence of nonstandard models for mathematics. (By a "nonstandard" model I mean a model in which, in addition to the integers 0, 1, 2, and so on there are also, viewed from the outside, "infinite integers." I say "viewed from the outside" because within the formal system itself there is no way to single out these nonstandard elements, or even to say that they exist. It is only in the metalanguage in which we prove the existence of the nonstandard model that we can say that there are "foreign elements" in the model, elements other than the "real" natural numbers.) Abraham Robinson showed, in fact, that by using such models one can carry out Leibniz's dream of a true calculus of infinitesimals, and the resulting branch of mathematics, which has been called nonstandard analysis, already has significant applications in many areas—for example, to the theory of Lie groups, to the study of Brownian motion, and to Lebesgue integration. What gives the subject its power is that, because the "infinite numbers" all belong to a model for the theory of the standard (finite) integers and real numbers, a model within which they are not distinguished by any predicate of the language from the standard numbers, we are guaranteed from the start that they will obey all the laws that standard numbers obey.

Kripke's Proof

In the past few decades various workers, the most famous being Paris and Harrington, have begun to use the existence of nonstandard models to give independence proofs in number theory itself.[4] The very existence of independent (or "undecidable") propositions of elementary number theory was proved by Gödel in 1934 by syntactic, not model-theoretic, means. The proposition proved independent by Paris and Harrington is a statement of graph theory (a strengthened version of Ramsey's theorem). What I want to tell you about now, however, is a different model-theoretic proof of the existence of undecidable sentences. This proof is due to Saul Kripke, and I wrote it up in a paper published in 2000 that is collected as the next chapter of the present volume.

Because Kripke's theorem does not aim at establishing the independence of a statement that is nearly as complicated as the proposition Paris and Harrington wished to prove independent, the proof is much simpler than theirs. Also, because the independence proof is semantic rather than syntactic, we do not need to arithmetize the relation "x is a proof of y," as Gödel had to do for his proof.[5] We do not need the famous predicate $Bew(x)$ (x is the Gödel number of a theorem), or the famous self-referring sentence that is true if and only if its own Gödel number is not the Gödel number of a theorem. In short, Kripke gave us a different proof of the Gödel theorem, not just a different *version* of Gödel's proof.

As the reader who peruses the next chapter will see, Kripke's construction of a nonstandard model in which a statement that is true under the intended interpretation of arithmetic is *false* is actually very short and elegant. Thus we have the remarkable result that post-Gödelian model theory can be used to replace recursion theory in the proof of the Gödel incompleteness theorem!

---

4. Jeff Paris and Leo Harrington, "A Mathematical Incompleteness in Peano Arithmetic," in Jon Barwise, ed., *Handbook of Mathematical Logic* (Amsterdam: North-Holland, 1977), 1133–1142.

5. To "arithmetize" a relation *Rab* between sentences (or other syntactic objects) means to define an arithmetic predicate *Fxy* that holds (for any x, y) if and only if x is the Gödel number of an expression *a* and y is the Gödel number of an expression *b* and *Rab*. Gödel actually showed how to define a *primitive recursive F* such that *Fxy* holds if and only if x is the Gödel number of an expression *a*, y is the Gödel number of an expression *b*, and *a* is a proof of *b*.

Prime Numbers

Here is yet another result that milks Gödel's theorem. This is actually an easy corollary of the results about Diophantine equations that I mentioned earlier, but number theorists expressed amazement when I pointed out in 1960 that (1) the prime numbers are exactly the positive values taken on by an exponential-polynomial function (i.e., a function that resembles a polynomial, except that the variables are allowed to have variable exponents) when the arguments are restricted to whole numbers; and (2) if it could be shown that Hilbert's tenth problem is recursively unsolvable (as it later was), "exponential polynomial" could be replaced by "polynomial" in this statement. (Both statements are proved in my "An Unsolvable Problem in Number Theory".)[6]

The resulting proof that the prime numbers are represented by a polynomial is sufficiently simple for me to give it in full:

The primes are a recursively enumerable set. So by the Davis-Matiyasevich-Putnam-Robinson theorem, there is a polynomial P with integral coefficients such that the equation

(1) $$P(n, x_1, x_2, \ldots x_k) = 0$$

has a solution in integers when and only when n is a prime number.

It is a theorem of number theory that every positive integer is the sum of four squares. I claim that the following equation has an integral solution with positive n if and only if n is a prime number:

(2) $$n = (y_1^2 + y_2^2 + y_3^2 + y_4^2) \cdot [1 - P(y_1^2 + y_2^2 + y_3^2 + y_4^2, x_1, x_2, \ldots x_k)]^2),$$

where P is the polynomial in (1).

---

6. Hilary Putnam, "An Unsolvable Problem in Number Theory," *Journal of Symbolic Logic* 25 (1960): 220–232. J. P. Jones, Daihachiro Sato, Hideo Wada, and Douglas Wiens, "Diophantine Representation of the Set of Prime Numbers," *American Mathematical Monthly* 83, no. 6 (1976): 449–464, have written: "Martin Davis, Yuri Matiyasevich, Hilary Putnam and Julia Robinson have proved that every recursively enumerable set is Diophantine, and hence that the set of prime numbers is Diophantine. From this and the work of Putnam [i.e., "An Unsolvable Problem in Number Theory"], it follows that the set of prime numbers is representable by a polynomial formula," (449).

Proof. First, suppose that n is a prime number. Then by (1) there are $x_1, x_2, \ldots$ x_k such that $P(n, x_1, x_2, \ldots x_k) = 0$. Choose such $x_1, x_2, \ldots x_k$ and let y_1, y_2, y_3, y_4 be such that $n = y_1^2 + y_2^2 + y_3^2 + y_4^2$ (by the four-squares theorem there are such y_1, y_2, y_3, y_4). Then $P(y_1^2 + y_2^2 + y_3^2 + y_4^2, x_1, x_2, \ldots x_k) = 0$, and the second factor in (2) is equal to 1. So (2) reduces to $n = y_1^2 + y_2^2 + y_3^2 + y_4^2$, which is correct, by the choice of y_1, y_2, y_3, y_4.

Second, suppose that (2) is true for natural numbers n, y_1, y_2, y_3, y_4, and $x_1, x_2, \ldots x_k$. If $P(n, x_1, x_2, \ldots x_k) \neq 0$, then $[P(y_1^2 + y_2^2 + y_3^2 + y_4^2, x_1, x_2, \ldots x_k)]^2 \geq 1$, and $(1 - [P(y_1^2 + y_2^2 + y_3^2 + y_4^2, x_1, x_2, \ldots x_k)]^2)$ is zero or negative. In either case $n = (y_1^2 + y_2^2 + y_3^2 + y_4^2)$. $(1 - [P(y_1^2 + y_2^2 + y_3^2 + y_4^2, x_1, x_2, \ldots x_k)]^2)$ is nonpositive. And if $P(y_1^2 + y_2^2 + y_3^2 + y_4^2, x_1, x_2, \ldots x_k) = 0$, then by (1), $y_1^2 + y_2^2 + y_3^2 + y_4^2$ is a prime number, and because $(1 - [P(y_1^2 + y_2^2 + y_3^2 + y_4^2, x_1, x_2, \ldots x_k)]^2) = 1$, (2) reduces to $n = y_1^2 + y_2^2 + y_3^2 + y_4^2$, so n is a prime. Thus the prime numbers are all and only the *positive* integers taken on as values by the polynomial $(y_1^2 + y_2^2 + y_3^2 + y_4^2) \cdot (1 - [P(y_1^2 + y_2^2 + y_3^2 + y_4^2, x_1, x_2, \ldots x_k)]^2)$.

15

Nonstandard Models and Kripke's Proof of the Gödel Theorem

My subject will be elementary number theory, that is, the theory of the nonnegative integers as formalized in standard (first-order) quantificational logic. As primitives I will simply take the primitive recursive predicates. (Alternatively, I could have just taken plus and times, or any other set of basic predicates or functions from which all the primitive recursive predicates can be defined.) As axioms I will take the standard first-order version of the Peano axioms.

One of the surprising facts that was not noticed by nineteenth-century mathematicians, but was observed after Gödel proved the completeness of first-order logic in 1930, is the fact that Peano arithmetic has *nonstandard models.* By a nonstandard model is meant a model in which, in addition to the integers 0, 1, 2, . . . , there are also (viewed from the outside) "infinite integers." I say "viewed from the outside" because within Peano arithmetic itself there is no way to single out these nonstandard elements, or even to say that they exist. It is only in the set-theoretic mathematical language in which we prove the existence of the nonstandard model that we can say that there are "foreign elements" in the model, elements other than the "real" natural numbers.

These models exist not only for elementary number theory but for mathematics as a whole (as represented by, say, Zermelo-Fraenkel set theory). In the metalanguage in which we talk about such a model, we can distinguish between "finite" and "infinite" real numbers, even though within the object language it is a theorem that "all real numbers are finite." And there are reciprocals of infinite real numbers in the model, that is, *infinitesimals.* Abraham Robinson showed, in fact, that by using such models one can carry out Leibniz's dream of a true calculus of infinitesimals, and the resulting branch of mathematics, which has been called nonstandard analysis, already has significant applications in many areas—for example, to the theory of Lie groups, to the study of Brownian motion, and to Lebesgue integration. What gives the subject its power is that because the "infinite numbers" all belong to a model for the standard (finite) integers and real numbers, a model within which they are not distinguished by any predicate of the language from the standard numbers, we are guaranteed from the start that they will obey all the laws that standard numbers obey.

As the title of this paper indicates, the honor of inventing this proof belongs to Saul Kripke. He has not published it yet, in part because he is still investigating what can be done using the "bounded ultrapower construction" that appears in the most constructive version of the proof. The version I am going to discuss is a quicker and less constructive version that is also due to Kripke.

I mentioned at the beginning of this paper that the existence of nonstandard models was first observed as a corollary to Gödel's 1930 work on the completeness of first-order logic. But today we know techniques that are purely algebraic rather than logical—the techniques of ultrapower construction—that can be used to give direct proofs of the existence of nonstandard models in a style with which mathematicians who are not trained logicians feel perfectly comfortable. In fact, these techniques do not even require knowledge of the Gödel completeness theorem or that logic itself be axiomatized. In short, the proof I am about to show you is one that establishes independence by means that could, in principle, have been understood by nineteenth-century mathematicians. I will exhibit a statement of number theory—one that is not at all "self-referring"— and construct two models, in one of which it is true and in the other of which it is false, thereby establishing "undecidability" (independence).

Consider a finite monotone increasing series ₛf natural numbers, say,

182, 267, 349, 518, ... , 3987654345.

And consider a formula A of number theory, say "$(x)(Ey)Rxy$" (with primitive recursive R). (I shall identify the series s with its Kleene Gödel number where convenient in what follows.) I shall say that s fulfills A if the second player (the "defending player") has a winning strategy in the game I shall describe.

The Game G

The game is played as follows. The first player (the "attacking player") picks a number less than the length of the given sequence s, say, 3. The sequence s is examined to determine the third place in the sequence (to determine "$(s)_2$," in the Kleene notation, because the members of a sequence with Gödel number s are $(s)_0$, $(s)_1$, $(s)_2$, ... in this notation). The same player (the attacker) now picks a number less than this number (less than 349, in the case of this example.) Let us suppose that he picks 17. We assume that the number picked by the first player was less than the length of the sequence (otherwise the first player has lost). If so, the second player (the "defending player") gets to look at the next number in the sequence (at $(s)_3$ or 518, in the case of the example). He must pick a number less than this number (less than $(s)_{n+1}$ if the first player picked the place $(s)_n$). Let us suppose that he picks 56. We now evaluate the statement $R(17, 56)$ (the statement $R(n, m)$, where n is the number picked by the first player and m is the number picked by the second player). Because R is primitive recursive, this can be done effectively. If the statement is true, the defending player has won; if it is false, the attacking player has won.

The statement that a sequence s fulfills this statement A (that there is a winning strategy for the defending player) can be written out in number theory as follows:

(I) $(i) \leq (\text{length}(s) - 1)(n \leq (s)_{i-1})(Em \leq (s)_i)Rnm.$

Similarly, if we are given a statement A with four, or six, or however many alternating quantifiers in the prefix, we can define "s fulfills A" to mean that there is a winning strategy for the defending player in a game that is played very much like the game G: a game in which the attacking player gets to choose a new place in the sequence each time it is his turn to play. The attacking player must also choose a number less than $(s)_{i-1}$, where $(s)_{i-1}$ is the number in the position he chose in the sequence. (The number in the ith position is called "$(s)_{i-1}$" and not "$(s)_i$" because Kleene—whose notation I am employing—calls the first position "$(s)_0$" and not "$(s)_1$.") Each time he plays, the attacking player has to choose a place that is to the right of the place in the sequence he chose before (unless it is his first turn to play) and not the last place in the sequence (unless he has no legal alternative, in which case he loses), and the defending player must then pick a number less than $(s)_i$ (less than the number in the next place in the sequence). The game ends when as many numbers have been chosen as there are quantifiers in the prefix of the formula. (We assume that all formulas are prenex, and that quantifiers alternate *universal, existential, universal, existential* . . .). The numbers chosen are then substituted for the variables in the matrix of the formula A in order (first number chosen for x_1, second number chosen for x_2, and so on, where x_1 is the variable in the first argument place, x_2 the variable in the second argument place, and so on). The resulting primitive recursive statement is evaluated, and, as before, the defending player wins if the statement is true and the first (attacking) player wins if the statement is false. Once again, for any fixed formula A we can easily express the statement that s fulfills A arithmetically (primitive recursively in s). And for any fixed recursively enumerable sequence of formulas A_1, A_2, . ., the statement that s fulfills A_n can be expressed as a primitive recursive relation between s and n, say, (Fulfills s, A_n). Note that we can also speak (by an obvious extension) of an ordinary infinite monotone increasing sequence "fulfilling" a formula (this means that if one picks any number less than a given number in the sequence to be the value of the first universal quantifier, it is always possible to pick a number less than the next place to the right in the sequence to be a value for the succeeding existential quantifier, so that no matter what number less than the number in an arbitrarily selected place still further to the right in the sequence one picks for the *next* universal quantifier, it is possible to pick a number less than the number in the place in the sequence immediately to the right of the last "universal quantifier place" chosen for the last existential quantifier so that the statement A comes out true). And note that a statement that is fulfilled by an infinite monotone increasing sequence is true (because the restriction that one must pick numbers as values for the universal quantifiers that are bounded by the numbers in the sequence is, in effect, no restriction on the "attacking player" at all—the numbers in the sequence get arbitrarily large, so he can pick any number he wants by going out far enough in the sequence).

Henceforth I shall confine attention to sequences with the following two properties (call them "good sequences"):

(1) *The first number in the sequence is larger than the length of the sequence.*

(2) *Each number in the sequence after the first is larger than the square of the number before.* (This is to ensure that the sum and product of numbers $\leq (s)_i$ are $\leq (s)_{i+1}$.)

Finally (this is the last of the preliminaries!), let P_1, P_2, P_3, . . . be the axioms of Peano arithmetic.

I will say that a statement is *n*-fulfillable if there is a good sequence of length n that fulfills the statement. The following is the statement that I will show to be independent of Peano arithmetic:

(II) *For every* n *and every* m, *the conjunction of the first* m *axioms of Peano arithmetic is* n-*fulfillable.*

Or (this is easily seen to be equivalent)

(III) *For every* n, *the conjunction of the first* n *axioms is* n-*fulfillable.*

What does this actually say? Well, if for "*n*-fulfillable" we substitute "fulfilled by an increasing infinite sequence," (III) becomes the statement that Peano arithmetic (or whatever consistent extension we take P to be) is *true*. Of course, truth is not expressible in P itself. So (III) is a kind of weak substitute for the statement that Peano arithmetic is a true theory. In fact, "*n*-fulfillable" is a Σ_1 property, so the above is only a Π_2 sentence. What it says, however, is that Peano arithmetic has a weak kind of *correctness*.

From here on I will only outline Kripke's proof; the details are not hard to verify.

First, we observe that if a formula A is true, then so is the statement that for every n, A is n-fulfillable. This is true because we can take any number larger than n we please for $(s)_0$ and then choose a number that is larger than the maximum values assumed by the "Skolem functions" corresponding to the existential quantifiers in A (as the arguments of those functions range through numbers $\leq (s)_i$), and also larger than the square of $(s)_i$, to be $(s)_{i+1}$, where $i = 0$, 1, . . . , $n-1$. This choice guarantees that a suitable value for the next existential quantifier in the formula can always be found without going more than one place to the right of the place chosen by the "attacking" player when he picked a number for the preceding universal quantifier. Moreover, for any fixed formula A, this argument can be formalized in Peano arithmetic, that is:

(IV) *It is a theorem of P that if A, then A is n-fulfillable (for n = 1, 2, 3, . . .).*

Obviously, because it is a theorem of P that if P_i, then P_i is *n*-fulfillable, and P_i is itself an axiom of P, then (for each i, n) it is a theorem of P that P_i is *n*-fulfillable. So if P is Σ_2-sound (if all Σ_2 statements implied by P are true), then it is not a theorem of P that there exists an n such that the conjunction of P_1, P_2, $\ldots$ up to P_n is not n-fulfilled. In other words, the negation of (III) is not provable in P unless P fails to be Σ_2-sound. (Note that Σ_2-soundness—truth of all Σ_2 statements implied by P—is weaker than Gödel's hypothesis of "ω-consistency.")

Our problem is to show by model-theoretic means that (III) (which is a true statement if P is Σ_1-sound, because for each n, the conjunction of P_1, P_2, $\ldots$ up to P_n is n-fulfillable is a Σ_1-consequence of the conjunction (P_1, P_2, $\ldots P_n$), and hence of P) is not a theorem of P. This means that we must construct a model in which (III) is false. Obviously the standard model will not do, because (III) is true in the standard model. So we must construct a nonstandard model.

We know (by the ultrapower technique) how to construct nonstandard models. So let us construct one—construct *any* nonstandard model, say, M. In M the statement "The conjunction of P_1, P_2, $\ldots P_n$ is n-fulfillable" is true for the "standard" numbers $n = 0, 1, 2, 3, \ldots$ because for each of these numbers the statement is a theorem and M is a model. But a statement expressible in the language of P cannot be true of the finite numbers in M and false of all the infinite ones. (For, by a theorem of P, there is a least number of which the statement is not true, if there are any numbers at all of which it is not true. But there cannot be a least infinite integer in M, because if k is an infinite integer in M, so is $k-1$.) So there must be at least one infinite integer in M of which this statement is also true, say N; that is, N is such that in M the conjunction of the first N axioms of P is N-fulfillable. Hence there must be a "Gödel number" S (also an infinite integer, as we shall see) that is the Gödel number of a "sequence" (in the sense of the model M) of length N that fulfills P_N.[1]

Consider the ordinary infinite sequence (this is not an object of the model M, but what model theorists call an "external" object) $(S)_0$, $(S)_1$, $(S)_2$, and so on. (The members of this sequence are all infinite integers from the model M, because S is a "good" sequence, and so even $(S)_0$ is larger than the infinite

---

1. A former student asked me: "Why must there be a Gödel number S of the sequence in M? Is this because there is some first-order formula true of the finite integers?" That is indeed the reason; for every finite integer n, "the conjunction of $P_1, \ldots, P_n$ is n-fulfillable" is a true Σ_1 statement of PA; and hence the nonstandard model M must contain an infinite N such that the formula "the conjunction of $P_1, \ldots, P_x$ is x-fulfillable" is satisfied by N in M.

integer N.) Because S fulfills P_1, P_2, . . . , it is easy to verify that this "external" sequence also fulfills these statements; that is, this external sequence fulfills each axiom of P. Now let H be the submodel of M that contains all the members of M that are smaller than a member of the external sequence. The external sequence is "good," so H is closed under " $+$ " and "$\cdot$"; and the external sequence we constructed is cofinal with H. But each axiom of P is fulfilled by an ω-sequence that is cofinal with the structure H, and this means that each axiom of P is true in H. (The argument that a formula that is fulfilled by an ω-sequence that is cofinal with the integers is true in the integers carries over to any structure.) So M is actually a model of P.

Now let us assume that we carried out this construction choosing as S the smallest Gödel number of a sequence of length N that N-fulfills P_N. S, considered as an "infinite integer," will be larger than every number in the sequence S (the Gödel number of a sequence in the Kleene system is always larger than every number in the sequence), and, by the construction of H, every number in H is smaller than some number in S. So S itself (considered as an "integer") is not in H. Is there any Gödel number of a sequence of length N that N-fulfills P_N in H? The answer is "no." For the statement "X N-fulfills P_N" is a Σ_1 statement, and Σ_1 statements "persist upward": if they are true in a substructure, they are true in the bigger structure. So if this statement were true in H, then it would also be true (of the very same X) in the original structure M. But then X would be a Gödel number smaller than S with the property of S (being the Gödel number of a sequence of length N that . . .)—contrary to the choice of S as the smallest integer with the property. So H contains no Gödel number that is a "witness" to the statement "*There is an X that N-fulfills P_N*"; that is, this statement is not true in H. Hence (III) is not true in H. We have succeeded in producing a model in which (III) is false.

One last remark: If P is any consistent finitely axiomatizable extension of Peano arithmetic, then if A is the conjunction of the axioms of P, A implies that $(n)(A$ is n-fulfillable) is a theorem of Peano arithmetic, and hence $(n)(A$ is n-fulfillable) is a theorem of P. So, if we let T_A be the theory each of whose axioms P_1, P_2, . . . is just A, the statement $(n)(Pn$ is n-fulfillable) is just (up to logical equivalence) $(n)(A$ *is n-fulfillable)*, and this is a theorem of T_A. But we just showed that this is undecidable in a consistent recursively enumerable extension of Peano arithmetic; hence Peano arithmetic has no consistent finitely axiomatizable extensions.

16

A Proof of the Underdetermination "Doctrine"

Before the publication of "On What There Is" in 1948, W. V. Quine was known to most philosophers (if they had heard of him at all) as a "logician." But now it is universally acknowledged that the series of Quine's philosophical papers and books that began with that famous essay are among the most important writings to have been published in (approximately) the second half of the twentieth century. One of the most famous theses in that series of works is "the doctrine that natural science is empirically underdetermined."[1] But I have observed that Quine's "doctrine" is widely misunderstood, and its logical status is, in fact, murkier than most readers imagine. The purpose of the present essay is to remove the murk and to prove the doctrine.

The principal misunderstanding I have observed is the idea that underdetermination follows immediately from "the Quine-Duhem thesis," by which is meant the thesis that theoretical claims cannot be either conclusively falsified or conclusively confirmed by isolated experiences. If those experiences appear to falsify a theory, it is always possible to revise parts of the theory

1. W. V. Quine, "On Empirically Equivalent Systems of the World," *Erkenntnis* 9 (1975): 313–328; quotation from p. 313.

(including what the logical empiricists called the "coordinating definitions"), or, if worst comes to worst, to reject the observation sentences that we use to report those experiences. However, this claim has literally nothing to do with the "underdetermination" doctrine. That doctrine, to which Quine gave considerable weight in his writings,[2] says that even if we do not reject any observation sentences that we take to have been verified, and even if we do not allow any changes to the theory, still, if there is at least one theory that has a given set of observational consequences, then there will always be more than one.

Another misunderstanding that I have observed, even among those who do not confuse the underdetermination doctrine with the Quine-Duhem thesis, is that the underdetermination doctrine is, somehow, logically obvious. But it is not obvious, and in her brilliant book on conventionalism, Yemima Ben-Menahem reminds us that in a little-known lecture Quine surprisingly confessed that the doctrine risks being either trivial or, at best, unprovable. As Ben-Menahem has pointed out, in the original version of "On Empirically Equivalent Systems of the World," Quine suggested a way of stating the doctrine that would avoid triviality, but concluded that so restated, the truth of the doctrine "is an open question."[3] However, all this rethinking was excised from "On Empirically Equivalent Systems of the World" before that essay was collected in *Pursuit of Truth.*[4]

As Ben-Menahem explains the problem:

Underdetermination crumbles under a problem that seems utterly trivial at first. Quine invites us to consider two theories that are identical except for a permutation of terms, for example "electron" and "proton." What is the relation between these theories? Taken at face value, they are clearly incompatible, for each affirms sentences the other denies, for instance, "The negative charge of the electron is . . ." It is likewise clear that these incompatible theories are empirically equivalent—they have exactly the same empirical import. The question is whether this would count as an

---

2. For example, in W. V. Quine, "On the Reasons for the Indeterminacy of Translation," *Journal of Philosophy* 67 (1970): 178–183, Quine gives the underdetermination of theory by data as one of the two principal reasons for his famous (and controversial) thesis of the indeterminacy of translation.

3. Quine, "On Empirically Equivalent Systems of the World," 327.

4. W. V. Quine, *Pursuit of Truth* (Cambridge, Mass.: Harvard University Press, 1990).

example of underdetermination, that is, whether such a minor permutation of terms suffices to render the two theories empirically equivalent but incompatible alternatives. Rather than taking them at face value, is it not more reasonable to regard them as slightly different, though perfectly compatible, formulations of the same theory?[5]

She goes on to explain that Quine endorsed the latter alternative. He wrote, "So I propose to individuate theories thus: two formulations express the same theory if they are empirically equivalent, and there is a construal of predicates that transforms one theory into a logical equivalent of the other."[6] Here is what I believe Quine meant by these words:

First, Quine certainly did not mean to allow us to reconstrue the observation terms, because that could change the empirical import of the theory. Second, in modern mathematical logic, we speak of an interpretation of one theory relative to another when there are possible definitions of the predicates of the first theory in the vocabulary of the other theory that are such that the "translations" of the theorems of the first theory by means of those definitions are all theorems of the second theory. In this terminology, what Quine's criterion comes to is this: two theories T_1 and T_2 ("two" in the merely syntactic sense) should be regarded as formulations that express the same theory just in case there is an interpretation of the theoretical (and possibly the mathematical?) predicates of the one in the language of the other (say, of T_1 relative to T_2) such that all the theorems of T_2 are consequences of the subset of theorems that are "translations" of theorems of T_1 (i.e., the "translation" of T_1 is, up to logical equivalence, just T_2). Note that if neither T_1 nor T_2 is interpretable relative to the other, or, in Quine's language, which I shall henceforth use, if neither is "translatable" into the language of the other by a suitable construal of its predicates (one that leaves the observation vocabulary fixed), then T_1 and T_2 certainly do not "express the same theory." And because failure of "construability" in both directions is a simpler (as well as a possibly stronger) criterion of difference than Quine's, I shall employ it in what follows.

Finally, by the "doctrine of the underdetermination of natural science by all observable events," Quine meant the claim that any scientific theory that accounts for a set of observations must always have an alternative that ac-

5. Yemima Ben-Menahem, *Conventionalism: From Poincaré to Quine* (Cambridge: Cambridge University Press, 2006), 246.

6. Quine, "On Empirically Equivalent Systems of the World," 320.

counts for the same set of observations.[7] But what happens to this doctrine if we now require that the "alternative" must be a really different theory, not just another "formulation that expresses the same theory"? Quine's answer in the 1975 lecture was: "This, for me, is an open question."[8]

Now let T be a consistent scientific theory, and let the language of T include the predicates and proper names contained in "observation sentences," under some understanding of that notion that is acceptable to those of empiricist sensibilities. (I will refer to those predicates and proper names as the O-vocabulary.) I shall assume that the theoretical terms of T are not definable in terms of the O-vocabulary plus logical-mathematical vocabulary. (Otherwise the theoretical terms were not needed in the first place.) I shall show that T is "underdetermined by the totality of true observation sentences" in Quine's sense of "underdetermined." This is a result that answers Quine's open question and vindicates his original intuition.

Theorem: Let T be as above. I shall show that there exists a theory T' that implies the same sentences in the O-vocabulary as T such that T is not "translatable into T' via construal of predicates" in Quine's sense, and T' is not "translatable into T via construal of predicates" in that same sense (holding the O-vocabulary fixed). In Quine's language, the choice between T and T' is thus "underdetermined" by the totality of true sentences in the O-vocabulary. The proof uses two famous pieces of logic: (1) Tarski's technique for defining "True-in-L" in any fixed formal language L; and (2) Gödel's second incompleteness theorem.

Proof: Let the vocabulary of T' be the O-vocabulary of T, plus enough arithmetic and set-theoretic vocabulary to define "true in L_O," where L_O is the first-order language whose predicates and proper names are just those in that observational vocabulary of T, say, by the formula $TR(x)$. I assume that the mathematical axioms of T' are correct and include the arithmetic needed for the Gödel incompleteness theorems, and that the definition of TR satisfies Tarski's criterion of adequacy, that is, all sentences of the form "$TR(\underline{n}) \Leftrightarrow Sn$" are provable in T',

---

7. What Quine means by "observations" is a subset of the observation sentences, those observation generalizations that he calls "observation conditionals." Here I will show that the "underdetermination doctrine" is correct even if we consider all the consequences of a theory statable in the observation vocabulary to count as part of its empirical import, and not only the "observation conditionals."

8. Quine, "On Empirically Equivalent Systems of the World," 327.

where Sn is a sentence in L_O and $\underline{n}$ is the formal numeral for its Gödel number. Let $N(x)$ be a formula of T' that represents the primitive recursive function whose value for any number n is the Gödel number of the result of prefixing the negation sign to the sentence with the Gödel number n. I also assume that the mathematical axioms of T' are strong enough to prove all sentences of the form "$Tr(N(n)) \Leftrightarrow \sim Sn$" (we might call these F-sentences because they say that "$TR(N(x))$" is an adequate falsity predicate for L_O.) The sole extralogical axiom of T' is (x) $((\exists n)\, \text{Proof}(n, x) \rightarrow TR(x))$, where "$\text{Proof}(n, x)$" represents the primitive recursive predicate "n is the Gödel number of a proof in T of the formula of L_O whose Gödel number is x." (Thus T' is a canonical formalization of the theory that "says," "All the consequences of T in its observational vocabulary are true.") I show that T' has the properties listed above.

1. T' is not translatable into T by construal of predicates. The idea of the proof is simple: if T' were translatable into T, then the obvious semantic proof of the consistency of T would also translate from T' into T, and T would be inconsistent. Here follow the details: Assume the contrary, and let the "translation" of the predicate TR be TR^*. Because "construals" in Quine's sense are not allowed to affect the O-vocabulary (because otherwise, the empirical import of the O-terms would not be held constant), each T sentence "$TR(n) \Leftrightarrow Sn$" of T' goes over into a "T-sentence" "$TR^*(n)) \Leftrightarrow Sn$" of T, and each F-sentence similarly goes over into an F-sentence "$TR^*(N^*(n)) \Leftrightarrow \sim Sn$." It follows that the predicate $TR^*(x)$ (respectively, $TR^*(N^*(x)))$[9] is also an adequate truth-predicate (respectively, falsity-predicate) for L_o. Trivially, $\sim(S_1 \land \sim S_1)$ is a theorem of T. Because the T-sentences and F-sentences (using the truth-predicate TR^* and the falsity-predicate $TR^*(N^*(x))$ for S_j and the negation of S_j) are theorems of T, it will be a theorem of T that $\sim(TR^*(\underline{1}) \land TR^*(N(\underline{1}))).$[10] But if T were inconsistent, then both "$(\exists n)\, (\text{Proof}(n, \underline{1}))$" and "$(\exists n)(\text{Proof}(n, N(\underline{1})))$" would be true and also would be theorems of T', because "Proof" represents provability-in-T in T', and hence "$(\exists n)(\text{Proof}^*(n, \underline{1}))$" and "$(\exists n)$ $(\text{Proof}^*(n, N^*(\underline{1})))$" would be theorems of T. Also, under the "construal," the axiom "$(x)((\exists n)(\text{Proof}(n, x) \rightarrow TR(x))$" of T' goes over into "$(x)((\exists n)(\text{Proof}^*(n,$

---

9. Of course, the numbers 1, 2, 3 . . . may become different "objects" under the construal, but all their relevant mathematical properties will be preserved, because the "translations" of the theorems needed for the Tarskian theorems and the Gödel-Rosser theorems will be theorems of T.

10. Recall that "$\underline{1}$" abbreviates the formal numeral for the sentence s_1.

x) → $TR^*(x)$))." So if both "$(\exists n)(\mathrm{Proof}(n, \underline{1}))$" and "$(\exists n)(\mathrm{Proof}(n, N(\underline{1})))$" were true, it would be a theorem of T that $TR^*(\underline{1}) \wedge TR^*(N(\underline{1}))$, contradicting $\sim(TR^*(\underline{1}) \wedge \sim TR^*(N(\underline{1})))$. Formalizing this argument in T (which is possible because T contains as much mathematics as T' does, via the assumed "translation" of T' into T), it is a theorem of T that T is consistent. But then, by the Gödel theorem, T is inconsistent, contrary to the assumption that T is a consistent theory.

2. T is not translatable into T' by construal of predicates. Because T' has no nonlogical vocabulary besides the O-vocabulary, such a translation would violate the assumption that the theoretical terms of T are not definable in terms of the O-vocabulary plus logical-mathematical vocabulary.

3. T and T' are empirically equivalent (imply the same sentences in the O-vocabulary).

Proof: T' obviously implies all the O-consequences of T, because for every theorem of T expressible in the O-vocabulary, say, Sj, "$(\exists n)(\mathrm{Proof}(n, j))$" is a theorem of Peano arithmetic and hence of T', so $TR(j)$ is a theorem of T' [by the one extralogical axiom of T', $(x)((\exists n)\mathrm{Proof}(n, x) \rightarrow TR(x))$, and the T-sentence $TR(j) \Leftrightarrow Sj$ is a theorem of T', and hence Sj is a theorem of T'. It remains to show that T' is a conservative extension of the theory (in the sublanguage L_O) whose axioms are the sentences in that sublanguage that are provable in T. Suppose the contrary; that is, suppose that some sentence S in the O-vocabulary is a theorem of T' but not of T. Because S is not a theorem of T, T has a model M in which S is false. Use the restriction of M to the O-vocabulary to interpret that vocabulary in both T and T', and interpret the mathematical predicates of T' in the standard way. Because T' implies S, T' is false under this interpretation, and hence its sole extralogical axiom $(x)((\exists n)$ $\mathrm{Proof}(n, x) \rightarrow TR(x))$ is false under this interpretation. But all the sentences in L_O that are provable in T are true in M, so this can be the case only if "TR" is not true of all and only the true sentences of L_O. But an "adequate" truth-definition in Tarski's sense can be wrong only if the mathematical part of the metalanguage in which that definition is given is incorrect,[11] and we assumed that the mathematical axioms of T' were correct, so this is impossible.

11. I pointed out the possibility that a truth-definition that is adequate in Tarski's sense can still be extensionally incorrect because the metalanguage is incorrect in Hilary Putnam, "Do True Assertions Correspond to Reality?" in Putnam, *Philosophical Papers*, vol. 2, *Mind, Language and Reality* (Cambridge: Cambridge University Press, 1975), 70–84; see p. 73. Note that consistency does not guarantee correctness, as the phenomenon of consistent theories that are ω-inconsistent illustrates.

Discussion

Obviously T' does not look like what we normally think of as a "scientific theory." For one thing, it has no "theoretical terms," no nonmathematical and nonobservational predicates. Can we produce a T'' that is more "normal" and is also empirically equivalent to T but not mutually relatively interpretable with it via Quinean "construals"?

Here is how to do that. Construct a theory T^* thus: Take the same logical-mathematical vocabulary and axioms as those of T' above and the same observational vocabulary as that of T and T', and add a new primitive "theoretical term" $H(x, y, z, t)$, where H is a predicate letter that does not belong to the vocabulary of T. "H" might be read "the electrical charge at space-time point (x, y, z, t) is positive," or as "space-time point (x, y, z, t) is magical," or however you like. Intuitively, H is a bivalent magnitude of space-time points. Instead of taking $(x)((\exists n)\text{Proof}(n, x) \rightarrow TR(x))$ as an extralogical axiom, this time we take the following two axioms:

(I) $\qquad\qquad\qquad (x)((\exists n)\text{Proof}(n, x) \rightarrow H(x, 0, 0, 0)),$

(II) $\qquad\qquad\qquad (x)(H(x, 0, 0, 0) \rightarrow TR(x)).$

(In positivist jargon, (I) is a "theoretical postulate" and (II) is a "bridge law" connecting the theory T^* with observational statements.) That T^* is "empirically equivalent" to T' and hence to T can be shown thus: Because the extralogical axiom of T' above follows immediately from (I) and (II), and T' has all the consequences of T in the "observational vocabulary" LO, T^* has all those consequences. That it has no further consequences can be shown by noting that T^* can be "translated into T' by construal of predicates": just define "$H(x, y, z, t)$" as "$(\exists n)\ \text{Proof}(n, x) \wedge y=y \wedge z=z \wedge t=t.$"

That T^* cannot be translated into T by construal of predicates can be shown exactly as T' was shown to have that property.

Finally, I prove that T cannot be translated into T^* by construal of predicates. If T could be translated into T^* by construal of predicates, then, because T^* can be translated into T' by construal of predicates, it would follow that T can be translated into T' by construal of predicates, which was shown to be impossible. QED.

17

A Theorem of Craig's about Ramsey Sentences

Frank Ramsey famously claimed that the whole empirical content of a theory can be expressed by a single second-order sentence *containing no theoretical terms* by conjoining the axioms of the theory and then replacing each of the theoretical terms in the resulting sentence by an existentially bound variable.[1] I recently became aware that many philosophers mistakenly believe that this second-order sentence (now known as the "Ramsey sentence" of the theory to which this technique is applied) has the same content as the whole set of observational consequences of the original theory. (I suspect that Ramsey himself believed this.) I give a simple proof of a result due to William Craig[2] that shows that this is false.

---

1. Frank Plumpton Ramsey, *Philosophical Papers,* ed. D. H. Mellor (Cambridge: Cambridge University Press, 1990), 131.
2. Craig communicated this result to me in a conversation, together with a proof that employed the fact that Peano arithmetic is not finitely axiomatizable, in the late 1960s or early 1970s. I communicated that proof to my student Jane English (see note 4).

Theorem (Craig): *It is possible for a theory to have only true consequences in its "observation vocabulary" (and hence a true "Craig translation"*[3]*) but a false Ramsey sentence.*[4]

Proof (mine): Consider a possible world W in which the individuals are an ω-sequence of lampposts going from east to west. The language L of the theory I shall construct contains the observation predicates Lx ("x is a lamppost"), xWy ("x is to the west of y"), and the individual constant a for the first lamppost. It also contains the single theoretical term P. The successor relation is defined (in "observation language") thus: $x\, S\, y =_{df} Lx \wedge Ly \wedge xWy \wedge (z)((Lz \wedge zWy) \Rightarrow (zWx \vee z=x))$. These lampposts are the only individuals there are in this world (although this is not essential; it should be obvious how to modify the proof if there are additional individuals). L_O (the "observation language") is just the language L minus the predicate P. Let T_O be the theory consisting of all the true sentences of L_O. Finally, let M be a nonstandard model of T_O. (Because T_O is consistent and has infinite models, such a model exists.) In M there are many "nonstandard elements," because in the intended model, the world W, there are only ω lampposts, while in M there are many more. (E.g., because "$(x)(Lx \Rightarrow (\exists y)(Ly \wedge yWx))$" belongs to T_O, and M is a model for T_O, every "nonstandard lamppost" in M has other "lampposts"—which must also be nonstandard—to its "west.") Interpret P to mean "x is a standard lamppost in M," thus extending M to a model M' of all of the language L.

Now consider the following theory T:

$$Pa \wedge (x)(y)(Lx \wedge Ly \wedge Px \wedge ySx \Rightarrow Py) \wedge (Ex)(Lx \wedge {\sim}Px).$$

T is true in M' (and hence consistent), but not true in W (the "actual" world) under the original interpretation of L_O and *any* interpretation of P. This is obvious, because T says that a has P and also says that P is hereditary, and hence all the standard lampposts would have to have P if there were such

3. For an explanation of the "Craig translation," that is, the axiomatizable theory in the "observational vocabulary" (whose existence was proved by William Craig) with the same observational consequences as a given scientific theory containing so-called theoretical terms, see Hilary Putnam, "Craig's Theorem," *Journal of Philosophy* 62, no. 10 (May 13, 1965): 251–259; reprinted in Putnam, *Philosophical Papers*, vol. 1, *Mathematics, Matter and Method* (Cambridge: Cambridge University Press, 1975), 228–236.

4. The only published proof I know of is in Jane English, "Underdetermination: Craig and Ramsey," *Journal of Philosophy* 70, no. 14 (August 16, 1973): 453–462

a P, and there are no nonstandard lampposts in W. Hence the Ramsey sentence of T is false in the world W. But T is true in M', and M' is a model of T_O, so T has no false consequences in L_O. QED.

COMMENT: I believe that this brings out how it is that a theory can have a true Craig translation and a false Ramsey sentence more clearly than Craig's original proof. Nothing in this phenomenon depends on second-order Peano arithmetic or the nonfinite axiomatizability thereof, which is what Craig used. But it does depend crucially on the existence of nonstandard models, which is what makes it possible for a theory to be consistent (and hence have models) even when we add to it the totality of true sentences in L_O (or even the graphs of all the predicates in L_O, adjoining for this purpose names of all the observable individuals of W), while not having any models that respect the intended interpretation of L_O.[5]

5. English says that the Ramsey sentence of the theory Craig constructs (the dual of Peano arithmetic) is "L-False." But this is only the case because in Craig's example the "observable things" are the natural numbers, and the truths of second-order arithmetic are (according to English) "L-truths." Nevertheless, Craig's T is not self-contradictory, and because it has models, of course its Ramsey sentence does too. It is just that they do not respect the intended interpretation of Craig's L_O (first-order arithmetic).

PART THREE

Values and Ethics

18

The Fact/Value Dichotomy
and Its Critics

As I have said more than once in these chapters, my favorite definition of philosophy is Stanley Cavell's: "education for grownups."[1] In this essay I shall be discussing an issue that is obviously a philosophical one and, at the same time, one on which many who certainly consider themselves "grownups," including many philosophers, economists, lawyers, and policy makers of all kinds, unquestionably *need* "education." That is the issue that I described in a book as "the fact/value dichotomy."[2]

The Fact/Value Dichotomy

The book I just mentioned begins thus:

Every one of you has heard someone ask, "Is that supposed to be a fact or a value judgment"? The presupposition of this "stumper" is that if it's a "value

---

1. Stanley Cavell, *The Claim of Reason: Wittgenstein, Skepticism, Morality, and Tragedy* (Oxford: Clarendon Press, 1979), 125.
2. Hilary Putnam, *The Collapse of the Fact/Value Dichotomy and Other Essays* (Cambridge, Mass.: Harvard University Press, 2002).

judgment" it can't possibly be a [statement of] "fact," and a further presupposition of this is that value judgments are subjective.[3]

I illustrated the way in which this idea can impact policy by citing the views of Lionel Robbins during the depths of the Depression. At that time Robbins argued against the whole idea of income redistribution on the philosophic ground that value judgments are (according to him) outside the sphere of reason altogether. I quote:

> If we disagree about ends it is a case of thy blood or mine—or live or let live according to the importance of the difference, or the relative strength of our opponents. But if we disagree about means, then scientific analysis can often help us resolve our differences. If we disagree about the morality of the taking of interest (and we understand what we are talking about), then there is no room for argument.[4]

Other influential economists in the 1930s, 1940s, and 1950s, while no less respectful of logical positivism and its insistence that value judgment totally lack what the positivists called "cognitive meaning," were not willing to follow Robbins in jettisoning welfare economics. Instead, they allowed the economist to appeal to values, *provided* it was made clear that all that she was doing was making means-ends judgments of the form "If you have such and such values, then such and such is the most feasible economic policy." That values themselves did not admit of rational argument was not challenged. Thus, for example:

> It is fashionable for the modern economist to insist that ethical value judgments have no place in scientific analysis. Professor Robbins in particular has insisted upon this point, and today it is customary to make a distinction between the pure analysis of Robbins *qua* economist and his propaganda, condemnations, and policy recommendations *qua* citizen. In practice, if pushed to extremes, this somewhat schizophrenic rule becomes difficult to adhere to, and it leads to rather tedious circumlocutions. *But in essence Robbins is undoubtedly correct.*[5]

---

3. Ibid., 7.

4. Lionel Robbins, *On the Nature and Significance of Economic Science* (London: Macmillan, 1932; reprinted 1952, quotation from the reprint edition, 150.

5. P. A. Samuelson, *Foundations of Economic Analysis* (original ed., 1947; enlarged version, Cambridge, Mass.: Harvard University Press, 1983), quotation from the enlarged edition, 219–220.

In this way the logical positivist claim that "why people respond favorably to certain facts and unfavorably to others is [merely] a question for the sociologist"[6] came to be regarded as "undoubtedly correct" by a policy science whose recommendations affect the lives of literally billions of our fellow human beings. Ethical questions are the questions that most of us think it most important to discuss rationally and not irrationally. But if the logical positivist view to which economists deferred for such a long time were indeed correct, then the very idea of discussing value questions rationally would be ("cognitively") *nonsense.*

Cavell's Place in This Discussion

I mentioned Stanley Cavell's definition of philosophy as "education for grownups," and one reason I did so is that he has an important place in the debate concerning facts and values. If this is not obvious at once, the reason, I believe, is the strange way in which one of Cavell's most important contributions has been neglected, both by his admirers and by his critics.[7] I am referring to the four chapters that make up part 3, titled "Knowledge and the Concept of Morality," of Cavell's book *The Claim of Reason.*

I just said that if the logical positivist view to which economists deferred for such a long time were indeed correct, then the very idea of discussing value questions rationally would be *nonsense.* But in my time and Cavell's many analytic philosophers became acquainted with (and, alas, often embraced) that idea as the result of reading Charles Stevenson's *Ethics and Language* rather than the logical positivists, and although Stevenson's affinity to logical positivism was recognized immediately, the irrationalist consequences of emotivism are intentionally played down in that work. In "Knowledge and the Concept of Morality" Cavell is concerned from the beginning to expose the points at which that irrationalism is nonetheless visible in Stevenson's book. Indeed, it is not very *well* hidden, because Stevenson devotes a whole chapter (chapter 7) to telling us that there is no such thing as a valid argument in ethics.[8]

---

6. Alfred Ayer, *Philosophical Essays* (London: Macmillan, 1954), 237.

7. An important exception is Cora Diamond, "Losing Your Concepts," *Ethics* 98 (January 1998): 255–277.

8. See Charles Stevenson, *Ethics and Language* (New Haven, Conn.: Yale University Press, 1944), 252–254.

Stevenson's "first question," as he calls it, is this:

> What is the nature of ethical *agreement* and *disagreement?* Is it parallel to that found in the natural sciences, differing only with regard to the subject matter; or is it of some broadly different source?[9]

His well-known answer is that "the disagreements that occur in science, history, biography" are "disagreements in belief," whereas "it is disagreements in attitude . . . that chiefly distinguish ethical issues from those of science."[10] Where a disagreement is in attitude, Stevenson does say that "reasons" can be offered for and against, but he says that these "reasons" are related (only) *psychologically* (since they are not related deductively and not inductively) to the judgments they support.

Stevenson assumes (as is too often assumed today) that all disagreements in "science" (or disagreements about "facts"—these notions are often simply equated) can be settled either deductively or inductively, and that such settlement results in agreement.[11] That not all moral disagreements can be so settled is another of Stevenson's reasons for concluding that these are not disagreements in "belief." For he argues that only on the assumption that "all disagreement in attitude is rooted in disagreement in belief" can moral disagreements be settled by rational proof, and this he calls a dubious "psychological generalization."[12]

One of Stevenson's early examples of "the methods used in moral arguments" is the following:

> A (SPEAKING TO C, A CHILD): To neglect your piano practice is naughty.
> B (IN C'S HEARING): No, C is very good about practicing. (Out of C's hearing): It's hopeless to drive him, you know, but if you praise him he will do a great deal.[13]

As Cavell remarks, one wonders why such examples "as much as seem to be examples of moral encounter."[14] In fact, the possibility that morality has char-

---

9. Ibid., 2.
10. Ibid., 13.
11. I criticize this assumption in Hilary Putnam, *Ethics without Ontology* (Cambridge, Mass.: Harvard University Press 2004), 75–78.
12. Stevenson, *Ethics and Language,* 136.
13. Ibid., 113.
14. Cavell, *Claim of Reason,* 253.

acteristic modes of *argument* that distinguish it from, inter alia, mere rhetoric and propaganda is (as we shall see) explicitly ruled out by Stevenson. But the investigation of such modes of argument (and the modes of description that they presuppose) is precisely what has always concerned the best moral philosophers, up to and including Cavell.

In addition, Cavell writes:

> But suppose that it is just characteristic of moral arguments that the rationality of the antagonists is not dependent on an agreement emerging between them, that there is such a thing as a *rational disagreement* about a conclusion. . . . Without the hope of agreement, argument would be pointless, but it doesn't follow that without agreement—and in particular, apart from agreement arrived at in particular ways, e.g. apart from bullying, and without agreement about a conclusion concerning what ought to be done—the argument was pointless.[15]

But if, as Cavell suggests here, there is such a thing as a rational argument that cannot be conclusively settled, then the whole argument from the existence of "irresolvable" disagreements in ethics to the absence of "cognitive meaning" in ethical judgments collapses. (In fact, as I point out in *Ethics without Ontology*,[16] there are "factual" issues, especially in the social sciences, on which it is difficult and perhaps impossible to get agreement.)

Summing up these opening observations in "Knowledge and the Concept of Morality," Cavell already lists four points on which he disagrees with Stevenson:

1. That all disagreement in attitude is *moral* disagreement;
2. that all disagreements which cannot be (rationally) *settled* (end in a conclusion which all parties agree is the right one) are irrational;
3. that a reason which is neither deductively nor inductively related to a judgment is "therefore" "only" "psychologically" related to it;
4. that what makes science rational is that it consists of beliefs about matters of fact—and hence consists of methods which rationally settle disagreements.[17]

15. Ibid., 254.
16. Putnam, *Ethics without Ontology*, 76.
17. Cavell, *Claim of Reason*, 260.

Cavell remarks that

> Stevenson's view requires, or contains, all of these ideas, and he must obviously take them to be obvious in themselves or to follow, obviously, from the fact that there are different "kinds" of disagreement. Given what I take to be the remorseless paradoxicality of his view, its wide acceptance— despite criticisms of *pieces* of his view which would have seemed essential to it (e.g., of his causal theory of "meaning," and in particular of emotive "meaning," and still more particularly of his analysis of the word "good")— must mean that these assumptions . . . are widely shared.[18]

They are indeed widely shared assumptions. In fact, they have become a sort of cultural institution—a most unfortunate one, which is why I have devoted a good deal of my writing in the past three decades to attacking it. That is why I began this essay by describing this as an issue where "education for grownups" is desperately needed. I have argued (sometimes in concert with the economist/philosopher Vivian Walsh)[19] that the intellectual legs on which the fact/value dichotomy stood are now in ruins. But to see that that is the case, one needs to bring together results from different parts of philosophy. Specifically, one needs to bring together the observations (by different philosophers) of the way in which so-called factual and so-called evaluative predicates are mutually "entangled"—the way in which it is a fantasy to suppose that the predicates we use to give sensitive and relevant descriptions of human beings and human interactions can be disentangled into two components, a purely descriptive component and an evaluative component—and the point, first argued at length by Morton White,[20] that Quine's demolition of the logical positivist dichotomy of theory and observational fact (or "fact" and "convention," as Quine sometimes put it) also destroyed the logical positivist arguments for the fact/value dichotomy. It was the prestige of those arguments that had exerted such a powerful influence on social scientists, such as the economists mentioned at the beginning of this chapter.

The logical positivist arguments in question depended on a *serious* effort, one continued over many years, to draw a clear line between factual proposi-

---

18. Ibid., 260.

19. Putnam and Walsh's joint papers are collected in Hilary Putnam and Vivian Walsh, eds., *The End of Value-Free Economics* (London: Routledge, 2011).

20. Morton White, *Towards Reunion in Philosophy* (Cambridge, Mass.: Harvard University Press, 1956; reprint, Princeton, N.J.: Princeton University Press, 2004); emphasis in the original.

tions, theoretical postulates (which they eventually came to regard as only "partially interpreted"),[21] mathematical-logical propositions (which they took to be analytic), and "pseudopropositions" (or "nonsense"), the last of which included, according to Carnap, "all statements belonging to Metaphysics, regulative Ethics, and (metaphysical) Epistemology."[22] The sequence of attempts to do this was summed up in Hempel's "Problems and Changes in the Empiricist Criterion of Meaning,"[23] a famous paper that closed with an idea that Carnap had proposed in "The Foundations of Logic and Mathematics" and later developed more fully in "On the Methodological Character of Theoretical Concepts."[24]

In brief, the idea was that "cognitively meaningful" language could contain not only observation terms (and terms defined in terms of these) but also "theoretical terms," terms referring to unobservables and introduced by systems of postulates, the postulates of the various scientific theories. As long as the system as a whole enables us to predict our experiences more successfully, such "theoretical terms" were now to be accepted as "empirically meaningful."

However, the acceptance of this idea led to a serious problem: to *predict* anything means (on the logical positivists' account) to *deduce observation sentences from a theory.* And to deduce anything from a set of empirical postulates, we need not only those postulates *but also the axioms of mathematics and logic.* And, according to the logical positivists, these do not state "facts" at all. They are *analytic* and thus "empty of factual content." In short, "be-

21. For an explanation and criticism of this rather unclear view, see Hilary Putnam, "What Theories Are Not," in Ernest Nagel, Patrick Suppes, and Alfred Tarski, eds., *Logic, Methodology, and Philosophy of Science* (Stanford, Calif.: Stanford University Press, 1962), 240–252; reprinted in Putnam, *Philosophical Papers,* vol. 1, *Mathematics, Matter and Method* (Cambridge: Cambridge University Press, 1975), 215–227.

22. Rudolf Carnap, *The Unity of Science* (London: Kegan Paul, Trench, Trubner and Co., 1934), 26.

23. C. G. Hempel, "Problems and Changes in the Empiricist Criterion of Meaning," *Revue Internationale de Philosophie* 4 (1950): 41–63.

24. Rudolf Carnap, "The Foundations of Logic and Mathematics," in Otto Neurath, Rudolf Carnap, and C. W. Morris, eds., *The International Encyclopedia of Unified Science* (Chicago: University of Chicago Press, 1939), vol. 1, pt. 1, 198–211; Rudolf Carnap, "On the Methodological Character of Theoretical Concepts," in Herbert Feigl and May Brodbeck, eds., *The Foundations of Science and the Concepts of Psychology and Psychoanalysis,* Minnesota Studies in the Philosophy of Science, vol. 1 (Minneapolis: University of Minnesota Press, 1956), 38–76.

longing to the language of science" was (after the acceptance of Carnap's idea) a criterion of *scientific* significance, but not everything *scientifically* significant was regarded by the positivists as a statement of *fact:* within the scientifically significant there are, according to Carnap and his followers, *analytic* as well as *synthetic* (i.e., factual) statements. Thus the search for a satisfactory demarcation of the "factual" became the search for a satisfactory way of drawing the analytic-synthetic distinction.

At this point, in "Two Dogmas of Empiricism,"[25] Quine demolished the positivists' metaphysically inflated notion of the "analytic" to the satisfaction of most philosophers. But Quine did not suggest that every statement in the language of science should be regarded as a statement of "fact" (i.e., as "synthetic"). Instead, he argued that the whole idea of classifying such statements as the statements of pure mathematics as "factual" or "conventional" (which the logical positivists equated with "analytic") was hopeless. As he later put it:

> The lore of our fathers is a fabric of sentences. In our hands it develops and changes, through more or less arbitrary and deliberate revisions and additions of our own, more or less directly occasioned by the continuing stimulation of our sense organs. It is a pale gray lore, black with fact and white with convention. But I have found no substantial reasons for concluding that there are any quite black threads in it, or any white ones.[26]

But if we lack any clear notion of "fact," what happens to the fact/value dichotomy? As Vivian Walsh has written, "To borrow and adapt Quine's vivid image, if a theory may be black with fact and white with convention, it might well (as far as logical empiricism could tell) be red with values. Since for them confirmation or falsification had to be a property of a theory *as a whole,* they had no way of unraveling this whole cloth."[27]

If the logical positivists' arguments are now generally regarded as failures (which does not mean that the *conclusions* they drew from those arguments

25. W. V. Quine, "Two Dogmas of Empiricism," *Philosophical Review* 60, no. 1 (January 1951): 20–43; reprinted in Quine, *From a Logical Point of View* (Cambridge, Mass.: Harvard University Press, 1961), 20–46.

26. W. V. Quine, "Carnap and Logical Truth," in Quine, *The Ways of Paradox and Other Essays* (New York: Random House, 1966), 107–132; quotation from p. 132.

27. Vivian Walsh, "Philosophy and Economics," in John Eatwell, Murray Milgate, and Peter Newman, eds., *The New Palgrave: A Dictionary of Economics,* vol. 3 (London: Macmillan, 1987), 862; emphasis in the original.

have ceased to exert a powerful influence), it must be said to their credit that those arguments were the product of years of careful effort, as the successive reformulations of "the empiricist criterion of meaning" charted by Hempel testify. Although it was his high regard for logical positivism that inspired Stevenson to defend their "emotivist views, his own arguments for a fact/value (or "belief/attitude") dichotomy rest on no such hard work. For him it is self-evident, as Cavell pointed out in his list of Stevenson's assumptions, that genuine "beliefs" can be proved or refuted by deduction or induction, and that is the only criterion of cognitive meaning that he thinks he needs.

Fact/Value Entanglement

Facts and values are entangled in at least two senses. First, factual judgments, even in physics, depend on and presuppose epistemic values. One would think that this ought to be uncontroversial, but in fact all the leading positivists—joined here by Popper, in spite of his frequently touted disagreements with Carnap and Reichenbach—made what I regard as pathetic attempts to evade this fact.[28] What the logical positivists were shutting their eyes to, as so many today who refer to values as purely "subjective" and science as purely "objective" continue to do, is obvious: the fact that judgments of coherence, simplicity (which is itself a whole bundle of different values, not just one "parameter"), "beauty," "naturalness," and so on are presupposed by physical science. But *coherence, simplicity,* and the like are *values.* All of the standard arguments for noncognitivism in ethics could be repeated without any change whatsoever for noncognitivism in *epistemology;* for example, Hume's argument that ethical values are not "matters of fact" (because we do not have a "sense impression" of goodness) could be modified to read "epistemic values are not matters of fact because we do not have a sense impression of simplicity or a sense impression of coherence." Disagreements about the beauty or "inner perfection" (Einstein's term) of a theory could certainly be described as "differences in attitude." And when it comes to fields less subject to experimental control than physics, fields like history or economics, for example, it is utterly simplistic to suppose that such disagreements can always

28. For details of these attempts, see Putnam, *Collapse of the Fact/Value Dichotomy,* chap. 8.

be settled by "induction and deduction." In fact, after the publication of Nelson Goodman's "The New Riddle of Induction," the idea that there is such a thing as *the* method of "induction" has been seen by philosophers of science to be extremely problematic.[29]

A second way in which values and facts are entangled might be described as "logical" or "grammatical." What is characteristic of "negative" descriptions like "cruel," as well as of "positive" descriptions like "brave," "temperate," or "just" (note that these are the terms that Socrates keeps forcing his interlocutors to discuss), is that to use them with any discrimination, one has to be able to understand an *evaluative point of view*. That is why someone who thinks that "brave" simply means "not afraid to risk life and limb" would not be able to understand the all-important distinction that Socrates kept drawing between mere *rashness* or *foolhardiness* and genuine *bravery*. It is also the reason that, as Iris Murdoch stressed,[30] it is always possible to *improve one's understanding* of a concept like "bravery" or "justice." If one did not at *any* point feel the *appeal* of the relevant ethical point of view, one would not be able to acquire a thick ethical concept, and sophisticated use of it requires a continuing ability to identify (at least in imagination) with that point of view.

My description of this phenomenon as "entanglement" in *The Collapse of the Fact/Value Dichotomy* and thereafter was suggested to me by John McDowell's use of the phrase "disentangling manoeuvre."[31] He describes as follows a move made by emotivists somewhat more recent than Stevenson:

> Typically, non-cognitivists hold that when we feel impelled to ascribe value to something, what is happening can be disentangled into two components. Competence with an evaluative concept involves, first, a sensitivity to an aspect of the world as it really is (as it is independently of value experience), and, second, a propensity to a certain attitude—a non-cognitive state that constitutes the special perspective from which items in the world seem to be endowed with the value in question.[32]

29. Nelson Goodman, "The New Riddle of Induction," in Goodman, *Fact, Fiction, and Forecast* (Cambridge, Mass.: Harvard University Press, 1955), chap. 3.

30. Iris Murdoch, *The Sovereignty of Good* (London: Routledge, 1970).

31. John McDowell, "Non-cognitivism and Rule-Following" (1981); reprinted in McDowell, *Mind, Value, and Reality* (Cambridge, Mass.: Harvard University Press, 1998), 198–218.

32. Ibid., 200–201.

He remarks:

> Now, it seems reasonable to be skeptical about whether the disentangling manoeuvre here envisaged can always be effected; specifically, about whether, corresponding to any value concept, one can always isolate a genuine feature of the world—by the appropriate standard of genuineness: that is, a feature that is there anyway, independently of anyone's value experience being as it is—to be that to which competent users of the concept are to be regarded as responding when they use it: that which is left in the world when one peels off the reflection of the appropriate attitude.[33]

To appreciate why McDowell believes that this claim is so dubious, he asks us to consider any specific conception of a moral virtue, and he continues:

> If the disentangling manoeuvre is always possible, that implies that the extension of the associated term, as it would be used by someone who belonged to the community, could be mastered independently of the special concerns that, in the community, would show themselves in admiration or emulation of actions seen as falling under the concept. That is: one could know which actions the term would be applied to, so that one would be able to predict applications and withholdings of it in new cases—not merely without oneself sharing the community's admiration (there need be no difficulty about that) but *without even embarking on an attempt to make sense of their admiration.*[34]

Later in this essay McDowell connects this discussion with a discussion of Cavell's views, and of Wittgenstein's views as interpreted in Cavell's *Must We Mean What We Say?*, an interpretation that McDowell likes. (Cavell expands on that interpretation in part 2 of *The Claim of Reason*.) Indeed, "entangled" terms are spoken of repeatedly in "Knowledge and the Concept of Morality." One of the earliest such places is the following:

> If . . . we take the case of some specific action, then we might take a case in which the "action" in question is described in ethically prejudicial terms (e.g., "Ought he to have murdered him" rather than ". . . killed him?," or "Was he wrong to betray him?" rather than ". . . to refuse to do what he said?"), or else we might feel that any agreement about the morality of an act will turn on some agreement about how the act is to be described. Was

---

33. Ibid., 201.
34. Ibid. (emphasis added).

it really breaking a *promise?* Is it fair *just* to say he lied when what he did
was lie *in order to* . . . or *as a way of* . . . (Socrates: Then [i.e., in moral dis-
putes] they [i.e., mankind, men] do not disagree over the question whether
the unjust individual must be punished. They disagree over the question
who is unjust, and what was done and when, do they not?)[35]

Apparently, what the "case" in question is *forms part of the content of the
moral argument itself.* Actions, unlike envelopes and goldfinches, do not
come named for assessment, nor, like apples, ripe for grading.[36]

An Important Difference between Cavell and McDowell

Despite all that I have just said, there is an important difference between
Cavell's and McDowell's views. Although each of them accepts the idea that
a competent use of what I have called "entangled" terms requires that one
understand an evaluative point of view—a "Wittgensteinian idea" that dates
back to Philippa Foot and Iris Murdoch in the 1950s, if not earlier[37]—their
understandings of the nature and function of the moral life seem to me quite
different. What makes "Knowledge and the Concept of Morality" so important, in my opinion, is the originality and profundity of the picture of morality that informs it.

McDowell, as is well known, defends a philosophical view of perception
with Kantian roots, a view in which all perception, indeed, all *experience,* is
conceptualized. His account of morality is dependent on that view of perception. In McDowell's account, ethical judgments are justified (as all judgments,
in the end, are justified on his view) by conceptualized experiences. The concepts involved are just the entangled concepts we have been speaking of, and
with their aid we are able to *perceive* that certain actions are cruel or consider-
ate, honest or morally dubious, and so on. The possession of those concepts is
a large part of what McDowell calls (following Aristotle) our "second nature."
And that second nature is one we come to have via a proper moral education.

35. Plato, *Euthyphro* 8c–8d.

36. Cavell, *Claim of Reason,* 264–265; emphases in the original.

37. Bernard Williams reports that "the idea that it might be impossible to pick up an
evaluative concept unless one shared its evaluative interest is basically a Wittgensteinian idea.
I first heard it expressed by Philippa Foot and Iris Murdoch in a seminar in the 1950s." Williams, *Ethics and the Limits of Philosophy* (Cambridge, Mass.: Harvard University Press), 218.

One feature of that view that stands out is that moral *disagreement* is far in the background. But what is most distinctive about Cavell's description of morality in part 3 of *The Claim of Reason* is precisely the centrality of disagreement.

In a passage I am especially fond of, Cavell describes the function of morality thus:

> Morality must leave itself open to repudiation; it provides *one* possibility of settling conflict, a way of encompassing conflict which allows the continuance of personal relationships against the hard and apparently inevitable fact of misunderstanding, mutually incomprehensible wishes, commitments, loyalties, interests and needs, a way of mending relationships and maintaining the self in opposition to itself or others. Other ways of settling or encompassing conflict are provided by politics, religion, love and forgiveness, rebellion, and withdrawal. Morality is a valuable way because the others are so often inaccessible or brutal; but it is not everything; it provides a door through which someone, alienated or in danger of alienation from another through his action, can return by the offering and the acceptance of explanation, excuses and justifications, or by the respect one human being will show another who sees and can accept the responsibility for a position which he himself would not adopt. We do not have to agree with one another in order to live in the same moral world, but we do have to know and respect one another's differences.[38]

This is a passage that demands many readings to be fully appreciated. What Cavell is getting at cannot be stated in a nutshell. It has obvious points of connection with democratic theory. (Like morality, democracy is "a valuable way" because "the others are so often inaccessible or brutal.") It is Kantian in the emphasis placed on mutual respect, but Cavell criticizes Kant because "the most serious sense in which Kant's moral theory is 'formalist' comes not from his having said that actions motivated only in *certain* ways are *moral* actions but in his having found too little difficulty in saying *what* 'the' maxim of an action is in terms of which his test of its morality, the Categorical Imperative, is to be applied."[39] And Cavell's obvious emphasis on knowing and respecting "differences" needs to be tempered by the reading of part 4 of *The Claim of Reason,* in which what he refers to as the "truth of

38. Cavell, *Claim of Reason,* 269.
39. Ibid., 265; emphasis in the original.

skepticism, or what I might call the moral of skepticism,"[40] turns out to problematize—problematize existentially, not just intellectually—the notion of "knowing" one another's differences.

But—in a sense this is the whole point of *The Claim of Reason*—to say that there is a truth *of* skepticism is not to say that the skeptic is right or that skepticism is flat out "true." Entangled terms do have extensions, and we do often get their uses right.

On this point the opposition between Cavell's view, which links evaluation, cognition, rationality, responsibility, and the inevitability of disagreement, and Stevenson's, which links evaluation, "feeling," "causing," and getting others to perform the actions we want them to, is stark. According to Stevenson, for example, the terms "propaganda" and "persuasion" have the *same* extension, if we discount the emotive meaning, and thus "when the terms are *completely* neutralized, we may say with complete equanimity that all moralists are propagandists, or that all propagandists are moralists."[41] He also tells us that "one may be said to exert a peculiarly 'moral' influence if he influences only those attitudes which are correlated with a sense of guilt, sin, or remorse, and so on."[42] Cavell's response is biting:

> So, for example, a mother who plays upon ("influences") her child's sense of guilt in order to have him give up the girl he wants to marry is acting the role of a moralist; and Kate Croy, interpreting her project to attract a legacy from Millie Theale in ways which muddle and blunt ("influence") Merton Densher's perception of his guilt, is a spokesman for morality. I hope it will not be thought that I would deny that parent and Kate Croy the title of moralist merely on the ground that they are morally wrong; moralists can, with the best will in the world, take morally wrong positions, positions which they themselves, could they see their positions more fully, would see to be culpable. But in the case of that parent and of Kate Croy, there is not so much as the intention to morality. . . . To propagandize under the name of morality is not immoral; it denies morality altogether.[43]

40. Ibid., 241.
41. Stevenson, *Ethics and Language,* 252; emphasis in original.
42. Ibid., 251.
43. Cavell, *Claim of Reason,* 287–288.

In Closing

I sometimes imagine myself confronted with an objector who says to me: "I grant you that Cavell's work is education for grownups. But so is the work of any significant writer. And Cavell's work is hardly typical of what is called 'philosophy' by most professors and graduate students in philosophy departments. Can philosophy in this more conventional sense ever be what Cavell means by that phrase?"

I have approached this challenge indirectly, and I have done so by looking at what I have referred to as the "collapse" of the fact/value dichotomy, and at Cavell's own contribution to that collapse. But—and this is important—I did not *confine* my attention to that contribution, although I did describe it in some detail. (Nonetheless, I had to leave out of my account a great deal that is important, because discussing all of it would have made this essay much longer than it is.) I also described what I see as Quine's contribution (even if he was willfully ignorant of the fact that he was undermining a dichotomy that he loved). A metaphor that Walsh and I use in this connection is the following:[44]

One may think of the logical positivists' fact/value dichotomy (and of the "emotivist" account of ethical language that goes with it) as the top of a three-legged stool. The three legs are (1) the postulation of theory-free "facts," leading to their dichotomy of fact and theory (or "experience" and "convention"); (2) the denial that fact ("science") and evaluation are entangled; and (3) the claim that science proceeds by a syntactically describable method (called "induction"). The fact that even theoretical physics presupposes *epistemic values* means that if value judgments were really "cognitively meaningless," all science would rest on judgments that are, in the language of Carnap's *The Unity of Science, nonsense.* That is why both Carnap and Reichenbach tried so hard to show that science proceeds by an *algorithm,* and why Popper tried to show that science needs only deductive logic. Thus the failure of the third leg is also a failure of the second leg. But the second leg also broke because, as we have seen, facts and values—ethical values—are entangled at the level of single predicates. And the first leg broke because the "two dogmas" on which it was based were refuted by Quine.

44. Hilary Putnam and Vivian Walsh, "A Response to Dasgupta," *Economics and Philosophy* 23 (2007): 359–364.

I wish to emphasize that the destruction of the fact/value dichotomy was a task that it took many brilliant women and men and many years of the past century to accomplish (I say "accomplish" and not "complete" because philosophical tasks are never really "completed"). Those women and men are associated in the textbooks (with their unfortunate love of such classifications) with many different kinds of philosophy. Quine was a high analytic philosopher if there ever was one, and close to the logical positivist movement, even if he turned out to be its severest critic. Morton White was sympathetic both to Quine's brand of analytic philosophy *and* to Oxford ordinary-language philosophy as practiced by Gilbert Ryle, among others. That there is no "algorithm" for doing science was stressed by Ernest Nagel and also by the most celebrated "philosopher-scientist," Albert Einstein. The failure of the "disentangling manoeuvre" that was supposed to split up thick ethical predicates into a value-free "cognitive" component and a cognition-free "emotive" component was first seen by Philippa Foot and Iris Murdoch and then further discussed, as we have seen, by Stanley Cavell and more recently by John McDowell and myself.

The moral I wish to draw is a simple one. It is not from any one "type" or "school" of philosophy that enlightenment comes. Enlightenment can come from any type of philosophy. Further, it is important to see how the different sorts of enlightenment that come from different philosophical sources can be *related* to one another. In "Knowledge and the Concept of Morality" Cavell shows how a vision of the function of morality can be related, on the one hand, to all the issues I have just mentioned, and, on the other hand, to the many other issues discussed in his book *The Claim of Reason*. But philosophy in "the more conventional sense" can also be "education for grownups." Philosophy stops being that only when it starts thinking of itself as a collection of "specializations" (like medical "specializations"). But philosophy, even in "the more conventional sense," need not and must not think of itself in that way. It is when different insights from different sources are connected with each other that philosophy truly educates us.

19

Capabilities and Two Ethical Theories

Explanation of the Terms "Expressivist" and "Kantian"

Let me begin by explaining two terms I am going to use. The term "expressivist" is probably unfamiliar to some readers. The heart of this position (formerly best known under the rubric "emotivism") is Charles Stevenson's idea, described in the preceding chapter, that "the disagreements that occur in science, history, biography" are "disagreements in belief," whereas "it is disagreements in attitude . . . that chiefly distinguish ethical issues from those of science."[1] Although the theories of truth and meaning employed by Stevenson's present-day successors are far more sophisticated than Stevenson's, this sentence represents a point on which they all still agree. Stevenson himself was an admirer of the logical positivists, and the conclusion that both Stevenson and the positivists drew from the doctrine that ethical sentences

---

1. Charles Stevenson, *Ethics and Language* (New Haven, Conn.: Yale University Press, 1944), 13.

do not state any facts but only "express attitudes" is that there is no such thing as a valid argument in ethics.[2]

This doctrine was enthusiastically embraced by Lionel Robbins, as pointed out in the preceding chapter, and was also endorsed by Paul Samuelson, two of the most influential economists in the world. There are many varieties of expressivism on the scene today, but to keep this chapter to manageable length I will pick out just one for examination, the version due to Simon Blackburn. (He calls it "Quasi-realism.") Blackburn denies that his version of expressivism has the irrationalist consequences that Robbins and Samuelson so enthusiastically endorsed, and he even explicitly endorses the capabilities approach to welfare economics.[3]

The other term I use is "Kantian," and here I need to say at once that I use this term not in its historical sense, as referring to the doctrines of Immanuel Kant, but rather as referring to a number of contemporary positions in ethics that share one particular idea of Kant's, however much they may differ in other respects. That idea is that what makes a course of action ethically right is that it is dictated by a rule on which all rational human beings (or, in some versions, all rational and morally concerned human beings) could agree. Rawls's ethical writings clearly have "Kantian" elements, and Habermas's idea that ethical problems are to be resolved (via "communicative action" in an "ideal speech situation") in a way on which *all* who are affected by the decision will agree is, it seems to me, likewise a Kantian idea. As with expressivism, there are many varieties of Kantianism on the scene today, but once again I will pick out just one for examination, the version due to Thomas Scanlon in *What We Owe to Each Other*.[4] It is not obvious that Kantianism does not comfortably fit the capabilities approach—I expect that we may not all agree about that—but I will explain in a few minutes why I say that it does not.

2. Ibid., especially 252–254. For a discussion of Stevenson's view, see the previous chapter.

3. The capabilities approach originated with Amartya Sen, *Commodities and Capabilities* (Oxford: Oxford University Press, 1985). See also Sen, "Human Rights and Capabilities," *Journal of Human Development* 6, no. 2 (2005): 151–166. I mention the capabilities approach here because this chapter originated as a lecture to the Human Development and Capabilities Association.

4. T. M. Scanlon, *What We Owe to Each Other* (Cambridge, Mass.: Belknap Press of Harvard University Press, 2000).

Expressivism and the Rationality of Ethical Judgments

In a lecture titled "Disentangling Disentangling" Simon Blackburn lists a number of things that, he says, "many of us wish to applaud," namely, "the demise of positivism and its contempt for value theory, the resurgence of ethics as a subject, the parallel resurgence of political philosophy, and, as Putnam stresses, the demise of *homo economicus* and the resurgence of pluralistic accounts of the good in writers such as Martha Nussbaum and Amartya Sen."[5] Blackburn goes on to argue that expressivists can have an "attachment to these civilized things" just as well as philosophers like myself (call those of us who believe in the possibility of moral knowledge "cognitivists"). Given all this, why do I say that the capabilities approach does not fit very comfortably with Blackburnian expressivism?

The core thesis of expressivism, in all its versions, is that the distinctive function of ethical utterances is to express "attitudes" (Blackburn also uses the term "stances"). They may also express beliefs, but, according to Blackburn, those beliefs vary from speaker to speaker. Contrary to my position in *The Collapse of the Fact/Value Dichotomy,*[6] Blackburn insists that the expressive "load" and the descriptive component of a thick ethical term can be "disentangled." Moreover, he claims that the descriptive component of such a term is extremely variable. According to him, although there may be a fact about whether or not the descriptive component associated with the term by a particular speaker on a particular occasion fits the world, there is no such thing as an "attitude" fitting or failing to fit the world. "Representations of how things stand must fit the world whereas it is the world that must fit, or be desired to fit, or regretted for not fitting, our attitudes."[7]

This leads to a problem that Blackburn himself recognizes and responds to, but unsuccessfully, in my opinion. The problem is that it is not clear how "attitudes" can be rationally evaluated if they are completely noncognitive. What is Blackburn's response?

5. Simon Blackburn, "Disentangling Disentangling," paper delivered at the "Colloque sur les défis de Hilary Putnam," Sorbonne, I, and École Normale Supérieure, March 22, 23, and 24, 2005. This paper has not yet been published as far as I know. Citations in what follows are to the version distributed at the conference.

6. Hilary Putnam, *The Collapse of the Fact/Value Dichotomy and Other Essays* (Cambridge, Mass.: Harvard University Press, 2002).

7. Ibid., 8.

He begins by admitting that

> Putnam is right, of course, that Sen's approach, like any substantial ethic, can only be supported if there is space for substantial, rational, discussion of value. He reminds us of the melancholy fact that some positivists, particularly in the nineteen thirties, denied that there was space for such discussion. His particular example is the economist Lionel Robbins who indeed made hair-raising pronouncements of just this kind. But for more than fifty years since then, and two hundred years before, expressive theorists have taken great care to acknowledge that discussion of attitudes and stances is fundamentally important. . . . It is important because when you change someone's mind about a value, you change their stance towards the world, and that will typically change what they do and what they support and what they regret and what they campaign for.[8]

This response may seem a little strange, because neither the positivists nor Lionel Robbins ever denied that ethical discussion is important in terms of its real-world effects. What they denied is that it is rational. But Blackburn immediately addresses this issue:

> Not only is discussion important, but there are better and worse ways of conducting it, and it is the good ways that are collected under the umbrella "rational." Manipulation, concealment, evasion, fantasies, arguments ad hominem, ad baculum, and the rest are bad. . . . Only some means are compatible with respect for the other person. Changing the other person's mind—changing their stance towards the world—is a fine art, but expressivists as much as anyone else can distinguish between my bringing it about that someone wants something by deception and manipulation, and bringing it about by revealing truths about it that, in my eyes, ought to impress the subject favorably.[9]

This is hardly reassuring. To begin with, the explanation just quoted is "loaded," to use Blackburn's own favorite term, with evaluative words: "deception," "manipulation," and even "good" and "bad." If the question is how, on the expressivist account, evaluative judgments can be rational, a string of evaluative judgments is hardly responsive to the question. Blackburn's account of the good ways that are collected together under the term "rational"

8. Ibid., 7.
9. Ibid.

faces an age-old problem here. The justification of any particular set of desiderata for rational belief fixation is normally that beliefs fixed in those ways are more likely to be true. Blackburn, however, is a "deflationist" about "true": that is analytic philosophers' jargon for someone who holds that all there is to say about truth is that to call a statement true is equivalent to simply making the statement.

According to Blackburn, even cognitively meaningless statements can be called "true" if one accepts them. Evaluative judgments are not possible "cognitions" according to Blackburn's view, but if one has an "attitude" that leads her to make one, say, *p*, then she is perfectly in order to say, "It is true that *p*," whatever the evaluative judgment *p* may be. Thus for Blackburn to say that evaluative judgments that we are persuaded to accept in one of the "good ways" are "more likely to be true" is just his way of indicating that they are more likely to be ones that he would accept. And this is what he does say, in effect, when he writes that in his eyes they "ought to impress the subject favorably."

The logical positivists would have said that "value judgments are neither true nor false," and it is to avoid saying this that Blackburn resorts to his deflationist account of truth. But he does say, as I mentioned, that they do not "represent" anything, where "representing" is equated with "fitting the world."

Here, it seems to me, Blackburn suffers from a severe impoverishment of categories. In *The Collapse of the Fact/Value Dichotomy* I argued that not all objectively correct judgments are representations. Truths of logic and mathematics are not true just because the world, as it contingently happens to be, does not falsify them, but because the world could not falsify them. They are not "representations" in Blackburn's sense. (However, Blackburn's skepticism extends to skepticism about objective mathematical and logical necessity, as he has made clear in a number of publications.)

Here it may be useful to contrast Blackburn's view with Scanlon's. According to Scanlon's famous formula in *What We Owe to Each Other*, the moral motivation par excellence is the desire to avoid an action if that action is such that any principle allowing it would be one that other people could reasonably reject. I will discuss the strengths and what I see as the limitations of Scanlon's contractarianism shortly. But Scanlon is surely right that the notion of what is reasonable is crucial to ethics, and indeed to science and to all of life. And judgments of what is reasonable, as I argue in *Ethics without*

Ontology (where I refer to them as "methodological" judgments[10]) do not fit the procrustean bed of "either it is a representation of the world or it is an expression of an attitude." That metaphysical dichotomy, which descends from Hume, is precisely what prevents Blackburn from giving any account of why we should care about arriving at our ethical convictions in the "good ways" that he "collects together" by calling them "rational."

For Blackburn, apparently it is enough if those ways change the minds of those he reasons with in ways that "in my eyes [Blackburn's], ought to impress the subject favorably." But we who are gathered to talk about human capabilities and development had better be able to say something better to the billions of people whose lives will be affected by our recommendations, if and when some or all of them are accepted, than "in our eyes, they should impress you favorably." To say this is just the late Richard Rorty's move of endorsing discourse ethics and simultaneously saying that the notion of rationality it assumes is justified because it is what "we" Westerners like. Like Stevenson's older emotivist view, Blackburn's new version of expressivism turns the choice of an ethical argument into a mere choice among the ways that will be likely to cause the change we desire in the behavior of those to whom it is addressed. To say that one does not approve of certain ways of doing this ("bad" ways), and that one does approve of others ("good" ways"), does not change the expressivist position at all—Stevenson and Lionel Robbins would have said that too—and "collecting together" the ways Blackburn likes under the label "rational" is not something Blackburn has shown that he is entitled to do.

Scanlon and Capabilities

In my view the most attractive of the "Kantian" approaches to ethics now under serious discussion by moral philosophers is the version of "contractarianism" that Thomas Scanlon developed in his wonderful book *What We Owe to Each Other*. What makes it attractive, at least in my eyes, is its modest and pluralistic approach to morality, in contrast with the rigorism and overweening ambition that often go with Kantianism. And I called it a "wonderful" book because its discussions are simultaneously broad, deep, and original.

10. Hilary Putnam, *Ethics without Ontology* (Cambridge, Mass.: Harvard University Press, 2004), 65.

If there were a version of "Kantianism" that capability theorists should be happy with, it would be Scanlon's. Nevertheless, at the end of the day, I do not think that it is what we want. In this section I explain why I think this.

According to Scanlon's theory, what makes an action wrong is that "its performance under the circumstances would be disallowed by any set of principles for the general regulation of behavior that no one could reasonably reject as a basis for informed, unforced general agreement."[11] The theory is supposed to simultaneously provide an account of moral motivation:

> According to the version of contractarianism I am advancing here our thinking about right and wrong is structured by . . . the aim of finding principles that others, insofar as they too have this aim, could not reasonably reject. This gives us a direct reason to be concerned with other people's point of view: not because we might, for all we know, actually be them, or because we might occupy their positions in some other possible world,[12] but in order to find principles that they, as well as we, have reason to accept. . . . There is on this view a strong continuity between the reasons that lead us to act in the way that the conclusions of moral thought require and the reasons that shape the process by which we arrive at those conclusions.[13]

I believe that Scanlon has well described how one sort of ethical claim can have motivating force in any community that shares one of the basic interests of morality. But I also believe that morality has a number of basic interests, including respect for the humanity in the other, equality of moral rights and responsibilities, compassion for suffering, and concern to promote human well-being, and not only the desire to be governed by principles for which we can give reasons to one another, although that too is one of them. Even though those interests sometimes conflict, I believe that on the whole and over time, promoting any one of them will require promoting the others. Precisely for that reason, a philosopher who succumbs to the temptation to see ethics as standing on a single "foundation" can always write a book "showing" that all of ethics "derives" from that interest—indeed, many such books have already been written. But I believe that this temptation should be resisted.

11. Scanlon, *What We Owe to Each Other*, 153.
12. Scanlon is referring to John Rawls's "veil of ignorance" and "initial position."
13. Scanlon, *What We Owe to Each Other*, 191

It should be resisted because if we do package together the various basic ethical interests under the single wrapper of "principles it is reasonable to accept," the package will be in constant danger of bursting. The strain on the ribbon "reasonable to accept" will be just too great.

To see that this is so, let us consider just two kinds of objections to the idea that morality requires the idea of equality (which I take to be the idea that all human beings are worthy of respect as moral agents). These objections represent principles that governed human societies millennia before the idea of moral equality was ever formulated.

The first objection I will imagine as coming from a society of slave-owning aristocrats. Such an aristocrat will doubtless object that Scanlon's formula ought to read: "An action is wrong if its performance under the circumstances would be disallowed by any set of principles for the general regulation of behavior that no slave owner could reasonably reject as a basis for informed, unforced general agreement among slave owners." Scanlon will doubtless reply that this formula is not an expression of morality at all. And indeed it is not. But my question is whether the Kantian ambience of Scanlon's book, an ambience I can best describe by saying that morality is constantly portrayed as something required by reasonableness, really tells us why morality requires that we be concerned with everybody, including the poorest and weakest, and not only with the rich and powerful. After all, knowing what reasonably follows from certain interests, or how certain aims may be reasonably attained, is required by every human activity, moral, amoral, or immoral. It is true that morality requires equal concern for others, the sort of concern I formulated in *The Many Faces of Realism* by the following three principles:

I. There is something about human beings which is of incomparable moral significance, some valuable aspect with respect to which all human beings are equal, no matter how unequal they may be in talents, achievements, social contribution, etc.

II. Even those who are least talented, or whose achievements are the least, or whose contribution to society is the least, are deserving of respect.

III. Everyone's happiness or suffering is of equal prima facie moral importance.[14]

14. Hilary Putnam, *The Many Faces of Realism* (La Salle, Ill.: Open Court, 1987), 45.

I noted that the respect of equality (the "something about human beings" mentioned in the first principle) has been differently interpreted by different traditions and different philosophers.[15] (For example, in the Kantian tradition, as I interpreted it in that book, it is the ability to think for ourselves about moral matters that is the respect of equality.) I also noted that the third principle came to be emphasized later than the first two.[16] It is precisely because equality (or "respect for the humanity in the other," as Kant put it) is a fundamental interest of morality that the slave owners' definition of "wrong" is not a moral one. But note that this does not show that the central concern of morality is reasonableness (pace Scanlon). After all, one could just as well argue that the central concern is equality, in this Kantian sense, and that morality requires that we look for principles that others with an interest in morality could not reasonably reject because every concern requires that we look for principles that others who share that concern could not reasonably reject.

Now let me imagine an objection to equality coming from a society dedicated to a warrior ethic (what I called a "macho" ethic in *Ethics without Ontology*). Members of such a society will doubtless object that Scanlon's formula ought to read: "An action is wrong if its performance under the circumstances would be disallowed by any set of principles for the general regulation of behavior that no courageous male could reasonably reject." Once again, Scanlon will reply that this formula is not an expression of morality at all. But this formula was at least implicitly accepted by many societies for probably something like thirty thousand years. The reason that we reject it is that we have come to feel the appeal of a way of life that does not regard the warrior as the very peak of human excellence. But that is not because we have became more able to see what is reasonable in given circumstances if we have certain given interests; it is because we have come to have fundamentally different interests.

What justifies those interests? The justification in each case has to come from within morality, not from outside or from a foundation prior to morality. We have come to reject warrior ethics ("macho" ethics) because we have come to appreciate that the idea that the warrior represents the ideal type of human was an extremely limited view of human excellence. (And "limited"

15. Ibid., 48.
16. Ibid., 45.

is, indeed, a "value-loaded" term, as Blackburn would say.) We have come to reject aristocratic ethics and patriarchal ethics partly because we reject the claims to certain sorts of intellectual superiority and superior reasonableness traditionally advanced by aristocrats and males as empirically false, but also because we have come to appreciate the superiority of what Dewey called "the democratic way of life." (And yes, "superiority" is a value-loaded term.) We have come to value universal moral norms, as, for instance, the idea of universal human rights, in large part because universal norms, somewhat like constitutions, contribute to the rule of law in democratic societies (that Kantian ethics emerged at the time of the first democratic revolutions is not accidental). And all the interests that the values of democracy, equality, compassion, and universal legal norms serve lead to and serve a pluralistic and compassionate vision of human well-being with such wide appeal that, in our time, even many totalitarian regimes have had to pay lip service to them.

In addition to what I have argued to be the theoretical weakness of Scanlon's theory—the arbitrariness of privileging the interest in having our behavior be governed by reasonable principles over all the other fundamental ethical interests—I have a further reason for claiming that his theory does not fit happily with the "capabilities approach," namely, Scanlon's obvious unwillingness to give the notion of human well-being much of a role, if any, in his account of ethics. Because the notion of capabilities that humans have reason to value would seem prima facie to be a notion of human well-being, one would naturally expect Scanlon obviously to mention that approach, but he does so only in passing. Most of the pages of *What We Owe to Each Other* devoted to well-being[17] are addressed to arguing that well-being is not a "Master Value," that is, a value from which all of morality can be derived, and that, moreover, the concept of well-being has boundaries that are not well defined (as opposed to, say, the concept "reasonably"?). But he does say that moral justification can appeal to (1) "the well-being of particular individuals with whom we interact, [those] whose well-being is determinate and can be known"; and (2) "more specific forms of opportunity, assistance and forbearance that we all have reason to want, rather than to the idea of well-being abstractly conceived."[18] Clearly this is a case of too little and too late.

17. Scanlon, *What We Owe to Each Other,* 108–146.
18. Ibid., 140.

But What Is the Alternative?

Before describing the "Deweyan" alternative to expressivism and (Scanlon's contractarian version of) "Kantianism", I want to say a word about the Kantian spirit in ethical theory. "Kantians," in my sense, are haunted by the idea of "universal agreement." This is explicit in the case of Habermas, who talks about rational discussions (concerning norms) that are supposed to continue until the "consent of all" is secured,[19] but it is also there, just under the surface, in Scanlon's talk of principles that no one (with an interest in being governed by such principles) could reasonably reject.

The problem with the Habermasian form of Kantianism is that in the real world it never happens that everyone agrees, if "everyone" means literally everyone and the norm in question affects a very large number of people. This is not a trifling objection. On the contrary, it cuts very deep in the sense that no simple modification will rescue Habermas's idea of the consent of everyone who is affected.

For example, the late lamented Richard Rorty, who strangely had his Habermasian moments, once proposed that the rejection of a claim *p* is justified if everyone except a few "dubious characters notorious for making assertions even stranger than *p*" think that those who defend *p* "must be a little bit crazy." But, as I am sure Habermas would agree, it may happen that it is the few "dubious characters notorious for making assertions even stranger than *p*" whose views are reasonable, and it is the majority that is unreasonable. One cannot "repair" Habermas's "consent of all" formula by allowing a norm to count as valid even if this or that minority is not included in the consensus of "all." And if, as I claim, the consent of literally "all" is not something we find in real life, then Habermas's universalization principle belongs to a utopian fantasy. This principle reads: "For a norm to be valid, the consequences and side-effects of its general observance for the satisfaction of each person's particular interests must be acceptable to all."[20]

This problem does not arise for Scanlon because (1) he does not require that everyone actually consent to ethical norms, or would consent if discus-

---

19. See Jürgen Habermas, *Moralbewusstsein und kommunikatives Handeln* (Frankfurt am Main: Suhrkamp, 1983); English translation, *Moral Consciousness and Communicative Action* (Cambridge, Mass.: MIT Press, 1992); quotation from p. 197 of the English translation.
    20. Ibid.

sion went on long enough, but instead that no one could "reasonably" reject them; and (2) he does not pretend to offer necessary and sufficient conditions for an objection being "reasonable," although he does argue that the notion of "well-being" has no role to play. Moreover, Scanlon wisely allows ethical principles themselves to be "open-ended" and to contain clauses of the form "unless there are overriding moral reasons." But when there is disagreement about what is reasonable (as in practice there always is in a democracy), how are we to decide?

So How Do We Decide?

The "Deweyan" answer I wish to defend is not metaphysical but practical. It can be summed up in two words: "democracy" and "fallibilism."

Democracy

What makes Habermas's philosophy so unrealistic is that it has so little to do with actual democratic politics. To be sure, Habermas offers justifications of freedom of speech and freedom of inquiry and everything necessary for these. But when it comes to actually deciding on policies, he gives us only the utopian idea of continuing discussion until everyone affected is in agreement. On the other hand, if, with Scanlon, we seek agreement of all those whose arguments are "reasonable," then we throw all the weight on a concept that, I argued, is unable to bear the load. As we all know, the alternative we settle on in practice, instead of seeking the consent of "all" or the consent of the "reasonable," is to seek arguments that convince substantial majorities, arguments that we hope will produce an "overlapping consensus."

That is the alternative as far as voters are concerned, but the alternative as far as experts are concerned—for example, academics, members of nongovernmental organizations, and others involved in suggesting and/or implementing policies—is similar. "Experts" may not take formal votes, but if their policy recommendations are to be acceptable, they must be arrived at by informed discussion that respects "discourse ethics" and that tries to understand and make explicit the concerns of all affected. Thus far, Habermas is surely right. But it is not to be expected that the result will usually be unanimous agreement, among the experts any more than among the voters.

So, am I saying that a decision reached in this way by a majority of the "experts," or a majority of the voters, is necessarily right? Or necessarily reasonable? Certainly not. This is where fallibilism enters.

Fallibilism

A Deweyan pragmatist does not propose necessary and sufficient conditions for "right" and "wrong," "reasonable," "well-being," or any other important value concept. What Deweyans possess is the "democratic faith" that if we discuss things in a democratic manner, if we inquire carefully, if we test our proposals in an experimental spirit, and if we discuss the proposals and their tests thoroughly, then even if our conclusions will not always be right, nor always justified, nor always even reasonable—we are only human, after all—still, we will be right, we will be justified, and we will be reasonable more often than if we relied on any foundational philosophical theory, and certainly more often than if we relied on any dogma or any method fixed in advance of inquiry and held immune from revision in the course of inquiry. In sum, what Winston Churchill said about democracy applies to inquiry as well: fallibilistic democratic experimentalism is the worst approach to decision making in the public sphere that has ever been devised—except for those others that have been tried from time to time.

20

The Epistemology of Unjust War

My friend Steven Wagner, a philosopher I very much admire, once wrote me that he finds "just-war theory" in its present form wholly untenable. With his permission, I shall quote part of what he wrote.

> Here's what I meant about just war theory (JWT) and ontology. The formulations of JWT effectively identify three distinct objects: a population, a nation, and the high-level decision-makers in the government. Therefore, even if JWT is invoked in the cause of peace, it surrenders the larger battle by buying into an authoritarian political ontology. So it's an irremediably spoiled tool for justice.
>
> A philosopher can easily work out epistemological costs of this ontological sin. Here, though, is a cost that is related to but does not immediately come out of the pernicious ontology:
>
> JWT imposes requirements of justified belief regarding the outcomes of alternative courses of military action. Applying elementary considerations about evidence, we argue that the relevant beliefs can be justified

This chapter originated as a lecture given in 2005; the examples used reflect our (partial) knowledge of the situation in Iraq at the time the lecture was given, but the philosophical points it makes apply to many other conflicts.

only if their source is a professional agency strictly independent of the decision-making sector and disinterested relative to the outcomes.

These clauses need more careful formulation. E.g., "professional" will imply membership standards no less rigorous (and applicable to the subject matter) than those governing, say, physicians and philosophy professors. "Disinterested" must mean *at least* "no more interested than is the population in general." E.g., if the professionals stand to gain from a war, then no more so than the run of the people. So, e.g., the agency must be strongly separate from the military, the war industries, etc. . . .

But in no nation, ever, has such an agency existed. Not even at distant approximation. Therefore, by the standards of JWT no nation has ever made war with good reason. *Q.e.d.*[1]

I am not prepared to go as far as Wagner, as will soon appear. For one thing, as a student of American pragmatism, I am not prepared to say that only judgments made by a body that is "disinterested relative to the outcomes" are justified. (Even the judgment that I have been aggressed against when I have just been hit on the nose is not normally made by such a body, but it is clearly justified.)[2]

With respect to Wagner's more political (as opposed to epistemological) reasons, I would agree that decisions in our imperfect approximations to democratic polities are frequently, indeed normally, made by bodies influenced by all sorts of special interests, but I still think that we live in what *are* approximations to democracies. (I suspect that it is because Wagner would find this idea naïve that he would require a body "strictly independent of the decision-making sector" to review a government decision to go to war before he would be willing to call it justified; indeed, I suspect that it is not only decisions to go to war that he would regard as epistemologically suspect when made by a "hierarchical" government.) Nevertheless, I would not quote Wagner's letter if I did not think that it contained an important idea.

1. Steven Wagner, personal communication.
2. Steven Wagner will remind me, I know, that he requires only that the agency that is the "source of the beliefs" concerning the justice of war (the Epistemic Bureau?) be "no more interested than is the population in general." Although I would agree that an agency that decides on war should not do so *because* it stands to gain from a war "more so than the run of the people," I do not think that apart from the case of material gain, there is any general rule about how "interested" deciders are allowed to be. Moreover, I do not agree that the epistemic justification of war requires an agency separate from the government—unless that "agency" be simply civil society as a whole.

The idea that I take from Wagner's reflections is that instead of thinking about a positive list of conditions for "just war," we might instead think of epistemic conditions for justified belief that resort to war is called for. This idea appeals to me because it generalizes an argument against certain wars that I first heard from my colleague at Harvard (whose life was tragically cut short by a stroke in 1987), Roderick Firth. To be precise, I heard Firth's argument during the Vietnam War, and I have reflected on it and, sadly, have had occasion to use it very often since he died. Firth's argument does not depend on suspicion of the representative character of what we call democratically elected governments, as it seems to me that Wagner's does, nor does it follow from Firth's argument that "no nation has ever made war with good reason."

But like Wagner's more radical argument, it avoids appealing to either a grand metaphysical story or a grand ethical theory. In essence both Wagner and Firth suggest that the fruitful question is the epistemological question about justified war rather than the metaphysical question about the "nature" of "just war."[3] And even a partial answer to the epistemological question is important if it identifies a class of cases in which we are definitely *not* epistemically justified in going to war (or, to put it less abstractly, in killing and maiming people as an instrument of state policy). Rather than seek an ontology of just war, these philosophers are saying, let us seek an epistemology of unjust war. But it is time for me to say what Firth's argument was.

Firth's Argument

Firth's argument was that even if we assume (as he himself, as a Quaker, did not) that war is justified under certain circumstances, those circumstances must be very special to override the clear moral presumption against inflicting suffering on such a large scale. What Firth claimed—this is the content of what I shall call "Firth's principles"—was that mere probabilities (e.g.,

---

3. In contrast to this epistemological approach, traditional just-war theory begins by requiring that the war be declared by a rightful authority. To interpret this would require the *whole* of political philosophy. And even then, wars that are not declared by an "authority" (e.g., wars that start with spontaneous popular resistance) are not even envisaged, as Wagner pointed out in his letter to me.

game-theoretic reasoning) are not enough.[4] One must *know,* not just have some opinions by members of a particular administration, that

1. the Bad Thing that the war one is thinking of waging is supposed to prevent really *will* happen if one does not wage it; and
2. the Bad Thing will actually be prevented (and not simply replaced by a different equally Bad Thing) if one does wage war.

I want to explain and discuss Firth's argument in a concrete context. That context is not today's, but it is immediately relevant to today's situation because it brought it about: the context of the decision on the part of the American government (one supported by the British government, as you all know) to wage a "preemptive war" against the then government of Iraq, headed by the dictator Saddam Hussein.

I want us to think ourselves back to the time when the decision to destroy the Iraqi army and institute "regime change" was arrived at. And although President Bush and Prime Minister Blair said at the time that the one overriding reason that justified "preemptive war" was the "weapons of mass destruction" that Saddam Hussein allegedly possessed, and we all know now that *this* "reason" was a mistake based on a mixture of faulty intelligence and wishful thinking, the mistake about weapons of mass destruction is *not* going to be the only matter on which I focus. I am equally interested in the question whether the decision to employ all the means of modern war, including, as we now know, napalm bombs,[5] would have been justified even if the intelligence had been correct (including, however, the intelligence, which was also available from United Nations inspectors, that Hussein's regime was far from ready for war, not because it lacked aggressive desires, but because it was successfully kept off balance by the whole series of United Nations actions after the Gulf War of 1990).[6]

4. This kind of argument is called an "antiutilitarian argument" by moral philosophers.

5. Authentic footage taken by embedded reporters of the coalition forces using napalm bombs is included in Michael Moore's film *Fahrenheit 911.*

6. This was Secretary of State Colin Powell's estimate on February 4, 2001, when in the course of press remarks with Foreign Secretary Amre Moussa of Egypt, he reported that "we had a good discussion, the Foreign Minister and I and the President and I had a good discussion about the nature of the sanctions—the fact that the sanctions exist—not for the purpose of hurting the Iraqi people, but for the purpose of keeping in check Saddam Hussein's ambitions

As I am sure you all remember, at the time I am writing about, there were serious *moral* disagreements about the decision to invade Iraq (in addition, of course, to the empirical disagreements about the likely results of the invasion). What could a philosopher, of all people, possibly say *as a philosopher* about this sort of deep moral disagreement? At that time I believed—and still believe—that Firth's argument was the most useful one for a philosopher to make.

Firth made his argument during the Vietnam War. Because some of you were not yet born at the time of the Vietnam War, and many of you may have been children at that time, let me briefly recall the situation. I am describing it because that was the situation in which Firth made his argument; I am *not* analogizing the present situation to America's or the United Kingdom's at that time, because obviously the Vietcong and the North Vietnamese were no direct threat to America, which was, indeed, over 10,000 miles away, while terrorism is certainly a direct threat to any government and any people the terrorists choose to single out for attack. Or rather, I am analogizing it only in the respect that Firth's principles are ones that I believe apply to both situations. America employed very harsh measures in an attempt to defeat the Vietcong and their North Vietnamese allies. (The most horrific was the dropping of napalm bombs—a weapon our nations have also used in Iraq—but the action that originally turned me into a protester

---

toward developing weapons of mass destruction. We should constantly be reviewing our policies, constantly be looking at those sanctions to make sure that they are directed toward that purpose. That purpose is every bit as important now as it was ten years ago when we began it. And frankly they have worked." Three years later, on March 22, 2004, CNN carried the following story: "WASHINGTON (CNN) The United Nations' top two weapons experts said Sunday that the invasion of Iraq a year ago was not justified by the evidence in hand at the time." "I think it's clear that in March, when the invasion took place, the evidence that had been brought forward was rapidly falling apart," Hans Blix, who oversaw the agency's investigation into whether Iraq had chemical and biological weapons, said on CNN's "Late Edition with Wolf Blitzer." Blix described the evidence Secretary of State Colin Powell presented to the UN Security Council in February 2003 as "shaky" and said that he related his opinion to U.S. officials, including National Security Adviser Condoleezza Rice. "I think they chose to ignore us," Blix said. In the same story, Mohamed ElBaradei, director general of the International Atomic Energy Agency, is quoted as speaking to CNN from the agency's headquarters in Vienna, Austria, saying, "Well, Wolf [Blitzer], I think I'd like to, for a moment, say that, to me, what's important from Iraq is what we learn from Iraq. We learned from Iraq that an inspection takes time, that we should be patient, that an inspection can, in fact, work."

against the war when I learned about it from the writings of the journalist David Halberstam was the destruction of the rice crop of the South Vietnamese peasants to keep it out of the hands of the Vietcong.) These harsh measures, which everyone admitted caused immense suffering to millions of Vietnamese people, were supposed to be necessary to prevent a Bad Thing. The Bad Thing was described using the metaphor of a "row of dominos." It was claimed that if South Vietnam fell to the Communists, so would Laos and Cambodia (which did fall), and then the Philippines and Indonesia would fall (which did not happen), and finally Japan would fall to the Communists.

Firth argued that in the case of the Vietnam War we did not actually *know* either that the Bad Thing (the fall of the dominos) that the war was supposed to prevent really *would* happen if we did not wage it or that the Bad Thing would actually be prevented if we did wage war; and, of course, he was right. What I want to claim is that this is the test that American and British citizens should have applied in considering whether the invasion of Iraq was morally permissible. If we take the "Bad Thing" to be increased terrorism, the growth of extremist Islamism, and the consequent weakening of the security of our nations and the "free world," what Firth's principles tell us is that what we (and morally concerned human beings everywhere, for that matter) should have asked are exactly the sorts of questions that Firth posed:

1. Did we actually *know* that the Bad Things that preemptive war and "regime change"[7] were supposed to prevent really would have happened if we desisted from them and tried other means, especially ones that are approved by the international community as a whole, or at least by the industrial democracies?
2. Did we actually *know* that the terrorist acts and the growth of extremist Islamism that we wished to prevent—the Bad Things in the present case—would not continue and even increase if we invaded Iraq and used military force to bring about "regime change"?

7. Of course, Firth spoke of war in general because this was the time of the Vietnam War, but I have substituted "preemptive war and regime change" because that is the application I am making of his (more general) principle.

Did we, for example, know (as President Bush alleged) that Iraq was a sponsor of al-Qaeda?[8] Or that invasion would not *bring it about* that Iraq became a seedbed of international terrorism on a huge scale?

I do not deny for one moment that one can offer arguments on both sides for the necessity of the problematic measures. But that is the problem that Firth's principles highlight. "One can offer arguments on both sides." To me it seemed clear even at the time that there is an enormous difference between saying that we *know* that invasion will be necessary and effective and "arguments on both sides." I am a Firthian here: if we do not *know*, then what we are doing is immoral. That is the philosophical question I invite us to think about, and what I will say from here on is, I am all too well aware, only the beginning of such a discussion.

But first, let me ward off a misunderstanding. Quite a few months ago a Republican friend of mine (yes, I do have some Republican friends) said to me in a tone of awe, "You were the only person I knew who opposed the war because you thought it wouldn't work." While I am not one to turn down compliments, even undeserved ones, this particular compliment involved a misunderstanding that I must ward off if the whole philosophical point of this lecture is not to be missed. My friend heard me as making an *empirical estimate*—one that, to his surprise, had turned out to be correct—about the future course of events in Iraq. In effect, he was complimenting me for political savvy that I do not pretend to possess. My point, like Firth's, and like Wagner's hermeneutic of suspicion, was an epistemic one. I did not deny (at the time the war started, anyway) that the rosy estimates of the Bush and Blair administrations *might* turn out to be right; my estimate was that they were not epistemically justified. And killing and maiming people on grounds that are not epistemically sound is *morally wrong,* not just practically unwise. To me this seems self-evident, as it did to Firth, but

8. "WASHINGTON—The Bush administration's assertion that Iraqi leader Saddam Hussein had ties to al Qaeda—one of the administration's central arguments for a preemptive war—appears to have been based on even less solid intelligence than the administration's claims that Iraq had hidden stocks of chemical and biological weapons. "Nearly a year after U.S. and British troops invaded Iraq, no evidence has turned up to verify allegations of Hussein's links with al Qaeda, and several key parts of the administration's case have either proved false or seem increasingly doubtful.

Senior U.S. officials now say there never was any evidence that Hussein's secular police state and Osama bin Laden's Islamic terrorism network were in league. At most, there were occasional meetings." *Miami Herald*, March 3, 2004.

I have discussed Firth's argument with enough people to know that it is far from being generally accepted. In the rest of the present lecture I will, accordingly, discuss objections to Firth's argument—both ones I have heard from others and ones that have simply occurred to me.

The Skeptical Objection to "One Must Know"

Both of Firth's principles employ the notion of "knowledge." According to the principles, justification of war requires that we *know* that the Bad Things will not happen if we resort to war (resort to maiming and killing) and that we *know* that the Bad Things will not continue (or be replaced by even worse Bad Things[9]) if we do resort to war. One of the most common objections that I have encountered to Firth's principles is simply that in matters like war and the Bad Things that a war is supposed to prevent, *knowledge* properly so called is simply impossible.[10] The conclusion that is drawn is that ordinary probable reasoning, faulty as it is, is what we must rely on.

As is well known, I am a "fallibilist" in epistemology, but this argument seems to me a misuse of fallibilism. One lesson that I have long insisted we should all have learned from the so-called classical pragmatists, that is, Peirce, James, and Dewey, is that *fallibilism does not entail skepticism*. Skeptics have always pointed to (or more often simply imagined) cases in which some judgment turns out to be false—Descartes famously imagined that he was not sitting in a chair in front of a fireplace, but only dreaming that he was—and have gone on to conclude (or, in Descartes' case, to worry) that we possess *no* empirical knowledge at all. But, as Peirce insisted, *real* doubt, as opposed to paper doubt, requires a context-specific *reason for doubting*—a reason with practical bearing—and the general fact that we are not infallible is, in any normal context, not such a reason. I know that I am in the United Kingdom, *and* I know that human beings are sometimes mistaken about which country they are in, and there is no contradiction between these two claims.[11]

9. This is the version I heard from Firth, rather than the more specific version I gave above. (Cf. note 7.)

10. However, I often hear this skeptical objection from people who, in other contexts, claim to have an amazing amount of political "knowledge."

11. However, Barry Stroud has defended Cartesian skepticism (or at least held out the possibility that it is correct) in *The Significance of Philosophical Scepticism* (Oxford: Clarendon Press, 1984). I criticize Stroud's defense in "Skepticism, Stroud, and the Contextuality of

To come to the case at hand, that there was no evidence of a connection between al-Qaeda (an extreme Islamist group if there ever was one) and the secular Baathist regime of Saddam Hussein is something we knew or should have known. That there was no good evidence that Saddam Hussein's regime was in a position to make aggressive war in the near future, or would be in such a position if the United Nations continued its inspections, flyovers, and other measures is also something we knew or should have known. There is such a thing as empirical *knowledge* in such matters.

However, lest this become a lecture on pragmatist epistemology, let me *give* the word "knowledge" to the skeptics. It is still possible to make Firth's essential point without using it. The essential point is that when it comes to the decision to use killing and maiming as instruments of state policy, the persons who make that decision must, if they are morally alive at all, accept the onus of an especially high burden of proof. There needs to be a "firebreak" between ordinary policies of building roads, raising or lowering taxes (within normal limits, anyway), and the like and policies that involve killing and maiming—at least if all talk about the value of human life is not to be exposed as sheer hypocrisy.

(Steven Wagner will remind me that it often *is* sheer hypocrisy in the mouths of the rich and powerful. But it is one thing to acknowledge this painful fact, and another to *accept* it as the way the world must be, and I urge that we resist such acceptance, even when it is presented as "worldly wisdom." Indeed, there is much more worldly wisdom in Firth's principles than in the *Realpolitik* of any of the world's present leaders.)

At this point some of you may be reminded of a famous notion of Ronald Dworkin's, the notion of "rights as trumps."[12] By this he meant that considerations of utility (in particular, considerations of wealth maximization) must not be allowed to "trump," that is, override, the moral rights of individuals. For example, the benefits that a majority might gain from discriminatory behavior against a minority cannot justify the violation of the inherent moral right of the members of the minority to be treated as free and equal citizens. Moral rights may have to be overriden in real emergencies, Dworkin recog-

---

Knowledge," *Philosophical Explorations* 4, no. 1 (2001): 2–16; reprinted in this volume as Chapter 29.

12. Ronald Dworkin, *Taking Rights Seriously* (London: Duckworth; Cambridge, Mass.: Harvard University Press, 1977), 153.

nizes, but such overriding requires strong moral justification, not just cost-benefit analysis.

There is good reason for you to be so reminded. For the right not to be maimed or killed, not to have one's children and other relatives maimed or killed, not to have one's house destroyed over one's head, and the like are prima facie moral rights in Dworkin's sense—indeed, if they are not recognized as such, I repeat, talk of "human rights" is meaningless hypocrisy. Thus Firth's argument can be regarded as an epistemological refinement of the idea of "rights as trumps." No general skepticism about the possibility of political "knowledge" should be allowed to efface or conceal the fact that both Dworkin and Firth are appealing to fundamental ideas of what our ideals of human equality and dignity require of us.

Am I forgetting, then, that the very human rights I am appealing to were violated repeatedly and on a large scale by Saddam Hussein and his regime? Not at all. That was, in the sense of Firth's principles, indeed a Bad Thing, and bringing that Bad Thing to an end was one of the goals that the war was intended to achieve (although not even the Bush—or the Blair—administration claims that it is right for our countries to invade any and all countries that have dictatorial regimes, regardless of whether they pose any threat to ourselves or our allies). But remember that according to Firth's second principle, to justify the decision to invade, it is not enough that we knew that the violation of human rights in Iraq would continue if Hussein were left in power; we needed to know that the result of the invasion would not, in the end, be an equally bad regime. And can we honestly claim to know *that* even today? Indeed, even if Iraq ends up with an elected government accepted by at least the Shi'ite majority of the country, do we know (or even have any basis for a reliable estimate of probability, for that matter) that *that* government will not, once the coalition forces leave, turn Iraq into an Iranian-style theocracy? Even to raise the questions that Firth's principles require us to address reveals the weakness of the justifications that were accepted by our regimes as justifying the invasion of another nation.

Exceptions to Firth's Principles

Firth's principles do have exceptions, however. When I talked to students about my reasons for opposing the Vietnam War, I often explained Firth's principles and applied them to the case of that war. Sooner or later an inge-

nious student would think of a conceivable situation in which Firth's principles gave the wrong result. The fact that these conceivable situations frequently had no relevant similarity to the Vietnam War was itself irrelevant
in the eyes of the ingenious student: only an appeal to an *exceptionless* moral
principle was felt to have any weight. And if there is no such thing as an *exceptionless* moral principle (or an exceptionless moral principle that applies to
cases more complicated than murdering someone who has not injured you or
anyone else simply for monetary gain)? In that case, perhaps, these students
thought that there is no such thing as an objective moral judgment, and that
relativism and/or subjectivism are right.

But that is not the way good moral thinking works. Good moral thinking, as Kant said, requires "mother wit" (or "healthy human understanding"), and there is no algorithm for healthy human understanding.[13] There *are*
rules of judgment that can help us. Kant famously listed three:[14] (1) think for
yourself; (2) think from the standpoint of humanity in general; and (3) be
sure your thinking is consistent. (The last, he said, is the hardest.) But who is
to decide how to apply these rules in any specific case?

The answer, as the great moralists have always said, is "Each one of us is."
Total skepticism about the normative is self-refuting, and half skepticism is a
cop-out. Each responsible human being has to decide the hard moral questions in the light of her own best judgment.

Still, even if the fact that Firth's principles have exceptions means that
they cannot be followed blindly—moral judgment has to be complex because
the world is complex and life is complex—those principles are still a valuable
guide. But to see their limits, let me now turn to some of the most common
exceptions.

The most common exception, I believe, is to Firth's second principle, the
principle that requires one to know that waging war will prevent the Bad
Thing (and not cause a Worse Thing to happen) in order to justify waging a
war. The exception I am thinking of is simply that when one's country is di-

---

13. See Juliet Floyd, "Heautonomy and the Critique of Sound Judgment: Kant on Reflective Judgment and Systematicity," in Herman Parret, ed., *Kants Ästhetik/Kant's Aesthetics/L'Estétique
de Kant* (Berlin: de Gruyter, 1998), 192–218, for an excellent discussion of this aspect of Kant's
thinking.

14. Immanuel Kant, *Critique of Judgment* (Berlin, 1790; English translation by Werner S.
Pluhar, Indianapolis: Hackett, 1987), 160–161.

rectly attacked by an aggressor, one may be justified in fighting back *even if* the resistance has no certainty of succeeding. In 1939 Poland's decision to use its army to resist the German Wehrmacht was, I believe, the right and honorable course even though Poland knew that the assistance it hoped for from France and Britain might not come in time. I will not argue this here, because I expect that this is something all of us who are not absolute pacifists agree on.[15] But ex hypothesi this exception does not apply to "preemptive war"— that is what makes such a war "preemptive" and not simply "defensive."

A related (and much more often cited) exception or possible exception to the second principle is the subject of the often-appealed-to *1939 analogy*. After all, I have heard people say, when Britain declared on war on Germany in 1939, Firth's second principle was violated; Britain did not know that the Bad Thing would be prevented if it waged war. Indeed, as the recent best-selling biography of the unhappy Lord Londonderry reminds us, there were voices raised in Britain who opposed the war on that ground, or partly on that ground (though not, by 1939, Londonderry himself).

I believe, nonetheless (and I hope you believe), that Britain's decision to declare war in 1939 was an admirable one, even more so in view of the fact that the odds appeared to be against success. A full discussion of why it was a justifiable decision would, I believe, require—or at least deserve—a whole book. Here I can touch on only a few of the principal considerations.

First, the "defensive-war" exception to Firth's second principle must not be confined to defense of one's own territory if one has entered into binding treaties of mutual defense with other states. Indeed, to limit the idea of just war to defense of one's own territory narrowly construed would be to reject the whole idea of mutual defense and collective defense, an idea that it is vital to strengthen, as the European Community has recognized, and as Kant long ago urged in *Perpetual Peace*. In the case of 1939, Britain was bound by such a treaty to come to the aid of Poland (and, earlier, of Czechoslovakia, although it regrettably failed to stand by *that* obligation). But beyond the narrow legal fact of the treaties, there is the fact that a German victory

15. Hegel argued—plausibly, I think—that the willingness of citizens to give their lives for the defense of the nation's territory is one of the preconditions for the existence of the modern nation-state: see Georg Wilhelm Friedrich Hegel, *Elements of the Philosophy of Right*, ed. Allen W. Wood and Hugh B. Nisbet (Cambridge, UK; New York: Cambridge University Press, 1991), 102.

would, as Churchill saw in 1939, have meant that the civilization of Europe would become a fascist and racist one. After the invasion of Russia and the attack on the United States at Pearl Harbor, what, indeed, turned out to be at stake was whether not just Europe but Europe and Asia, at the very least, would suffer the fate in question.

But the Bush doctrine of "preemptive war" is *not* justifiable on similar grounds. Hussein's regime, bad as it was, was not a real threat to the free world, and, provided one had continued to inspect and monitor and take other such measures as the United Nations was already doing, it was hardly likely to become one in the coming decade, at least.[16] A rush to war was hardly called for on defensive grounds, which is why a large part of the international community refused to support it. Today Bush's supporters defend the war as part of a "war on terror." But "war on terror" is a confused concept.

It is confused because, unlike a state, "terror" has no fixed boundaries, no fixed population, and no fixed army, and there is no such thing as a "surrender" (or, for that matter, a "peace agreement"). Like "the war on drugs," the "war on terror" is a metaphor to provide an open-ended license to the government to do whatever it wants in a particular area.

In practice, the Bush administration has identified the "war on terror" with a war on "state-sponsored terror." But this too is problematic. Any degree of tolerance of or assistance to terrorist organizations might count as "sponsorship"; and because all the radical Islamic states (and some non-Islamic states, e.g., North Korea) do tolerate and/or assist terrorist organizations, in principle, the idea of a war on "state-sponsored terror" could be used to justify war (simultaneously or seriatim) with a large number of states (including some of our "allies," such as Saudi Arabia). Some neoconservatives indeed welcome this, saying that the "war on terror" should have been called a "war on radical Islamism." But in the light of Firth's principles one must ask: do we actually know that, say, al-Qaeda was either created by or depends for its continued existence on particular states? Do we know that a crusade against such a significant number of Islamic states would not inflame the whole Islamic world, even the part that is now living under moderate regimes?

I do not deny that some regimes may, even in the near future, really begin to develop weapons of mass destruction, and that at some point military action might have to be taken, although occupying more Islamic states hardly

16. See the statements by Colin Powell, Hans Blix, and Mohamed ElBaradei cited in note 6.

seems wise or likely to be effective. And in any case, the volatile character of the Islamic world is clearly related to poverty and underdevelopment, which the current policies of the G7 nations and the World Bank may be aggravating rather than alleviating.

Another function of "war on" talk is, of course, to make it seem that anything *other* than a military response is Munich-style "appeasement." But talk of avoiding "appeasement" is a way of ignoring the fact that we are not fighting a power-mad dictator but a hydra-headed movement with complex roots. To say, as I am sure Firth would have, that we, together with the United Nations and other democratic nations, should try to pressure "extremist" nations not to develop nuclear weapons or provide material support to terrorism and, at the same time, begin to address some of the root causes of violence in the third world, and not simply send in the armed forces to "smash the "terrorists," is not "appeasement" but elementary morality and common sense.

Exceptions to the First Principle

But, it may be objected, "You have only considered objections to the second principle. What about the first principle?"

Yes, there are cases when the first principle should be overridden. If the Bad Thing is terrible enough, in some cases just knowing that there is a significant probability that the Bad Thing will happen if one does not act militarily may justify war. But at this point I think that I should reconsider my criticism of Steven Wagner's position.

Recall that, as I quoted him saying at the outset, Wagner holds that "the relevant beliefs can be justified *only* if their source is a professional agency strictly independent of the decision-making sector and disinterested relative to the outcomes." And I said initially that I do not think that we need to require that the relevant beliefs be vetted by an agency independent of the decision-making sector. But when the beliefs on the basis of which we are asked to maim and kill other human beings are beliefs about *probabilities,* then I would ask that those beliefs be ones that reasonable and well-informed judges in other democratic countries also regard as sound. Where Wagner wants an impartial "agency" to survey beliefs on which warfare is based because he is afraid that those beliefs will be disguised justifications for imperialism, I want the "agency" of well-informed people everywhere to support such morally momentous decisions because I am afraid that they will be ideologically based, that is, based on what is ultimately *fantasy*—whether that fantasy does or does not serve this or that economic interest. If Wagner will accept this as a friendly amendment to his view, we may not be so far apart.

Conclusion

The present situation is one of occupation rather than war, and I do not—any more than anyone else, apparently—have any clear principle to suggest about how to bring that occupation to an acceptable end. That we cannot simply "pack up and leave" is, I think, clear. I am inclined to think that the best outcome (one suggested some months ago by Peter Galbraith[17]) would be one that led to a federation (even if it was not called that) of three largely autonomous communities—the Shi'ite, the Sunni, and the Kurdish—in Iraq. But that is only an opinion.

So why do I ask us to reexamine the justification of the decision to wage war *now?*

My reason is not just the abstract importance of Firth's principles, although I think that they do have enormous importance (as does Dworkin's idea of "rights as trumps"). That we stop treating killing and maiming one another as just "policies" subject to a cost-benefit analysis is vital if respect for human rights and international law is ever to have so much as a *chance* to take hold and grow. (In *Perpetual Peace* Kant estimated that it would take four hundred years.) But writing as an American, I am also concerned because "neoconservative" voices in the administration are reported to have urged and to still be urging further "preemptive wars." If that is right, then no "philosophical problem" is more urgent than the problem of thinking wisely and morally about the justification of war. I urge that Firth's principles are the place at which we should start that thinking.

17. Peter Galbraith, "How to Get out of Iraq," *New York Review of Books* 51, no. 8 (May 13, 2004), retrieved at the URL http://www.nybooks.com/articles/17103.

21

Cloning People

When Ian Wilmut and his coworkers at the Roslin Institute in Edinburgh announced in March 1996 that a sheep, "Dolly," had been successfully cloned, there was an amazing spontaneous reaction. People all around the world felt that something morally problematic was threatening to happen. I say "threatening to happen" because most of the concern centers not on the cloning of sheep, guinea pigs, and other animals, but on the likelihood that sooner or later someone will clone a human being. And I say "spontaneous reaction" because this is probably not a case in which a moral principle that had been formulated either by traditional ethical sources, religious or secular, was clearly violated or would be violated if we succeeded in cloning people, or someone would be deprived (if the feared possibility materialized) of what is already recognized to be a human right. Of course, some proposed uses of cloning do violate the great Kantian maxim against treating another person only as a means—for example, cloning a human being solely so that the clone could be a kidney donor or a bone-marrow donor—but these are probably not the cases that came immediately to people's minds.[1] I will argue that cloning humans (if and when that happens) may indeed violate human

---

1. In a few cases, I understand, people have already had a child precisely for such a purpose.

dignity even when the purpose is not as blatantly instrumental as producing an organ donor; but I do not think that the spontaneous reaction I described resulted from a considered view about *how* this would be the case.

The reaction was rather the kind of reaction that gives rise to a morality rather than the product of worked-through reflection.[2] Such reactions can be the source of moral insight, but they can also lead to disastrous moral error. I do not conclude from the fact that a ban or restriction on or condemnation of the practice is not easily derived from already-codified moral doctrines that it should be presumed that cloning human beings presents no moral problems; on the contrary, I shall argue that it poses extremely grave problems. What I want to do is say *why* the issue is a grave one—for I believe that the spontaneous reaction is justified—and to begin the kind of reflection that I believe we need to engage in, the kind of dialogue that we need to have, to make clearer to ourselves just what issues are at stake.

The scenario with which I shall be concerned—because it is, I believe, precisely the sequence of events that people fear may transpire—is the scenario in which (1) we learn how to clone people; and (2) the "technology" becomes widely employed, not just by infertile couples, but by ordinary fertile people who simply wish to have a child "just like" so-and-so. Why do so many of us view this scenario with horror?

Although this will be my central question, it must be mentioned that quite apart from the spontaneous horror I described, there are additional grounds, including some obvious utilitarian grounds, for being worried about the possible misuse of the cloning technology. (And anyone who relies on "market mechanisms" or "consumer sovereignty" by themselves to prevent the misuse of *anything* must have his head in the sand.) For one thing, even techniques of "bioengineering" that seem utterly benign, such as the techniques that have so spectacularly increased the yields of certain crops, have the side effect, when used as widely as they are being used now, of drastically reducing the genetic diversity of our food grains and thus increasing the probability of a disaster of global proportions should a disease strike these high-yield crops. If cloning were to be used not just to produce animals for the production of

2. Cf. William Gass's provocative essay, "The Case of the Obliging Stranger," *The Philosophical Review* 66, no. 2 (1957): 193–204, for a brilliant defense of the significance of such spontaneous reactions, accompanied (unfortunately) by a radical pessimism about the powers of ethical reflection.

medically useful drugs, but to produce "twins" of, say, some sheep with especially fine wool, or some cow with especially high milk yield or beef yield, and the practice were to catch on in a big way, the resulting loss of genetic diversity might well be even more serious. And if cloning ever became a really popular way of having babies, then the question of the result of the practice on *human* genetic diversity would also have to be considered. But although these issues are obvious and serious, it is not my purpose to address them further at this time, for I do not believe that *they* are what lie behind the spontaneous reaction to which I alluded.

The Family as a Moral Image

When I say that I want to engage in a reflection that will help us become clearer about the cloning issue, I do not mean that I want in this lecture to propose a specific set of principles or a methodology for deciding whether cloning people could ever be justified (or deciding when it would be justified). In a book I published in 1987,[3] I argued that moral philosophers who confine themselves to talk about rights or virtues or duties are making a mistake: that what we need first and foremost is a moral image of the world. A moral image, in the sense in which I use the term, is not a declaration that this or that is a virtue or a right; rather, it is a picture of how our virtues and ideals hang together with one another, and what they have to do with the position we are in. I illustrated what I meant by the notion of a moral image by showing that we can get a richer appreciation of the Kantian project in ethics if we see the detailed principles that Kant argued for as flowing from such a moral image (one that I find extremely attractive): an image of human equality. I claimed that the respect in which human beings are equal, according to Kant, is that we have to think for ourselves concerning the question of how one ought to live, and that we have to do this without knowing of a human telos, or having a clear notion of human flourishing. I argued that this notion of equality, unlike earlier ones, is incompatible with totalitarianism and authoritarianism. But although I find this image very appealing, I argued that it is not sufficient, and I spoke of us as needing a plurality of moral images. Commenting on this, Ruth Anna Putnam has written:[4]

3. Hilary Putnam, *The Many Faces of Realism* (LaSalle, Ill.: Open Court, 1987).

4. Ruth Anna Putnam, "Moral Images and the Moral Imagination," published in Spanish as "Imágenes morales y imaginación moral," *Dianoia* 38 (1992): 188.

I think that this reference to pluralism must be understood in a twofold way. On the one hand, the image of autonomous choosers is too sparse; it needs to be filled out in various ways, we need a richly textured yet coherent image. On the other hand, we need to recognize that people with different moral images may lead equally good moral lives. I do not say, of course, that all moral images are equally good; there are quite atrocious moral images; I am saying that there are alternative moral images by which people have led good lives, and that we can learn from their images as they can learn from ours. One of our very deep-seated failings is, I suspect, a tendency to have moral images that are too sparse.

One more remark about moral images before I return to our primary topic. In stressing the importance of moral images I do not mean to suggest that in some way moral images are a *foundation* from which moral principles, lists of rights and virtues, and so on are to be *derived*. I have never been a foundationalist in ethics, in philosophy of science, or in any other area of philosophy. We can raise the question of justification with respect to a moral image of the world (or of a part or aspect of the world or of life) just as we can raise it about any particular and partial value.[5] As I put it in *The Many Faces of Realism,* "The notion of a value, the notion of a moral image, the notion of a standard, the notion of a need, are so intertwined that none of them can provide a 'foundation' for ethics."[6] Commenting on this, Ruth Anna Putnam has rightly observed "It follows, it seems to me, then when he says that a moral philosophy requires a moral image he must not be understood to say that a moral philosophy requires a foundation, or an ultimate justification for whatever values it espouses, a foundation that can only be provided by a moral image. Rather we must understand him to say, and with this I agree, that without a moral image any moral philosophy is incomplete."[7] What I want to do now is to describe some moral images of the *family,* moral images that turn out to influence how we think not just about the family but about communal life in general.

That we do use images derived from family life in structuring our whole way of looking at society and our whole way of seeing our moral responsibilities to one another is not difficult to see. I became keenly aware of this many

5. This is a point that Ruth Anna Putnam stresses in "Imágenes morales y imaginación moral." I am enormously indebted to her reflections for helping me see more clearly just what I was trying to do with the notion of a moral image in *The Many Faces of Realism.*

6. Putnam, *The Many Faces of Realism,* 79.

7. Ibid.

years ago as the result of a conversation with my late mother-in-law, Marie Hall. In her youth Marie had lived a committed and dangerous life, being active in the anti-Hitler underground in Nazi Germany for two years before escaping the country and eventually making her way to the United States, and she continued to show an admirable commitment to a host of good causes until she died (she lived to be eighty-six). I admired her and loved her and adored not just her commitment but the vitality and humor that accompanied everything she did, and I constantly "pumped her" about her attitudes and activities during various periods in her life. On one occasion I asked her why she was inspired to make such efforts and run such risks for a better world in spite of all the setbacks, and I was amazed when she answered by saying simply, "All men are brothers." What amazed me was simply that she *meant* it. For her, "All men are brothers" was not a cliché; it was an image that informed and inspired her whole life. It was at that moment that I understood the role that a moral image can play.

The particular image that inspired Marie Hall, the image of us as all brothers (and sisters), played a huge role in the French Revolution and after, of course: the great slogan "Liberty, Equality, Fraternity" listed, perhaps for the first time, fraternity as an ideal on a par with equality and liberty. And to this day union members and members of oppressed groups frequently refer to one another as brothers and sisters. Of course, we all know that in "real life" siblings frequently do *not* get along. But we do have images of what *ideal* familial relations should be, and it is clear that these images are enormously powerful and can move large numbers of people to do both wonderful and (as the French Revolution also illustrates) terrible things.

The use of the image of an ideal family as a metaphor for what society should be is not confined to the West. Confucian thought, for example, has an elaborate picture of an ideal harmonious (and also hierarchically ordered) family that it consciously uses as a guiding metaphor in thinking about what an ideal society would be. I now want to turn to the following questions: what *should* our image of an ideal family be, and what bearing does that image have on whether we do or do not view the "cloning scenario" with horror?

Moral Images of the Family

I began this lecture by describing a scenario that is, I claimed, evoked by the prospect of cloning people. In that scenario, (1) we learn how to clone people; and (2) the "technology" becomes widely employed, not just by infertile couples, but by ordinary fertile people who simply wish to have a child "just like" so-and-so. I want now to see how we will evaluate this scenario from the standpoint of different possible moral (or immoral) images of the family.

Let us begin with an image that we would not regard as a *moral* image at all. Imagine that one's children come to be viewed simply as parts of one's "lifestyle." Just as one has the right to choose one's furniture to express one's personality, or to suit one's personal predilections, or even (even if one does not wish to admit it) to "keep up with the Joneses," so, let us imagine that it becomes the accepted pattern of thought and behavior to "choose" one's children (by choosing whom one will "clone" them from—from among available relatives, or friends, or, if one has lots of money, persons who are willing to be cloned for cash). In the brave new world I am asking us to imagine, one can have, so to speak, "designer children" as well as "designer clothes." Every narcissistic motive is allowed free rein.

When I say that one can have "designer children," I do not mean that in scientific fact, clones (or identical twins) *will* (or do) have the same abilities, attitudes, and other attributes. Well-informed writers on the cloning issue are aware that identical twins do not (and clones would not) have, for example, identical *brains* even if, *per impossible,* the environments were exactly the same (most of the brain cells that get produced are eliminated in the course of the embryo's development; and which will survive is not preprogrammed). Moreover, what cell assemblies get formed in the brain—the detailed "wiring"—is heavily influenced by "neuronal group selection," which has different results even when the genes are the same.[8] Still, clones will *look* like the person they are cloned from, and looks are likely to be what matter the most to the narcissistic parent.

What horrifies us about this scenario is that in it, one's children are viewed simply as objects, as if they were commodities like a television set or a new carpet. Here what I referred to as the Kantian maxim against treating another person only as a means is *clearly* violated.

[In an article published in the *New York Review of Books*,[9] Richard Lewontin, himself a great evolutionary biologist and an outspoken radical, has ar-

8. See G. M. Edelman, *Neural Darwinism: The Theory of Neuronal Group Selection* (New York: Basic Books, 1987).

9. Richard Lewontin, "Confusion about Cloning," *New York Review of Books,* October 23, 1997, 18–23.

gued that it is hypocritical to worry about this as long as we allow capitalist production relations to exist:

> We would all agree that it is morally repugnant to use human beings as mere instruments of our deliberate ends. Or would we? That's what I do when I call in the plumber. The very words "employment" and "employee" are descriptions of an objectified relationship in which human beings are "things to be valued according to externally imposed standards."[10]

But Lewontin is confused about what the Kantian maxim means. Even when someone is one's employee, there is a difference between treating that someone as a mere thing and recognizing his humanity. That is why there are criteria of civilized behavior with respect to employees but not with respect to screwdrivers. An excellent discussion of what these criteria require of one and how they are related to the Kantian principle can be found in Agnes Heller's book *A Philosophy of Morals.*[11]

Let me now describe the moral image I want to recommend, and then consider some alternatives that are not as blatantly narcissistic as the one in which children are treated as simply adjuncts to one's so-called lifestyle. To begin with, of course, if our image is to be a *moral* image at all, it should conform to the Kantian maxim. In an ideal family the members regard one another as "ends in their own right," as human beings whose projects and whose happiness are important in themselves, and not simply as they conduce to the satisfaction of the parents' (or anyone else's) goals. Moreover, it should be inspired by the Kantian moral image I described at the outset, the image that assigns inestimable value to our capacity to *think for ourselves in moral matters.* Here, perhaps, one should think not only of Kant but of Hegel, who, in *The Philosophy of Right,* argues that the task of good parents is precisely to *prepare children for autonomy.* The good parent, in this image, looks forward to having children who will live independently of the parents, not just in a physical or an economic sense, but in the sense of thinking for themselves, even if that means that they will inevitably disagree with the parent on some matters and may disagree on matters that both parents and child regard

---

10. Ibid., 21. Here Lewontin is (unfortunately) mocking the language of *Cloning Human Beings: Report and Recommendations of the National Bioethics Advisory Commission* (Rockland, Md.: June 1997).

11. Agnes Heller, *A Philosophy of Morals* (Oxford: Oxford University Press, 1990).

as important. Rather than regarding the tendency of one's children to think for themselves, to disagree with one when they reach the age of reflection, and to have tastes and projects one would not have chosen from them, one should welcome each of these. I note that although the Confucian image of the good family to which I referred earlier would agree that all the members of a good family should value one another as ends and not as mere means, the valorization of autonomy is foreign to classical Confucianism, with one exception: having the capacity to stand up for the right, when the alternative is clearly evil, is valued in the Confucian tradition. What is missing is the value of independent thinking about what morality requires of one. (Contemporary neo-Confucian thinkers are struggling with the problem of incorporating such Enlightenment values as autonomy and equality, including gender equality and equality among younger and older siblings, into a broadly Confucian framework, with interesting results.)

If one does accept the values that I have described and incorporates them into one's moral image of the family, then, I think, there is one further value that it is important (and very natural) to add, and that is the value of *willingness to accept diversity.* As things stand now (I write as a parent of four children and a grandparent as well), the most amazing thing about one's children is that they come into one's life as different—*very* different—people seemingly from the moment of birth. In any other relationship one can choose to some extent the traits of one's associates, but with one's children (and one's parents) one can only accept what God gives one to accept. And, paradoxically, that is one of the most valuable things about the love between parent and child: that, at its best, it involves the capacity to love what is very different from oneself. Of course, the love of a spouse or partner also involves that capacity, but in that case loving someone with *those* differences from oneself is subject to choice; one has no choice in the case of one's children.

But why *should* we value diversity in this way? One important reason, I believe, is precisely that our moral image of a good family strongly conditions our moral image of a good society. Consider the Nazi posters showing "good" Nazi families. Every single individual, adult or child, male or female, is blond; no one is too fat or too thin; all the males are muscular; and so on. The refusal to tolerate ethnic diversity in the society is reflected in the image of the family as utterly homogenous in these ways.

I am not claiming that a positive valuation of diversity *follows from* a positive valuation of autonomy. On the contrary, there have been societies that valued autonomy and moral independence while disvaluing diversity to the extent of believing in their own "racial superiority" and engaging in widespread sterilization of those who were seen as "unfit." I am thinking of the social-democratic Scandinavian countries that passed sterilization laws in the 1930s "with hardly any secular or religious protest."[12] Here is Daniel Kevles's description:

> Eugenics doctrines were articulated by physicians, mental-health professionals and scientists, notably biologists who were pursuing the new discipline of genetics. They were widely popularized in books, lectures and articles to the educated public of the day, and were bolstered by the research that poured out of institutes for the study of eugenics or "race biology" that were established in a number of countries, including Denmark, Sweden, Britain and the United States. The experts raised the spectre of social "degeneration," insisting that "feeble-minded" people—to use the broadbrush term then commonly applied to persons believed to be mentally retarded—were responsible for a wide range of social problems and were proliferating at a rate that threatened social resources and stability. Feebleminded women were held to be driven by a heedless sexuality, the product of biologically grounded flaws in their moral character that led them to prostitution and illegitimacy. Such biological analyses of social behavior found a receptive audience among middle-class men and women, many of whom were sexually prudish and apprehensive about the discordant trends of modern urban, industrial society, including the growing demands for women's rights and sexual tolerance.
>
> The Scandinavian region's population was relatively homogenous, predominately Lutheran in religion, and Nordic in what it took to be its racial identity. In this era, differences of ethnicity or nationality were often classified as racial distinctions. Swedish analysts feared that the racial purity of their country might eventually be undermined, if only because so many Nordics were emigrating. Swedish speakers in Finland feared the proliferation of Finnish speakers, holding them to be fundamentally Mongols and as such a threat to national quality.[13]

Our moral image of the family should reflect our tolerant and pluralistic values, not our narcissistic and xenophobic ones. And that means that we

12. Daniel L. Kevles, review of Gunnar Broberg and Nils Roll-Hanson, eds., *Eugenics and the Welfare State, Times Literary Supplement,* January 2, 1998, 3–4.

13. Ibid., 3.

should welcome rather than deplore the fact that our children are not us and not designed by us, but radically Other.[14]

Am I suggesting, then, that moral images of the family that depict the members of the ideal family as all alike, either physically or spiritually, may lead to the abominations that the eugenics movement contributed to? The answer is, "Very easily." But my reasons for recommending an image of the family that rejects the whole idea of trying to "predesign" one's offspring, by cloning or otherwise, are not, in the main, consequentialist ones. What I have been claiming is that the unpredictability and diversity of our progeny are intrinsic values and that a moral image of the family that reflects them coheres with the moral images of society that underlie our democratic aspirations. Marie Hall was willing to risk her very life in Hitler's Germany for the principle that "all men are brothers." She did not mean that "all men are identical twins."

Let me conclude by saying that if "rights" talk has not figured in my discussion, it is because, as I explained, I believe that our conceptions of rights, values, duties, and the like need to cohere with a moral image that is capable of inspiring us. But perhaps one novel human right *is* suggested by the present discussion: the "right" of each newborn child to be a complete surprise to its parents.

14. I use the Levinasian expression as a tribute to the philosopher who, more than any other, has taught us the moral importance of valuing alterity.

PART FOUR

Wittgenstein: Pro and Con

22

Wittgenstein and Realism

Writing about Wittgenstein is dangerous, I find. One is liable to be attacked by both sides, by Wittgenstein-*Schwärmer,* on the one hand, and by Wittgenstein haters (of whom there are a great many nowadays), on the other. But in my philosophical life I have always discovered in Wittgenstein's texts the sort of *depth* that makes them worth thinking about whether one agrees or disagrees with them on first (or even subsequent) reading. So I shall once again take the risk.

The title of this essay is "Wittgenstein and Realism," but I could also have titled it "Wittgenstein, Carnap, and Reichenbach on Realism" because, as will shortly become evident, I have found it necessary to say something about the views of all three of these philosophers.

I shall begin by discussing a famous passage in the *Tractatus Logico-Philosophicus* in which Wittgenstein appears to address the issue of realism versus solipsism, namely:

> Here we see that solipsism strictly carried out coincides with pure realism. The I in solipsism shrinks to an extensionless point and there remains the reality co-ordinated with it.[1]

---

1. Ludwig Wittgenstein, *Tractatus Logico-Philosophicus* (London: Kegan Paul, Trench, Trubner and Co., 1922), §5.64.

What Sort of "Solipsism" Does Wittgenstein
Have in Mind?

It was reading Brian McGuinness's brilliant essay "Solipsism" that made me realize just how many different meanings "solipsism" had in Wittgenstein's writings (including not just published writings but letters, discussions reported by Russell in 1912, the *Notebooks,* and successive drafts of the *Tractatus*). In particular, Wittgenstein used the term to characterize Schopenhauer's philosophy. A striking fact is that that philosophy is not "egocentric" in the sense in which Russell was concerned with escaping from an egocentric predicament. As McGuinness writes, after pointing out that Boltzmann's physicalism was one of Wittgenstein's early influences,

> If Boltzmann was his first idol, how does Wittgenstein still contrive to value the thinker who seemed to Boltzmann to represent philosophy at its most sterile and ridiculous? Once again the answer lies in Wittgenstein's wish to transcend the old philosophy. He uses Schopenhauer's terms, or ones like them, to make philosophical moves that confirm Boltzmann's hostility to philosophy.[2]

In this section, as a preliminary to my own discussion, I summarize McGuinness's interpretation, with which I largely agree.

McGuinness points out that the very first proposition of the *Tractatus* both parallels (in style) and opposes (in content) Schopenhauer's famous dictum that the world is my idea. "The world," as McGuinness puts it, "is not my idea, but is all that is the case (i.e. regardless of what is known or thought to be the case). This contrast remains even if there is also a contrast with the idea that the world consists of objects. Later in the book we realize that Wittgenstein nevertheless wants to accommodate Schopenhauer's insight within his own. 'The world is my world' is the point of reconciliation within the two."[3]

The "reconciliation" of which McGuinness speaks is not a reconciliation in the usual sense of that term, however. It is not that Wittgenstein finds that Schopenhauer's and Boltzmann's "insights" are both right, but that he finds that they are both *empty*. In fact, the remarks about the world being my world were originally intended to be part of the discussion of

2. Brian McGuinness, *Approaches to Wittgenstein* (London: Routledge, 2002), 133.
3. Ibid., 133–134.

pseudopropositions,[4] but as they became longer, they got moved to a section of their own, the 5.6's.[5] As McGuinness explains:

> Originally the point was to say that the world is my world has as little real content as saying that a thing is identical with itself. It could be paraphrased by saying meaninglessly—"The *world* we describe is the world *we* describe." . . . It is as if he reacts to the tendency he finds in Mach and Russell by saying to them, Look, you're not saying anything different from what the realist or physicalist (Hertz or Boltzmann) wants to say. No insight is being conveyed when the solipsist says, "The world is what I experience (or could experience)," unless one goes on to say with Wittgenstein, "But I am nothing," and then one conveys both a logical lesson (as here) and (in the diary and *Notebooks*) a moral lesson which we have yet to consider.
>
> The logical lesson is that just as there is no a priori science of identity and of the nature of the proposition (two of his examples in this context), so there is not an a priori experience of the world's relation to the subject, no a priori order of the world. Thus Wittgenstein takes a further step in the rejection of the a priori, and hence of philosophy itself, which was his heritage from Boltzmann, and yet at the same time is able to do justice to the hidden stream, with its origin in Schopenhauer and Tolstoy, that has all along accompanied his devotion to logic.[6]

In understanding what McGuinness means, it is important to keep in mind that for Wittgenstein, as for Frege in the same period, there is only one language, of which all natural languages are different realizations. That is, there is a fixed totality of possibilities, or, as one might also put it, of coherent thoughts to think, which are simply expressed in different signs in different natural languages. If we put aside the famous self-destruction of the *Tractatus* (the fact that the ladder is to be "thrown away" at the end), a self-destruction that does not seem to have been in Wittgenstein's mind in the *Prototractatus* and the *Notebooks,* then the picture is this: The only world that the "subject," that is, the utterly impersonal speaker of "the" language, can speak of or think of is the world that all speakers of the language necessarily

---

4. §§5.534–5.535 in both Ludwig Wittenstein, *Prototractatus,* ed. by B. F McGuinness, T. Nyberg, and G. H. von Wright (London: Routledge and Kegan Paul, 1971), and *Tractatus Logico-Philosophicus.*

5. McGuinness, *Approaches to Wittgenstein,* 135.

6. Ibid., 135–136.

share by virtue of sharing a language—not a particular natural language, such as English or Chinese, but the only possible language (that is what makes "the" subject, in a sense, impersonal). Professing to be a "physicalist" or a "solipsist" cannot change or add anything intelligible to the content of the propositions (*Sätze*) of the language. And those are the same propositions no matter what metaphysical gloss one may attempt to put on them.

A way I find it helpful to think of all this is via a contrast with Kant. The so-called realism of the *Tractatus*, the realism that consists in taking at face value the totality of possibilities represented in the language (which are also the propositions of science) is what Kant called "empirical realism." But Kant thought that the possibility of empirical realism could be seen only from the perspective of his "transcendental idealism." Wittgenstein (in the *Tractatus* and pre-*Tractatus* writings) is saying that "transcendental idealism" is unintelligible nonsense. I wrote a moment ago of taking the propositions of "the" language— which is also "ordinary language," according to the *Tractatus*, and also the language of science—at "face value"; but taking them at face value is the only way there *is* to take them! *That's* the point.

In line with this deflationary reading of the supposed "solipsism" of the *Tractatus*, McGuinness reads "the much-discussed proposition"—"That the world is my world gets shown in the fact that the limits of language (of language which *alone* I understand) mean the limits of *my* world [*Die Grenzen der Sprach* (*der Sprache die* allein *ich verstehe*) *die Grenzen* meiner *Welt bedeuten*]"[7]—as meaning simply that "there are no possibilities other than those guaranteed to be such, permitted to be such, by language, and since anyone can envisage everything language allows (and cannot envisage anything else), everyone has the same relation to the whole world."[8]

McGuinness anticipates the inevitable question: "Why then call it my world?" and answers, "Because it follows from the above that everyone is, and I in particular am, a measure of the world. We define its possibilities by being a completely neutral point of view. We are dual with it, as language is and for the same reason. It is for this reason that Wittgenstein exclaims, 'It is true. Man is the microcosm.'"[9]

---

7. Wittgenstein, *Tractatus Logico-Philosophicus*, §5.62.

8. McGuinness, *Approaches to Wittgenstein*, 136–137.

9. Ibid., 138; Ludwig Wittgenstein, *Notebooks, 1914–1916*, ed. G. H. von Wright and G. E. M. Anscombe, 2nd ed. (Chicago: University of Chicago Press 1979), 84; a note dated October 12, 1916.

Enter Carnap

It is eighty-five years since the *Tractatus* was published by Routledge and Kegan Paul in its series titled The International Library of Psychology, Philosophy and Scientific Method, but its issues continued to reverberate for decades and, in a way, continue to reverberate today. What I want to do now is to trace some of those reverberations.

It is well known that the members of the Vienna Circle spent many months reading and discussing the *Tractatus*. Among the ideas that they found congenial in that short book, in addition to the idea of clarifying meaning with the aid of modern logical notation, with which they were also familiar from Russell's writings, was certainly the identification of what could meaningfully be said with the propositions of science. (Recall: "The only strictly correct method in philosophy would be this. To say nothing except what can be said, *i.e.* the propositions of natural science, *i.e.* something that has nothing to do with philosophy: and then when someone else wished to say something metaphysical, to demonstrate to him that he had given no meaning to certain signs in his propositions. This method would be unsatisfying to the other—he would not have the feeling that we were teaching him philosophy—but it would be the only strictly correct method."[10]) But they also felt that the *Tractatus* fell far short of "the strictly correct method"; in fact, they thought that it was full of metaphysics. (Wittgenstein felt the same way about the views of the Vienna Circle, but that is a story for another occasion.)

Part of what they saw as "metaphysical," that is, as nonsensical, was the idea that the world has one fixed logical structure, or, alternatively, that propositions have one fixed logical structure. (In *Philosophical Investigations* Wittgenstein also famously rejects this view.) But before I consider Carnap's alternative view, I need to say a word about Carnap's notion of "metaphysics."

Usually that notion is explained in terms of the Vienna Circle's famous verifiability theory of meaning: what is metaphysical (or "nonsense") is what cannot be (scientifically) verified or refuted, and during much of his life Carnap did try to find a satisfactory statement of that theory. But in the *Logical Syntax* period appeals to the verification principle are entirely replaced by the idea that philosophy, which Carnap identifies with "logic of science," "is

---

10. Wittgenstein, *Tractatus Logico-Philosophicus*, §6.53.

nothing other than the syntax of the language of science."[11] With the verification principle conspicuous by its absence, what then justifies the rejection of metaphysics? Wittgenstein! I quote: "Wittgenstein has shown that the so-called sentences of metaphysics and of ethics are pseudo-sentences."[12] The idea was that logical syntax would separate the nice scientific sheep from the nasty metaphysical goats. (Of course, Carnap was soon to turn to embrace, and even formalize, the very semantics that he had rejected in *Logical Syntax*.)

If there is one thing that is constant in Carnap's description of "metaphysics," however, it is that he gives assertions of "reality" as paradigmatic examples of the sort of metaphysics that he repudiates. Thus in "The Old and the New Logic," Carnap writes, "We speak of 'methodological' positivism or materialism because we are concerned here only with methods of deriving concepts, while completely eliminating both the metaphysical thesis of positivism about the reality of the given and the metaphysical thesis of materialism about the reality of the physical world."[13] Twenty years later, in "Empiricism, Semantics and Ontology,"[14] ordinary empirical questions of existence, which Carnap regards as meaningful "internal questions" provided a scientific language has been selected, are sharply separated from "external questions" (e.g., "Do material objects exist?" "Do sense data exist?"), which are meaningful only when they are reconstrued as questions concerning the utility of selecting one or another form of language, and are meaningless if they are construed as theoretical questions ("Do material objects *really* exist?").

That there are problems with this stance goes without saying. One problem is raised by Quine's attack on the intelligibility of Carnap's "internal/external" distinction in the famous paper "Carnap and Logical Truth."[15] One can read

11. Rudolf Carnap, "Testability and Meaning, Continued," *Philosophy of Science* 4, no. 1 (January 1937): 1–40; quotation from p. 7.

12. Rudolf Carnap, *The Logical Syntax of Language* (London: Routledge and Kegan Paul, 1937), 282.

13. Rudolf Carnap, "Die alte und die neue Logik," *Erkenntnis* 1, no. 1 (1930): 12–26; English translation, "The Old and the New Logic," in A. J. Ayer, ed., *Logical Positivism* (Glencoe, Ill.: Free Press, 1959), 133–146; quotation from p. 144.

14. Rudolf Carnap, "Empiricism, Semantics and Ontology," *Revue Internationale de Philosophie* 4 (1950): 20–40; reprinted in Leonard Linsky, ed., *Semantics and the Philosophy of Language* (Urbana-Champaign: University of Illinois Press, 1952), 208–228.

15. W. V. Quine, "Carnap and Logical Truth," in P. A. Schilpp, ed., *The Philosophy of Rudolf Carnap* (LaSalle, Ill.: Open Court, 1963), 385–407.

Quine as asking, "If I say that chairs exist and that chairs are material objects, does that not imply that material objects exist?" Carnap would, of course, reply that as an "internal" sentence, "chairs exist" is a trivial empirical proposition, while as an "external" sentence, "material objects exist" is a metaphysical pseudoproposition, and very likely he would add that "chairs are material objects" likewise has two roles: as an analytical internal sentence and as a metaphysical external sentence. But Quine finds both the internal/external distinction and the (inflated) analytic/synthetic distinction Carnap appeals to suspect. Quine is telling us that when we acquiesce in our home language, there is no intelligible difference between "chairs exist" and "chairs really exist," or between "material objects exist" and "material objects really exist." Quine asks us to give up the idea of a fixed "scientific/metaphysical" distinction along with the other positivist baggage—the verifiability theory of meaning, the (inflated) analytic/synthetic distinction, and the internal/external distinction.

Apart from the last step—giving up the scientific/metaphysical distinction—it seems to me that Quine's attitude is close to the attitude of Wittgenstein in the *Tractatus*. What Carnap is trying to do in "Semantics, Empiricism and Ontology," it would seem to both Quine and Wittgenstein, is to find an *external standpoint from which to condemn external questions as meaningless.* To distinguish a sense that "chairs exist" or "material objects exists" has *within the language,* an "internal" sense, from a pseudosense that can be characterized only from outside the language is precisely to miss the point that there is no standpoint available to us outside the language—precisely the point that Wittgenstein was making in the *Tractatus,* as we have seen. What is "outside the language" is only what has not been given a sense at all, "nonsense" in the most ordinary sense of the term.

Carnap in the *Aufbau*

In the book that first made him famous, *Der logische Aufbau der Welt,*[16] Carnap reveals a significant disagreement with Wittgenstein. As I already mentioned, in the *Tractatus* Wittgenstein believes (like Frege before him) that there is a unique correct analysis of the propositions of science (the propositions

---

16. Rudolf Carnap, *Der logische Aufbau der Welt* (Berlin-Schlachtensee: Weltkreis Verlag, 1928).

of the one and only *Sprache,* that is, "my" language). If Wittgenstein is really arguing in the 5.6's of the *Tractatus* (as, following McGuinness, I have claimed that he is) that the dispute between "physicalists" and "solipsists" is *empty,* then the assumption of the uniqueness of the logical form of the *Sätze* (propositions) of science (and the further assumption that there is nothing particularly "metaphysical" about that logical form) is an essential premise of the argument. But this is an assumption that Carnap disputes.

There is an irony here, in that Carnap's position, that the language of science admits of *both* a ("methodological") solipsist reconstruction *and* a physicalist reconstruction, is one that Wittgenstein himself had held as a student of Russell's. This is something we know from a letter of Russell's to Lady Ottoline Morrell dated March 23, 1912:

> I argued about Matter with him [Wittgenstein]. He thinks it is a trivial problem. He admits that if there is no Matter then no one exists but himself, but he says that doesn't hurt since physics and astronomy and all the other sciences could still be interpreted so as to be true.[17]

To be sure, neither Carnap nor (presumably) the young Wittgenstein who argued this with Russell in 1912 thought that the existence of different formulations of the language of science meant that the *metaphysical* issue whether solipsism is "really true" was thereby reinstated. Rather, Carnap thought (and it sounds as if Wittgenstein thought in 1912) that these formulations (solipsist and physicalist) are in some way "equivalent." But there are serious problems with this "equivalence" claim. (McGuinness thinks that this was also Wittgenstein's position in the *Tractatus,* but I disagree.)

For one thing, the equivalence cannot be ordinary logical equivalence, because the models of the physicalist language and the models of the phenomenalist language are not even isomorphic. In fact, the physicalist language, if it includes quantification over space-time points or even regions, will have only nondenumerably infinite models, while the phenomenalist language, if it quantifies only over the *Elementarerlebnisse* (elementary experiences) of one human being (the "subject"), as does the phenomenalist language of Carnap's *Aufbau,* will have only *finite* models.[18]

17. Quoted in McGuinness, *Approaches to Wittgenstein,* 132.
18. I develop this point in more detail in Hilary Putnam, "Logical Positivism and Intentionality," in Allen Phillips Griffiths, ed., *A. J. Ayer Memorial Essays,* Royal Institute of Phi-

When Carnap wrote the *Aufbau,* presumably the "equivalence" between the possible reconstructions of the language of science, of which the phenomenalist reconstruction given in that work was supposed to be only one, was supposed to be cognitive equivalence as measured by a "methodologically solipsist" version of the verifiability theory of meaning. According to Carnap at that time, any sentence has the same consequences as far as the experiences of the subject are concerned, whether the sentence is formalized in the physicalist version of the language or in the phenomenalist version of the language. But this is a highly problematic criterion of cognitive equivalence.

To see why, let us recall Reichenbach's argument against the choice of "the egocentric language" in *Experience and Prediction* (1938).

Enter Reichenbach

Following Carnap's recommendation, Reichenbach describes the question dividing phenomenalists and physicalists as simply a question of "choice of a language." As Carnap later put it, "Our mistake was that we did not recognize the question as one of decision regarding the form of language; we therefore expressed our view in the form of an assertion—as is customary among philosophers—rather than in the form of a proposal."[19] But the argument Reichenbach offers against the choice of an "egocentric language" (i.e., a "methodologically solipsist" language, like that of the *Aufbau*) actually cuts against the idea that nothing is at stake but a pragmatic "choice."

To be sure, the way in which Reichenbach describes the "choice" may make it seem that he is not talking about Carnap at all. The two languages between which we are supposed to choose in the first part of *Experience and Prediction* are described as "egocentric language" and "usual language."[20] And "egocentric language" is not a sense-datum language (a language of *Erlebnisse*). It is a language in which I can speak of things as existing only when

---

losophy Supplement 30 (Cambridge: Cambridge University Press, 1991): 105–116; reprinted in Putnam, *Words and Life* (Cambridge, Mass.: Harvard University Press, 1994), 85–98.

19. Rudolf Carnap, "Testability and Meaning, Continued," 5.

20. Hans Reichenbach, *Experience and Prediction* (Chicago: University of Chicago Press, 1938), 140–141. For a fuller discussion, see Hilary Putnam, "Reichenbach's Metaphysical Picture," in Putnam, *Words and Life* (Cambridge, Mass.: Harvard University Press, 1994), 99–114.

they are observed by me. But if "things" are supposed to exist only when they are observed by me, and are assigned only the properties that they appear to me to possess, then these "things" will be just as phenomenal as Carnap's *Erlebnisse* or the empiricists' "impressions." And I have no doubt that Carnap was Reichenbach's intended target in this discussion.

The justification Reichenbach offers for the "choice" of usual language is, it seems to me, extremely deep. Instead of making the expected empiricist argument that "usual language" enables us to formulate more successful scientific theories—that is, that it enables *me* to make better predictions concerning *my* future experiences—he offers what we might call a Kantian argument (thinking of the Kant of the third *Critique* rather than the first): namely, that the choice of an egocentric language would leave us unable to formulate the justifications of a great many ordinary human actions (and here, revealingly, he calls it "the strictly *positivistic* language"):

> The strictly positivistic language contradicts normal language so obviously that it has scarcely been seriously maintained; moreover, its insufficiency is revealed as soon as we try to use it for the rational reconstruction of the thought-processes underlying actions concerning events after our death, such as [purchasing] life insurance policies. . . . We find here that the decision for the strictly positivistic language would entail the renunciation of any reasonable justification of a great many human actions.[21]

I described Reichenbach's argument as "deep," and indeed it is deep in reminding us of what is too often forgotten in discussions of "realism," "solipsism," and the like: that language is not used only to predict but also to *justify our actions* (and, of course, much else besides—for example, to explain ourselves, to exculpate ourselves, to make it possible for us to share a moral world: someone who, say, can only explain why he buys life insurance in terms of the *sensations* it gives him does not share our moral world). Further developed, this insight could have led Reichenbach to reflect, with the Wittgenstein of *On Certainty* and with the Stanley Cavell of *The Claim of Reason*,[22] on the fact that our relation to the world is not rightly conceived as one of *knowledge*. But it remains alone and isolated in Reichenbach's *oeuvre*.

21. Reichenbach, *Experience and Prediction,* 150.

22. Ludwig Wittgenstein, *On Certainty* (Oxford: Blackwell, 1969); Stanley Cavell, *The Claim of Reason* (Oxford: Clarendon Press, 1979).

A Problem with Reichenbach's "Egocentric Language"

Although this may not seem important if we think of Reichenbach's "egocen-tric language" as simply a stand-in for Carnap's full-blown phenomenalist sys-tem, it must be noted that, taken on its own, that language seems unintelligi-ble. Supposedly that language speaks about "things" (trees and chairs and other human beings), but what concept of a "thing" is in play here? It is not as if the ideas that things exist *independently* of the perceiver, that many of them ex-isted before their current perceivers did, that many will exist after those per-ceivers die, and that many of them are not perceived by us at all are *superficial* features of our thing concept. To just say, "Subtract all that from the notion of a thing" leaves us with—just *what* notion? Reichenbach has given us a *moral* argument for using "usual language" (though *not,* obviously, for regarding that use as just a "choice of a language," as he pretends), but deep as that argu-ment is, he would have gone still deeper had he asked whether there really is such a thing as an "egocentric language" to *choose* (or refuse to choose).

Enter the Later Wittgenstein

In closing, I want to suggest that a useful approach to the philosophy of the later Wittgenstein, and to the so-called private-language argument in partic-ular, is to see it as raising precisely that question. (I will break this last sec-tion, explaining why I think this, into three subsections for clarity.)

How I Understand "Grammar" in Wittgenstein

My look at the *Tractatus* may, I hope, have clarified the sense in which Witt-genstein accuses his earlier self of having conceived (or rather "preconceived") of logic as having a "crystalline purity." Logic is the very structure of the world and imposes demands: No vagueness! Concepts must make sense in all possible situations! What is possible is absolutely fixed in advance! Hence what thinkers *can* meaningfully think is also fixed (and, as we saw, must be the same for all thinkers)! These are all demands that the later Wittgenstein comes to see as chimerical. I hear pathos in *Philosophical Investigations* when he asks: "But what becomes of logic now? Its rigour seems to be giving way here.—But in that case doesn't logic altogether disappear? For how can logic lose its rigour? Of course not by our bargaining any of its rigour out of it.—The *preconceived idea* of crystalline purity can only be removed by turning our whole examination around. (One might say: the axis of reference of our examination must be rotated, but around the fixed point of our real need.)"[23]

The "new axis of reference" is called "grammar" by Wittgenstein. But what exactly is Wittgensteinian "grammar"?

In a conversation John McDowell suggested to me the way of thinking of that notion that I find most helpful. What he suggested was that I should think of the sort of examination of concepts that Elizabeth Anscombe undertook in *Intention* as a "grammatical" investigation in Wittgenstein's sense;[24] for example, pointing out the differences in the way we answer the questions "Why is water trickling down the window pane?" and "Why are you cutting the bread?" is clarifying the "grammar" of intentional explanation.

I like a number of things about McDowell's explanation; particularly, I like the fact that it does not turn grammar into a procrustean bed. There is no suggestion here that grammatical investigation must be *all* of philosophy (even if that may have been Wittgenstein's preferred use of the word "philosophy" after a certain period), nor any suggestion that constructive philosophical work and grammatical investigation must be incompatible, that is, that philosophy must consist merely of "therapy." As McDowell went on to say, what Wittgenstein does when he is at his best is convince one that certain philosophical "theses" (e.g., no one ever directly perceives a material object") do not, in the end, say anything intelligible, or better, do not say anything that I any longer feel I can express any insight by saying—but why should not showing that *not* require constructive work, provision of an alternative philosophical picture? But I will stop here because my aim is not to develop an interpretation of Wittgenstein's later philosophy as a whole, but simply to indicate a way of thinking that helps me learn from that philosophy.

Why I Say "an Approach"

I said earlier that I want to suggest that a useful approach to the philosophy of the later Wittgenstein, and to the so-called private-language argument in particular, is to think of it as taking up the question Reichenbach failed to ask, whether the idea of an "egocentric language" really makes sense. I take it that

---

23. Ludwig Wittgenstein, *Philosophical Investigations* (Oxford: Blackwell, 1953), §108.
24. Elizabeth Anscombe, *Intention,* 2nd ed. (Ithaca: Cornell University Press, 1963).

this question calls for a "grammatical" investigation in the sense I just explained. I speak only of "suggesting an approach" and not of "interpreting," both because it would be absurd to attempt anything as ambitious as interpreting "the private-language argument" (or arguments) in a few pages, and because what I want to accomplish in this essay is less to "interpret" Wittgenstein than to suggest a useful approach to his work, one that enables one to learn from it and build on it in one's own way. The private-language argument, to the extent that there is *a* private-language argument in *Philosophical Investigations,* is, like all of Wittgenstein's writing, a collection of aperçus that touch on many complex philosophical issues. Still, when Wittgenstein wrote the famous paragraph that contains the words "private language,"[25] I find it plausible to suppose that *one* of the issues he had in mind was the issue that concerned him in the 5.6's of the *Tractatus,* the issue of "solipsism." But (apart from this remark) I shall not try to read his mind, but rather to apply ideas from *Philosophical Investigations* to that issue in my own way.

"Private Language" and Public Language

The question I accused Reichenbach of ignoring was whether there really is such a thing as an "egocentric language," as opposed to "usual language" (our public language). If we take the "egocentric" language *not* to be Reichenbach's strange thing-language-that-is-not-about-things (he himself identifies his egocentric "things" with "impressions" in much of his discussion in part 1 of *Experience and Prediction*), but to be, say, Carnap's "methodological solipsist" language in the *Aufbau,* then that is precisely the question that launches the private-language argument:

> Now, what about the language which describes my inner experiences [*Erlebnisse*] and which only I myself can understand. *How* do I use words to stand for my sensations?—As we ordinarily do?—Then are my words for sensations tied up with my natural expressions of sensation? In that case my language is not a "private" one. Someone else might understand it as well as I.—But suppose I didn't have my natural expressions of sensation, but only had the sensations? And now I simply *associate* names with sensations and use these names in descriptions.[26]

25. Wittgenstein, *Philosophical Investigations,* §256.
26. Ibid., §256.

I certainly do not want to speculate that Wittgenstein actually had Carnap's *Aufbau* in mind here (especially because Wittgenstein himself had conceived of the possibility of phenomenalist reconstruction of the whole language as early as 1912 and was to flirt with phenomenalism again in the so-called early middle period [1929–?]). But I will apply what he says here to that work because such an approach fits surprisingly well.

The question Reichenbach ignored was whether our ordinary notion of a "thing" does not involve many attributes incompatible with "egocentric" existence. But it is not hard to see that there is a similar question to be asked about sensations and "experiences" in the sense of Carnap's *Erlebnisse*. Knowing what a pain is certainly involves knowing that pains are "located" in parts of the body, and that they cause me to move the part of the body when I think or instinctively feel that that will cause the pain to stop (e.g., when I discover that my hand is in water that is too hot), that severe pain will cause me to scream or cry, and so on. All our sensations talk enters into explanations of *physical movements* and of responses to physical events. In sum, sensation language and thing language are parts of a single public language. If someone did not know that, he could not have our "grammar" of sensation talk. So, what concept of a "sensation" could he have? That is the question that Wittgenstein poses in *Philosophical Investigations* §256. The question does not presuppose either behaviorism or verificationism or any other "ism." It presupposes that (normally, anyway) to ascribe a concept to someone is to ascribe some mastery of its grammar.

But why, you may wonder, did Wittgenstein say that the private language is one that only I myself can understand? If we suppose that a position like Carnap's is being targeted, this makes perfect sense. To see why, recall two features of the *Aufbau:*

1. The purpose of the *Aufbau* is not merely to present *a* reconstruction of the language of science (or the language of cognition, because for Carnap there is no cognition outside science), but *the* reconstruction that has *epistemic priority,* that is, the one we need when we do epistemology. (In 1936 Carnap was to renounce "epistemology" altogether, but in 1922 he was still under residual Husserlian influence.) But that means that the *Aufbau* is meant to depict all we can *know,* and to show how we know it. Because the point of view is constructionist and certainly not "holistic" in the Quine-Duhem sense, that means that we are supposed to have knowledge expressed by "protocol sentences," and hence to fully understand the primitive phenomenalistic concepts

they contain, *before* we ascend to understanding talk of physical objects (whose reduction to talk of *Erlebnisse* the *Aufbau* was, famously, unable to complete). So it is essential to this sort of *foundationalist phenomenalism* that sensation concepts (*Erlebnisse* concepts) should *not* presuppose thing concepts.

2. Talk of other people's sensations was supposed to be interpreted behavioristically once talk of their bodily motions had been reconstructed (i.e., after talk of physical objects had been reconstructed). So sensation concepts at the ground level, the concepts that figure in protocol sentences, must not presuppose the sense of "sensation" or "experience" in which *other people* have "sensations" or "experiences" (which is a totally different sense, viewed from the *Aufbau*). That is why the reconstruction has to be "solipsist."

In sum, the *Aufbau,* or any similar "solipsist" reinterpretation of the language, must start with a notion of experience *utterly* different from the one we actually have. And this is what Wittgenstein doubts we can understand.

Concluding Remarks

Of course, my account of Wittgenstein's long struggle with the possibility of a "solipsist" reconstruction of the language ends where most discussions of the private-language argument begin. At this point, as *Philosophical Investigations* §256 already starts to do, Wittgenstein considers the possibility that the solipsist will concede that his notion of a sensation is not the one we have in "usual language" (as Reichenbach called it) and will say that the private language refers to private objects (today some might call them "qualia"), and that this reference is made possible by private ostensive definition, and he tries to make us suspicious of both notions. I say "tries to make us suspicious" because I do not think that he was giving a "proof" of the nonexistence of private objects or the impossibility of private ostensive definition. But I think that we are well advised to ask ourselves whether either notion really makes sense, and that Wittgenstein has given us a valuable introduction to serious reflection on these topic—an introduction, but not a "last word," because there are no last words in philosophy.

Carnap, of course, did not try to defend the phenomenalist reconstruction by invoking private objects (which would have seemed to him too metaphysical) or private ostensive definition. Instead, he lost interest in the project and also renounced the idea of epistemology. Gradually, too, solipsism and phenomenalism ceased to be a worldwide preoccupation of philosophers—ceased to such an extent that today our graduate students hardly have an inkling of what the *Tractatus*, the *Aufbau*, and *Experience and Prediction* were about. I hope that I have convinced you that this ignorance should be rectified. It should be rectified because the period of the *Tractatus*, the *Aufbau*, and *Experience and Prediction* was one of the great periods in the history of philosophy, and to miss what its issues and debates were really about is to misunderstand writings that we can still profit from today.

23

Was Wittgenstein *Really* an Antirealist about Mathematics?

It seems clear that we understand the meaning of the question: "Does the sequence 7777 occur in the development of π?" It is an English sentence; it can be shown what it means for 415 to occur in the development of π; and similar things. Well, our understanding of that question reaches just so far, one may say, as such explanations reach.[1]

It is widely assumed that Wittgenstein was an "antirealist" about mathematics, and this assumption then undergirds claims by philosophers as different in other respects as Michael Dummett, Paul Horwich, Saul Kripke, and Richard Rorty that Wittgenstein was an antirealist across the board. Indeed, very often this is not recognized to be an assumption; the secondary literature often takes it to be self-evident. My aim in this essay is to challenge this assumption, not by presenting a lengthy textual analysis of Wittgenstein's extensive writing on the foundations of mathematics, but by offering a reaction of my own to the "realism problem" that is built on recognizably Wittgensteinian materials and does not smack of any sort of "antirealism." To be sure, even if I can bring this off, it will not show that a more realist (or less "antirealist") reading is the only possible reading that can be given to Wittgenstein's text; for I will claim that one can also build a case for an antirealist position out of Wittgensteinian materials, and that this is what Michael Dummett, Saul Kripke, and Paul Horwich claim to have done (Rorty does not cite texts but simply tells us what Wittgenstein allegedly thought). Moreover, as late as

1. Ludwig Wittgenstein, *Philosophical Investigations* (Oxford: Blackwell, 1953), §516.

the last sections of *Remarks on the Foundations of Mathematics* (1944),[2] Wittgenstein *was,* I believe, attracted to the idea that mathematical propositions that are not humanly decidable *lack* a truth-value[3]—an idea that I find deeply problematic for reasons that I will explain. However, by showing how Wittgenstein's work as a whole—not just the *Remarks on the Foundations of Mathematics*—can support a very different view, I hope to encourage us to take a second look at the *Remarks* itself, and to ask whether even this view—which is usually taken to be "antirealist" almost by definition—may not have sprung from very different considerations.

The plausibility of the "antirealist" readings of Wittgenstein's later philosophy rests on the existence in Wittgenstein's work of what look like "verificationist" ideas. Indeed, Wittgenstein was attracted to such ideas in the early and mid-1930s, although not, I believe, when he wrote *Philosophical Investigations* (which is not to deny that he found some insights in verificationism even in his last writings).[4] In addition, the idea that changes in mathematical method—for example, the change from proving a proposition in geometry by classical geometric means to proving the "same" proposition by means of algebraic geometry—constitute a change in the very identity of the proposition being proved exerted a certain fascination for Wittgenstein (but did not, I believe,

---

2. But not necessarily as late as January 1945. For the quotation from Wittgenstein that serves as the epigraph to this paper is compatible with the line I suggest in this paper as the one Wittgenstein should have taken, and it is silent on the claim that even "omniscience" cannot know whether a pattern occurs in the decimal expansion of π if there is no humanly available calculation to show that there is—a claim that is central to the discussion of the law of the excluded middle in part 4 of *Remarks on the Foundations of Mathematics.* It is, of course, important that Wittgenstein published *Philosophical Investigations,* while the material editors have put together as *Remarks on the Foundations of Mathematics*, G. E. M. Anscombe, R. Rhees, and G. H. Wright, eds. (Cambridge, Mass.: MIT Press, 1983) was unpublished.

3. In *Remarks on the Foundations of Mathematics,* he writes, "Suppose people go on and on calculating the expansion of π. So God, who knows everything, knows whether they will have reached '777' by the end of the world. But can his omniscience decide whether they *would* have reached it after the end of the world? It cannot. I want to say: even God can determine something mathematical only by mathematics. Even for him the mere rule of expansion cannot decide anything that it does not decide for us" (V, §34; this remark dates from 1944). To say that omniscience cannot decide a question is, I believe, Wittgenstein's way of saying that there is no answer to be known, i.e., it is an illusion that there must be a right answer.

4. Cora Diamond, "Realism and the Realistic Spirit," in Diamond, *The Realistic Spirit* (Cambridge, Mass.: MIT Press, 1991), 39–72. See also Hilary Putnam, "Pragmatism," *Proceedings of the Aristotelian Society* 95 (June 1995): 291–306.

convince him).[5] And this idea can be *seen* as an expression of verificationism. In addition, there are Wittgenstein's remarks on the law of the excluded middle in mathematics (in part 4 of *Remarks on the Foundations of Mathematics*), which, although they are clearly not as "revisionist" as Brouwer's views, have sometimes been seen as expressing quasi-intuitionist sentiments. But I have already said that my concern in this essay is not with this sort of textual question. To repeat, I shall be quite content if I can show that there is *a* way of looking at the questions in the philosophy of mathematics (in particular, a way of trying to dissolve the "realism problem") that is not "antirealist" but is "Wittgensteinian."

In seeking such a way, I shall first be guided by Cora Diamond's interpretation of the "rule-following" discussion in *Philosophical Investigations*. She has shown that this famous discussion can be read in a way that is neither metaphysically realist nor "antirealist,"[6] and that way of reading the "rule-following" discussion is one of my guides in my present enterprise.

Second, there is an attitude toward the philosophy of mathematics that is widely shared, and to which I myself contributed, according to which the problems in question are so hard that one should despair of seeing any way at all of resolving them. (In fact, I once wrote a report on the present state of the philosophy of mathematics with the subtitle, "Why Nothing Works.")[7]

---

5. For the reasons for not believing that Wittgenstein accepted this idea, see Juliet Floyd, "On Saying What You Really Want to Say: Wittgenstein, Gödel, and the Trisection of the Angle," in Jaakko Hintikka, ed., *Essays on the Development of the Foundations of Mathematics* (Dordrecht: Kluwer Academic Publishers, 1996), 373–426. In *Remarks on the Foundations of Mathematics,* VII, §10, Wittgenstein writes:

> Now how about this—ought I to say that the same sense can only have *one* proof? Or that when a proof is found the sense alters?
>
> Of course some people would oppose this and say: "Then the proof of a proposition cannot ever be found, for if it has been found then it is not a proof of *this* proposition." But to say this is so far to say nothing at all.—
>
> It all depends *what* settles the sense of a proposition, what we choose to say settles its sense. The use of the signs must settle it; but what do we count as the use?—
>
> That these proofs prove the same proposition means, e.g.: both demonstrate it as a suitable instrument for the same purpose."
>
> This last remark is quite compatible with saying that the same proposition can have both, e.g., an "elementary" proof and an "analytic" proof.

6. Diamond, "Realism and the Realistic Spirit."

7. Hilary Putnam, "Philosophy of Mathematics: Why Nothing Works." Originally published with the title "Philosophy of Mathematics: A Report", in P. D. Asquith and H. E. Kyburg

The second idea that will guide the present discussion is that *that* view—the view that the problems in the philosophy of mathematics are entirely *sui generis* and add up to something close to an antinomy of reason—cannot be right. In the end, I shall try to show, the conundrums that bother us in the philosophy of mathematics are intimately connected with, and have precisely the same roots as, familiar conundrums about the indeterminacy of reference, the objectivity of value judgments, etc. Of course, the fact that the difficulties are in a sense the same does not mean that they do not require special treatment in each case. It is a feature of philosophical difficulties that each time a difficulty arises anew, it seems to be in some way more serious, more real, and more intractable than it seemed before, and that appearance is always connected with particular aspects of the particular form that the difficulty takes, aspects that have to be carefully analyzed in each case. It would be profoundly contrary to the spirit of Wittgenstein to suppose that there is some one magic wand—called, as it might be, "the resolution of the realism/antirealism problem"—that one can wave to make all the protean forms of that difficulty vanish with a "Poof!" Nevertheless, it seems to me that what we have to eventually reach as we work through the difficulties in the philosophy of mathematics is a standpoint from which the problems here will seem not mysterious and charming but *boring*.[8]

I must also mention a third idea that will guide me in the present essay, the idea that *before* one attempts to dissolve the difficulties, one must give the difficulties a chance to be *felt;* and that means that I will have to try to understand the position of someone for whom those difficulties seem real and intolerable (that is not hard for me to do, because that is the position I occupied myself for many years).

Wittgenstein on Following a Rule

We are all acquainted with analyses of Wittgenstein's text according to which the point of Wittgenstein's discussion of rules is that the notion of following

---

Jr, eds., *Current Research in Philosophy of Science: Proceedings of the P.S.A. Critical Research Problems Conference* (East Lansing, Mich.: Philosophy of Science Association, 1979), 386–398; reprinted in Putnam, *Words and Life* (Cambridge, Mass.: Harvard University Press, 1994), 499–512.

8. I no longer agree with this view, which seems to me now [2011] too close to what I think of as the negative pole of Wittgenstein's thought in "Wittgenstein: A Reappraisal" (Chapter 28 in this volume), which concerns the treatment of metaphysics as a kind of illness.

a rule, or the normativity of rule following, is in one way or another to be explained in terms of the notion of conforming to the standards of a community.[9] Such a view may strike one as entailing a profound change in some of the things we think about rules, especially in mathematics, and indeed it does have startling (Kripke would call them "skeptical") consequences. For example (although Wittgenstein himself *never* says any such thing), it is a consequence of these interpretations that the notion of one individual following a rule in isolation from any linguistic community makes no sense in Kripke's version, unless we—in our imagination—"take him into our community" and apply our notions of rule following to him.[10]

If, however, we actually look at the discussion of following a rule in *Philosophical Investigations*,[11] it is striking that we find no hint of this claim. The closest Wittgenstein comes to addressing the question whether one individual in isolation could follow a rule is to ask, "Is what we call 'obeying a rule' something that it would be possible for only *one* person to do, and to do only *once* in his life?" and to answer (he describes this as "a note on the grammar of the expression 'to obey a rule'"):

> It is not possible that there should have been only one occasion on which a person obeyed a rule. It is not possible that there should have been only one occasion on which a report was made, an order given or understood, and so on.—To obey a rule, to make a report, to give an order, to play a game of chess are *customs* (uses, institutions).[12]

Now, if Wittgenstein had meant "It is not possible that one person in isolation could obey a rule *period*," would he not have said so? One wonders why Wittgenstein would have made such a weak statement if what he thought was so much stronger. Compare the question Wittgenstein puts to himself in *Remarks on the Foundations of Mathematics*, II, §67: "But what about this

9. Cf. Saul Kripke, *Wittgenstein on Rules and Private Language* (Cambridge, Mass.: Harvard University Press, 1982). In "Wittgenstein and Kripke on the Nature of Meaning," *Mind and Language* 5, no. 2 (Summer 1990): 105–121, Paul Horwich also defends such a reading, although he criticizes Kripke's notion of a "fact" and Kripke's description of the position as "skeptical." On Horwich's view, I understand a word correctly if I assign degrees of assertibility to sentences containing the word in a way that accords with community standards. Rorty's remarks on Wittgenstein in *Contingency, Irony, Solidarity* (Cambridge: Cambridge University Press, 1989) presuppose a similar interpretation.

10. Kripke, *Wittgenstein on Rules and Private Language*, 110.

11. Ludwig Wittgenstein, *Philosophical Investigations*, §§143–242.

12. Ibid., §199. I have retranslated the German.

consensus—doesn't it mean that *one* human being by himself could not calculate? Well *one* human being could at any rate not calculate just *once* in his life." Here too he refuses to simply answer yes.

In defense of the "community-standards" interpretation, it might be argued that Wittgenstein's use of the word "institution" implies the existence of a community. But it remains the case that (1) the explicit point of §199 is only to rule out the "possibility" that it should happen exactly once that a rule is obeyed, or an order given, and so on, and not to say that a community is necessary; and (2) the discussion that follows connects the notion of rule following not with "institutions" in the sociological sense, but with the notion of a "practice" and then with the notion of a "regularity," and certainly it is possible for there to be regularities in what one individual human or animal does. One individual can certainly make it a *custom* to do something whether or not his community does (the German word here is *Gepflogenheit,* by the way—a word that normally describes the way in which an individual is in the habit of doing something, unlike the English word "custom," which does normally refer to a communal practice). Does the whole textual evidence for this astounding interpretation rest, then, on the fact that Wittgenstein included "institutions" in his *parenthesis?*

Moreover, the context makes it quite clear why Wittgenstein made the statement just quoted. Although he certainly does not claim that we can *define* "rule" in terms of "regularity" or reduce the notion of a rule to the notion of a regularity, he does argue that following a rule is a practice, and that practices presuppose regularities. A world in which no one even thought of the notion of a rule or of following a rule except for one isolated individual, and that isolated individual had the concept for just one brief moment and followed a rule for just that moment—if, contrary to Wittgenstein's "grammatical point," there could be such a world—would be a world in which for one moment there was a rule even though none of the regularities in linguistic and extralinguistic behavior that give content to "rule" talk were ever in place. *That* is the supposition that Wittgenstein thinks is senseless, and it requires no commitment to any astounding philosophical views (no "philosophical revisionism") to accept what Wittgenstein said *here.*

Nor are the doubts and rejoinders that Wittgenstein anticipates in response to what he writes in §§199–207 the ones he should have anticipated if the claim in §199 were so radical. Having stated a condition for talk of rules to make sense (a condition that uses the notion of a regularity), what Wittgenstein quite sensibly worries about is that someone might suppose that even the notion of a regularity *(Regelmässigkeit)* already presupposes the notion of a rule *(Regel)*, and thus what Wittgenstein has said is simply *circular*. "Then am I defining 'order' and 'rule' by means of 'regularity'?—How do I explain the meaning of 'regular,' 'uniform,' 'same' to anyone?"

Let me now quote Wittgenstein's response to *this* question in full:

> I shall explain these words to someone who, say, only speaks French by means of the corresponding French words. But if a person has not yet got the *concepts,* I shall teach him to use the words by means of *examples* and by *practice.*—And when I do this I do not communicate less to him than I know myself.
>
> In the course of this teaching I shall show him the same colors, the same lengths, the same shapes, I shall make him find them and produce them, and so on. I shall, for instance, get him to continue an ornamental pattern uniformly when told to do so.—And also to continue progressions. And so, for example, when given . . . to go on. . . .
>
> I do it, he does it after me; and I influence him by expressions of agreement, rejection, expectation, encouragement. I let him go his way, or hold him back; and so on.
>
> Imagine witnessing such teaching. None of the words would be explained by means of itself; *there would be no logical circle.*[13]

Wittgenstein makes two principal points in this passage. First, he makes the point that we all know how to explain these notions, and he gives examples of the different ways in which we do explain them—all, notice, perfectly *ordinary* ways, that is, not philosophical ways, ways that presuppose "Platonism" or "mentalism" or any other philosophical account of the real nature of rule following. Second, he adds quite categorically, "And when I do this [teach the notions in this way] I do not communicate less to [the student] than I know myself." That is, Wittgenstein denies that there is some other kind of explanation of the notions involved in following a rule (notions of doing the "same" thing, of "continuing" a progression in "the same" way, of behaving "uniformly," and of doing such and such "ad infinitum" [which is also mentioned in §208], and even the notion "and so on") that one is in possession of either by virtue of being a philosopher or by virtue of having direct acquaintance with something ineffable.

Cora Diamond has pointed out that the target Wittgenstein has in mind is no straw man.[14] F. P. Ramsey, Wittgenstein's friend and interlocutor for several

---

13. Ibid., §208. Emphasis added in the last sentence.
14. Diamond, "Realism and the Realistic Spirit."

years, believed that although indeed there is no other *explanation* that one can give of what one does when one follows a rule than the ordinary one (the "banal" one, so to speak), nevertheless, following a rule consists in following "psychological laws"—not laws of theoretical psychology, but laws of which each of us has (and can only have) implicit knowledge. This view combines a very special brand of mentalism about the nature of rule following with the idea of incommunicable knowledge that is a well-known target of *Philosophical Investigations*. Thus it makes perfect sense that Wittgenstein would reject that view here, as well as contrast the ordinary teaching of the concept of rule following with the various mentalist and Platonist accounts that he has discussed before.

What I am suggesting is the following: if the discussion of rule following in *Philosophical Investigations* expresses skepticism at all, the "skepticism" is directed at *philosophical accounts* of rule following, and not at rule following itself. What readers like Kripke and Horwich have done is to take Wittgenstein to oppose not only *metaphysical realism* about rule following but also our commonsense realism about rule following, when what Wittgenstein actually doubts is the need for and the possibility of a philosophical *explanation* of rule following that will justify the commonsense things we say (e.g., our talk of "right" and "wrong" ways to follow rules) and the ways in which we actually teach people to follow rules. Indeed, the point of §208 is that even the concept of a *regularity* does not have or need a philosophical *explanation* in this sense.

But why do we not need a noncommonsensical account? What is wrong with explanations of rule following in terms of special powers of "grasping concepts" and so on? The answer to this question is, of course, the whole of the discussion in *Philosophical Investigations* itself (which is, perhaps, the whole of *Philosophical Investigations*). I do not intend to repeat that discussion, but I shall remind you of certain features of it.

Wittgenstein repeatedly tries to show that when philosophers offer accounts of rule following—accounts that are supposed to explain the possibility of rule following or to tell us the metaphysical nature of rule following—although the words used in these explanations may sometimes be in place (may sometimes hit it off in the sense of giving us a description of an impression we have when we follow a certain kind of rule, a description of the "phenomenology," one might say, of following a rule), the minute we take them seriously as *explanations,* the minute we suppose that some *entity* has been introduced whose *existence* clarifies the supposed question about how we are able to follow rules, then we slip into thinking that we have discovered a "supermechanism";[15] moreover, if we try to take the supposed supermechanism seriously, we fall into all sorts of absurdities. At certain points Wittgenstein indicates the general character of such "explanations" by putting forward parodistic proposals, for example, the proposal that what happens when we follow a rule is that the mind is guided by invisible train tracks, "lines along which [the rule] is to be followed through the whole of space."[16]

Even this proposal, please note, may be right as a description of the "phenomenology" on at least some occasions. If I am computing the sine of a number that I have never computed before, or taking the square root of a number, then perhaps it will seem to me that I am being "taken" to a series of "new places" quite automatically, as if I am arriving at a series of train stations that I have never visited before while sitting in a train as it continues on its predetermined route. Wittgenstein remarks that my description makes sense only if it is to be understood symbolically—that is, metaphorically—"I should have said: *This is how it strikes me.*"[17] But what if people were to believe that the existence of invisible train tracks and the mind's going along like a train from one station to the next is the *explanation* of what it is to follow a rule? It would be as if they confused the grammar of a myth with the grammar of a scientific explanation. "My symbolical expression was really a mythological description of the use of a rule."[18]

Of course, the answer to the question "Well, what is wrong with wanting to supplement the ordinary account of rule following with another account?" is that there is nothing wrong if it is simply a desire to understand the empirical facts, to know, for example, what the causal preconditions—cultural, biological, psychological, or what have you—of following this, that, or the other sort of rule are. But that is not the question the philosopher tries to investigate.

15. Jerry Fodor's talk of "nomic connections" as what enable us to have concepts may be a present-day example of this. According to Fodor, what enables me to refer to things that may not exist, and that may not even be objects of any science—e.g., witches—are nomic connections to the property of being an object of the kind in question. Cf. Fodor, *A Theory of Content and Other Essays* (Cambridge, Mass.: MIT Press, 1992), 89–136. A nomic connection to the property of being a witch? Talk about supermechanisms!

16. Ludwig Wittgenstein, *Philosophical Investigations,* §218–219.

17. Ibid., §219.

18. Ibid., §221.

What bothers Kripke, for example, is really the question, "How can we grasp rules that do not in themselves contain any errors with our finite and all-too-fallible brains that inevitably do make errors?" And the trouble with that question is that, in general, it seems genuinely pressing only because one supposes what is not the case, that a fallible brain could understand an infallible rule only by being connected with a mechanism (a supermechanism) that was itself infallible. Of course, most philosophers will protest that such a mechanism is not what they are seeking at all, but at that point, I believe, Wittgenstein's strategy is to show that they cannot say *what* it is they are seeking; that in the end they have a desire for *they know not what.*

I do not mean to suggest that Wittgenstein's discussion ends here. Stanley Cavell has reminded us that Wittgenstein has a deep interest in what in us—I have heard Cavell call it our "perversity"—drives us to seek answers that cannot make sense to questions that have no clear sense. But my purpose here has been more limited; to remind you that what Wittgenstein is trying to bring us back to is a standpoint from which notions like "learning to add two," "understanding the rule keep adding two," and "knowing what it means to add two" do not seem problematic at all. At least as a first step in his investigations, he wants to make it seem problematic that these notions have become problematic; rather than wondering how it is that we can add two, we should be wondering at the fact that it can come to seem to us the most puzzling thing in the word that we can add two. In the same way, at the end of my discussion I want to be able to say that what is puzzling in the philosophy of mathematics is not that we are able to use and learn and understand and acquire the concepts of, say, number theory, but that it has come to seem to so many philosophers the most puzzling thing in the world that we should be able to do this. Unfortunately, there is a large gap between commonsense realism about adding two and anything that could be called "commonsense realism" about *number theory.* I still have a number of issues to discuss in this connection. One of them is the so-called ontological question.

Do the Quantifiers in Mathematics Range over "Intangible Objects"?

For Willard V. Quine, the quantifiers—that is, the logical particles "every" and "some" (or "there exist")—are our way of talking about objects; because we use them in mathematics, mathematics too consists of generalizations about *objects.* "The words 'five' and 'twelve,'" Quine writes, "name two intangible objects, numbers, which are *sizes* of sets of apples and the like."[19] This attitude raises two broad questions. One question, the question on which Quine himself has focused almost exclusively in his writing on the philosophy of mathematics, is the epistemological question: what *justification* is there for positing the existence of these "intangible objects"? Quine's answer (as is well known) is that the justification is analogous to the justification for positing the existence of electrons and other such unobservable entities in physics: such posits[20] increase our ability to predict observation conditionals and generalized observation conditionals, statements of the form "If *A,* then *B,*" where *A* and *B* are observation sentences, and sentences of the form "At any place and any time, if *A,* then *B,*" where *A* and *B* are observation predicates. The other question, which has been in the foreground of discussion since the publication of Paul Benacerraf's famous paper "Mathematical Truth,"[21] is the semantic question. It is widely accepted in analytic philosophy nowadays that reference to objects presupposes that those objects are ones that we can observe or otherwise causally interact with, or at least describe in terms of properties and objects with which we can causally interact. We describe quarks, for example, in terms of their charge, their "charm," and other properties; and we have access to charge and "charm" because these are magnitudes that determine the outcomes of physical interactions: quarks, like photons and electrons, are the quanta of fields with which we interact. But we do not have physical interactions with numbers, and the properties in terms of which we describe numbers and sets—say, being the successor of a number, or containing certain sets or numbers—are not properties that characterize physical objects or fields. Mathematical objects, if there be such, are causally inert. But then how can we so much as refer to them?

19. W. V. Quine, "Success and Limits of Mathematization" (originally given as a talk in 1978 at the Sixteenth International Congress of Philosophy), in Quine, *Theories and Things* (Cambridge, Mass.: Harvard University Press, 1981), 148–155; quotation from 149.

20. Quine employed this talk of "posits" in his celebrated essay on the ontological question, "On What There Is," *Review of Metaphyics* 2 (1948): 21–38; reprinted in W. V. Quine, *From a Logical Point of View* (Cambridge, Mass.: Harvard University Press, 1953), 1–19.

21. Paul Benacerraf, "Mathematical Truth," *Journal of Philosophy* 70, no. 19 (1973): 661–679; reprinted in Paul Benacerraf and Hilary Putnam, eds., *Philosophy of Mathematics: Selected Readings,* 2nd ed. (Cambridge: Cambridge University Press, 1983), 403–420.

The attitude Wittgenstein expresses, for example, in his *Lectures on the Foundations of Mathematics* is very different.[22] He clearly pooh-poohs the idea that talk of numbers—either ordinary numbers or the so-called transfinite numbers—is in any way analogous to talk of objects. If we think that way, then of course we will think that set theory has discovered (or posited the existence of) not just objects, and not just intangible objects, but an *enormous*—an *unprecedentedly* large—quantity of intangible objects. And the very vastness of the universe that set theory appears to have opened up for our intellectual gaze will then be part of its charm; but Wittgenstein thinks that this reason for being charmed is a bad one. Wittgenstein believes—and I think he is right—that it does not make the slightest sense to think that in pure mathematics we are talking about objects. The difference between the use of the quantifiers here and the use of the quantifiers when we are literally formulating generalizations about (what we ordinarily call) objects is just too great. In *Remarks* he is equally scornful:

> The comparison with alchemy suggests itself. We might speak of a kind of alchemy in mathematics.
>
> It is the earmark of this mathematical alchemy that mathematical propositions are regarded as statements about mathematical objects,—and so mathematics is the exploration of these objects?[23]

To be sure, someone might argue that there is no real opposition between Wittgenstein's view and Quine's view. One might claim that Quine and Wittgenstein are both trying to show the senselessness of metaphysical realism with respect to mathematics; Wittgenstein shows it by means of a certain sort of philosophical therapy, while Quine shows it by means of what one might call philosophical irony. Quine has reinterpreted the statement that in mathematics we are talking about objects in a way that totally robs it of its sup-

---

22. Cf. Ludwig Wittgenstein, *Lectures on the Foundations of Mathematics,* ed. Cora Diamond (Chicago: University of Chicago Press, 1976), XXV and XXVI, particularly 251–256.

23. Wittgenstein, *Remarks on the Foundations of Mathematics,* IV, §16. Of course, if one wants simply to extend the notion of "object" by speaking of "mathematical objects" in connection with mathematical propositions, that is something different; but then the application of "intangible" makes no sense, because the tangible/intangible distinction goes with the ordinary, unextended, use. On the idea that we continually do extend the use of "object," see Hilary Putnam, "The Dewey Lectures: Sense, Nonsense, and the Senses; An Inquiry into the Powers of the Human Mind," *Journal of Philosophy* 91, no. 9 (September 1994): 445–517.

posed metaphysical significance. Although the subject of this essay is Wittgenstein's philosophy of mathematics and not Quine's, I should like to make a couple of remarks about this as a possible interpretation of Quine's doctrine of ontological commitment.

First, even if it were correct—and I soon discuss why I do not think that it is—it clearly does not represent the way in which realistically inclined philosophers of mathematics have understood Quine's criterion of ontological commitment. The paper by Paul Benacerraf I cited, for example, clearly assumes that if we take the quantifiers in classical mathematics to be "objectual" quantifications in Quine's sense,[24] then we are committed to regarding them as ranging over objects in a sense perfectly analogous to the sense in which quantifiers over the books in my office range over objects. It is even more obvious that the work on the philosophy of mathematics by Benacerraf's colleague at Princeton, David Lewis,[25] takes the idea that the quantifiers of classical mathematics need to be interpreted as ranging over some "*stuff,*" some *things* or other, perfectly seriously. But—and this is my second point—Quine himself is clearly not free of the tendency to think of "abstract objects" as perfectly analogous to physical objects, as his epistemology of mathematics shows.

Still, one might ask, what is wrong with speaking of anything that *exists* as an object? And what better criterion have we for deciding whether talk genuinely commits us to the *existence* of something than how that talk looks when it is "regimented" in the notation of symbolic logic?[26] My answer would be

24. "Substitutional" quantifiers are interpreted in terms of the notion of truth of substitution instances of open sentences, rather than in terms of (nonlinguistic) "objects."

25. David Lewis, *Parts of Classes* (Oxford: Blackwell, 1991).

26. At times Quine plays down the significance of symbolic logic as such, writing, e.g., "The artificial notation '∃x' of existential quantification is explained merely as a symbolic rendering of the words 'there is something x such that.' So whatever more one may care to say about being or existence, what there are taken to be are assuredly just what are taken to qualify as values of 'x' in quantification. The point is thus trivial and obvious."

("Trivial and obvious" if one assumes that the ordinary-language [?] expression "There is an *x* such that" has just one use.) W. V. Quine, "Ontology and Ideology Revisited," *Journal of Philosophy* 80, no. 9 (September 1983): 499–502; quotation from p. 499. At other times he seems to deny that ordinary language *has* ontological commitments; e.g., in "Facts of the Matter," in R. W. Shahan and K. R. Merrill, eds., *American Philosophy from Edwards to Quine* (Norman: University of Oklahoma Press, 1977), 176–196, he writes, "It is only our somewhat regimented and sophisticated language of science that has evolved in such a way as really to raise ontological questions" (183), and "The ontological question for such a language [one with

that there are two different attitudes that one can have toward mathematical logic. One can think of symbolic logic as the skeleton of an *ideal* language— this is the view that drives Quine's philosophy[27]—and, of course, it is part of the ideal-language view that the real standard for what *any* sentence means is its translation (or, as Quine says, its "regimentation") in the ideal language. But one can also think of symbolic logic as simply a useful canon of rules of inference, and of formalization as simply a technique of idealization that facilitates the statement of those rules and their representation as (recursive) calculating procedures.[28] Indeed, as Wittgenstein reminds us in *Philosophical Investigations,* the word "ideal" is problematic here. "We are dazzled by the ideal," he writes, criticizing our tendency to think that "we are *striving after* an ideal, as if our ordinary vague sentences had not yet got a quite unexceptionable sense, and a perfect language awaited construction by us."[29] In a formal language the relation "*B* is a deductive consequence of A" has a syntactic representation in the sense of being coextensive with a relation that is definable in syntactic (in fact, in computational) terms; but that does not mean that a formal language is in some way a "better" language than a natural language.

I believe that the second of these attitudes is the reasonable one. But if we refuse to allow ourselves to think of symbolic logic as "sublime," it becomes difficult to see why we should take seriously the idea that every statement whose symbolization involves the symbol "$\exists x$" implies the existence of an object in some univocal sense of "object" and some univocal sense of "exis-

---

certain kinds of 'foreign notations'], *as for ordinary language generally,* makes sense only relative to agreed translations into ontologically regimented notation. . . . *Translation of ordinary language into the regimented idiom is not determinate*" (185–186; emphases added).

27. Cf. Hilary Putnam, "Convention, a Theme in Philosophy," *New Literary History* 13, no. 1 (Autumn 1981): 1–14, reprinted in Putnam, *Philosophical Papers,* vol. 3, *Realism and Reason* (Cambridge: Cambridge University Press, 1983), 170–183, particularly 182–183, for an analysis of Quine's scientism. Note that my reading of Wittgenstein has changed, however, since that paper was written.

28. A classic statement of this view was Ernest Nagel, "Logic without Ontology," in Y. H. Krikorian, ed., *Naturalism and the Human Spirit* (New York: Columbia University Press, 1944), 210–241; reprinted in Nagel, *Logic without Metaphysics* (Glencoe, Ill.: Free Press, 1959), chap. 4.

29. Wittgenstein, *Philosophical Investigations,* §98, §100. Compare §103: "The ideal, as we think of it, is unshakable. You can never get outside of it; you must always turn back. There is no outside; outside you cannot breathe.—Where does this idea come from? It is like a pair of glasses on our nose through which we see whatever we look at. It never occurs to us to take them off."

tence." Indeed, there are also purely mathematical reasons not to think this. Quine himself has pointed out that existential quantification can be interpreted in at least two different ways, the ways he calls "substitutional" and "objectual." He argues that if one accepts the classical theory of real numbers at face value—that is, without intuitionist or other constructivist reinterpretation of some kind—then one cannot construe *its* quantifiers as "substitutional," and so he concludes that they must be "objectual"; but, as we shall see shortly, there are still other *known* interpretations of the quantifiers over and beyond the objectual and the substitutional. The axioms of first-order logic are simply not categorical in the sense of determining a *unique* interpretation of the quantifiers themselves. There is thus no *mathematical* reason to think that there is some one thing that we must be saying whenever we use the symbol "∃x."

But Does Rejecting the Notion of "Ontological Commitment" Resolve the Realism/Antirealism Problem?

So far I have ascribed two attitudes to the later Wittgenstein: commonsense realism about rule following and contempt for the idea that mathematics presupposes an "ontology," that is, contempt for the idea that any branch of mathematics is literally a description of the behavior of objects of some kind. But I do not wish to suggest that these two attitudes by themselves suffice to give clear content to the notion of commonsense realism with respect to, say, number theory. And even though at the end of the day I hope to arrive at a standpoint from which the "realism problem" does not genuinely arise with respect to number theory,[30] it is clear that many bright people have tried and failed to find such a standpoint—too many to encourage the idea that it is easy to find. I said at the outset that one of my guiding thoughts will be that we have to see *why* the problems in the philosophy of mathematics seem so hard; that we have to do *justice* to the sense that they may actually be insoluble before we can begin to unravel them. In this connection, it is particularly important to understand why the problems look so much *worse* after Gödel's theorem.

Entertain, for the moment, the fantasy that Gödel's theorem has not been proved, and imagine that we all believe (as Hilbert did) that it is or will in the

---

30. When I speak of "mathematics" in this essay, I shall be thinking of number theory and analysis (e.g., calculus, theory of functions), not of set theory.

future be possible to find axioms of mathematics that are *complete;* that is, that either every sentence in, for example, properly axiomatized number theory is a theorem or else its negation is a theorem, if not in number theory itself, then in a suitable extension of number theory. In that case truth in number theory would be coextensive with provability in that extension; and to be provable in a formal system, we know, is just to be a sentence that is obtainable as the last line of a longer or shorter proof (or as the bottom of a larger or smaller proof tree) by following certain rules. Now, our commonsense realism with respect to rules (which I have already seen Wittgenstein as defending) could be seen as implying commonsense realism about provability; and if truth can be identified with provability in a suitable sense of "provability," then it seems that we have an easy way of being commonsense realists about mathematical (number-theoretic) truth.

Another advantage to this way of thinking is that it bypasses the whole question of the "objectual" or "substitutional" character of the quantifiers in analysis. If analysis is complete, or some extension of analysis is complete with respect to the propositions of analysis (as Hilbert also believed or, better, hoped), then we can explain the truth of whole sentences in analysis in the same way, by identifying it with their provability in an appropriate formal system. And we can then argue that even if the quantifiers in analysis are not substitutional, that does not mean that they have to be understood in terms of intangible objects. We could say (speaking, at this point, like good Fregeans, even if he would have been horrified at the view we were defending) that to understand the quantifiers—or, indeed, to understand any symbol in analysis—is just to understand the contribution that those symbols make to the truth-value of whole claims; and we understand the contribution that the various symbols make to the truth-value of whole claims in analysis by understanding the *proof procedure* of analysis. (We would then have a nice example of the way in which a quantifier may be neither objectual nor substitutional.)

But alas, Gödel did prove his theorem, and we do know that there are undecidable sentences in number theory, and that some of them will remain undecidable no matter how we consistently extend number theory and analysis. What are we to do?

One possibility would be to try to hold on to the line just suggested, and to take the existence of undecidable sentences in number theory as showing that there are sentences in number theory that are neither true nor false. To be sure, some of those sentences become provable and others become disprovable if we add further axioms to number theory—for example, if we add the sentence expressing the formal consistency of the given system of number theory as an additional axiom to obtain a stronger system, then, in particular, that very sentence CON(NT)[31] that was undecidable in the original system becomes decidable in the extended system and hence *true in the extended system*. We could say—even if it seems a most unattractive thing to say at first blush—that we have here a case of *ambiguity;* that CON(NT) was one statement (and was neither true nor false) in the original system NT and is a different statement in the extended system, and is, moreover, true in the extended system. (This is the line that Wittgenstein entertained in his famous [or notorious] remarks about the Gödel theorem.[32] Elsewhere I shall try to show that this does *not* mean that Wittgenstein accepted the rest of the antirealist story—e.g., that he believed, as Michael Dummett does, that understanding mathematical propositions is just a matter of understanding the proof procedures connected with them.)[33]

But this move flies in the face of another Wittgensteinian insight—one about which Cora Diamond has written so beautifully[34]—that it is a fundamental feature of our mathematical lives that we do not experience every change in our mathematical language games as a change in the very meaning of the sentences.[35] As Diamond puts it, we sometimes "see the face" of one mathematical language game in another mathematical language game. What was wrong with the line that Wittgenstein was tempted by in 1937 in connection with the Gödel theorem is that every mathematician in the world sees the face of number theory based on Peano's axioms in number theory based on Peano's axioms plus CON(NT).[36]

31. CON(NT) is here used to stand for the particular number-theoretic proposition of which Rosser (improving Gödel's result) proved that (1) number theory is consistent if and only if that proposition is true; and (2) if number theory is consistent, then that proposition cannot be proved in it.

32. Wittgenstein, *Remarks on the Foundations of Mathematics,* I, appendix I, §§8–19.

33. See Chap. 25 in this volume.

34. See Cora Diamond, "The Face of Necessity" (written in 1968), in Diamond, *The Realistic Spirit,* 243–266, esp. 246–249.

35. Wittgenstein's remark (cited in note 5) "that these proofs prove the same proposition means, e.g.: both demonstrate it as a suitable instrument for the same purposes" explicitly allows that we *can* see two different proofs as proving the "same" mathematical proposition, contrary to many interpretations of Wittgenstein

36. Wittgenstein's example (*Remarks on the Foundations of Mathematics,* I, appendix I) was a different one: Russell's *Principia,* and Principia + CON*(Principia).* But the same rejoinder applies.

Still, even if CON(NT) is not provable in Peano arithmetic, it *is* provable by metamathematical reflection *on* number theory, as Gentzen showed. Can we identify truth in number theory with the disjunction of formal provability and—however vague the notion may be—provability by metamathematical reflection on number theory? But we have no reason to think that every sentence of number theory is decidable even by metamathematical reflection. What reason is there to think that, for example, the number-theoretic sentence expressing the consistency of Zermelo-Fraenkel set theory is decidable by metamathematical reflection?[37]

Following Michael Dummett, one might *propose a revision in the logic of classical mathematics itself;* in particular, to adopt *intuitionist logic.* This *is* compatible with identifying truth in mathematics with (informal) provability; the logic was motivated by that very identification. And because we would no longer accept the law $p \vee \neg p$, we would no longer be required to assert

$$CON(ZF) \vee \neg CON(ZF).$$

Indeed, for any sentence S of number theory, we would assert $S \vee \neg S$ when and only when we had succeeded in proving S or in proving $\neg S$.

Alternatively, if we do want to preserve the formula

(1) $p \vee \neg p,$

we can do that too. For Gödel[38] pointed out that there is a "translation" of classical first-order number theory into intuitionist first-order number theory under which all the laws of classical logic and first-order number theory are intuitionistically valid. The formula $p \vee \neg p$, for example, gets "translated" as $\neg(\neg p \wedge \neg \neg p)$, where "$\neg$" is intuitionist negation. Thus one gets (1) above as a theorem, though not, of course, under its classical interpretation.

In short, if one distinguishes between the validity of (1) with respect to number theory and the semantic principle that every sentence of number theory has a determinate truth-value (the law of the excluded middle), then

---

37. Indeed, to date all attempts to find a Gentzen-style proof for the consistency of even *ramified analysis* have been utter failures.

38. Kurt Gödel, "Zur intuitionistischen Arithmetik und Zahlentheorie," *Ergebnisse eines mathematischen Kolloquiums* 4 (1933): 34–38; English translation, "On Intuitionistic Arithmetic and Number Theory," in Martin Davis, ed., *The Undecidable* (New York: Raven Press, 1965), 75–81.

we may say that syntactically speaking, under Gödel's "conjunction-negation translation" of classical number theory into intuitionist number theory we end up retaining (or, if you prefer, regaining) the law (1) of propositional calculus without thereby committing ourselves to the law of the excluded middle understood as "bivalence."

In addition, under Gödel's translation, the existential quantifier "$\exists x$" of classical number theory gets "translated" as "$\neg(x)\neg$," where "(x)" is the intuitionist universal quantifier. This is, by the way, the example I promised a little while ago of yet another interpretation of the existential quantifier (over and beyond the substitutional and objectual interpretations that Quine mentions). Unfortunately, however, intuitionism is not the way out of our perplexities.

A Logicist Insight

It is a logicist insight that part of the problem is to understand how mathematical statements function as part of science, and not just to understand talk about proofs and theorems. Science is not separable into a part that is empirical and free of all mathematical concepts and a part that is the mathematics and is free of all empirical content. Empirical science itself contains "mixed statements," statements that are empirical but speak of, for example, functions and their derivatives, as well as of physical entities.

Even as simple a statement as Newton's law of gravitation is "mixed." That law, of course, asserts that there is a force *fab* exerted by any body *a* on any other body *b*. The direction of the force *fab* is toward *a*, and its magnitude *F* is given by:

$$F = \frac{gMaMb}{d^2}$$

where *g* is a universal constant, *Ma* is the mass of *a*, *Mb* is the mass of *b*, and *d* is the distance that separates *a* and *b*. The law presupposes the existence of real numbers corresponding to forces, distances, and masses, and the operations of multiplication and division of real numbers. A "regimented" version of the law would thus require quantifiers over real numbers, or at the very least over rational numbers.[39]

---

39. For a detailed discussion, see Hilary Putnam, "Philosophy of Logic," in Putnam, *Philosophical Papers*, vol. 1, *Mathematics, Matter and Method* (Cambridge: Cambridge University Press, 1975), 337–341 (but see also 347).

Philosophical accounts of our mathematical concepts face the challenge of explaining how we are able to assign a truth-value to such mixed statements. If our account of mathematical language is intuitionist—that is, if the only notion of "truth" we have in connection with mathematical statements is the notion of provability (and a restricted kind of provability at that)[40]—while our physical language is incompatible with the assumption that truth coincides with verifiability,[41] then the problem of interpreting the mixed statements seems unsolvable. On the one hand, their interpretation has to be realistic, or at least not antirealistic, if we are to derive from them empirical statements that are themselves understood realistically; but on the other hand, they contain concepts whose *only* meaning was supposed to be in terms of the contribution they make to the provability of statements.

In some cases this problem can indeed be finessed, and in a way that Wittgenstein might well have approved of. If I say that a stick is three feet long (where it is clear from the context that what is meant is three feet plus or minus the inherent inaccuracy of some measurement procedure), then I can reformulate this as "If the stick is correctly measured, the result will be '3' "; and this does not assume that there is a logically necessary and sufficient condition for a measurement's being correct that is statable in "observation terms."[42] Similarly, if I say that a certain formula gives the position of the moon at any time, up to a given accuracy, nothing about the way we understand that statement commits us to a notion of mathematical truth that outruns calculability. But the case I describe below cannot be finessed in this way.

I believe that Wittgenstein failed to see the seriousness of this problem, precisely because his examples of the empirical application of mathematics are limited to *calculation* and to comparing the results of calculations to experiments. (It is important to recall that Wittgenstein was trained as an *engineer*.) Thus he may well have thought that the mixed statements of physics all have

40. Intuitionists reject nonconstructive proofs of existence and impredicatively defined sets, for example.

41. Cf. Hilary Putnam, "On Not Writing Off Scientific Realism," Chapter 4 in this volume, in connection with the issue of verificationism.

42. The idea that the use of such a notion as "correct" measurement does not commit us to a necessary and sufficient condition for the notion is connected with Stanley Cavell's insistence in *The Claim of Reason* (Oxford: Clarendon Press, 1979), chaps. I and II, that "criteria," in the sense in which Wittgenstein uses the term, are not necessary and sufficient conditions.

this form: the values of such and such parameters can be *calculated* in such and such a way given such and such initial data. But this is simply not the case.

Here is an example that may help bring out the problem. In mathematical physics the value of a physical magnitude generally depends on the solution of one or another differential equation—say, for example, the wave equation. But those solutions may in certain cases *not be recursively calculable,* even if the initial conditions are recursively describable[43]—and in such a case we may very well not be able to *prove* that the solution to the differential equation in question is such and such, even to a finite number of decimal places worthy of accuracy. But if we take our physics seriously, there *is* an answer to the question, what the solution of the differential equation in question is to such and such a number of decimal places.

One way out, Dummett's, is to widen the notion of provability, to interpret the logical connectives[44] not in terms of mathematical provability alone but in terms of the wider notion of *verifiability,* and to collapse the notions of truth and verifiability for empirical statements, as well as for mathematical statements. But not only for metaphysical realists but even for commonsense realists, for all of us who are anti-antirealists, that way out is unacceptable.[45]

Of course, if we construe mathematical quantifiers (in purely mathematical statements and mixed statements alike) as expressing generalizations about intangible objects ("abstract entities"), and at the same time construe truth for all three kinds of statements—statements that do not quantify over any abstract entities, pure mathematical statements, and mixed statements—as verification transcendent, then we can preserve our realism with respect to the statements of the first class, but at the cost of endorsing a form of metaphysical

43. Marian Boykan Pour-El and Ian Richards, "The Wave Equation with Computable Initial Data Such That Its Unique Solution Is Not Computable," *Advances in Mathematics* 39 (1981): 215–239. It is not known whether the Newtonian three-body problem is such a case; there may not be, in general, any way to determine recursively whether the bodies will *collide* or not.

44. I speak of "interpreting the logical connectives" because that is one way in which intuitionism is frequently presented; e.g., it may be said that for intuitionists, $p \wedge q$ means "p is (constructively) proved and q is (constructively) proved," $p \vee q$ means "there is a (constructive) proof of p or there is a (constructive) proof of q, and a way of determining which of the two statements is proved," $p \Rightarrow q$ means "there is a construction that applied to a hypothetical (constructive) proof of p would yield a constructive proof of q," and so on.

45. A position similar to Dummett's but not committed to an atomistic conception of verification is also possible. My position in Hilary Putnam, *Reason, Truth and History* (Cambridge: Cambridge University Press, 1981), was of this kind.

realism for the pure mathematical and the mixed statements, and thereby having to face the familiar epistemological and semantic problems. This is the situation that Paul Benacerraf described so well in the paper I referred to.

Indeed, the epistemological and semantic problems may not even be the most serious difficulties for this sort of metaphysical realism. Consider, for example, the form of metaphysical realism in the philosophy of mathematics developed by David Lewis.[46] Lewis proposes that we postulate that reality contains not only the matter we know about in this world, and similar matter in other logically possible worlds,[47] but also a vast amount of additional stuff—stuff that need not be material and need not be thought of as causally interacting in any way with matter, but need only be "atomless"[48] and multitudinous. There has to be a *lot* of this stuff, so much that the number of its parts is a very large cardinal (a cardinal *so* large that one does not need to suppose that there are any more sets than that even in Cantor's paradise).

If the atomless stuff is that multitudinous, then, Lewis shows, one can reinterpret the notion of a set and the notion of membership in a set in terms of mereology (in terms of the notions of part and whole) in such a way that one obtains what amounts to a standard model for set theory. At first blush, this form of metaphysical realism might seem to provide us with a way of understanding mathematics fully realistically without postulating any sort of supermechanism. After all, the role of the atomless stuff in Lewis's philosophy of mathematics is not to act on us in any way, or to act on our minds in any way, or to be perceived by us in any way; it is only to provide an "interpretation" of the sentences of set theory (including the mixed sentences) that is, in a suitable sense, "standard."[49] So the problem with Lewis's metaphysics cannot be that he is somehow confusing conceptual and causal questions, because he does not introduce anything that even resembles causality into his story. There need not be any cause-and-effect connections at all involving the atomless stuff.

But this very causal inertness of the atomless stuff brings to light another problem,[50] one that, I believe, affects all forms of metaphysical realism in the

---

46. Cf. Lewis, *Parts of Classes.*

47. Lewis was notorious for his "realism" about possible worlds.

48. "Atomless" means that every part of the "stuff" has proper parts in turn.

49. The model must be one that a Platonistically inclined set theorist like Gödel could regard as making the same sentences true as the sentences he believes are true "in the universe of sets."

50. I am indebted to Jody Azzouni for the idea of the argument that follows, if not for its precise form. See his stimulating *Metaphysical Myths, Mathematical Practice* (Cambridge: Cambridge University Press, 1994).

philosophy of mathematics, though not as transparently.[51] Suppose that Lewis is wrong, and that his atomless stuff actually does not exist (supposing for the sake of the argument that it makes sense to suppose either that it exists or that it does not). *Would mathematics fail to work?*[52] Because the success or failure of mathematics in any of its real-world applications depends only on how the empirical objects behave, and that would not be any different if the atomless stuff did not exist (because by hypothesis it exerts no causal influences on the ordinary empirical objects), it follows that if the mysterious extraphysical "atomless stuff" did not exist at all, then *mathematics would work exactly as it actually does.* But does not this already show that postulating immaterial objects to account for the success of mathematics is a useless shuffle?

It is, I think, because we philosophers of mathematics have a tendency to rehearse precisely the sort of dialectic that I have just laid out before you (with, of course, our own individual variations) that we suffer from the occupational disease of tending to believe that at the end of the day the problems must be both *sui generis* and hopeless. After all, have not the arguments I laid before you shown that commonsense realism about rule following does not suffice to yield the desired commonsense realism about even about the elementary theory of numbers? And antirealism and metaphysical realism likewise do not work. So, in short, neither the opposing metaphysical positions nor the Wittgensteinian therapy designed to rid us of the problem is successful—or, at any rate, that is how it can look to one. But is there another way to look at these problems?[53]

---

51. I no longer [in 2011] agree with this claim. I believe that the problem does not arise for realism with respect to *modality,* which I argue for in "Set Theory: Realism, Replacement, and Modality" (Chapter 11 in this volume), and which I believe suffices for a realistic interpretation of set theory.

52. David Lewis would doubtless respond that this counterfactual makes no sense, given his own theory of counterfactuals. *Tant pis* for that theory of counterfactuals! The counterfactual "If Lewis's atomless stuff did not exist, everything we observe in our world—including the success of mathematics—would still occur" is what I have called a "strict" counterfactual (see Hilary Putnam, *Renewing Philosophy* [Cambridge, Mass.: Harvard University Press, 1992], 51–52), that is, one in which the consequent is a *logical consequence* of the antecedent, and such counterfactuals ought to be true in any adequate theory of counterfactuals. (Note also that even if the antecedent is "metaphysically impossible" in Lewis's sense—if it is a sense—it still is not logically contradictory. There is nothing degenerate about this particular strict counterfactual.)

53. I have intentionally not discussed Hartry Field's nominalism (see his *Science without Numbers* [Oxford: Oxford University Press, 1980]). Although it was an important contribution

A Revisiting of the "Commonsense Realism about Rules" Issue

I said that it was an insight of the logicists (one inherited by Wittgenstein and Quine) that an adequate account of our mathematical practice must include the whole of that practice, and that means in particular the application of mathematics, the way in which mathematics is used in accounting, measuring, empirical science, and other areas. I went on to argue that, in particular, antirealist and formalist interpretations of pure mathematics do not fit well with commonsense realism—or just anti-antirealism—with respect to, for example, empirical science. But this argument may have another consequence that is less welcome for the position I wish to defend, namely, that I was wrong to praise Wittgenstein for what I described as his defense of commonsense realism with respect to rule following. The problem with commonsense realism, it may seem, is not that commonsense realism is a *wrong,* or bad, or incoherent philosophical position, but that it is not a philosophical position at all. Of course, Wittgensteinians will agree that this is the case; it is a part of most, if not all, interpretations of Wittgenstein's later work that it was not his *intention* to put forward a "philosophical position" in any conventional sense of that term, and this is something he himself tells us. Still, analytic philosophers of a more conventional mind-set will ask (and indeed do ask) whether Wittgensteinians are not too *complacent* about this state of affairs. After all, not having a philosophical position may be just another name for refusing to acknowledge a real problem. And is that not what is going on here?

I have been arguing that it does not help to suppose that what makes the sentences of number theory true is some bunch of objects, some *stuff.* In the

---

to the discussion, it is widely realized that it did not work. Unfortunately, most of the critics have directed their fire solely at the mathematical errors in that work. In my view, the more serious problems with Field's nominalism are philosophical, and these remain even when the mathematical errors are repaired. In particular, (1) Field's position requires that we be realists about the classical space-time continuum. If we do not take the nondenumerable infinity of that continuum seriously as a *physical* hypothesis, then the construction immediately collapses. (2) Even if we waive this objection, the construction is not "robust"; it does not apply to quantum mechanics, for example, not even to as classical and "realist" a version of quantum mechanics as Bohm's hidden-variable theory, nor to theories in which space-time is discrete. But the success of mathematics surely does not depend on the particular Newtonian form of empirical science that Field considers. (Would mathematics stop working if Bohm's theory were true, or if space-time were discrete?)

same vein, I suggested that Wittgenstein argued against the very *intelligibility* of the view that what makes it the case that one is following a rule correctly, when one is, is some special relation that one has to some intangible objects, be they mental or Platonic. But does not that at once raise the question "Then what *does* make these sorts of claims true?"? Is not that quite obviously a *real* question? And am I not praising Wittgenstein for denying that a real question is a real question?

But what exactly *is* the "real question" that Wittgenstein is supposed to have swept under the rug? The formulation suggests that the question is this: what *makes* the statements of mathematics true? For that matter, what *makes* *true* the statement that such and such is the correct way to follow the rule "add two"? ("Remember that when the last digit is a nine you have to change it to a one and carry a one.") But one does not need to be a "Wittgensteinian" to find that question suspect. The question, after all, rests on a particular picture of truth: truth is what results when a statement stands in a particular relation (it "is made true by," or more simply, it "corresponds to") something else, say, a "reality."

But although that picture may fit *some* statements—for example, the statement that a sofa is green may "correspond" to a certain green sofa on a particular occasion—it does not seem to fit other equally familiar statements without intolerable strain. Consider ordinary negative existentials. What makes the statement that there is no red book in this room true? The "negative fact" that there are no red books in this room?[54] There is an ancient metaphysical problem about the existence or nonexistence of "negative facts." Indeed, are "facts" objects, entities that make propositions true? Today many, perhaps most, philosophers would say that "it is a fact that there is nothing red in this room" is just a syntactic and semantic variant of "It is true that there is nothing red in this room" (compare "It is the case that there is nothing red in this room"—why has no one proposed an "ontology" of *cases*?). To speak of the fact that there is nothing red in this room is not to cite an object that stands in a mystery relation called "making true" to the statement that there is nothing

54. Cf. Wittgenstein, *Lectures on the Foundations of Mathematics,* 248: "We have certain words such that if we were asked, 'What is the reality which corresponds?' we should all point to the same thing—for example 'sofa,' 'green,' etc." And (247) "If you say, 'Something corresponds to the word 'red,' namely the colour'—how does it correspond if you say (truly) 'There's nothing red in this room?'"

red in this room; it is simply to refer to the *truth* of the claim that there is nothing red in the room. Indeed, if we are going to think of the statement "There is nothing red in this room" as being made true by an object, why not take that object to be "the truth" of the statement that there is nothing red in the room? Or why not just say that the statement is true *because* there is nothing red in the room? This last, at least, has the virtue of being trivially true.

But is it an explanation? Is "the statement *p* is true because *p*" an *explanation?*

Of course, we could *deny* that speaking of "the fact that there is nothing red in this room" is just a reformulation of "It is true that there is nothing red in this room."[55] Certain philosophers would say that there are such "entities" (i.e., intangible objects) as facts, and that facts make statements true. But one does not need to be a positivist to point out that positing ghostly counterparts of sentences to explain the truth of statements is mere verbiage. Such talk meets none of the standards that we require a serious explanation to meet in everyday life. As an explanation of anything, the idea that "facts" are literally intangible objects is in infinitely bad shape.

Of course, the emptiness of this particular explanation of the truth of the statement that there is nothing red in a certain room, namely, that it is "made true by a negative fact," is especially easy to see. But once we have broken free of the grip of the particular picture that if a statement is true, there must be a *something* that "makes" it true, then we can also see how that picture is responsible for the feeling that there is a "deep problem" whose existence Wittgenstein is somehow refusing to face. The fact that Wittgenstein does, however, use such homely examples as the words "sofa" and "green" and the sentence "There is nothing red in this room"[56] illustrates very well, I think, the fact that he is not at all trying to sweep a genuine problem under the rug; rather, what he is trying to do is to see just what picture "holds us captive"—that is, to find the roots of our conviction that we *have* a genuine problem, and to enable us to see that when we try to state clearly what the genuine problem is, it turns out to be a nonsense problem. "The results of [Wittgenstein's] philosophy are the uncovering of one or another piece of plain nonsense and of bumps that

---

55. I prescind from the fact that the most common use of the locution "it is a fact that" is *epistemic* and not ontological at all. "It is a fact that" has much the same force as "It is quite certain that" in this use.

56. See note 58.

the understanding has got by running its head up against the limits of language. These bumps make us see the value of the discovery."[57]

Kripke's Interpretation of Wittgenstein Again

This is, I believe, a good point at which to say something about Saul Kripke's brilliant attempt to state a supposed genuine problem in this area, one to which he thinks (mistakenly, as I already indicated) Wittgenstein offered a "skeptical solution."

In order to explain what I think is wrong with Kripke's interpretation of Wittgenstein, I need first to say more than I did above about the nature of what I have been calling "commonsense realism." It is a feature of commonsense realism (as opposed to metaphysical realism) that it always seems to ignore (or beg) the philosophical problem rather than respond to it. Consider, for example, commonsense realism about perception. The commonsense answer to a question like "How do you know that Joan has a new car?" might well be "I saw it." But when John Austin writes that the fact that we hear a bird and recognize the bird "by its booming"[58] is sufficient reason for saying that we know that there is a bittern at the bottom of the garden (or if I say that the fact that I saw it is sufficient reason for saying that I know that my neighbor has a new car), the objection is always that the response has entirely ignored the "problem," namely, that perception, after all, gives us only "direct" acquaintance with our own "sensations" (and that "the real problem" is how we are justified in so much as speaking of perceiving material objects when all we have directly or immediately before the mind is the sensations). And indeed, commonsense

---

57. Wittgenstein, *Philosophical Investigations*, §119. I now [as of 2011] strongly disagree with the above repudiation of the question "what makes certain statements true." In "From Quantum Mechanics to Ethics and Back Again" (Chapter 2 in this volume) I argue that we should be willing to say that empirical statements do correspond to states of affairs that obtain, and that this has explanatory value; for example, it enables us to see how there can be what I call "equivalent descriptions" with startlingly different "ontologies" in physics; and in "Set Theory: Realism, Replacement, and Modality" (Chapter 11 in this volume) I argue that mathematical statements (including mixed statements) correspond to facts about possibility and impossibility (the "modal-logical" version of "structuralism"). In general, as I explain in "Wittgenstein: A Reappraisal" (Chapter 28 in this volume), I now reject the rejection of metaphysics.

58. John Austin, "Other Minds," *Proceedings of the Aristotelian Society,* suppl. 20 (1946): 148–187; reprinted in Austin, *Philosophical Papers,* ed. J. O. Urmson and G. J. Warnock, 2nd ed. (Oxford: Oxford University Press, 1970), 44–84; see 79.

realism by itself is not a metaphysical position, or even an antimetaphysical position. The philosophical work that Austin does and that Wittgenstein also does of undermining the picture on which the supposed difficulty rests, the picture of experience as consisting of "sensations" in a private mental theater, still has to be done. But notice the nature of the strategy that I am attributing to both Austin and Wittgenstein, and that I have discussed at more length elsewhere.[59] The strategy is not to *counterpoise* an alternative thesis to the various theses of the traditional epistemologists, be they realist theses or idealist theses or empiricist theses or whatever. Rather, it consists in, first, taking perfectly seriously our ordinary claims to know about the existence of birds and automobiles and what Austin referred to as "middle-sized dry goods" and our ordinary explanations of how we know those things, and, second, meeting the objection that these ordinary claims simply ignore a philosophical problem by challenging the very intelligibility of the supposed problem.[60]

With this now in mind, let us consider what the parallel dialectic about following a rule might be. Kripke defines a function "quus" by the stipulation that a quus b (where a and b are whole numbers) is 5 when either a or b is greater than or equal to 57, and is equal to a plus b otherwise. He asks us how we can know that we ourselves did not in the past mean "quus" when we said or thought "plus." (His ultimate aim is to cast doubt on my confidence that I *now* mean anything when I say or think "plus.")

If someone were to ask me, how do I know that someone—call her "Joan"—does not mean "*quus*" by "plus"? my response would be that the questioner is using the question as a "ploy" to start a philosophical discussion (all the more so if the question is not about someone else but about my own past self!). If Joan were an intelligent adult and a fluent speaker of English, I would be at a loss what to say. Perhaps I would just say, "She speaks English." And this seems unresponsive to the philosophical problem, because of course Kripke does not deny that in *some* sense that is the right answer to give.

But let me pretend that the question is a serious one. Then I may say that the "hypothesis" that Joan means "*quus*" by "plus" is one that can, in fact, be empirically refuted by asking Joan what $2 + 57$ is.[61] To be sure, it is "logically

59. See Hilary Putnam, "On Not Writing Off Scientific Realism," Chapter 4 in this volume.

60. I am now [2011] unhappy with this kind of "challenging," as I explain in "Wittgenstein: A Reappraisal" (Chapter 28 in this volume).

61. The "right" answer, on the hypothesis that Joan means "quus" by "plus," is, of course, "5."

possible" that Joan would still answer the question "What is $2 + 57$?" by saying "59" rather than by giving the answer that is correct on the "quus" interpretation of "plus," namely, "5," because Joan might make a mistake in "quadition." The inference from the response "$2 + 57 = 59$" to "Joan does not mean '*quus*' by 'plus'" is not a *deductive* inference. But so what? As Austin famously reminded us, "Enough is enough: it doesn't mean everything."[62]

But Kripke's problem is deeper. Even if we can rule out the particular hypothesis that Joan means "*quus*" by "plus," we cannot in the same way rule out every possible hypothesis of this kind. Thus, for each n define a function $quus_n$ as follows:

$$a \text{ quus}_n b = a + b \text{ if } a, b < n;$$
$$a \text{ quus}_n b = 5 \text{ if } a \text{ or } b \geq n.$$

If what Joan means by "plus" is $quus_N$, for some very large N, Joan's response to the sums we actually encounter will be "normal." Still, if she is mathematically sophisticated, could we not ask her "Is there a number N such that a 'plus' $b = 5$ whenever either a or b is $\geq N$?" If Joan answers "No," does this not show that whatever Joan may mean by "plus," she *does not* mean $quus_n$, for any n whatsoever?[63] At this point we encounter the really *deep* move in Kripke's argument.

Joan, Kripke tells us, might not understand only "plus" in a different way from the rest of us; she might conceivably understand almost every word in the language differently, but in such a way that she speaks exactly like the rest of us in all actual circumstances. She would speak differently from the rest of us if we could utter or write down the number N in decimal notation and ask

62. Austin, "Other Minds," 84.

63. Someone might object that I have misunderstood Kripke's problem, and that the problem is just this: The underdetermination of, say, a hypothesis about quarks by laboratory evidence is not a problem for us because we grant that there are *facts* about quarks ("considered in isolation," if you please), but the underdetermination of the supposed "fact" that Joan means "plus" by her speech dispositions is an underdetermination by "all the facts" that there *are*. This is, of course, just Quine's problem of "the underdetermination of reference"; and perhaps Kripke fails to see that it is only because he thinks that Quine's problem turns on behaviorist assumptions rather than on the notion of a "fact of the matter." (See his remarks about Quine on p. 57 of *Wittgenstein on Rules and Private Language*). However, Kripke's emphasis on the fact that the plus function is defined for infinitely many cases, so that it is (supposedly) puzzling how my "finite mind" can hold it all, introduces a different and more interesting line of argument, even if in the end that line too turns on the notion of a "fact," and it is that line of argument that I have tried to reconstruct here.

her, "What is $2 + N$?" or some such question. But suppose that the number N is so large that it is *physically impossible* to write it down.[64] Then there might be no *physically possible* situation in which Joan would speak differently from the rest of us, and yet she would "mean something different" from the rest of us by "plus."

Kripke is right to think that this is a possibility that Wittgenstein would dismiss as unintelligible. The idea that someone might mean something different by a word although the supposed difference in meaning does not show up in any behavior at all, indeed, in any possible behavior, linguistic or extra-linguistic, is a "possibility" that Wittgenstein would reject. But must Wittgenstein's rejection be based on a community-standards view of what correctness in rule following consists in? (This is close to Wittgenstein's supposed "skeptical solution" as Kripke interprets him.)

Here I wish to bring out the *distance* between what we know Wittgenstein thought and what Kripke says he thought. All interpreters agree that we know that Wittgenstein thought that we can speak of understanding a word (or of understanding a word one way rather than another) only against the background of a whole system of uses of words, and we know, moreover, that when Wittgenstein speaks of the "use" of words, he means also the actions and events with which those uses of words are interwoven. Descriptions of language and descriptions of the world, including what speakers do in the world, are interwoven in Wittgensteinian accounts. Moreover, we know that Wittgenstein thinks that this observation—that the notion of simply understanding a word in isolation from anything one might do with the word, or from the presence of an appropriate background of other uses and actions, makes no sense—is itself a "grammatical" one, that is, in some sense a conceptual observation.

According to Kripke, however, Wittgenstein has much more radical beliefs and much more metaphysical beliefs in addition to these. Specifically, Kripke's Wittgenstein, Kripkenstein, holds that the only condition we possess for the truth of the claim that someone understands a word the way the other members of the relevant linguistic community do is that that someone be disposed to correctly answer certain specific questions, or, more broadly, to give certain specific linguistic responses in certain situations.[65] Giving those

---

64. It would also suffice to suppose that N is so large that it is physically impossible for someone with a human brain to calculate with it.

65. However, even this condition—the only one we possess—may not actually determine that the claim *is* true, but only that we agree in calling it true: in *Wittgenstein on Rules and*

responses in those situations, and the other members of the community then saying that he or she has the concept, itself constitutes a kind of metalinguistic language game, which Kripke calls the "concept attributing game."

In addition, Kripke ascribes to Wittgenstein a certain concept of a "fact": there are no facts about "a person considered in isolation" except (1) physicalistic facts, that is, facts about his or her brain states and other such materialistic facts, and (2) (possibly) mentalistic facts, that is, facts about his or her sensations, mental images, and the like (described without reference to any intentional content we might ascribe to them).[66] In short, Kripke assumes—or rather Kripkenstein assumes—that either materialism or a certain limited form of dualism is the only possible account of what a "fact" is. Given this view of what a "fact" is, Kripkenstein goes on to argue that there are no "facts about a person considered in isolation" that *constitute* the fact that that person understands a word the way other people in the community do. Of course, it is well-known that there is no textual basis at all for attributing these views about facts, or even for attributing the concept of "a fact about a person considered in isolation," to Wittgenstein as opposed to Kripkenstein.

Now suppose that Joan uses the word "plus" and the other arithmetical words and mathematical signs the way the rest of us do. (I do not mean just that Joan gives the same answer to questions of the syntactic form "What does $a + b$ equal?" that the rest of us do, but also that she talks *about* addition the way the rest of us do, and that she talks and behaves the way normal people do about matters that involve the *application* of addition. If you are worried about certain science-fiction possibilities, assume also that there are no funny "molecular" facts about Joan that would cast the "sincerity" of any of these responses into question—for example, funny facts about her polygraph

---

Private Language, 112, Kripke has Wittgenstein deny that the community's responses determine the truth of *any* sentence, and assert instead that all that the facts about the community's responses determine is that the community does not doubt certain sentences (e.g., does not doubt that $a + b \neq c$). On this interpretation of Wittgenstein there is no fact of the matter about whether any sentence in the language is true or false; yet Wittgenstein still thinks that he is able to employ the notion of a fact.

66. Kripkenstein (Kripke, *Wittgenstein on Rules and Private Language*, 52–54) does consider the possibility that there are other sorts of mentalistic facts, but he argues that a mental state that determined the value of the plus function in infinitely many cases would be a "finite object" (because we have "finite minds") that could not determine the value of the plus function "in an infinity of cases." I leave it to the reader to judge whether this argument, with its image of the mind as a finite space and its picture of mental states as objects, is Wittgensteinian at all.

results, her blood pressure, or her brain waves.[67] If you are worried about the possibility that Joan might be able to conceal her true thoughts even from the polygraph, suppose—with Kripke—that there are no funny facts about her interior monologue as well.) The grammatical remark I ascribe to Wittgenstein is that we cannot understand talk of "meaning something different" by a piece of language when there is no connection between the alleged difference in meaning and anything the person to whom the difference in meaning is ascribed does or says or undergoes or would do or say or undergo. It follows from that "grammatical" observation that in such a case we should not be able to make heads or tails of the suggestion that Joan *really* means something other than "plus" by the word "plus" or the sign "+." The thesis Kripke ascribes to Wittgenstein, in contrast, is that there are no *facts that determine* that an arbitrary sentence *S*'s being true under particular circumstances is incompatible with the meaning the community assigns to the words except the verdicts the community actually renders or would actually render in the concept-attributing game.

Note that this is a *general* thesis—not a thesis only about the word "plus," but a thesis about every word in the language. Let us consider what Wittgenstein would be committed to if he held that thesis. I begin with mathematical examples and move to nonmathematical ones.

First, let *a, b,* and *c* be three numbers too large (and too "complicated" considered as sequences of digits) for human beings to actually add, say, strings of more than 264 more or less random digits. Then on Kripke's interpretation there will be *no fact that determines that one particular answer to the question "Is it the case that a + b = c?" is correct.* Our communal understanding of the words *is* just our disposition to give certain responses to certain questions, plus the relation of that disposition to the dispositions of the community.

---

67. I take *Philosophical Investigations,* §270—"Let us now imagine a use for the entry of the sign S in my diary. I discover that whenever I have a particular sensation a manometer shews that my blood pressure rises"—not only to go against the view that for Wittgenstein, only "criterial" behavior is relevant to the occurrence or nonoccurrence of sensations, and, more generally, of propositional attitudes, but also to explicitly allow the (possible) relevance of scientific facts about a person (facts about so-called molecular behavior). These issues were, of course, much discussed during the controversies about logical behaviorism that took place in the 1950s and 1960s. See Hilary Putnam, "Brains and Behavior," in R. J. Butler, ed., *Analytical Philosophy,* 2nd ser. (Oxford: Basil Blackwell, 1963), 211–235; reprinted in Putnam, *Philosophical Papers,* vol. 2, *Mind, Language and Reality* (Cambridge: Cambridge University Press, 1975), 325–341.

Those dispositions *may*, of course, determine that it is *wrong* to say $a+b=c$, even for very large a, b, and c. For example, if we are told that the numbers a and b end with, respectively, 2 and 3, and that the number c does not end with 5, then no matter *how* long a, b, and c are, we can say that $a+b \neq c$. But these negative tests (tests by which we can say on the basis of a limited amount of information about the numbers that $a+b \neq c$) will not suffice to justify a *positive* statement to the effect that $a+b=c$ (or to employ such a statement in considering whether or not a speaker passes the concept-attributing game in connection with the sign "+"); indeed, nothing will ever justify such a statement if a, b, and c are very long numbers that are not given as values of functional expressions that are short enough to be written down and understood and that we can prove theorems about. For infinitely many triples of numbers a, b, and c, there will be no "fact" about whether we understand the word "plus" in such a way that a plus b equals c, on Kripke's interpretation of Wittgenstein.[68]

Of course, Kripkenstein's view does not imply that we cannot *say* things like "No matter how long the numbers a and b are, there is a number c such that c is the sum of a and b." Kripkenstein would say that it is part of the language game we play that we *do* say things like that. But I think that Kripkenstein would have to say that when we say, "There is a number c such that $a+b=c$," what we say does not necessarily imply that there is a fact about *which* number c is such that, as a matter of "fact," "$a+b=c$" is true.

Just to make the point clear: the point I am making is not just the point (which indeed follows immediately from what Kripke writes) that there is nothing about the understanding of any individual considered in isolation that determines the correct answer to all addition problems. It is that when we consider the necessary finiteness of the responses *of a whole community*, Kripkenstein's argument yields the result that *there is no fact about the whole community*

68. I mentioned at the beginning of this essay that Wittgenstein was attracted to the idea that undecidable propositions lack a truth-value at various times. Does this not support Kripke's interpretation here? No, for several reasons. First of all, Kripke is interpreting *Philosophical Investigations* (in fact, just the "rule following discussion"). His interpretation is not supposed to depend on evidence from unpublished writings (Kripke subtitles his book *An Elementary Exposition*). Secondly, as pointed out in note 1 above, *Philosophical Investigations* refrains from mentioning any controversial views Wittgenstein may have held on undecidable statements in mathematics. Moreover, as I go on to argue above, other discussions in *Philosophical Investigations* are inconsistent with the views Kripke attributes to Wittgenstein.

that determines the answer to *all* the indefinitely many possible addition problems.

Consider a nonmathematical kind of case that Wittgenstein himself discusses in *Philosophical Investigations*. It can happen that there are disagreements that we are unable to resolve—that is, that the community is unable to resolve—concerning the genuineness of people's expressions of their feelings.[69] I feel that someone's expression of emotion is genuine, but I cannot convince other people. But all of us pass the relevant concept-ascribing tests. If the disagreement is about whether someone's avowal of love is sincere, then in the typical case no one would say that some of us lack the concept, or at least no one would say this on the basis of Kripkean criteria. (Of course, we do in fact ask whether people really know what love is, but this sort of discussion is one for which Kripke's view seems to leave no room—or perhaps Kripke would say that this is simply a metaphorical way of talking.)

Now Wittgenstein explicitly says that such judgments, judgments on which the community does not come into agreement, are judgments that nevertheless may be right. There are people who are better *Menschenkenner,* people who are better at understanding people, and such people can make "correct judgments." It is true that Wittgenstein says that these better *Menschenkenner* are also, in general, better at making prognoses; but there is nothing in Wittgenstein's text that implies that *each individual correct judgment* of the genuineness of an emotion will eventually be confirmed behaviorally in a way that will command the assent of the whole community—this would be an extraordinary view for Wittgenstein to hold, writing as he does in the same pages that such judgments are typically made on the basis of "imponderable evidence."

Of course, Kripke has a possible answer here. Kripke ascribes a deflationist account of truth to Wittgenstein, and no doubt he would reply, "Of course, one can say that my judgment or anyone else's judgment of an emotion is correct, or, for that matter, that a judgment concerning the sum of two huge numbers is correct, *meaning by that* simply to endorse, or repeat, the judgment in question. To say that someone is a superior *Menschenkenner* is, thus, simply to endorse that person's verdicts, but it is not to say that there is a fact about that person considered in isolation that is the fact that he is a better *Menschenkenner* than others, and, for that matter, if the community does not agree that he is a better *Menschenkenner,* then there is not even a fact about the commu-

---

69. Wittgenstein, *Philosophical Investigations,* part II, 207. See Hilary Putnam, *Pragmatism* (Oxford: Blackwell, 1995), chap. 2, for a discussion.

nity as a whole—or indeed a fact about the universe as a whole—that is the *fact* that the person is a better *Menschenkenner*." In short, Kripkenstein's view is a combination of a "deflationist" account of truth and a metaphysical realist account of fact.[70]

But what could have led Kripke to attribute such an extraordinary combination of views to Wittgenstein? One hypothesis—which I am sure is wrong—would be that Kripke himself believes that there are only the two sorts of fact that I mentioned, and that this seems so self-evident to Kripke that he cannot but believe that Wittgenstein thinks this too. But I am sure that this is wrong because it is quite clear that Kripke himself believes that there is another kind of fact about a person considered in isolation, namely, the fact that the person *grasps a certain concept*.[71] What makes Kripke think that Wittgenstein would *deny* that it can be a fact about someone ("considered in isolation") that he or she grasps the concept of addition?

I think that the answer must lie in a peculiar ambiguity in Kripke's cumbersome phrase "a fact about a person considered in isolation." Consider the following two statements:

1. Susan means by "plus" what most English speakers mean by the word.
2. Susan has the concept of addition.

70. It is here that Paul Horwich, who agrees with Kripke in ascribing deflationism about truth to Wittgenstein—and whose view of what constitutes understanding, like Kripke's, implies that there is no determinate right answer to the question "Is *p* true?" in cases in which the community's standards do not require that one say yes or no—parts company with Kripke. (For references, see my discussion of Horwich's views in "Dewey Lectures," lecture 3.) Horwich's Wittgenstein is an antirealist about both truth and fact. Given the views that Horwich ascribes to him, if the community says that "there is a number *c* such that *c* is the sum of *a* and *b*," but the community's standards do not require that one say that any particular *c* is the sum of *a* and *b*, then—rather than say that there is such a *c* but no fact about which it is (the position that I claim Kripkenstein should hold, given the views ascribed to him by his creator)—Horwichstein should say that there is some particular *c* (which we cannot know) such that $a+b=c$ is true, but no *c* such that $a+b=c$ is determinately true. (The idea that statements can be "indeterminately" true—whatever that means—is introduced by Horwich in *Truth* [Oxford: Blackwell, 1990], 114.) For a criticism of Horwich and Kripke's identification of Wittgenstein's view of truth with contemporary versions of deflationism, see Putnam, "Dewey Lectures," 510–516.

71. Kripke confesses that "personally I can only report that in spite of Wittgenstein's assurances, the 'primitive' interpretation ['that looks for something in my present mental state to differentiate between my meaning addition or quaddition'] often sounds rather good to me." Kripke, *Wittgenstein on Rules and Private Language*, 67.

Now, on *any* view, philosophical or nonphilosophical, statement 1 does not express "a fact about Susan considered in isolation." What it says is that there is a certain *relation* between Susan's understanding of a word and a particular linguistic community's understanding of that word. But because the fact that there is no fact about *Susan considered in isolation* that is the fact that she grasps the concept "plus" is supposed to be a shocking thesis (Kripke calls it a "skeptical" thesis and compares it with Hume's celebrated theses about causation, induction, the external world and so on), this cannot be what Kripke means to point out. Evidently, Kripke, or rather Kripkenstein, means to make the surprising claim that statement 2 is not a fact about Susan considered in isolation. But it is not clear why not, *even on Kripkenstein's view of what is involved in having a concept.*

To see why not, let us suppose that Susan does have those speech dispositions that enable her to pass the concept-attributing game when the concept ascribed is the concept of addition, or more specifically "plus," and the people playing the game are the members of the English-speaking linguistic community. Now, that Susan is disposed to make such and such responses under particular circumstances would appear to be as much a fact about "Susan considered in isolation" as any dispositional fact about her—say, the fact that she is fond of sweets, or the fact that she despises Erich Segal's *Love Story*. It is true that English speakers would not express that particular fact about Susan by saying that she has the concept of addition if *their* concept of addition were different; but it is equally true that the fact that Susan likes sweets would not be described by English speakers in those words if their concept of "liking sweets" were different.

Indeed, one might wonder why Kripke chose to express his view in this puzzling way rather than by saying—what he might straightforwardly have said—that, on his interpretation, what Wittgenstein thinks is that having a concept ("addition" or any other concept) is simply having those linguistic dispositions that enable a speaker to pass the appropriate concept-attributing game. Of course, if he had put his interpretation that way, then the interpretation would not have been called a skeptical interpretation. Rather, it would have been called a *behaviorist* interpretation, for it amounts to saying that possessing a concept is just possessing a certain behavioral disposition; which disposition is determined by the appropriate concept-attributing game.[72]

---

72. Kripke does argue that (on his interpretation of Wittgenstein) our behavioral dispositions do not determine the truth and falsity of our mathematical claims; but that is because there is no fact about their truth or falsity on this interpretation (see note 71 above). This does

But I do not mean to suggest that it was just for "packaging" purposes that Kripke expressed his interpretation in the way he did. I think that, rather, the way in which Kripke chose to express his interpretation is an indication of what Kripke's own view must be. On Kripke's own view, I suspect, that I grasp the concept of addition is not *just* "a fact about me considered in isolation," but rather a very special kind of fact, a fact that is not reducible to facts about my behavior dispositions, or my behavior dispositions-cum–bodily states, or the sameness and difference of the foregoing from the behavior dispositions and bodily states of others. Kripke, I believe, sees Wittgenstein as *denying that*.

But even this is not enough to explain what is going on here, because it is pretty easy to see that Wittgenstein would not wish to deny that, say, I (or Susan) possess the concept of addition or understand the word "plus"; and many have seen that Wittgenstein is antireductionist, and that he would not wish in any way to reduce the statement that I understand the concept "plus" to any set of statements about my behavior dispositions. Kripke, of course, knows this; so why, in the end, does he—if not in those very words—ascribe to Wittgenstein a view that makes having a concept come to no more than possessing certain behavior dispositions, or, alternatively, a view that implies that there is no such fact as the fact that anyone has a concept; it is just that we sometimes talk that way?

The answer, I think, must be this: although Kripke knows that Wittgenstein was concerned to deny being a behaviorist or a reductionist about the possession of concepts, he finds these denials somehow *unsatisfying*. They do not come, as it were, with the right metaphysical emphasis. To have the right metaphysical emphasis, Wittgenstein would have to say that the fact that I grasp a certain concept is a fact about me somehow on the same metaphysical *plane,* one that possesses the same metaphysical *reality,* as the fact that my neurons do such and such, or the fact that I have such and such a mental image when I think of chocolate ice cream. *That* is what Wittgenstein must be denying. If my guess is right, Kripke speaks of "facts about a person considered in isolation" not because of anything in *Wittgenstein's* writings, but because that is a description of what *Kripke* himself believes in, and because he sees himself as having a disagreement with Wittgenstein.[73]

---

not show that (on the same interpretation) having a concept is either more or less than being able to pass the concept-attributing tests.

73. I know that Kripke insists that it is not his purpose to talk about his agreement or disagreement with Wittgenstein, but he openly ridicules Wittgenstein's claim that the "primitive"

At the same time, Kripke is obviously honest in telling us that Wittgenstein created a problem for him. This comes out in a remarkable misreading of §195 of the *Investigations*. Kripke writes, "Yet (§195) 'in a *queer* way' each such case [of the addition table] is 'in some sense already present.'"[74] And after ruminating on how "mysterious" a supposed mental state of understanding would have to be (he writes that it would be a "finite object" that contained an infinite amount of information), Kripke quotes "the protest in §195 more fully," namely, "But I don't mean that what I do now (in grasping a sense) determines the future use *causally* and as a matter of *experience,* but that in a *queer* way the use is itself somehow present."[75] There is no hint in this that this sentence is set off in quotes in Wittgenstein's text (i.e., that the voice is the voice of an interlocutor), much less a hint of Wittgenstein's brusque response: "Really all that is wrong with what you say is the expression 'in a queer way.' The rest is all right; and the sentence only seems queer when one imagines a different language-game for it from the one in which we actually use it." In short, Kripke takes the voice of the interlocutor to be Wittgenstein's voice, and he cannot hear Wittgenstein's response at all!

It is because he sees Wittgenstein as showing that it is "queer" that we can understand a rule that Kripke is virtually forced to see him as a skeptic. But if we grant that this is a misreading, then what is it that Wittgenstein fails to do that Kripke would have him do, other than, as it might be, pound the table when he says that we do grasp concepts and follow rules?

If grasping concepts is (as Kripke hints that he thinks[76]) a matter of special "facts about a person considered in isolation," then we need to be told more about what the causal or other powers of "facts about a person considered in isolation" are supposed to be. It is striking that the alternative to Wittgenstein's view turns out to be *no clear alternative at all.* Indeed, I suspect that it must consist in saying what practically all of us, including Wittgenstein, would wish to say, but with a special stamp of the foot ("It is something more than physical facts and mental imagery" versus "The statement that someone

---

interpretation is a philosopher's imposition, writing, "Personally I think that such philosophical claims are almost invariably suspect. What the claimant calls a 'misleading philosophical construal' of an ordinary statement is probably the natural and correct understanding." Kripke, *Wittgenstein on Rules and Private Language,* 65.

74. Ibid., 52.
75. Ibid., 53.
76. Is this not the "primitive interpretation" that "looks rather good" to Kripke?

grasps a concept is not replaceable by a set of statements about physical facts and mental imagery"—but what does "something more" add?).

Mathematics

Almost everything that I have said so far (which is to say, most of this essay) is preliminary to discussing my announced topic—the case of mathematics, or rather the case of number theory, which is the only part of mathematics I wish to discuss here.[77] And I have to begin this final section with yet another "preliminary"; for if we are to find an analogue to our commonsense realism about tables and chairs, and our commonsense realism about rule following, in this mathematical case, it will be necessary first to understand what one is to mean by the "ordinary use" of mathematical notions.

Here I am guided by a remark of Stanley Cavell's,[78] that the notion of the ordinary in Wittgenstein has nothing to do with the distinction between the vernacular and technical language, or the language of the "man (or woman) on the street" and scientific language, or anything like that. The notion of the ordinary in Wittgenstein's later philosophy is meant to contrast only with the philosophical. In this sense of "ordinary," for example, a mathematician who proves a theorem about the zeros of the Riemann zeta function is employing *ordinary language.* Nor is the ordinary use of mathematical language confined to *proof* (here one must remember Wittgenstein's remark that "mathematics is a motley"[79]). Mathematicians not only prove theorems, they also make conjectures, they apply the theorems that have been proved both inside and outside mathematics proper, etc.

An interesting case is the following. Wittgenstein imagines that a mathematician gives a nonconstructive proof that the pattern 777 exists somewhere in the decimal expansion of π, where the nonconstructivity of the proof resides precisely in the fact that the proof gives no indication *where* in the decimal expansion of π the pattern occurs.[80]

---

77. It seems appropriate that this is the case because, in a sense, Wittgenstein's philosophy always consists of "preliminaries."

78. This is a remark Cavell made in a course he and I co-teach on Wittgenstein's later philosophy.

79. Wittgenstein, *Remarks on the Foundations of Mathematics,* III, §46.

80. Wittgenstein, *Remarks on the Foundations of Mathematics,* V, §27: "A proof that 777 occurs in the expansion of π without shewing where, would have to look at this expansion

Let us now imagine that a mathematician living at the time Wittgenstein wrote these remarks offered the following probabilistic argument (such arguments are a commonplace in number theory and often guide research): as far as we know, the decimal expansion of π passes all the standard statistical tests for randomness. But if that is so, and the digits occur with equiprobability (as they appear to), then, with a probability greater than .95, we should expect the pattern 777 to occur by such and such a place in the expansion.

We can imagine that with the development of the computer this conjecture is later confirmed by carrying out the expansion that far. (Indeed, I have learned that a computer search shows that 7777 occurs in the expansion.)

I want to say that although such a probabilistic argument is not a proof, nevertheless, giving such arguments and applying them are part of the present-day activity of mathematics, although it was probably not a part of the activity of mathematics one hundred or two hundred years ago, and the technique by which we (informally, of course) assign a "probability" to certain mathematical assertions gives a sense to saying things like "It is probable that 777 occurs in the expansion of π before such and such a place."

The application I wish to make of this remark is the following: consider a mathematical conjecture that we have not succeeded in deciding, for example, the celebrated conjecture that there are infinitely many pairs of twin primes (pairs of numbers p and $p + 2$ that are both primes). A great deal of the discussion in the philosophy of mathematics has centered on the question whether in the case of such a statement—one that is undecided and, for all we know, possibly undecidable—it makes sense to say that the statement is either true or false. If someone says that it is either true or false that there are infinitely many twin primes, then I think that what a Wittgensteinian philosopher should ask is, "In what context does the assertion occur?" If it occurs in the context of a philosophical discussion about whether undecided propositions are "really" true or false, then a Wittgensteinian may well doubt that it makes any sense because of the peculiar metaphysical emphasis that has been put on the notion of being *really* true or false. But if it occurs as part

---

from a totally new point of view, so that it showed e.g. properties of regions of the expansion about which we knew only that they lay very far out. Only the picture floats before one's mind of having to assume as it were a dark zone of indeterminate length very far on in π, where we can no longer rely on our devices for calculating; and then still further out a zone where in a *different* way we can once more see something."

of a mathematical argument, then it would be regarded as a perfectly trivial remark. If there is—in that context—a commonsense answer to the question "But how do you *know* that it must be either true or false?" it would be the rhetorical question, "What other possibility is there?"[81]

In short, I want to suggest that just as Wittgenstein points out that there is a perfectly ordinary way of learning the concepts "do so-and-so *ad infinitum*," "keep doing the *same* thing," "do this *uniformly*," ". . . " (as in "Do A, B, C, . . ."), so there is an ordinary way of learning the use of such expressions as "either *S* is true or *S* is false" in mathematics; and just as I take the point of Wittgenstein's discussion of rule following not to be the expression of skepticism of any kind about *rule following,* but rather the expression of a "skepticism" about philosophical discussions of rule following (or, better, the expression of a conviction that such discussions contain a great deal of nonsense), so I would urge that a Wittgensteinian attitude toward the use of the law of the excluded middle in mathematics would involve not a skepticism about the applications of that law within mathematics, but a "skepticism" about the very *sense* of the "positions" in the philosophy of mathematics—logicism, formalism, intuitionism, Platonism, nominalism, and so on.

I would like to remind you at this point that, as I said at the beginning, my aim is not to produce Wittgenstein's own view (although I believe he would agree with what I have just written), but to produce a view that builds on genuinely Wittgensteinian materials. I do find the things Wittgenstein says about the law of the excluded middle in *Remarks on the Foundations of Mathematics* troubling.

At any rate, to shift from the twin-prime conjecture to an example that Wittgenstein actually uses (*Remarks,* V, §9)—contrary to what Wittgenstein writes about the question (he calls it "queer"), I think that we should regard the question whether 770 (note that Wittgenstein shifts from "777" to "770" here) ever occurs in the expansion of π as a perfectly sensible mathematical question. For example, if mathematicians were to make the conjecture that this is the case, and to try to prove this conjecture in various ways, they would

---

81. Today [April 11, 2011] I blush at this claim. What I wrote here amounted to denying that we understand the question whether sentences of pure mathematics are really bivalent. The answer to "What other possibility is there?" is, of course, "That it lacks a truth-value." I have long argued that that answer should be rejected (my "indispensability argument"). But neither that answer nor the question it responds to is senseless.

be doing something that makes perfect sense. What Wittgenstein should have said is that mathematicians do understand the question whether 770 ever occurs in the decimal expansion of π, and that they have learned to understand such questions by learning to do number theory; and that something that they have also learned by learning to do number theory is that either 770 will occur in the expansion or 770 will never occur in the expansion. (Indeed, this seems to be what Wittgenstein *does* say in *Philosophical Investigations* §516—the paragraph I chose as the epigraph to this paper.)

Of course, objections to what I have just proposed as a "Wittgensteinian" strategy (even if it is not Wittgenstein's strategy in *Remarks*) will come from supporters of all the traditional positions: from metaphysical realists, antirealists of all varieties, intuitionists, formalists, and others. The previous discussion of how one answers these sorts of objections in the case of rule following is meant to indicate the general strategy to be followed in responding to each of these objections, although, of course, to even begin to carry it out in detail would require a work as long as Wittgenstein's *Remarks on the Foundations of Mathematics* (which is, as we know, not a "work" but a selection of material from Wittgenstein's *Nachlass*). In particular, the following question will be raised, both by metaphysical realists and by a certain sort of antirealist: If you say, "Either it is true that 770 will occur in the decimal expansion of π or it is true that 770 will never occur in the decimal expansion of π," then you must be doing one of two things. Either (1) by saying that this assertion is perfectly all right (provided it is not meant as a "philosophical" assertion), you merely mean that we can utter the *words* "Either it is true that 770 will occur in the decimal expansion of π or it is true that 770 will never occur in the decimal expansion of π" in the context of a mathematical argument, while leaving the *philosophical interpretation* of those words entirely open—which is something that no philosopher denies; or (2) you are claiming that the mathematician is entitled to say that one of the two disjuncts is *determinately* true. But if one of the two statements is determinately true, then there must be some *fact* that makes one of those statements true; and because the true disjunct may be undecidable, that fact is not (or is not necessarily) any fact about the existence of a proof. But what sort of a fact could this be?

Once again, exactly as in the above discussion of rule following, a number of suspect philosophical moves are being made here; for example, the distinction between its being the case that either A is true or B is true and its being the case that one of the two propositions is *determinately* true has been assumed, and, more important, it is being assumed that if a proposition is true, then there is something that *makes* it true (I have already discussed *that* one).

One might also consider a different objection, which I will not identify with a person or position, and that is this:

"Is not mathematical truth supposed to be in some way *conceptual* truth?[82] Now, if all mathematical truths were provable, then it would be clear (or at least *clearer*) in what way all mathematical truths were conceptual truths; but after the Gödel theorem it seems that we are stuck with the idea that some number-theoretic propositions are true even though they have no proof (if we are willing to take mathematical instances of the law of the excluded middle as plain mathematical common sense, in the way you are recommending). But then, what sort of thing *is* this 'mathematical truth'? If there is anything that Wittgenstein insists on, it is that it is not empirical truth, and if it is not conceptual truth either, then what is it?"

Again, in this case, a special philosophical idea, in fact, a dichotomy, is in play; namely, the dichotomy between *empirical* and *conceptual* truth is being given a metaphysical emphasis. And we know that this metaphysically emphasized dichotomy is vulnerable to both Quinean and Wittgensteinian criticisms. The right answer to "What sort of truth is it?" is, of course, "A *mathematical truth.*" And that really is a "grammatical" remark.

Of course, we might say that what "makes" it true (if it is true) that the pattern 770 occurs in the decimal expansion of π is that there is a number n such that the nth, $n+1$st, and $n+2$nd places in the expansion are 7,7, and 0, respectively, just as what makes it true that snow is white is that snow *is* white—but, of course, this would not be to give an *informative* answer to the

82. Indeed, Wittgenstein does say (in *Lectures on the Foundations of Mathematics*, 251) that "$2+2=4$" is a *grammatical* truth. For an account of what this might mean, see Hilary Putnam, "Mathematical Necessity Reconsidered," in Putnam, *Words and Life*. 245–263. Wittgenstein's frequent observation that we use (established) mathematical assertions as rules (e.g., for determining the coherence or incoherence of empirical statements containing numerical expressions) is often taken to constitute an explanation of mathematical truth; it should be evident that if the present essay is on the right track, it is no such thing. Wittgenstein is simply describing certain plain facts about the use of these assertions, not advancing a theory of mathematical truth. I am indebted to Yemima Ben-Menahem's "Explanation and Description: Wittgenstein on Convention" (unpublished), a fine study of the issue of "conventionalism" in Wittgenstein's writing that employs the distinction between reading the passages in question as descriptions of our practices and reading them as explanations to shed light on Wittgenstein's discussions.

question, what "makes" either statement true, but to reject the metaphysical picture of statements as being "made true" by . . . (what, *intangible* objects?).

But, it may be objected, is not the statement that snow is white like the statement that the sofa is green? Even if he did not use the language of "making true," Wittgenstein was willing to say (in *Lectures on the Foundations of Mathematics*) that "green" and "sofa" "correspond" to "realities." Could we not say that that white stuff that we see falling from the sky (and have seen through the ages) is the reality to which the statement that snow is white corresponds?

But there is a problem. "Snow is white" is, after all, a universal generalization—a negative existential. And Wittgenstein was unwilling to extend the language of "realities" corresponding to statements so far as to introduce *negative realities* for negative existentials to correspond to. (Here I am repeating a point made above, of course.) But the objector might respond, "Okay, but even if you do not accept negative facts or negative realities (or universal facts or universal realities), one could say that the statement that there is no yellow book in this room corresponds not to 'a reality' but simply to *reality;* that is, *it says something about reality.* Does the statement that 770 never occurs in the decimal expansion of π, if true, correspond to reality as a whole?"

In this connection, I would like to remind you of Wittgenstein's remarks (in *Lectures on the Foundations of Mathematics*) on G. H. Hardy's (supposed) claim that "to mathematical propositions there corresponds—in some sense, however sophisticated—a reality":[83]

> We have here a thing which constantly happens. The words in our language have all sorts of uses; some very ordinary uses which come into one's mind immediately, and then again they have uses which are more and more remote. . . . A word has one or more nuclei of uses which come into everyone's mind first. . . . So if you forget where the expression "a reality corresponds to" is really at home.—What is "reality"? We think of "reality" as something we can *point* to. It is *this, that.* Professor Hardy is comparing mathematical propositions to propositions of physics. This comparison is extremely misleading.[84]

83. Wittgenstein, *Lectures on the Foundations of Mathematics,* 239. What Hardy actually wrote in the article Wittgenstein mentions is that mathematical theorems are "theorems concerning reality" Hardy, "Mathematical Proof" in *Mind* 38, no. 149 (1929): 1–25; quotation, 4. Hardy does not speak of a "correspondence to reality" anywhere in this article.

84. Wittgenstein, *Lectures on the Foundations of Mathematics,* 239–240.

Wittgenstein is not forbidding us ever to use the expression "corresponds to reality" in connection with mathematics;[85] on the contrary, he devotes the next few lectures to exploring different things that one might mean by expressions of the form "statements of such and such type correspond to—or are responsible to—a reality" when this expression is applied to mathematical statements and when it is applied to other sorts of statements. Wittgenstein's point is, rather, that Hardy is unable to free himself from the illusion that when he speaks this way, he is still employing the expression in a way that is closely akin to its paradigmatic use, as if mathematical expressions were *descriptions* of a region of reality, only not an ordinary region. As James Conant has recently put it, "The use to which Hardy attempts to put the expression is to be seen in the end as a confused attempt to amalgamate several of these possibilities of use in such a way as to fail in the end to be saying anything at all."[86]

In addition, the idea that each true mathematical statement has to *describe* reality—an exotic region of reality—is in tension with the fact that mathematical truths are not *contingent*. If mathematical statements *describe* an exotic reality, must they not *depend* on that reality for their truth? And if so, does that not mean that *they would be false if that reality were otherwise?* From as early as 1913, as Burton Dreben and Juliet Floyd have pointed out,[87] Wittgenstein resolutely opposed the idea that any statement could simultaneously be *about* reality and yet true independently of how reality is; this is, indeed, what lay behind his insistence that logical truths are "tautologous." Of course, if we could make sense of Kant's "synthetic a priori," we might be able to understand how a statement could describe reality and yet not be in any way *contingent*. But even the German idealists had trouble in making sense of that notion.

But what of Benacerraf's question in his famous "Mathematical Truth" paper? Am I saying that there is a kind of "realist semantics" for the language

---

85. This represents a change in my reading of this passage from the shorter version of the present essay published as "On Wittgenstein's Philosophy of Mathematics," *Proceedings of the Aristotelian Society,* suppl. 70 (1996): 243–264. James Conant's response (see the next note) convinced me that my former reading was wrong.

86. James Conant, "On Wittgenstein's Philosophy of Mathematics: II," *Proceedings of the Aristotelian Society* 97 (1997): 195–222.

87. Burton Dreben and Juliet Floyd, "Tautology: How Not to Use a Word," *Synthese* 87 (1991): 23–49.

of classical mathematics that is part of the activity of classical mathematics itself? And if so, how do I account for the notion of reference involved in that semantics? Is the causal theory of reference false for mathematical entities? And how do I answer all the other philosophical conundrums about mathematics?

But again, what one needs to do is to take these questions apart patiently. As Jamie Tappenden once remarked to me in conversation, "Benacerraf's paper showed so well why this way of posing the question leads to an impasse, that it should really have been the *end* of that discussion rather than leading people to put additional epicycles on views that don't work."

When people speak of a "semantics" for a piece of language, whether they be linguists or philosophers, all they give us (and all they *can* give us) is a formalism that is ultimately interpreted in *ordinary* language. In the sense of having a translation into ordinary language, of course, mathematics has a semantics—a "realist semantics," if you please. Or rather, mathematics as it is actually done, "unregimented" mathematics, already *is* in ordinary language. On the other hand, if a "semantics" is something like a metaphysical relation of "correspondence," then I would say that it is wholly unclear what a "semantics" is, or why the alternative to having a semantics in that sense has to be antirealism.[88] But I am beginning to repeat points that I have already made—not, however, unintentionally, but because I wish to emphasize that the assumption that the notion "semantics" has already been given a special philosophical sense that we all, of course, understand is like the assumption that we understand (and have to accept) the claim that if a statement is true, there is "something" that makes it true.

Why the Problems in the Philosophy of Mathematics Are Not *Sui Generis*

One thing that makes our understanding of mathematical notions seem paradoxical is that we tend to be attracted not only to the metaphysical realist picture—in fact, it is only since Quine's talk of "ontological commitment" has become so accepted that that has become a very popular picture—but also to a formalist picture in which understanding mathematical concepts is something like "internalizing" rules for syntactic manipulation. We have a

88. For a fuller discussion of the idea that commonsense realism does not depend on a mystery relation of "correspondence," see Putnam, "Dewey Lectures," particularly lecture 3.

tendency to think that somehow (perhaps unconsciously) what we are doing—and all we are doing—when we do elementary number theory is deriving formulas from other formulas, that we have, as it were, a proof procedure in our brains for doing that, and the *syntactic* capacity to operate that proof procedure is what constitutes our *understanding* of number theory; and then we encounter the Gödel theorem, and we are all at sea. But *that* picture is already inadequate as a picture of what our understanding of first-order logic consists in. As I pointed out above, it is just not the case that our understanding of first-order logic is correctly described as our having a set of axioms in our brains and a technique for deriving the valid sentences of first-order logic from those axioms. The *interpretation* of the quantifiers themselves is not simply exhausted by the axioms and rules of derivation. The problem is an instance of a more general problem that one might call the problem of *reductionism*. Our understanding of our concepts and our employment of them in our richly conceptually structured lives are not mystery transactions with intangible objects, transactions with something *over and above* the objects that make up our bodies and our environments; yet as soon as one tries to take a normative notion like the understanding of a concept or Wittgenstein's notion of the use of a word and equate that notion with some notion from stimulus-response psychology ("being disposed to make certain responses to certain stimuli"), or a notion from computational psychology, or a notion from the physiology of the brain, then the normativity disappears, and hence the concept itself disappears. Cora Diamond's very lovely example[89] of picture faces is, I think, a useful one here. The smile in a picture face is not an "object" over and above the charcoal and the paper before us, but that does not mean that statements about picture faces are simply reducible to statements about molecules of carbon and molecules of paper. The expression is something we see *in* the charcoal and the paper; it is not the charcoal and the paper, and it is not something "besides" the charcoal and the paper. In the same way we understand, say, the notion of truth, the logical connectives, the notions of number theory, and so on. That understanding is not something "immaterial" in the sense of being something "over and above" the systems in our brains, the stimuli and our reactions to them, and so on; but it is something we see *in* the practices of certain of the embodied beings that we are. "If you want to understand what it means 'to follow a rule,' you have already to be

---

89. The example is used by Cora Diamond in *Realistic Spirit,* 249.

able to follow a rule," Wittgenstein reminds us.[90] And if you want to understand what it means "to grasp generalizations about the natural numbers," you have to be able to grasp generalizations about the natural numbers.

At the beginning of this essay I said that one of my goals would be to show that the problems in the philosophy of mathematics are not as *sui generis* as they appear; and in closing I wish to say a little more about why that is the case. Consider, for purposes of comparison, the "problem of the indeterminacy of translation." The problem, as Quine poses it, is precisely that we cannot point to trajectories of particles (or of fields) that make it true that *A* is the correct interpretation of a certain discourse, even when we all agree that *A* is; and Quine concludes from this that "there is no fact of the matter" about whether *A* is the correct interpretation. Or consider the so-called realism problem about ethical statements, which is again that one cannot point to causal processes or physical objects that make it true that something is good. The discussion that we have just reviewed in the philosophy of mathematics is not, fundamentally, a different discussion. The problem in all of these cases—as, indeed, Simon Blackburn has seen,[91] although I believe that he has taken exactly the wrong view—is that we wish to impose a *pattern* of what it is to be true, a pattern derived largely from the successes of physical science, on all our discourse. We believe that there is such a phenomenon as magnetism, and it is also the case that our best explanation of the experiences that have led us to believe this involves an appeal to the theory of (electro)magnetism. Magnetism causally, scientifically, *explains* those experiences. But what on earth, one might ask, does this have to do with a discussion of "value judgments," or a discussion of "translation," or a discussion of mathematical truth?

For a certain kind of "scientific realist," the answer is "everything." For "scientific realism" is scientific imperialism. *Any* belief that claims "objectivity" must conform to this pattern—the pattern of causal explanation in a natural science. But that is not how scientific realists see themselves.

The reason that scientific realists do not see that what they are doing is simply trying to force all belief that claims to be objective into a single procrustean bed is that they have prestructured each of these debates in such a way that one *must* choose between two options: in the metaethical case,

90. Wittgenstein, *Remarks on the Foundations of Mathematics,* V, §32.
91. Simon Blackburn, *Spreading the Word* (Oxford: Oxford University Press, 1984).

either concede that "value judgments" possess "only the illusion of objectivity" or produce metaphysical objects standing behind value judgments and guaranteeing their objectivity; in the semantic case, either concede that judgments concerning the correctness of a translation possess only the illusion of objectivity or defend "the museum myth of meaning"; in the philosophy of mathematics, either concede that mathematical statements possess a truth-value only when they are decidable or defend the existence of a Platonic realm of mathematical objects. Either there are "intangible objects" corresponding to value terms, to interpretations, to mathematical statements—objects that are causally efficacious in the way in which magnetism is causally efficacious—or else value judgments, interpretations, and undecidable mathematical conjectures are (if they are taken to have a truth-value) as misguided as belief in *ghosts*. In contrast, the Wittgensteinian strategy, I believe, is to argue that although there is such a thing as correctness in ethics, in interpretation, and in mathematics, the way to understand that is not by trying to model it on the ways in which we get things right in physics, but by trying to understand the life we lead with our concepts in each of these distinct areas. The problems in the philosophy of mathematics are not simply the *same* as the problems in metaethics or the problems about the indeterminacy of translation, because the way the concepts work is not the same in these different areas; but what drives the sense that there is a problem—a problem that calls for either a "skeptical solution" or an absurd metaphysics—is usually the very same preconceptions about what "genuine" truth or "genuine" objectivity or "genuine" reference must look like.[92]

92. I [2011] still believe that teaching us to describe our language games and the actions and forms of life with which they are interwoven was an important contribution of Wittgensteinian philosophy (although I do not think that Wittgenstein himself had enough acquaintance with what mathematicians actually do to carry that task out in the case of mathematics). I do not agree that it is a mistake for philosophers to try to *explain* either the purposes or the success of those "language games"; nor, in particular, do I believe that the important distinction between *using the law of the excluded middle in a mathematical argument and accepting the idea that truth in mathematics is bivalent* is metaphysical nonsense. This chapter probably represents my closest approach to being a "Wittgensteinian"; other chapters in this volume illustrate why and how I have moved further away.

24

Rules, Attunement, and "Applying Words to the World": The Struggle to Understand Wittgenstein's Vision of Language

Wittgenstein's writing produces two sorts of controversy: first, there are controversies between those who, like myself, think that Wittgenstein was one of the greatest philosophers of the twentieth century and those who think him the most overrated philosopher of the century (Saul Kripke represents an interesting middle position: Wittgenstein was great but misguided, if his reading is correct). These are not the controversies I wish to discuss here.

The second sort of controversy is a controversy *among* philosophers of the first kind. Such controversies are a familiar phenomenon in the history of philosophy. They arise because different "lines of thought" can arguably be supported by various statements and arguments in the text of a great philosopher. On the surface, the question about which of these lines of thought best represents what the great philosopher meant to teach us may seem to be a purely "textual" one, but it almost never is. Because interpreters quite properly apply the principle of charity, each side attributes to the great philosopher the line of thought that it finds strongest in its own right. I believe that that is what was going on in a wonderful exchange between Steven Affeldt and Stephen Mulhall in the pages of the *European Journal of Philosophy*.[1] And it is proper that it should be, for the important question raised by these papers is whether one or another view of how language and the world connect is correct, and secondarily whether the preferred view is really Wittgenstein's.

For the most part I shall focus on Mulhall's paper because it represents what he himself regards as an "orthodox" interpretation of Wittgenstein, and I believe that seeing what is wrong with the "orthodox" view as a view not only can prepare us to entertain the possibility that an interpretation like Stanley Cavell's[2] (which is the interpretation Steven Affeldt defends) does more justice to the subtlety and originality of Wittgenstein's later philosophy than the "orthodox" view (which Mulhall associates with the names of Baker and Hacker), but it can also help us appreciate deep issues about our linguistically mediated intercourse with one another and the world that do not even come into sight if one accepts the "orthodox Wittgensteinian" view as the right view in philosophy.

I used the laudatory epithet "wonderful" to describe this exchange not just because the intellectual quality of both papers is so high, but because in many ways the exchange was a model of what a philosophical criticism and a response to it ought to be, but too rarely is. Both papers are learned, thoughtful, serious, and inspired by a search for truth and not rhetorical advantage. Moreover, both papers are courteous—Mulhall's fairness and courtesy toward someone who is, after all, criticizing a book he wrote are exemplary. I hope that I shall manage to display the same traits in this response.

Although I shall be defending substantially the view that Affeldt defended, I shall approach the issues in my own way, and, as I just indicated, I shall focus on the defensibility of the "orthodox Wittgensteinian" view as a position in philosophy, rather than on the textual evidence for and against it as a reading of Wittgenstein.[3] I have chosen Mulhall's paper not only because I was so impressed by its quality, but because it contains some extremely clear

---

1. Steven Affeldt, "The Ground of Mutuality: Criteria, Judgment and Intelligibility in Stephen Mulhall and Stanley Cavell," *European Journal of Philosophy* 6, no. 1 (April 1998): 1–31; Stephen Mulhall, "The Givenness of Grammar: A Reply to Steven Affeldt," ibid., 32–44.

2. Stanley Cavell, *The Claim of Reason* (Oxford: Clarendon Press, 1979).

3. That it is the correct interpretation of Wittgenstein's later philosophy is argued by Cavell, *Claim of Reason,* and (on independent grounds) by Charles Travis in *The Uses of Sense* (Oxford: Oxford University Press, 1989).

and concise statements of the claims that the debate is all about. In the next
section I shall quote a few of these statements, and in the subsequent sections I shall argue that, taken at face value anyway, they lead to a disastrous
epistemology.

The "Orthodox" View of Criteria and Rules

The word that occurs again and again in Mulhall's account of the "orthodox"
view is "rule," for example, in "criteria as rules," [4] "uncovering a framework of
rules," [5] and "orthodox rule-based accounts of grammar and criteria." [6] Although Mulhall is the author of a sympathetic (in fact, highly laudatory)
account of Cavell's interpretation of Wittgenstein, *Stanley Cavell: Philosophy's
Recounting of the Ordinary* (henceforth *SC*),[7] he does not accept Cavell's criticisms of the idea that criteria provide a "framework of rules" or Cavell's rejection of the idea (beloved of "orthodox" Wittgensteinians) that philosophical
nonsense is to be diagnosed as the result of a misguided attempt to make sense
outside the framework. In fact, he writes:

> I was of course aware [when writing the book] of the two texts of Cavell's . . .
> articulating his hostility to the idea of grammar as a framework of rules,
> but I gave little detailed attention to either since neither seemed to me to
> provide any clear and detailed justification for this hostility.[8]

And

> It became clear to me that Cavell was deeply suspicious from a very early stage
> of his work of any such talk of Wittgensteinian criteria as rules, or of gram-
> matical investigations as uncovering a framework of rules; but it was not at
> all clear to me what the grounds of this suspicion were, and it was equally
> unclear to me that anything significant in Cavell's reading of criteria and
> grammar was threatened by reformulating it in the Baker & Hacker termi-
> nology and turns of phrase with which my writing has been inflected.[9]

4. Mulhall, "Givenness of Grammar," 33.
5. Ibid.
6. Ibid., 40.
7. Stephen Mulhall, *Stanley Cavell: Philosophy's Recounting of the Ordinary* (Oxford:
Oxford University Press, 1984).
8. Mulhall, "Givenness of Grammar," 33.
9. Ibid.

Of course, we cannot rule out in advance the possibility that the disagreement between Cavell, on the one hand, and "Baker & Hacker,"[10] on the other—or even the disagreement between Affeldt and Mulhall—is a purely verbal one; the possibility, that is, that there is an understanding of "rule" on which the claim that criteria are rules of grammar says nothing that Cavell need disagree with. To avoid being caught in what might be a purely verbal controversy, I shall, therefore, avoid taking "Are (Wittgensteinian) criteria rules?" to be *the* question at issue. Instead, I want first to look at what Mulhall thinks rules *do*.

Rules, in Mulhall's sense, do not tell us how to process "marks or features" that are themselves not already conceptualized. (Mulhall says that Affeldt misunderstood him on this point.)[11] One must already be within the schema to proceed on the basis of criteria. Mulhall writes as follows:

> So my saying that criteria constitute the marks or features on the basis of which we judge whether something counts as a chair is not meant to suggest that, whenever we encounter chairs (whether familiar or exotic), we first recognize the presence of criteria for something's being a chair and then go on to call it a chair. . . . *My claim concerns the order of justification, not that of perception or judgment:* the point is that if my judgment that something is a chair were to be subject to question or contestation, then I must be able to, and would, justify it by reference to certain features of the object itself, and of the ways in which it is intended to be or can be employed.[12]

The idea that criteria figure only in the "order of justification" is not repeated, however, and is difficult to square with what follows. (It is also difficult to square with Cavell's point, with which Mulhall seems to agree in *SC,* that the judgment that something is a chair or an inkwell or any other generic object, is not ordinarily a *claim,* and that in an ordinary "nonclaim context" the question of justification does not so much as arise.) For example, on the very next page Mulhall writes:

> Neither is my claim meant to imply—as Affeldt at another point suggests— that criteria are "assertability conditions" (p. 3). To be sure, someone might

10. G. P. Baker and P. M. S. Hacker, *Scepticism, Rules and Language* (Oxford: Blackwell, 1984).

11. Cf. Mulhall, "Givenness of Grammar," 35: "As Affeldt has it in his footnote 31."

12. Ibid.; emphasis added

gloss my idea in such terms, and thereby invoke a complex machinery of meaning-theoretic analysis of the kind that informs disputes between Davidsonians and Dummettians; but any such gloss would be entirely foreign to the spirit of SC as a whole, and is certainly not built in to *the simple idea that criteria are what we go on when we apply words to the world.*[13]

Here (and throughout Mulhall's paper, apart from the one sentence about criteria and "the order of justification") it certainly sounds as if all talk about the world *employs* criteria, whether a "claim" has been "contested" or not. Indeed, a few pages further on, Mulhall seems to explicitly contradict his assertion that "my claim concerns the order of justification, not that of perception or judgment." Thus he writes, responding to Affeldt's view (which is itself an interpretation—a correct one, I believe—of Cavell's view) that talk of our criteria arises only in connection with specific philosophical or empirical confusions (that the question "What are the criteria for the use of such and such a concept?" has no sense apart from a specific philosophical or empirical confusion):

> These statements conjure up a sense of criteria as forged when, and only when, we encounter specific confusions or crises in going on with our words—as if criteria are absent in the absence of such problems, *as if our uncontested or unconfused linguistic judgments are not already shaped or informed by criteria, as if in such circumstances we have judgments without criteria.*[14]

Not only is Mulhall here claiming that we never have "judgments without criteria" (not even uncontested and unconfused ones), but also he is putting forward what he sees as a serious dilemma for the Cavell-Affeldt view: if Cavell and Affeldt reject the idea that criteria are a "framework of rules" that we "go on" in judgment, then must not they think of them as *created* by the philosopher's investigation? But let me quote the passage in full (this will be the last of these quotations from Mulhall's paper, for it contains a crucial argument—one that it will be my aim to rebut). I shall repeat the previously quoted sentence to remind you of the context:

> These statements [Affeldt's] conjure up a sense of criteria as forged when, and only when, we encounter specific confusions or crises in going on with our words—as if criteria are absent in the absence of such problems, as if

13. Ibid., 36; emphasis added.
14. Ibid., 39; emphasis added.

our uncontested or unconfused linguistic judgments are not already shaped or informed by criteria, as if in such circumstances we have judgments without criteria. This would not only make it hard to comprehend Cavell's and Wittgenstein's frequent talk of being recalled to or reminded of our criteria by philosophical and nonphilosophical confusions—talk which seems to imply that while criteria may be discovered through such confusion, they are not created thereby. It would also leave little room for talk of our everyday judgments as normative, as open to evaluation as correct or incorrect. For such talk presupposes the existence of standards of correctness, of norms; it must be possible for us to justify how we go on,[15] and as Wittgenstein tells us "justification consists in appealing to something independent of what is being justified."[16] *It is that justification that, on my account, criteria provide;* but its very possibility seems threatened by some of Affeldt's more unguarded remarks.[17]

Going On without Rules

Evidently, on the "orthodox" (or "Baker & Hacker") view, going on without criteria—criteria construed as rules that belong to a framework of rules that is independent of the particular judgment that those rules "justify"—is making sounds to which no "normativity" attaches, in effect, mere babble. Let us see whether this is so.

Probably I do not need to remind the reader that I began my philosophical career as a philosopher of science, and for the next few minutes I will return to philosophy of science. From very early on, what impressed me about the great events in science in the first third of the twentieth century was the way in which what once were taken to be "a priori" truths, perhaps even "conceptual" truths, had to be given up one after another. In "It Ain't Necessarily So," a paper I published almost forty years ago (but one I still agree with),[18] I tried to explain just how important this fact is for all of epistemology. Imagine, for example, that in, say, 1700 Jones had said, "There is a triangle

---

15. Here again, Mulhall assumes that the question of justifying how we go on always makes sense, whereas the claim that it does not is central to Cavell's discussion of skepticism in *The Claim of Reason*.

16. Wittgenstein, *Philosophical Investigations* (Oxford: Blackwell, 1953), §265.

17. Mulhall, "Givenness of Grammar," 39–40; emphasis added.

18. Hilary Putnam, "It Ain't Necessarily So," *Journal of Philosophy* 59, no. 22 (October 1962): 658–671; reprinted in Putnam, *Philosophical Papers,* vol. 1, *Mathematics, Matter and Method* (Cambridge: Cambridge University Press, 1975), 237–249.

both of whose base angles are right angles." Would these words have been intelligible? At best this would have been taken to be a riddle. Perhaps Smith would have replied, "Oh, I get it. You mean a *spherical* triangle." But let us imagine that Jones says, "No, I do not mean a triangle on a sphere. I mean a triangle on a plane, on the locus of all straight lines that intersect two given straight lines." Perhaps Smith tries again to "guess the riddle." "Are you perhaps considering a finite line segment as a degenerate case of a triangle, one whose third angle is zero degrees?" But Jones says, "No, I mean a triangle all three of whose angles are positive, and two of whose angles are right angles." At this point, Smith would doubtless say, "I give up. What is the answer?" And if Jones could say no more than "I just mean that there is a triangle whose base angles are both right angles and whose third angle is positive," then he would have been utterly unintelligible. He would not have provided us with a context in which we knew how to assign a determinate sense to his words.

Another example I used in the same paper is the following: it is now conceivable that space is finite. Again, if Jones had said this in 1700, Smith might have said, "Oh, you are going back to the Aristotelian view. You believe that if we could travel far enough we would encounter a boundary, a sphere that surrounds the whole cosmos, and that the question 'What lies beyond the sphere?' makes no sense." (Note that strangely enough, by Kant's day—and perhaps already in 1700—this view, which had been accepted for two millennia, already seemed inconceivable. What seems to makes sense can *stop* making sense.) But if Jones says, "No, there is no boundary. Space is *finite but unbounded*," Smith—that is to say, our former selves—would have said, "You are talking gibberish." Or, perhaps, more charitably, Smith might have first asked, "What do you mean by 'finite and unbounded?'" We suppose that Jones gives our present-day answer: "By 'finite' I mean just the obvious thing: that there are only finitely many distinct nonoverlapping places the size of, say, this room to get to, travel as one may (even if one were allowed to travel instantaneously from any one place to any other). And by 'unbounded' I mean that no matter which direction one travels in, one never encounters an impassible barrier to continuing to travel in that direction." Again, if Jones had been unable to say more than this, if he could only repeat this explanation without satisfactory elaboration, then he would have been utterly unintelligible. He would not have provided us with a context in which we knew how to assign a determinate sense to his words.

Today every educated person knows at least the outlines of what happened to make these strange assertions intelligible. In the usual quick story, which is indeed correct as far as it goes (except for overlooking Thomas Reid's remarkable anticipation of non-Euclidean geometry in 1764),[19] at the end of roughly the first quarter of the nineteenth century a German mathematician, Bernhard Riemann, and a Russian mathematician, Lobachevsky, independently discovered two different sorts of "non-Euclidean" geometries. Each of them, moreover, at once concluded that a non-Euclidean geometry (rather than the traditional Euclidean geometry) might well be the one to describe physical space—that is, the space in which all physical objects are located—correctly. The propositions "The sum of the angles in *any* triangle is always *greater* than two right angles" and "Space is finite but unbounded" both hold true in any "Riemannian" space, any space described by Riemann's (original, ungeneralized[20]) non-Euclidean geometry. Indeed, in Riemannian space, if one constructs two straight lines both perpendicular to a third straight line and prolongs them sufficiently, they will eventually meet (there are no parallels in a Riemannian world; any two straight lines meet), and when they meet, they will form a nonzero angle. Thus the kind of triangle Jones described does indeed exist if space is Riemannian.

The extent to which a space deviates from Euclidean space is measured by a quantity called the "curvature" of the space (note that "curved" space does not literally *bend;* the intrinsic curvature of a space might be called its "non-Euclideanness" rather than its "curvature"). The more the sum of the angles of a triangle of a given size is greater (or smaller) than 180°, the greater is the "curvature" of the space. Already in the nineteenth century, models of spaces in which there is "variable curvature" (i.e., the space approximately obeys Euclidean geometry more closely in some places than in others) had been constructed, and W. K. Clifford had even advanced the speculation that physical space might have variable curvature.

19. Thomas Reid, *Inquiry into the Human Mind on the Principles of Common Sense (1764),* Derek R. Brooks, ed. (University Park: Pennsylvania State University Press, 1997). The story of this anticipation is beautifully told in Norman Daniels, *Thomas Reid's Inquiry: The Geometry of Visibles and the Case for Realism* (Stanford, Calif.: Stanford University Press, 1989).

20. Riemann's generalized geometry is a mathematical formalism for representing arbitrary geometries, including ones with "variable curvature," while "Riemannian geometry" *sans phrase* usually refers to the geometry of constant curvature in which there are no parallels (constant positive curvature).

But, some "Wittgensteinian" philosophers might suggest, the fact that some scientists talked this way does not show that they were (fully) making sense.[21] Perhaps these nineteenth-century speculations only made the kind of sense that a science-fiction story makes; the kind of sense that we can indeed enjoy but might nevertheless find to be incoherent if someone were to "take it seriously."

In the twentieth century, however, the idea of applying non-Euclidean geometry to physical space was elaborated into a highly successful physical theory by Einstein in his general theory of relativity. (The main paper was published in 1916.)[22] And in the subsequent decade the scientific community came to accept this theory (with minor modifications). Although there were holdouts against this consensus for a number of years (Whitehead went so far as to propose a rival theory), today the theory is regarded as well confirmed by virtually every competent astrophysicist. This theory implies that the two propositions I imagined Jones uttering in 1700 *may* be true (whether they are depends on the average mass density of the universe, a quantity that has proved difficult to estimate.) That "Jones's" propositions may be true, and that they "make sense," is something that every astrophysicist today believes.

The title of the present section of this paper, I remind you, is "Going On without Rules," and it is time to connect all this to Wittgenstein. Perhaps the following quotation from Stanley Cavell's *Claim of Reason* can serve as a connector:

> This is how, in my illiteracy, I read Thomas Kuhn's *The Structure of Scientific Revolutions:* that only a master of the science can accept a revolutionary change as a natural extension of that science, and that he accepts it, or proposes it, in order to maintain touch with the idea of that science, with its internal canons of comprehensibility and comprehensiveness, as if against the vision that under altered circumstances the normal progress of explanation and exception no longer seems to him to be science.[23]

In accepting the general theory of relativity as "a natural extension" of physics, physicists were treating assertions like "Jones's" as intelligible ways of

---

21. Cf. Hilary Putnam, *The Threefold Cord: Mind, Body, and World* (New York: Columbia University Press, 1999), 98–100, the section titled "On Lacking Full Intelligibility"; and for my present view, see "Wittgenstein: A Reappraisal," Chapter 28 in this volume.

22. Albert Einstein, "Über die Grundlage der allgemeinen Relativitätstheorie," *Annalen der Physik* (August 1916): 769–822.

23. Cavell, *Claim of Reason*, 121.

"going on," indeed, as the right, the justified, ways of going on, given the totality of data-cum-theory to date. Indeed, the possibility of this kind of scientific revolution was already implicit in Einstein's earlier (1905) special theory of relativity. For accepting the special theory involves giving up the idea of an "absolute" simultaneity, as we all know. And what is it to "give up" absolute simultaneity? It is precisely to allow that in certain circumstances there is literally no fact of the matter about whether *A* happened before *B* or *B* happened before *A* or they happened simultaneously (and not just because they happened so close together that our watches are not good enough to distinguish which happened first, as might happen when two horses reach the finish line in a race). Even putting aside uncertainties about the precise second when something happened, there are enormously many cases, if special relativity is correct, when there just is not a *fact* about which happened first or whether they happened simultaneously. And Smith would have no more been able to understand such an assertion—that is, it would not have been an intelligible assertion before Einstein told his story and showed how to apply it—than we would have been able to understand "Space is finite but unbounded" before Riemann and Einstein told their stories and showed us how to apply them.

I remind you that for Wittgenstein a rule *(Regel)* is a subspecies of regularity *(Regelmässigkeit)*. Now it is certainly a *regularity*, a *Regelmässigkeit*, in the behavior of physicists that under our twentieth-century conditions (where the relevant "conditions" include both the data that were collected and the space of available theories to interpret the data), physicists accepted special and general relativity. And it is a regularity that has normative significance; a physicist who deviates from it is regarded as irrational, as at best an unreasonable reactionary. But is it a *rule?*

If Stephen Mulhall is prepared to say that it is, then the disagreement between him and Steven Affeldt may well be in large part a verbal one. I myself think that there is a natural understanding of the notion of a "rule" on which it would be decidedly odd to say that physicists who accepted Einstein's theories and who persuaded other physicists to accept them were "following a rule." The word "rule" suggests something one could state (perhaps after reflection). I recall Paul Ziff protesting[24] against the tendency to postulate "rules" whenever there is a question of right and wrong. "There are right and

---

24. In a wonderful seminar on his forthcoming book *Semantic Analysis* (Ithaca: Cornell University Press, 1960) at Princeton in 1959.

wrong ways to use a screwdriver," he said, "but there isn't a *rule* for using a screwdriver." But is this perhaps just a bit of "ordinary-language philosophy" of the kind we should set aside?

I am inclined to think that we should not set it aside. What happened in this case was that scientists—eventually an overwhelming majority, though at first only a few—discovered that they were in what Cavell calls "attunement." Discovering an attunement is phenomenologically quite unlike being reminded of a rule to be followed. Indeed, Mulhall himself stresses that "rules" in his sense are *independent* of what they justify. But if the regularity "In *such* conditions good scientists will eventually prefer the special and general theories of relativity" *is* a rule, it is so very particular! We can, of course, give it a pseudogenerality by saying that this is an instance of the rule "Choose the simpler theory," or something of that kind, but by what *prior* standard of "simplicity" was it "simpler" to abandon the maxim (which had always been regarded as a priori) that there is a fact of the matter about whether events precede one another in time or are simultaneous (setting aside borderline cases, such as the close finish in a race)? By what *prior* standard of "simplicity" was it "simpler" to abandon Euclidean geometry? By what *prior* standard of simplicity was it "simpler" to think of space as "finite but unbounded"?

Indeed, it is not even quite right to speak of a regularity *(Regelmässigkeit)* here, let alone a rule. To be sure, once large numbers of physicists were won over, there was a "regularity" in their scientific judgments and practice for philosophers and historians of science to observe. But we now regard the first physicists to accept the theory (that is, to accept it as at least a strong candidate for acceptance) as *rational,* indeed, as displaying a high order of scientific insight, and their decisions were not *yet* even instances of a regularity. In this case masters of a science "accept[ed] a revolutionary change as a natural extension of that science," in Cavell's phrase, but not on the basis of anything it seems right to call a *rule* or even a regularity.

I have chosen these cases because they are exemplary for the difference between Cavellian and "orthodox" readings of Wittgenstein. Cases like these illustrate the difference between "going on" in Wittgenstein's sense (or "applying words to the world," in Mulhall's phrase) on the basis of a prior and independent rule and going on without any such basis, but in a way that is fully rational (if revolutionary), a way that would not be comprehensible without our attunement, often unforeseeable, with one another. Note that I do not speak of going on *on the basis* of an attunement. For reasons made clear below, I do not think that our attunements are a *foundation*, or a *basis*, or a *justification*. Rather, they are the *preconditions of intelligibility* of our utterances. Recall Wittgenstein: " 'So you are saying that human agreement decides what is true and what is false?'—It is what human beings say that is true and false; and they agree in the language they use. That is not agreement in opinions but in form of life."[25]

One possible reaction, of course (I hope that it would not be Mulhall's, or Baker and Hacker's), would be to say that what we have here is simply a string of cases in which the *words* (i.e., the phonetic shapes) "triangle," "right angle," "straight line," "plane," "finite," "unbounded" (and perhaps "space"?) are committed to new and different concepts. When we are told that straight lines can behave in these "non-Euclidean" ways, the old grammar is not being contradicted but simply abandoned; in fact, the concept of a straight line has been altered. Perhaps it has been; but not *arbitrarily* altered. For to assimilate these cases to cases in which there is a *mere* change of meaning would be quite wrong. As I pointed out in "It Ain't Necessarily So," what one should ask anyone who took this line is: "Pray, then, which are the straight lines *in the old sense?*" What was literally inconceivable in Jones's and Smith's day was not only that straight lines, properly so called, should not exhibit "Euclidean" behavior; it was also equally inconceivable that *there should be no straight lines in that sense* in space.

Moreover, if we were to insist on regarding scientific revolutions as disguised redefinitions of words, or on saying that whenever we "go on" in a way that forces us to modify or abandon previous criteria, we are really "changing the meaning of words," we would, in fact, have gone back to exactly the Carnapian view that I and others spent our efforts attacking in the 1960s. In that case I would have to say that despite Mulhall's insistence that the "orthodox" view is not a version of logical positivism, it seems to have exactly the same disastrous consequences.

Attunement and Ordinary Language

Another possibility is that someone might say that scientific revolutions are a "special case," and that none of this shows anything much about ordinary language. Besides misunderstanding Wittgenstein's notion of ordinary language

25. Wittgenstein, *Philosophical Investigations*, §242.

(in which "ordinary" contrasts with "philosophical," not with "scientific," "technical," or the like), such a reply would simply be dead wrong. This sort of projection of old concepts into new situations—projection that reveals attunements that have not previously been made manifest—is fundamental to all use of language. Let me begin with an everyday case: the case of jokes.

Here is a (presumably true) story I heard last year. There was, I was told, a professor of philosophy at a Catholic university in the United States who had lost his faith. In these liberal post–Vatican II days he had, however, kept his position at that university. This professor was about to give a paper at a Catholic philosophy conference where all the participants knew of this philosopher's unbelief. When he stood up to read his paper, he smiled and said, "I guess I am the lion being thrown to the Christians."

The amazing thing is that this witticism is instantly intelligible to us (given the background, of course, which includes knowing about gladiatorial games in ancient Rome—this is an instance of what Cavell calls the "systematicity" of our attunements). But this particular metaphor had never been employed by anyone before, as far as we know. The regularity that people understand this metaphor and that they regard it as amusing is extremely strong (I have tested it on a number of occasions). And there are *appropriate* and *inappropriate* ways to understand this joke (which is surely normative). But *rules?* Come on!

I once was talking to Adolf Gruenbaum about his well-known attacks on psychoanalysis, and I said, "I grant you that Freud was mistaken in thinking that psychoanalysis is a *science.* But does that show that it is all just *suggestion,* or *hypnosis,* or something like that, as you maintain? Look, is *philosophy* a science?"

Gruenbaum looked a little crestfallen (to my surprise—this was not at all the reaction I expected) and finally answered slowly, "Yes, it is a shame that we have not yet succeeded in writing down *the canons of rationality.*"

I confess that the idea that we distinguish appropriate from inappropriate metaphors, interpretations of jokes, and the like on the basis of *rules* seems to me much the same fantasy as Gruenbaum's fantasy of a set of "canons of rationality" waiting for us to write them down.

Yet another possibility is that someone will say that jokes and metaphors too are a "special case." (But was not the argument that normativity presupposes the possibility of justification, which in turn presupposes the existence of rules "independent" of what is to be justified, a perfectly *general* argument?)

However, as Charles Travis has brilliantly argued over many years,[26] if anything is central to Wittgenstein's vision of language, it is that the meaning of our words does not determine the precise truth-evaluable content they have in particular contexts. If I say, "That apple is green," even if you know what apple I am talking about, and that it is the color green that I am ascribing, you need also to understand what it would be for that apple to be green in this context. (E.g., am I saying that you should not eat the apple because it is "still green," that is, not ripe? Or that the peel is green? Or that it belongs to a *kind* whose peel is normally green? Each of these in turn permits of different "understandings" in different contexts, and each of those understandings permits of different possible further interpretations, and so on.)[27] Our ability to arrive—often instantaneously—at the proper understanding of what is said in a context is, again, a manifestation of our attunement with one another, not of "rules" (unless it be rules in Chomsky's peculiar sense, a sense that is certainly not Wittgenstein's, and one that I believe to be ultimately incoherent).[28]

Do Our Attunements Have Normative Significance?

I have already quoted Mulhall arguing that "it must be possible for us to justify how we go on" and saying that Wittgenstein tells us that "justification consists in appealing to something independent of what is being justified."[29] I am not claiming that our attunements with one another, the attunements we manifest in speaking in both familiar and novel ways, are a *justification* for what we say or how we say it. The very idea of a *general* problem here, a general question about how *any* of the things we say is justified (or about how we are justified in saying them), whether we are chatting at the dinner table, arguing about the next elections, advising a client, performing an experiment,

26. Cf. Charles Travis, "Annals of Analysis," *Mind* 100, no. 398 (April 1991): 237–263; Travis, "Pragmatics," in Bob Hale and Crispin Wright, eds., *A Companion to the Philosophy of Language* (Oxford: Blackwell, 1997), 87–107; and Travis, *Uses of Sense.*

27. See Travis's works cited in the preceding note. See also the works listed in the appendix to Kent Bach, "Semantic Slack: What Is Said and More," in S. L. Tsohatzidis, ed., *Foundations of Speech-Act Theory: Philosophical and Linguistic Perspectives* (London: Routledge, 1994), 267–291.

28. See my criticism of Chomsky's idea of a "semantic component" of "UG" (universal grammar) in *Threefold Cord*, 123–125.

29. Wittgenstein, *Philosophical Investigations*, §265.

reproving someone for his behavior, or whatever, seems to me one that Wittgenstein would certainly reject as senseless. Just as the things we say make sense (when they do) in particular contexts, so the demand for justification, when *it* makes sense (and very often it does not arise, and it would not make sense if someone were to say, "Justify that claim"), is met in *particular* ways, depending on the particular claim that is called into question. Our attunements enable us to understand "what is going on"; they are not facts that we appeal to in going on.

Missing this is, I think, responsible for much of the oscillation between apriorist and extreme relativist positions in philosophy. In ethics, for example, students often ask, "By what criteria can one tell when an ethical claim is justified?" and are startled when I reply, "By just the considerations that we advance in a good ethical argument." It is as if, over and above the things we say when we argue for or against an ethical judgment, there had to be a *more fundamental* consideration, a *philosophical* consideration, that we ordinarily neglect to give, but one that has to be given lest our ordinary arguments lack what? A foundation? (But Wittgenstein beautifully quips, "I have arrived at the rock bottom of my convictions. And one might almost say that these foundation-walls are carried by the whole house.")[30] Similarly, in the philosophy of mathematics the different positions often seem to be seeking so many different *foundations* for mathematical judgments, for reasons that particular mathematical judgments are true that the mathematician neglects to give, but that have to be given lest mathematics lack a foundation. As the aphorism I just quoted illustrates, Wittgenstein is no foundationalist. Affeldt is right that we do not need a "framework of rules" to serve as a foundation for the ways we go on. (But I am troubled by his concluding remark that "if there is a ground of intelligibility, then I am that ground. But picturing ground as given, I may not be."[31] Perhaps Cavell too sounds at times as if he were saying that we, or each of us individually, were a "ground"; but the metaphor is too easily taken as accepting [and providing an answer] to the question "What is the foundation?")

The Significance for Philosophy

If the view I have been defending is right, then we cannot be convicted of speaking nonsense just by showing that we have used a word in a case where

---

30. Ludwig Wittgenstein, *On Certainty* (Oxford: Blackwell, 1969), §248.
31. Affeldt, "The Ground of Mutuality," 23.

the "criteria" of its ordinary use are not fulfilled. Here is a nice example from *The Claim of Reason:*

> We learn the use of "feed the kitty," "feed the lion," "feed the swan," and one day one of us says, "feed the machine," or "feed his pride," or "feed wire," and we understand, we are not troubled. Of course we could, in most of these cases, use a different word, not attempt to project, or transfer, "feed" from contexts like "feed the monkey" into contexts like "feed the machine." But what should be gained if we did. And what should be lost?
>
> What are our choices? We could use a more general verb like "put," and say merely, "Put the money in the meter," "Put new material into the machine," "Put film into the camera," etc. But first, that merely deprives us of a way of speaking which can discriminate differences which, in some instances, will be of importance; e.g., it does not discriminate between putting a flow of material into a machine and putting a part made of some new material into the construction of the machine. And it would begin to deprive us of the concept we have of the emotions. Is the idea of feeding pride or hope or anxiety any more metaphorical, any less essential to the concept of an emotion, than the idea that pride and hope, etc., grow and moreover grow in certain circumstances? Knowing what sorts of circumstances these are and what the consequences and marks of overfeeding are, is part of knowing what pride is. And what other way is there of knowing? Experiment? But those are the very concepts an experiment would be constructed from.
>
> Secondly, to use a more general verb does not reduce the range of transfer or projection but increases it. For in order that "put" be a relevant candidate for this function, it must be the same verb we use in contexts like "Put the cup on the saucer," "Put your hands over your head," "Put out the cat," "Put on your best armor," "Put on your best manner," "Put out the cat and then put out the light."[32]

What one can add, however, is that if someone uses a word in a case where the criteria for its previously familiar uses are not fulfilled, then if we do *not* automatically project the new use (as Cavell imagines us naturally understanding "feed the machine" without any explicit explanation), we need to be told a "story" about how the word *is* to be understood. I have claimed that Lobachevsky, Riemann, and Einstein told a story that enabled us to understand how and why they said things about straight lines that defied the accepted

---

32. Cavell, *Claim of Reason*, 181.

criteria, and that enabled us to see what they were saying as "a natural exten-sion" of the geometric concepts, in Cavell's phrase. What kind of story will enable us to see a use of a word or concept as a "natural extension" and in what circumstances something for which there are no general rules.

It may seem as if this interpretation deprives Wittgensteinian grammati-cal investigation of all its philosophical power, of its critical bite. If I can-not show that the skeptic is talking nonsense by suggesting that our words may not apply to the world although the criteria for applying them are manifestly fulfilled, or that concepts like "dream" or "illusion" may apply even though the criteria for applying them are not fulfilled, if I cannot show the traditional philosopher that she "violates rules of language" (a lo-cution Wittgenstein never uses, by the way),[33] then *what good is Wittgen-stein's philosophy?*

The answer, I suggest, is that the philosopher's claim to be *justified* in us-ing the words in question outside or apart from their (Wittgensteinian) crite-ria *cannot* be rejected *a priori*. In each case one has to listen to the story the philosopher tells and show why and how it is incoherent. (I have tried to do just that in a Wittgensteinian spirit in some recent writing on skepticism and on the philosophy of mind.)[34] What is true, perhaps, is that once we strip "Wittgensteinianism" of the appearance of being a machine for refuting tra-ditional philosophy, then it may turn out to be much more continuous with philosophy as Socrates practiced it than it is customary to think.

33. James Conant has told me that as far as he can determine, Wittgenstein speaks of vio-lating rules of language only in "Some Remarks on Logical Form"—the one piece of writing he explicitly later disowned. "Some Remarks on Logical Form" was published in the *Proceed-ings of the Aristotelian Society Supplementary* 9 (1929): 162–171; it was reprinted in I. M. Copi and R. W. Beard, eds., *Essays on Wittgenstein's "Tractatus"* (London: Routledge and Kegan Paul, 1966), 31–37. Elizabeth Anscombe added a note there saying, "Wittgenstein disowned the following essay. . . . I have consented to the reprint of the essay because I suppose that it will certainly be reprinted some time, and if that is to happen there had better be a statement indicating how little value can be set upon it as information about W's ideas." Juliet Floyd informs me that the current most accessible place to find the essay is in K. Klagge and A. Nordmann, eds., *Philosophical Occasions* (Indianapolis: Hackett, 1993), 36–44 (see further citations of relevant correspondence there).

34. Cf. Hilary Putnam, "Skepticism," in Marcelo Stamm, ed., *Philosophie in synthetischer Absicht* (Stuttgart: Klett-Cotta, 1998), reprinted here with the title "Skepticism and Occasion-Sensitive Semantics" as Chapter 30; and Putnam, "Strawson and Skepticism," in *The Philoso-phy of P. F. Strawson*, Lewis E. Hahn, ed., (La Salle, Ill.: Open Court/Library of Living Phi-losophers, 1998), reprinted here as Chapter 31.

25

Wittgenstein, Realism, and Mathematics

Previously[1] I have argued that *Philosophical Investigations* and most of Wittgenstein's *Remarks on the Foundations of Mathematics* do not represent a generally antirealist philosophy of any kind.[2] That does not mean that there are no problems with Wittgenstein's remarks, and we shall soon see that at least one of them does flirt with a mathematical form of verificationism. In particular, it does seem that in 1944 (when he wrote the remark that even God's *omniscience* cannot decide whether people would have reached "777" in the decimal expansion of π "after the end of the world" if they have not reached it before that[3]) that there was a moment at which Wittgenstein thought that a mathematical proposition cannot be true unless we can *decide* that it is true on the basis of a proof or calculation of some kind. One might well ask, "How can such a view *not* spring from antirealism?"

---

1. Hilary Putnam, "On Wittgenstein's Philosophy of Mathematics," *Proceedings of the Aristotelian* Society, suppl. 70 (1996): 243–264.

2. Ludwig Wittgenstein, *Philosophical Investigations* (Oxford: Blackwell, 1953); and Wittgenstein, *Remarks on the Foundations of Mathematics,* ed. G. H. von Wright, Rush Rhees, and G. E. M. Anscombe (Oxford: Blackwell, 1978) (referred to in the main text as *PI* and *RFM*).

3. Wittgenstein, *Remarks on the Foundations of Mathematics,* V, §34. I quote the paragraphs in question in full later in this essay.

What I shall argue, in brief, is that the mistake in the remark in question (*RFM*, V, §34) represents a combination of genuine insight with an inadequate knowledge of actual mathematical practice and especially of sciences that depend on mathematical practice, in particular, mathematical physics.[4]

Indeed, a very striking fact about *RFM* from the point of view of a mathematical physicist or a mathematician is that in spite of the *philosophical* importance that Wittgenstein himself attaches to the application of mathematics *outside* mathematics, his examples of such applications are remarkably trivial. Apart from applications of arithmetic to results of counting and a few scattered examples of applications of geometry, there are very *few* examples of applications in *RFM*. In particular, as I shall shortly explain, crucial difficulties for the view that mathematical truth cannot transcend provability can be discovered from an examination of the ways in which mathematics is used in mathematical physics. But apart from a couple of examples from engineering,[5] the only example that might even be regarded as an application in mathematical physics in the whole of *RFM* that I was able to find is a trivial application to the orbit of a comet.[6]

In a way, of course, this is not surprising. Wittgenstein was trained as an engineer; and, indeed, one gets the feeling in all his writing on the philosophy of mathematics that he imagines that the application of mathematics to empirical material either consists in reading off geometric relations from pictures or comparing the results of calculations with the results of counting and measurement. These are exactly the ways in which an engineer applies mathematics. I do not mean to suggest that Wittgenstein did not know that there

---

4. One reason—by no means the only one—that I say that Wittgenstein's examples indicate an inadequate knowledge of mathematical practice is the assumption throughout that what I called "quasi-empirical methods" in Hilary Putnam, *Philosophical Papers*, vol. 1, *Mathematics, Matter and Method* (Cambridge: Cambridge University Press, 1975), chap. 4, play no role. The fact is that even *before* computers succeeded in showing that "777" and "7777" occur in the decimal expansion of π, every first-class number theorist was sure that they did on "quasi-empirical" grounds.

5. Examples from engineering in Wittgenstein, *Remarks on the Foundations of Mathematics*, III, §49; V, §51.

6. Wittgenstein, *Remarks on the Foundations of Mathematics*, IV, §23, refers to the proposition that a comet describes a parabola. I do not count the examples of the attractive force exerted on something by "an endless row of marbles of such and such a kind" in IV, §8, and, in the same section, the weight of a pillar composed of "as many slabs as there are cardinal numbers" as applications in *physics*, but rather as bits of pure mathematics.

are much more *complicated* uses of mathematics than these in mathematical physics. What I think likely is that he had no idea of the detailed nature of those applications, and that he assumed that although they might indeed be more complicated than the trivial ones that he used as examples, that is, indeed, all they are—more "complicated"; that is, nothing of philosophical interest could be lost by confining attention to the few and trivial applications that he did discuss. If so, I believe that he made an error, but an error that did not flow from a metaphysical position.

I want to emphasize, however, that Wittgenstein's remarks about mathematics involve numerous insights. I shall begin by describing just two of these insights.

Two Wittgensteinian Insights

1. Mathematical propositions would not be *propositions (Sätze),* that is, meaningful statements, if mathematics were not applied *outside* mathematics. (Note that this does not mean that every single mathematical proposition need have such applications to be meaningful.)

If mathematics were not applied outside mathematics, that is, if there were not "mixed" statements—empirical statements that contain mathematical terms—then there would be no reason to view it as more than a game (one in which, for example, we are allowed to write certain marks on paper when the marks are "axioms" or when other players have written certain other marks on paper—or in a "solitaire" version of the game, when we ourselves have written certain other marks on paper). The question of the "truth" of anything that is produced in the course of the game would be as silly as the question whether a move in a chess game is "true." Insight 1 is, of course, one that Wittgenstein has in common with Frege and Russell, who stressed the importance of the fact that numbers can be used to *count* things, and for that reason number words occur constantly in *empirical* statements.[7]

---

7. The logicists had other insights as well: they stressed the importance of the fact that one can count "abstract entities," e.g., numbers and equations, as well as emperors and cabbages, and also stressed that the things counted need not be adjacent in space or time or even exist in the lifetime of the person doing the counting; hence counting and forming sums of things counted are not physical operations. Wittgenstein's attacks on Russell's *Principia* in *Remarks* are attacks on the significance Russell attached to symbolic logic, not on *these* insights. One thing that I like about *Remarks on the Foundations of Mathematics* is that Wittgenstein

The fact that Wittgenstein shares this insight with the great logicists is the reason that it is a mistake to claim (as Michael Dummett does) that Wittgenstein believed that understanding mathematical propositions is just a matter of understanding the *proof procedures* by which we verify them. Indeed, at one point Wittgenstein suggests that a mathematical "proposition" might have a proof but no real meaning, precisely because we would have no idea how to *apply* it.[8]

2. Commonsense realism with respect to "rule following"—which, I argued in Chapter 23, Wittgenstein defends—does not, in and of itself, commit us to views about infinity (i.e., about the sort of problem that motivates Kripke's discussion).[9]

Kripke's celebrated "skeptical problem"[10] is initially formulated as the question, how can our understanding of a rule determine what is true in infinitely many cases, and so in more cases than it is physically possible for human beings to "get to"? But Wittgenstein is quite clear that when someone (say, a child) learns to follow the rule "add two," he learns to follow a practice that it is possible for human beings to engage in; but such a practice is not one that extends to infinitely many cases, because human beings cannot—not really—add two to infinitely many numbers.

Consider, for example, what Wittgenstein says about a child's understanding of infinity:[11]

---

treats empiricist views with a certain respect. It is true that, like Frege, he finds such views in the end completely inadequate, but he is willing to let us see the *appeal* of such views. He does not treat an empiricist view as simply a *dumb* view, as Frege does. On the contrary, empiricist views (and also finitist views) are treated as views to which one is naturally led by a desire for clarity, although Wittgenstein does agree with Frege that coming to see their inadequacy is an essential step to any further progress with the questions.

8. Wittgenstein, *Remarks on the Foundations of Mathematics,* IV, §25. I find this remark wrong, by the way, because it seems to me to forget something Wittgenstein himself elsewhere points out, which is that mathematical propositions also have applications *within* mathematics. Wittgenstein seems to connect having an application with having a *constructive* proof here, but this is unjustified unless one thinks of a very limited sort of application. Nevertheless, the fact that Wittgenstein made the remark shows that much more than the mastery of proof procedures is involved in the understanding of mathematical assertions, in his view.

9. I realize that it is an anachronism to refer to Kripke in explaining one of Wittgenstein's insights, but the anachronism may help a present-day reader.

10. Saul Kripke, *Wittgenstein on Rules and Private Language* (Cambridge, Mass.: Harvard University Press, 1982).

11. Wittgenstein, *Remarks on the Foundations of Mathematics,* IV, §14.

Suppose children are taught that the earth is an infinite flat surface; or that God created an infinite number of stars; or that a star keeps on moving uniformly in a straight line, without ever stopping.

Queer, when one takes something of this kind as a matter of course, as it were in one's stride, it loses its whole paradoxical aspect. It is as if I were to be told: don't worry, this series, or movement, goes on without ever stopping. We are as it were excused the labor of thinking of an end.

"We won't bother about an end."

It might also be said: "for us the series is infinite."

"We won't worry about an end to this series; for us it is always beyond our ken."

It is significant that Wittgenstein says, "Suppose *children* are taught"; that is, this last quotation is not a comment on how a sophisticated mathematician might understand these statements, but about how a child might understand the statement that a series (e.g., the series 1, 2, 3 . . .) is infinite, or that it "has no end."[12]

The point Wittgenstein makes here can be put this way: that when we speak of a human being as being able to follow a rule, in the ordinary applications of that concept we are not required to ascribe to the person (in this case the child) the mathematical notion of infinity, or the notion that the rule *determines* what it is correct to say in cases that it is not actually possible to get to.

Kripke, on the other hand, runs together two different questions: the question how our understanding of a rule determines what we count as correct in actual human practice, and the question of *what the mathematical consequences of the rule are.* In posing his "skeptical problem," Kripke takes it as evident that an *arbitrary* sum is correct if and only if the answer is *determined by the rule;* but this is a way of thinking that Wittgenstein criticizes from the very beginning of *RFM.* As Wittgenstein makes clear,[13] to criticize this is not to say that one cannot introduce a *mathematical* sense of "determine" according to which, if we take the "rule" of addition to be, for example, the definition of addition by primitive recursion, we can say that "every correct sum is

---

12. Speaking for myself, I do not think that the mathematical notion of the infinite is always completely absent in the child's grasp that "you can always go on counting." Wittgenstein's discomfort with mathematical talk about what happens in infinite sequences may be influencing his perceptions here.

13. See Wittgenstein, *Philosophical Investigations,* §189, for the distinction between these two uses of "determine."

determined by the rule." But to say this is to make a *mathematical* comment about addition; it is not to speak to any of the philosophical problems about mathematics, because just the notion that puzzled us, the notion of mathematical correctness, has been simply taken for granted.

Thus we might respond to Kripke as follows: "Professor Kripke, we propose to use the expression '*A* is determined as the correct answer by the rule *R*' only when it is possible for human beings to *actually* calculate *A* by using the rule *R;* and in all other cases we will use the different expression '*A* is a mathematical consequence of the rule *R.*' Now, one of the problems you raised, namely, how does a rule that we grasp manage to determine what is correct in infinitely many cases? simply does not arise; for if 'determine' means what we have just proposed that it ought to mean, then a rule does *not* determine what it is correct to say in infinitely many cases (although it is 'always beyond our ken' just where we will cease to be able to actually apply it, and hence where it ceases to determine an answer). And if you reformulate your puzzle by saying 'How can a finite rule have infinitely many different mathematical consequences?' then that seems to be a mathematical question, and one whose answer is trivial." I do not mean to say, of course, that this response completely defuses Kripke's worries—those worries have complex sources—but at least one strain in Kripke's complex argument does seem to depend on the conflation of these two senses of "determine."

A further remark in this connection: if one accepts what I described as Wittgenstein's defense of our commonsense realism about rule following, then one could be led to suppose that the notion of being "provable" is an unproblematic notion, because to be provable in a formal system, we might say, is just to be a sentence that is obtainable as the last line of a longer or shorter proof (or as the bottom of a larger or smaller proof tree) by following certain *rules*. But we can now see that that is too simple a response. In particular, we can object that Wittgenstein's commonsense realism about rule following cannot give us the *mathematical* notion of "provability" (if we take the notion of "following a rule" as referring to something that it is *actually possible* for humans to do, just as we took the notion of being "determined by a rule" to refer to something that it is actually possible for human beings to find out); at best it can give us the ordinary empirical notion of "provability." (Of course, in daily life we sometimes use the word "provable" in one way [corresponding to the notion of being "determined by the rules," in the terminology just introduced] and sometimes in the other [corresponding to the notion of being a mathematical consequence of the rules], which adds to the confusion.) When I said earlier that Wittgenstein may have thought that a mathematical proposition cannot be true unless we can decide that it is true on the basis of a proof or calculation of some kind, I meant the modal word "can" to refer to what is "humanly" possible; for to say that a mathematical proposition cannot be true unless it is "decidable" in the mathematical sense—the sense in which it might be decidable even though the shortest proof or disproof was longer than the number of elementary particles in the universe—would be open to the objection that I just made against Saul Kripke, the objection that to understand *that* notion of decidability (or of provability or of disprovability) requires being able to understand the sort of mathematical notion Wittgenstein wishes to investigate.

Dummett's Interpretation of Wittgenstein on Truth and Provability in Mathematics

Michael Dummett has claimed that Wittgenstein held the view that "being actually proved" is a necessary condition for mathematical *truth*.[14] But even though here and there in Wittgenstein's *Nachlass* one can turn up a note that shows that Wittgenstein "played with the idea" at certain moments in his life, I do not find anything in *RFM* that should be construed as *committing* Wittgenstein to such a radical view. For example, when we read the passage written in 1944 that I mentioned at the beginning of this lecture, let us pay close attention to what Wittgenstein says:[15]

> Suppose that people go on and on calculating the expansion of π. So God, who knows everything, knows whether they will have reached "777" by the end of the world. But can his *omniscience* decide whether they would have reached it after the end of the world? It cannot. I want to say: Even God can determine something mathematical only by mathematics. Even for him the rule of expansion cannot decide anything that it does not decide for us.[16]

---

14. See Michael Dummett, "Wittgenstein on Necessity: Some Reflections," in Peter Clark and Bob Hale, eds., *Reading Putnam* (Oxford: Blackwell, 1994), 49–65.

15. Wittgenstein, *Remarks on the Foundations of Mathematics*, V, §34.

16. Ibid. It is important to contrast what Wittgenstein published on this question—§516 *of Philosophical Investigations*—with this unpublished paragraph in *Remarks on the Foundations*

Of course, human beings cannot possibly calculate at all "after the end of the world"—especially if, as is reasonable to suppose, the "end of the world" means the end of space and time.[17] Thus the most reasonable interpretation is that Wittgenstein meant us to suppose that human beings go on calculating to the very limit of what is possible for such beings. That is, Wittgenstein is not here—contrary to Dummett's interpretation—saying that not even God could know what human beings would count as a correct calculation prior to their actually accepting it as correct (which would involve attributing a strongly "antirealist" attitude toward counterfactuals to Wittgenstein, on no evidence that I can see); he is saying—as he goes on to make clear—that not even God can decide whether the pattern does or does not occur in the expansion of π except by a calculation that is *actually*—not just "mathematically"—possible.

But even if Dummett's interpretation is wrong, it still remains to say how, given what I have been calling Wittgenstein's "commonsense realism about rule following," a commonsense realism that, we have just seen, does not in and of itself involve one of the sorts of problems that Kripke raises, because it does not require a notion of infinity beyond the notion Wittgenstein says a child might have ("For us the end is always out of reach"), and given a robust insistence on the fact that mathematical propositions are statements with sense only because mathematical concepts have applications in the realm of the *nonmathematical*—given that these are both genuine insights *and* genuinely commonsensical—Wittgenstein can have been led to flirt with the view that provability (in the ordinary empirical sense) is a necessary condition for mathematical truth.

My answer is that if one makes the mistake of supposing that the sorts of examples of the applications of mathematics outside mathematics that Wittgenstein uses *exhaust* the philosophically relevant sorts of examples there are, then the position I have ascribed to Wittgenstein can appear quite attractive,

---

of Mathematics. Philosophical Investigations, §516, reads: "It seems clear that we understand the meaning of the question: 'Does the sequence 7777 occur in the development of π?' It is an English sentence; it can be shown what it means for 415 to occur in the development of π; and similar things. Well, our understanding of that question reaches just so far, one may say, as such explanations reach." There is nothing here about what God can or cannot determine!

17. The passage could, of course, be read in another way: there is a Last Judgment in time, and that is "the end of the world," and there is still more time available for calculation after the Last Judgment—but then why should not human beings go on calculating in eternity?

for there is nothing in those sorts of applications that would *require* us to suppose that humanly[18] unprovable mathematical propositions can have a truth-value. The notion that a humanly unprovable mathematical proposition can have a truth-value can then appear (as I believe it did appear to Wittgenstein when he wrote this paragraph) to be just a piece of metaphysical fabulation. The situation is quite different, however, when we consider a very different sort of example, one taken from serious mathematical physics.

Where Wittgenstein Went Wrong

When we make the statement that a physical system obeys a certain equation (this is an example of what I earlier called a "mixed" statement, an empirical statement that contains mathematical terms), for example, when we say that the state vector of a physical system obeys the Dirac equation, or, in Newtonian physics, when we say that gravitational forces obey the Newtonian law of gravitation, or even when we say that a certain phenomenon obeys the wave equation, what is the situation? As long as we accept the correctness of Newton's law of gravity, we are committed to the statement that the evolution of an N-body system will be in accordance with the solutions to the appropriate system of differential equations. In general, any application of mathematical physics to a physical system involves treating the system as behaving in accordance with some equations—usually differential equations—or others. If a physicist believes that the equations are correct, and furthermore that they describe the behavior of the system at each time point, then she is committed to the claim that those equations have *solutions* (in real numbers, or, in certain cases, in complex numbers) for each real value of the time parameter t.

Now, suppose that the solution to the equations, in a particular physically given case, for a particular rational value of t is in a certain rational interval, say, between r_1 and r_2, where r_1 and r_2 are rational numbers. That is, suppose that the statement "The solution to such and such equations for the given

18. The fact that Wittgenstein is willing to consider the thought experiment of imagining that humans go on and on calculating until "the end of the world" may also indicate that he is willing to allow us to idealize human abilities to calculate to the uttermost limits of physical possibility. Thus the relevant line here is not between what *humans* can calculate and what, say, Martians might be able to calculate, but between what it is within physical possibility to calculate and mere "mathematical possibility" of calculation. If so, this would go even more strongly against Dummett's reading.

value of t is in the interval between r_1 and r_2," is true.[19] If truth is the same as provability (in the "ordinary empirical sense" of provability), then what the physicist is committed to, if she believes the view that Wittgenstein flirted with in *RFM*, V, §34, is that it is possible to calculate the solution to the equations in question for the specified value of t, and that the solution will be found to lie in the interval between r_1 and r_2. But, given present knowledge, this is something a sensible physicist better not commit herself to.

Because this point is at the heart of my argument, let me state it again for clarity: if truth is coextensive (in the mathematical case) with the empirical possibility of being proved by a calculation or a proof from axioms, then no mathematical statement can be true but not demonstrable by calculation or proof. Not even "God's omniscience" can know the truth-value of such a humanly undecidable statement, which is to say that it does not have a truth-value. If the statement is that the equations of motion of a system S have a solution, say, "$P(t)$ is in the interval between 3.2598 and 3.2599," where P is some physical parameter, then if it is not physically possible for human beings to compute $P(t)$, or to prove by some deduction from acceptable axioms that $P(t)$ is or is not in the interval between 3.2598 and 3.2599, then there is no fact of the matter about whether $P(t)$ is in that interval or not. But to accept this is precisely to be a verificationist in one's physics. It is to give up a claim that is part of our best physical theory of the world, the claim that the equations of that theory describe the behavior of certain systems accurately and completely, in the sense that those equations have *solutions* for each real value of the time parameter t, and those solutions give the value of the physical parameter P in question even if it is not feasible to verify that they do in certain cases. Systems of equations are, on a verificationist view, just prediction devices, and when it is not feasible to derive a prediction from them (even if we are allowed to go on calculating until "the end of the world"), then there is nothing that they *say* about the case in question.

But why do I say that the physicist "better not commit herself to" the claim that it is physically possible to calculate the solution to the differential equations of physics for arbitrary specified values of t? One important reason has to do with *chaos* (a phenomenon represented mathematically by certain

---

19. I have confined attention to rational values of t, r_1, and r_2 so as to avoid the objection that not all real numbers have names in the language (a remark for the mathematically sophisticated).

kinds of differential equations): when a phenomenon is sufficiently "chaotic," it may be "empirically" (though not "mathematically") impossible to actually calculate the time evolution of certain parameters *even though we know the equations they obey.* An additional (possible) reason has to do with nonrecursiveness: it is to this day quite unknown whether the solutions to, for example, the Newtonian gravitational equations are recursively calculable even when $N=3$.[20] In addition, it is known that there are cases in which it has been proved that the solutions to the wave equation are not recursively calculable, even given recursive initial data.[21] Indeed, we do not even know whether the values of physical magnitudes at specified future times are, in general, effectively calculable to even *one* decimal place when those magnitudes obey these equations.

In sum, short of being a verificationist about physics, one cannot consistently sustain the identification of mathematical correctness with provability and with calculation that seems to be asserted in *RFM,* V, §34.

How might what Wittgenstein wrote in *RFM,* V, §34, be defended against this criticism?

As far as I can see, there are only three ways in which a defense of *RFM,* V, §34, that one could take seriously might go. First, one might look for a different interpretation—one according to which when Wittgenstein asked, "But can his [God's] *omniscience* decide whether they *would* have reached it after the end of the world?" and answered "It cannot," he was not asserting that there is no fact of the matter about whether 777 occurs in the decimal expansion of π unless human beings are able to show that it occurs "before the end of the world" (or, presumably, to show that it does not).[22] I anticipate that some will suggest such possible interpretations. But I myself find the fact that Wittgenstein says not just that God could not decide this, but that his *omniscience* could not, makes it almost crystal clear that V, §34, means that the

---

20. At Gabriel Stolzenberg's suggestion, I emphasize that this is connected with the (possible) undecidability of the question whether the bodies will or will not collide.

21. Marian Boykan Pour-El and Ian Richards, "The Wave Equation with Computable Initial Data Such That Its Unique Solution Is Not Computable," *Advances in Mathematics* 39 (1981): 215–239.

22. As remarked in an earlier note, computers have shown that 777 *does* occur in the decimal expansion of π. (So does 7777, Wittgenstein's example in *Philosophical Investigations,* §516.)

statement that 777 occurs in the expansion is neither true nor false in the envisaged circumstance.

If we assume for the sake of the argument that my interpretation of V, §34, is the right one, the other two ways of defending it against my criticism that I can envisage are the following:

1. One can point out (correctly) that the physical theories we have today and are likely to have in the future are *idealizations* and not literally correct descriptions of the physical universe. One might argue that this means that the entire problem I have raised does not really arise.

2. One can bite the bullet and argue that one should be a "verificationist" in physics, at least to the extent of agreeing that what physical theories say about physical reality does not go beyond what it is (physically) possible for human beings to calculate on the basis of those theories.

In the next two sections I shall consider these two defenses in turn.

Physics Is Idealized. Does That Make a Difference to the Argument?

It is quite true that our most fundamental physical theories—quantum mechanics and general relativity—cannot be regarded as perfectly correct as they stand. For one thing, neither satisfactorily incorporates the other. ("String theory" might alter this situation if some version of it succeeded. However, the "measurement problem" of quantum mechanics would still be unsolved—in my opinion, at least—and it may well require a deeper theory than any we have today to resolve it.) For this reason, we can regard these "fundamental theories" only as idealizations, or approximately correct descriptions of the evolution of physical systems in time. Does this undermine the argument for interpreting the equations of those theories realistically?

I cannot see that it does. In a host of ordinary situations (situations in which the curvature of space-time plays a part only in a way that is well modeled by existing theories of "quantum gravity," such as the familiar terrestrial situations in which classical mechanics works quite well), the correctness of quantum mechanics has been confirmed to many decimal places. On the other hand, the phenomena I pointed to above—phenomena such as "chaos" and the possible nonrecursivities in the solutions to certain differential equations—prevent us in many cases from calculating the values of certain parameters to even the *first* decimal place. Thus even if we replace the claim that the wave equation or the gravitational equation or the equations that govern a complicated "chaotic" phenomenon are *precisely* correct (a claim that is certainly not true) with the claim that in such and such a situation the equations yield the right answer to, say, four or five decimal places (a claim we believe to be true), the same problem arises. On the one hand, our best theory of the world includes the claim that these equations do have solutions, and those solutions yield the values of the parameters, whether it is feasible to calculate those solutions or not. On the other hand, *RFM*, V, §34, as I am interpreting it, says that if it is not feasible to compute a solution, even if we are allowed to go on computing to the "end of the world," then there is not one. The statement that any sufficiently small rational interval contains the solution is neither true nor false if it is not feasible to determine that it does, on the view expressed by this paragraph. But then the equations of physics are not what we take them to be—a nomological account of how physical systems evolve in time. It is not, for example, that it is too hard (because of chaos or other factors) to determine how a complicated system evolves (if *RFM*, V, §34, is right): the equations do not predict how it evolves. What they predict depends on our human powers of computation.

If this had been Wittgenstein's considered view, as opposed to an isolated remark, then his claim that he offered no "theses" in philosophy would have been deceptive. But the claim about what *omniscience* cannot decide in *RFM*, V, §34, is, fortunately, one that Wittgenstein does not repeat when he touches on the same question in *Philosophical Investigations*, §516.

Should We Bite the Bullet?

Should we, however, say that one *should* be a "verificationist" in physics, at least to the extent of agreeing that what physical theories say about physical reality is exhaustively determined by what it is (physically) possible to calculate on the basis of those theories? Can one coherently be this sort of verificationist without being a verifications *tout court?*

Ian Hacking, for example, has proposed that we can and should be "realists" about the existence of positrons but not about the theories about them. In brief, Hacking claims that being a realist about the entities physical theories postulate does not have to mean being a realist about the theories. But this attempt to disjoin realism about physical entities and realism about theories does not work—it fails for a very Wittgensteinian reason. The notion of a "positron" (positrons were Hacking's example[23]) depends on the language game to which positron talk belongs, and it is the quantum theory that structures that very strange game. We speak of "spraying" positrons in a certain experiment Hacking describes, for example. But we may also (depending on the experiment) speak of "spraying a *superposition* of three positrons and five positrons"—which does not mean that "we sprayed three positrons and we sprayed five positrons on top of them" (that would be spraying *eight* positrons, something that the statement that we sprayed a superposition of three positrons and five positrons rules out), and which does not mean that "we sprayed three positrons *or* we sprayed five positrons," and which does not mean anything else that a classical physicist could understand.[24] Positron talk has an entirely novel grammar, and that grammar is *provided by the theory*. When Hacking says that he is "a realist about positrons," he doubtless means that he believes that "positrons exist." But that statement means literally nothing apart from a specification of what concepts do and what concepts do not apply to positrons. If, for example, Hacking thinks that positrons exist in the sense of being countable entities with a position in space at each time, then he is not an antirealist about quantum theory at all; he simply thinks that quantum theory is *false,* and that positrons are classical particles. But I am sure that that *is not* what he thinks. The fact is that to say that one is a realist about posi-

23. Ian Hacking, *Representing and Intervening* (Cambridge: Cambridge University Press, 1983), 23.
24. This response to Hacking was written in 2002, before the revision of the different interpretations of quantum mechanics I described in "A Philosopher Looks at Quantum Mechanics (Again)" (Chapter 6 of the present volume) and "Quantum Mechanics and Ontology" (Chapter 7 of the present volume); it reflects what most physicists probably believed at the time. Today [2011] I would have to say that the answer to the question "Is there a fact of the matter about the 'number of positrons' in an arbitrary quantum-mechanical state?" is one we do not now *know*. If the Bohm interpretation turns out to be right, for example, the answer is "yes." But if the Ghirardi-Rimini-Weber interpretation turns out to be right, the situation is still interpretation dependent; if the "flash" interpretation of Ghirardi-Rimini-Weber suggested by Bell and developed by Roderich Tumulka (this is described in Tim Maudlin, *Quantum Non-locality and Relativity: Metaphysical Intimations of Modern Physics*, 3rd ed. [Oxford: Wiley-Blackwell, 2011]) turns out to be right, then individual positrons rarely exist as actual entities, even in the situations that so impressed Hacking; if the nonrelativistic "mass-density" ontology suggested by Ghirardi turns out to be right, then positrons exist, but they are smeared out. And there are other possibilities, including ones connected with the phenomenon of "duality" referred to in Chapter 2 of this volume.

trons but not about quantum theory or any other physical theory would be to say *nothing*. One can no more take the existence of positrons seriously without taking the conceptual apparatus that goes with positron talk seriously than one can take the notion of *baldness* seriously while rejecting the concept of *hair*.

In sum, being a verificationist about the theory precludes any substantial sense of being a realist about the entities. This can be seen from the sort of example I used to put pressure on Wittgenstein's remark in *RFM*, V, §34. Suppose that the temperature of the gas in a certain small region in the sun cannot be measured directly or indirectly. To say that in that case the notion of the temperature of the gas in the region is meaningless would be classical verificationism. Now suppose that although the temperature of the gas cannot be measured directly, there is an equation whose solution is the temperature in question. Saying that whether the notion of the temperature of the gas in the region is meaningful or not depends on whether a certain calculation (which is mathematically well defined, we will suppose) is *feasible* or not would be just as verificationist, would it not?

Alternatively, suppose that we say (assuming that there is nothing relevantly wrong with our theory) that the notion of the temperature of the gas in region X is a perfectly meaningful one, and that, as commonsense realists with respect to temperature, we of course believe that there is a fact of the matter about whether the temperature of X is between A and B or between C and D (where $A, B, C,$ and D are possible temperatures for such a region). But why would we think that the statement that, say, the temperature is between A and B ($A < t(X) < B$) has a truth-value even if people could not find it out if we reject the idea that the purely mathematical statement "777 occurs in the decimal expansion of π" has a truth-value even if people could not find it? Both statements, after all, employ mathematical notions. Such a view does not seem coherent.

Well, why should not one bite the bullet and be a verificationist all the way? Enough people have argued that this is what Wittgenstein was, after all. I will not repeat the arguments against verificationism, which are by now familiar to everyone.[25] But I will say a word about my reasons for not reading Wittgenstein as a verificationist.[26]

---

25. My own reasons for giving up the liberal form of verificationism that I defended in the early 1980s are given in Hilary Putnam, "Pragmatism," *Proceedings of the Aristotelian Society* 95, no. 3 (1995): 291–306.

26. In a sense, all of Cora Diamond's essays in *The Realistic Spirit* (Cambridge, Mass.: MIT Press, 1991) can be regarded as a nonverificationist interpretation of Wittgenstein. She

On Not Reading Wittgenstein as a Verificationist

A strong reason for not reading the later philosophy of Wittgenstein as a form of verificationism has already been mentioned in passing: Wittgenstein is insistent that his aim is not to defend any "theses" in philosophy, but simply to teach us to expose disguised nonsense.[27] It is hard to see how any of the different forms of verificationism can be regarded as anything other than a philosophical thesis. Moreover, the philosophers who more than any other have defended a verificationist interpretation of Wittgenstein—Michael Dummett and Crispin Wright—are avowed critics of what Wright calls Wittgenstein's "quietism" and what Dummett has referred to in conversation as Wittgenstein's "opposition to theory." The fact that these verificationist philosophers, brilliant as they are, are so out of sympathy with what was so obviously at the center of the later Wittgenstein's philosophical stance should lead us to view with suspicion their attempts to foist a "theory" on such works as *Philosophical Investigations* and *Remarks on the Foundations of Mathematics*.

Doubtless one reason that Dummett and Wright wish to read verificationism into the later philosophy is that they believe that their own (different) versions of verificationism are right, and so, in accordance with the principle of charity, they naturally try to find passages in Wittgenstein that can be read to accord with those versions, and to dismiss passages that disclaim the ambition to propound philosophical theses as places where Wittgenstein was "weak" or confused about the nature of his own best contribution. But for those of us who, like myself, think that verificationism is wrong, the principle of charity works the other way: we think that if Wittgenstein does not have to be read as a (crypto)verificationist, then he *should not* be. And, in fact, almost all the famous passages that were at one time read in a verificationist way (or a "skeptical" way, in Kripke's case) have been given what seem to me much better readings of a very different sort. I myself have interpreted the "rule-following discussion" in *Philosophical Investigations* in a way that makes

---

has tackled this issue explicitly in "How Old Are These Bones? Putnam, Wittgenstein and Verification," *Proceedings of the Aristotelian Society,* suppl. 73 (1999): 99–134.

27. This was written in 2002. For my present [2011] critical attitude toward Wittgenstein's claim that the various positions he attacks are literally "nonsense," see Chapter 28. In addition, I would say that although Wittgenstein is right that there cannot be nonmathematical explanations of the truth of statements of pure mathematics, there can be better and worse philosophical reasons *for and against the view* that statements of pure mathematics have bivalent truth-values; offering such reasons is not talking nonsense.

clear that if the discussion of rule following expresses skepticism at all, the skepticism is directed at *philosophical accounts* of rule following, and not at rule following itself. What readers like Kripke have done, I claimed, is to take Wittgenstein to oppose not only metaphysical realism about rule following but also our commonsense realism about rule following, when what Wittgenstein actually doubts is the need for and the possibility of an account of rule following over and above the commonsense account.[28] Stanley Cavell and others have long since demolished the idea that Wittgenstein endorsed some version of "behaviorism" in *Philosophical Investigations*.[29] More and more, it seems that it is only passages in *Remarks on the Foundations of Mathematics* that are being used to support the ascription of verificationism to Wittgenstein.

But here too, at least on a second reading, there is precious little that *demands* a verificationist reading. The comparison of certain mathematical statements to *conventions,* for example, is meant as an analogy that highlights certain ways those statements are used, not as an *explanation* of their truth, much less an attempt to explain "the nature of mathematical truth," as Yemima Ben-Menahem has shown.[30] The "notorious paragraph" on the Gödel theorem,[31] which for a long time seemed to me conclusive evidence that Wittgenstein identified mathematical truth with proof, admits of a much more interesting (and better) interpretation, as Juliet Floyd and I have discovered.[32] Indeed, *RFM,* Part V, §34, seems to be the only place where, at least for a moment, Wittgenstein expresses what I have called a "verificationist" attitude (toward a possible mathematical truth, not physics, however).[33] Should we conclude, on such slight evidence, that this was Wittgenstein's considered opinion?

It seems to me that we should not. To see how to account otherwise for *RFM,* V §34, we need, I think, to do two things: we need to look at the paragraph in question again more carefully, and we need to look carefully at its

28. See Putnam, "On Wittgenstein's Philosophy of Mathematics."

29. Stanley Cavell, *The Claim of Reason* (Oxford: Clarendon Press, 1979).

30. Yemima Ben-Menahem, *Conventionalism* (Cambridge: Cambridge University Press, 2006).

31. Wittgenstein, *Remarks on the Foundations of Mathematics*, I, appendix III, §8.

32. Juliet Floyd and Hilary Putnam, "A Note on Wittgenstein's 'Notorious Paragraph' about the Gödel Theorem," *Journal of Philosophy* 97, no. 11 (2000): 624–632. Also see Chapter 27 in this volume.

33. At the time Wittgenstein wrote, it was not known that we can prove that 777 occurs in the decimal expansion of π.

more cautious twin, §516 in *Philosophical Investigations*. Here are the two paragraphs in question:

> Suppose that people go on and on calculating the expansion of π. So God, who knows everything, knows whether they will have reached "777" by the end of the world. But can his *omniscience* decide whether they *would* have reached it after the end of the world? It cannot. I want to say: Even God can determine something mathematical only by mathematics. Even for him the rule of expansion cannot decide anything that it does not decide for us.[34]
>
> It seems clear that we understand the meaning of the question: "Does the sequence 7777 occur in the development of π?" It is an English sentence; it can be shown what it means for 415 to occur in the development of π; and similar things. Well our understanding of that question reaches just so far, one may say, as such explanations reach.[35]

What I take to be Wittgenstein's considered opinion is that the very notion of a *nonmathematical explanation of the correctness of mathematics* is disguised nonsense. As I suggest we read it, what *PI*, §516, tells us is that just as our explanations of rule following described in §208 of *PI* ("I shall, for instance, get him to continue an ornamental pattern uniformly, when told to do so") suffice perfectly, and there is no "deeper" explanation for philosophy, or future neurology (*PI*, §158), or anything else to give, so our ordinary (mathematical) explanations of mathematical correctness suffice perfectly, and here too there is no "deeper" explanation to be given.[36] A brief look at one

---

34. Wittgenstein, *Remarks on the Foundations of Mathematics*, V, §34.

35. Wittgenstein, *Philosophical Investigations*, §516. See note 16.

36. Earlier I pointed out that what certain philosophers have done is to take Wittgenstein to oppose not only metaphysical realism about rule following but also our commonsense realism about rule following. They have taken Wittgenstein to have doubts about the objectivity of rule following when what Wittgenstein actually doubts is the need for and the very idea of an account of rule following over and above the ordinary account. The ordinary account of rule following is, in my view, contained in Wittgenstein's description of how one teaches someone such concepts as "go on like this," "and so on," and "and so on ad infinitum" at *Philosophical Investigations*, §208. It is especially important that Wittgenstein insists that when he teaches someone the concept of "going on like this" or "and so on," "I do not communicate less to him than I know myself." The attempt to explain these concepts—which function perfectly well in our lives—in terms of either Platonic entities (such as the invisible rails of *Philosophical Investigations*, §218) or mysterious mental powers is an attempt to explain something that we perfectly well understand in terms of something that Wittgenstein wants to

well-known attempt at such a "deeper" explanation, David Lewis's, may clarify what I mean.

David Lewis tries to explain why mathematics "works" by postulating a large cardinal's worth of invisible "stuff."[37] But in doing so, what he is embroiled in is a search for *nonmathematical reasons that the truths of mathematics are truths.* But such a search makes no sense. Showing in detail that it does not—which I do not pretend to do here—would be one way of carrying out the philosophical task that Wittgenstein describes when he writes, "The results of philosophy are the uncovering of one or another piece of plain nonsense" (*PI,* §119). But the following is how I would start.

I would ask the question: if the vast amount of invisible stuff that David Lewis postulates as part of his explanation of the truth of set theory ceased to exist, would mathematics stop working? If Lewis answers yes, he will be required to say that we *causally interact with the invisible "stuff"*—which he emphatically denies. (And if he *had* been willing to say *that,* he would have turned mathematics into a fictitious descriptive science—which is to say, it would have ceased to be mathematics.) But if mathematics would work just as well even if Lewis's "stuff" did not exist, what explanatory work does the "stuff" do? But I digress.

It seems to me that the reading just proposed perfectly fits *PI,* §516. But what of *RFM,* V, §34? *Part* of this paragraph fits the same reading: "Even God can determine something mathematical only by mathematics" fits well with the idea that there is no "foundation" for mathematics, and no explanation of mathematical concepts (including the concept of mathematical correctness) other than the mathematical ones. Where Wittgenstein went astray, I have been suggesting, is in the identification of mathematical correctness with *decidability.* But is it really an *identification?*

It is an identification only if we suppose that Wittgenstein thought that the thought experiment proposed in *RFM,* V, §34, is really an *intelligible* one; but the minute we stand back and reflect, this looks quite unlikely, I believe. Is it at all clear what it would mean for human beings to compute until "the end of the world"? Is it at all clear what they would and what they would not succeed

---

"uncover" as "plain nonsense." The need for something mysterious to "support" or provide a "foundation for" our grammar is what is always under attack in Wittgenstein's writing, not the grammar itself.

37. David Lewis, *Parts of Classes* (Oxford: Blackwell, 1991).

in calculating? To suppose that Wittgenstein is, say, *defining* mathematical truth, for example, thus:

> *S* (a sentence of mathematics) is true if and only if human beings would, were they to go on and on calculating and writing down proofs, succeed in proving *S* before the end of the world,

seems absurd. Thus the assumption on which I have proceeded throughout this lecture, that Wittgenstein thought that mathematical truth is *coextensive* with something I called the "empirical" possibility of (human) proof is enormously implausible. It is time to revise it.

What I suggest instead is the following: Wittgenstein at the moment he wrote *RFM,* V, §34, meant to say that the idea that there is such a thing as a mathematical truth that utterly escapes the possibility of proof is metaphysical nonsense. I think that he was wrong about this for the reasons given above. But to (mistakenly) reject the idea that mathematical truth transcends provability does not require one to think that the notion of "possibility of proof" is itself a completely clear one. (Mathematical truth may also not be a completely clear notion.) The point was not, if I am right, to take a supposedly quite clear notion of "possibility of proof" (in the paragraph Wittgenstein considers only a constructive proof, via a calculation) and use it to *explain* mathematical truth; the point was rather to express discomfort with the idea that mathematical truth can completely *transcend* possible recognition by "us." I have been arguing that the discomfort should be resisted: we have excellent reasons for thinking that mathematical truth *can* transcend recognition by "us." But if Wittgenstein succumbed momentarily to a discomfort that it is extremely easy to feel, that does not mean that he fell into the trap of developing a verificationist philosophy. For, as already mentioned, he simply excised the suspicious part of the remark when he came to write *Philosophical Investigations,* §516. In short, it looks as if he later felt at least some discomfort with the discomfort he felt when he wrote *Remarks on the Foundations of Mathematics,* V, §34.

26

Wittgenstein and the Real Numbers

A number of people have remarked that Wittgenstein has had a significant influence on my own philosophical work. If that is right (and I believe that it is), a lot of the credit (or blame, if you happen to be one of the unfortunate philosophers who "hates" Wittgenstein) should go to Cora Diamond. At Harvard, for many years the most vocal (in every sense of "vocal") advocate of Wittgenstein was my dear friend and longtime debating partner Burton Dreben, and Dreben was a firm advocate of what he called a "Jacobin" (and what the rest of us called an "end-of-philosophy") reading of *Philosophical Investigations*. I always resisted this view, although I had no doubts about Wittgenstein's genius. In early 1987 Cora Diamond sent me a copy of her paper "The Face of Necessity," a paper that I still regard as a masterpiece, and one that enabled me to see both what was right and what was wrong in the "Jacobin" interpretation.[1] More important, it enabled me to appreciate Wittgenstein's importance as a philosopher.

1. Cora Diamond, "The Face of Necessity" (originally written in 1968), in Diamond, *The Realistic Spirit* (Cambridge, Mass.: Bradford Books, MIT Press, 1991), chap. 9. I discuss the importance of the insights in this paper in my third Dewey Lecture, "The Dewey Lectures 1994: Sense, Nonsense, and the Senses; An Inquiry into the Powers of the Human Mind,"

However, I do not describe myself as a "Wittgensteinian." In part this is because I do not like sects in philosophy, and I do not like treating mere mortals as divinities.[2] But it is also because there are elements in Wittgenstein's writings that I cannot defend. In this paper I shall consider some of those elements, but it should be kept in mind that if this paper emphasizes what I see as the negative in some of Wittgenstein's thoughts, first, I still regard Wittgenstein's work as some of the most important philosophical work done in the twentieth century, and, second, the thoughts I shall be criticizing are unpublished thoughts. *Remarks on the Foundations of Mathematics,* in particular, is a selection made by others from notebooks Wittgenstein kept for his own use. None of these remarks made their way into *Philosophical Investigations.*[3] I find it appropriate to discuss them here because they clearly relate to both my own and Cora Diamond's philosophical work and interests.

A Troublesome Part of Wittgenstein's
Remarks on the Foundations of Mathematics

If there is a part of Wittgenstein's oeuvre that I find deeply troublesome, it is the brief second part of the revised edition of *Remarks on the Foundations of Mathematics,*[4] henceforth cited simply as "II." As I interpret Wittgenstein's later philosophy, and as the interpreters I most admire, including Cora Diamond, interpret it, a central feature of that philosophy is meant to be the

---

Journal of Philosophy 91, no. 9 (September 1994): 445–517; reprinted in Hilary Putnam, *The Threefold Cord: Mind, Body, and World* (New York: Columbia University Press, 2000).

2. For similar reasons, although I am often described as a "pragmatist" by other philosophers, I do not (not often, anyway) describe myself as one.

3. The "Editors' Preface to the Revised Edition" of Ludwig Wittgenstein, *Remarks on the Foundations of Mathematics,* ed. G. H. von Wright, R. Rhees, and G. E. M. Anscombe (Oxford: Blackwell, 1978), tells us (30) that "it must have been Wittgenstein's intention to attach appendices on Cantor's theory of infinity and on Russell's logic, as well as the appendix on Gödel's theorem . . . to *Philosophical Investigations.*" However, it also tells us (29) that "Wittgenstein did not return to this subject matter in the last years of his life."

4. In the 1956 edition, this material was in what was then I, appendix II. The present II includes a few remarks that were left out of the 1956 edition and, in addition, "The arrangement of sentences and paragraphs into numbered Remarks corresponds to the original text (which was not wholly the case in the 1956 edition)." ("Editors' Preface to the Revised Edition," 31). The numbers of the remarks are, therefore, different from those in the 1956 edition, as is the order of some of the remarks.

avoidance of "philosophical theses." The exchanges between the different "voices" in *Philosophical Investigations* bring out the different ways in which one may be led into confusion in thinking about philosophical problems. When this way of dealing with a problem is most successful, one comes to see that the "thesis" that created the problem (if, indeed, it began with someone's enunciating a thesis) is one that does not require a yes or no answer, but instead requires to be picked apart, disentangled. But in II one finds Wittgenstein himself enunciating what sound like a number of philosophical theses—in fact, I think that they *are* philosophical theses, and, moreover, ones that I strongly reject. In order to explain what I mean by this claim, however, I need to say something about the terminology Wittgenstein uses in II.

II deals with Cantor's diagonal proof, a proof that Wittgenstein regards as containing a confusion, and it represents an attempt to remove this supposed confusion. Although elsewhere in *Remarks on the Foundations of Mathematics* he talks about the "real numbers" and "the number line," in II Wittgenstein often finds it convenient (as indeed it is for expository purposes) to consider the diagonal argument not as applied to infinite series of real numbers, but rather as applied to series of "expansions," by which term he means decimal representations of real numbers, for example, the representation of π as 3.141592 . . . (The relevant difference is not that expansions are infinite series of digits with one decimal point somewhere and real numbers are, according to most philosophers, "abstract entities,"[5] but that some real numbers correspond to two different expansions and not one: e.g., the real number 3 corresponds both to the expansion 3.0000 . . , with infinitely many zeroes, and to the expansion 2.9999 . . , with infinitely many nines. Wittgenstein does not take note of this, however.) Applied to expansions, what Cantor's diagonal argument would be taken by mathematicians to show is that the set of all expansions is nondenumerably infinite. What does Wittgenstein say about this?

He tells us, amazingly, that it *means nothing* to say of any class X of numbers that it is nondenumerable. I quote the amazing paragraph in full:

> 10. It means nothing to say [after one has shown that any given infinite series of expansions can be diagonalized]: "*Therefore* the X numbers are not denumerable." One might say something like this: I call number concept

5. For a criticism of this "ontological" way of talking about numbers as "entities," see Hilary Putnam, *Ethics without Ontology* (Cambridge, Mass.: Harvard University Press, 2004), pt. 1. Here, I expect, Wittgenstein and I are in agreement, but my reasons are not his.

X non-denumerable if it has been stipulated that, whatever numbers falling under this concept you arrange in a series, the diagonal number is also
to fall under that concept.

The argument that comes with §10, and that indeed runs through all of II,
goes like this: (according to Wittgenstein) the very notion of an *expansion,*
that is, of an infinite sequence of digits (with a decimal point somewhere), is
an indeterminate notion. We have, indeed, decided that certain "developments"
are expansions, for instance, the decimal expansion of π, and we can even
produce infinite sequences of expansions. And Wittgenstein concedes that
when we produce such a sequence, Cantor's diagonal argument does indeed
give us a way of producing something different from all the expansions in
the particular sequence (or, to use Wittgenstein's terminology, the particular
"system") of expansions. But as §10 makes clear, Wittgenstein thinks that it
already involves a *stipulation*—it is not fixed in advance—to say that what we
get by the diagonal proof is itself an "expansion" in the original sense. For
Wittgenstein, the expansions are an *indeterminate class*—we have not fully
specified what it would mean to speak of "all expansions."

This does not mean, however, that Wittgenstein wants us simply to *reject*
the whole of Cantor's argument. Rather, he wants to separate what he sees as
the genuine mathematical content of that argument from what he sees as the
confused add-on (the add-on that led Cantor and others into what Wittgenstein saw as the pseudoparadise of set theory). The problem is that in deciding what is confused, he imports some rather strange convictions of his own,
convictions I want to describe.

In the case of Cantor's proof, it is fairly easy to see what Wittgenstein is saying. What Cantor proved, according to Wittgenstein, is that if you are given a
denumerable "system" of real numbers (or, alternatively, expansions), then you
can exhibit (assuming we have made the "stipulation" mentioned in §10) a real
number or an expansion that is not in the system. But what is misleading,
and what should be avoided, is expressing this by saying, "*The* real numbers
are nondenumerable," or "There are *too many* real numbers to enumerate."

That this is, in fact, what Wittgenstein means is not conjecture on my part.
He tells us as much in so many words. In §16, for example, Wittgenstein writes,

> That, however, is not to say that the question "Can the set R [the real num-
> bers] be ordered in a series [i.e., enumerated]?" has a clear sense. For this
> question means e.g. Can one do something with these formations corre-

sponding to the ordering of the cardinal numbers [Wittgenstein means the natural numbers] in a series? Asked, "Can the real numbers be ordered in a series?" the conscientious answer might be "For the time being I can't form any precise idea of that."—"But you can order the roots and the algebraic numbers for example in a series; so you surely understand the expression!"—To put it better, I *have got* certain analogous formations which I call by the common name "series." But so far I haven't any certain bridge from these cases to that of "all real numbers." Nor have I any general method of trying whether such-and-such a set "can be ordered in a series."

Now I am shown the diagonal procedure and told: "Now here you have the proof that this ordering can't be done here." But I can reply: "I don't know—to repeat—what it is that *can't be done* here."

And again:

19. The dangerous, deceptive thing about the idea: "The real numbers cannot be arranged in a series," or again "the . . . set is not denumerable" resides in its making what is a determination, formation, of a concept look like a fact of nature.

Wittgenstein continues:

20. The following sentence sounds sober: "If something is called a series of real numbers, then the expansion given by the diagonal procedure is also called a 'real number,' and is moreover said to be different from all members of the series."

And again:

28. Why should we say: the irrational numbers cannot be ordered?—We have a method of upsetting any order.

29. Cantor's diagonal procedure does not show us an irrational number different from all in the system, but it gives sense to the mathematical proposition that the number so-and-so is different from all those in the system.

And again:

31. Cantor gives a sense to the expression "expansion which is different from all the expansions in a system," by *proposing* [my emphasis—HP] that an expansion should be so called when it can be proved that it is diagonally different from the expansions in a system.

32. Thus can be *set* as a question: find a number whose expansion is different from all those in the system.

It is not just my "interpretation" that in these remarks Wittgenstein is giving a very weak constructive interpretation to the Cantor proof, namely, that it provides a method whereby, for any given "system" of expansions,[6] it is possible to exhibit an expansion—provided we are willing to *stipulate* that the term "expansion" applies to this new object—different from all the expansions in the system; but he is not willing to agree that Cantor showed anything that should be formulated as "the set of all real numbers is nondenumerable." Here is my proof text:

> 33. It might be said, besides the rational points there are *diverse systems* of irrational points to be found in the number line.
> There is no system of irrational numbers—but also no super-system, no "set of irrational numbers" of higher-order infinity.[7]

"*There is no system of irrational numbers—but also no super-system, no 'set of irrational numbers' of higher-order infinity.*" This from a philosopher who does not put forward philosophical "theses"?!!

Stop and think about this remark. Here Wittgenstein has taken one of the central theorems, not just of set theory but of all modern mathematics, namely, that there is a set of all irrational numbers, and that set is nondenumerable, and arrogantly asserted its negation!

Nor is it merely my conjecture that Wittgenstein's reasons for saying that "there is no set of irrational numbers" (and *a fortiori* no set of real numbers) is that the notion of an "expansion" (which he treats as interchangeable with "real numbers" in II) is not "determined."[8] On the contrary, here too the work of the interpreter is done by Wittgenstein himself:

---

6. By a "system" Wittgenstein seems to mean, roughly, any recursive series of expansions. However, if that is what he meant, his idea that we have to keep extending the notion of what counts as an "expansion" or as a "development" would seem to indicate that he was unaware of Church's thesis.

7. In §14, Wittgenstein seems to identify irrational numbers with certain methods of calculation, and the assertion that the irrational numbers are nondenumerable with the assertion that "these methods of calculation" cannot be ordered in a series. And he adds in a footnote: "And here the meaning of 'these' just gets vague."

8. In any case, the denial of the existence of a set of all real numbers and the denial of a set of all irrational numbers, along with his criticism of the notion of a Dedekind cut (discussed

30. Cantor shows that if we have a system of expansions it makes sense to speak of an expansion that is different from them all—but that is not enough to determine the grammar of the word "expansion."[9]

Thus in II we have a clear and unmistakable rejection by Wittgenstein of Cantorian set theory and everything that presupposes it. And we have a partial explanation of this attitude, namely, his belief that although the notion of an "expansion" does have meaning in certain contexts (we have many "developments" that we are willing to count as expansions), nevertheless, the grammar of the notion "all expansions" (all infinite series of digits) has not been "determined" (and presumably cannot be determined in a way that would suit Cantor's purposes). In order to understand *why* Wittgenstein thinks this, however, it is necessary to inquire into Wittgenstein's attitude toward the mathematical notion of infinity. Fortunately, that too is the subject of many of the remarks in II.

Wittgenstein's Account of Infinity

What Wittgenstein criticizes, not only in II but throughout the material we have as *Remarks on the Foundations of Mathematics,* is thinking of statements of the form "There are infinitely many _____" as meaning that there is a vast number of _____. His alternative is expressed very simply:

45. To say that a technique is unlimited does *not* mean that it goes on forever without stopping—that it increases immeasurably: but that it lacks the institution of the end, that it is not finished off. As one may say of a sentence that it is not finished off if it has no period. Or of a playing field that it is unlimited when the rules of the game do not prescribe any boundaries—say by means of a line.

It is also interesting to look at II, §26 and §27:

26. We say of a *permission* that it has no end.
27. And it can be said that the permission to play language games with cardinal numbers [integers] has no end. This would be said e.g. to someone

---

later), are equally rejections of the heart of what is today considered the theory of functions of real and complex variables.

9. See also §14; §20, "*If* something is called a series of real numbers"; §22, "One pretends to compare the 'set' of real numbers"; and §31, "Cantor *gives a sense* to the expression 'expansion which is different from all the expansions in a system.'"

to whom we were teaching our language and our language games. So it would again be a grammatical proposition, but of an *entirely* different kind from "$25 \times 25 = 625$." It would however be of great importance if the pupil were, say, inclined to expect a definitive end to this series of language game (perhaps because he had been brought up in a different culture).

That he means these rather strange analogies perfectly seriously is shown by §47 and §48:

> 47. What is the function of such a proposition as "A fraction has not a next biggest fraction, but a cardinal number has a next biggest cardinal number"? Well, it is as it were a proposition that compares two games. (Like: in draughts pieces jump over one another, but not in chess.)
>
> 48. We call something "constructing the next biggest cardinal number" but nothing "constructing the next biggest fraction."

This picture of what it means to talk of infinity in mathematics is utterly inadequate even if we confine our attention to elementary number theory. It is not, after all, just that number theory fails to "stipulate" that we are to say of any number that it is the last number, or that we are to say after counting to any number, "We are not allowed to go on." We make the positive assertion—even in intuitionist mathematics—that it is *always possible* to go on. And this is not merely a misleading statement in words,[10] a misleading metastatement that we are accustomed to make; it is an axiom of arithmetic that every number has a successor.

Perhaps Wittgenstein would have taken the same scornful attitude toward Peano's axiomatization of arithmetic that he took toward *Principia Mathematica* and toward set theory; but if so, his position was a radically "eliminationist" one indeed.

---

10. In §55 Wittgenstein writes: "If someone says I have proved the proposition that we can order pairs of numbers [positive integers] in a series, it should be answered that this is not a mathematical proposition, since one doesn't calculate with the words 'we,' 'can,' 'the,' 'pairs of numbers,' etc. The proposition 'one can . . .' is rather a mere approximate description of the technique one is teaching, say a not unsuitable *title*, a heading to this chapter. But a title with which it is not possible to *calculate*." Wittgenstein does not realize—or does not care—that mathematics is not simply calculation, and that what he calls "not a mathematical proposition" is, on the contrary, a paradigmatic mathematical proposition.

A Brief Summary up to This Point

To sum up what we have found so far in II: Wittgenstein unquestionably makes three flat (and in my view unbelievable) assertions:

1. The notion of an expansion (an infinite series of digits) is indeterminate and requires new stipulations as new techniques for developing such expansions are invented.
2. Because of this indeterminacy, "it means nothing to say, '*Therefore*, the *X* numbers are not denumerable.'"
3. To say that a mathematical technique produces an infinite series does not mean that it goes on without stopping. It means simply "that it lacks the institution of the end, that it is not finished off."

But before attempting to draw any moral from all this, we should look at the closing sections of II, in which Wittgenstein himself considers saying "There is nothing infinite" in mathematics and, I think, draws back from saying it:[11]

59. This way of talking: "But when one examines the calculus, there is nothing infinite there" is of course clumsy—but it means: is it really necessary here to conjure up the picture of the infinite (of the enormously big)? And how is this picture connected to the *calculus*? For its connection is not that of the picture | | | | with 4.

60. To act as if one were disappointed to have found nothing infinite in the calculus is of course funny; but not to ask, what is the everyday employment of the word "infinite," which gives its meaning for us; and what is its connection with these mathematical calculi?

61. Finitism and behaviorism are similar trends. Both say, but surely all we have here is . . . Both deny the existence of something, both with a view to escaping from a confusion.

62. What I am doing is, not to show that calculations are wrong, but to subject the *interest* of calculations to a test. I test e.g. the justification for

---

11. That talk of the infinite in mathematics does not have to be taken to mean that mathematics literally deals with a Platonic realm of "abstract objects" that is enormously big is something with which I agree. See Putnam, *Ethics without Ontology,* chap. 3. But §45, which I quoted earlier, disappoints us by making it seem as if comparison of the idea that there are infinitely many numbers with the fact that no boundary has been prescribed for the playing field was sufficient to show this.

using the word . . . here. Or really, I keep on urging such an investigation. I show that there is such an investigation, and what it is like to investigate there. Thus I must say, not: "We must not express ourselves like
this," or "That is absurd," or "That is uninteresting," but: "Test the justification of this expression in this way." You cannot survey the justification of an expression unless you survey its employment, which you
cannot do by looking at some facets of its employment, say a picture
attaching to it.

Section 62 seems to me completely unobjectionable—and when we remember that these are remarks in Wittgenstein's personal notebooks, it is
natural to take "Thus I must say, not: 'We must not express ourselves like this,'
or 'That is absurd, or 'That is uninteresting,' but: 'Test the justification of this
expression in this way'" as a warning by Wittgenstein *to himself.* Nevertheless,
I do not find in the remainder of *Remarks on the Foundations of Mathematics*
any indication that Wittgenstein ever gave up his utterly negative attitude
toward set theory or his rather simplistic interpretation of such statements as
"There are infinitely many natural numbers."

The Continuum in Physics

What Wittgenstein wanted was not to *change* our use of the word "infinite"
in what he was prepared to count as mathematics (or at least as "chapter head-
ings" in mathematics),[12] but rather to understand that use; but as we have just
seen, his *way* of understanding it is decidedly "minimalist," guided by the
analogy he sees between the infinite and the playing field to which no limit
has been stipulated. But there is also another line of criticism of set theory that
runs through much of *Remarks on the Foundations of Mathematics,* based on
Wittgenstein's conviction that set theory is *useless* in the sense of lacking
application. For example, about the proposition that two to the power aleph-
null is greater than aleph-null, he writes (in §35):

> In what practice is this proposition anchored? It is for the time being a
> piece of mathematical architecture which hangs in the air, and looks as if

12. Recall (see note 10) that the theorem that the ordered pairs of integers are denumerable is not "mathematics" in Wittgenstein's view (but only a "chapter heading"), although the
particular recursion by which we enumerate them (Wittgenstein gives one) is a "calculation"
and hence counts as mathematics.

it were, let us say, an architrave, but not supported by anything and supporting nothing.

And he follows with this remark:

> 36. Certain considerations may lead us to say that 10^{10} souls fit into a cubic
> centimeter. But why do we nevertheless not say it? Because it is of no use.
> Because, while it does conjure up a picture, the picture is one with which
> we cannot go on to do anything.

A similar attitude seems to underlie his well-known pooh-poohing of the
concern of logicians like Frege and Russell with achieving a formalization
of mathematics that is free of contradictions. Wittgenstein expresses the attitude that as long as contradictions did not keep cropping up all over the
place, mathematics could go on perfectly well even if some arguments did
lead to contradictions. We could, so to speak, just avoid those arguments.[13]
Thus Wittgenstein also sees the search for rigorous and consistent foundations for mathematics as useless.

But this is simply ignorance on Wittgenstein's part. The fact is that the
confusions about the correct formulation of the notion of a set in nineteenth-
century mathematics exactly parallel and are intimately related to confusions
about the correct formulation of the notion of a function, and those confusions not only affected pure mathematics but also interfered with the resolutions of difficulties having to do with the understanding of certain physical
systems. Documenting this in detail, however, goes beyond the scope of the
present essay.[14]

---

13. See, for example, Wittgenstein, *Remarks on the Foundations of Mathematics,* I, appendix III, §11 and §12; and III, §78 and §80.

14. I am indebted to Mark Wilson for the following examples: One of the great crises in
physical theory arose because Bernhard Riemann and Lord Kelvin (William Thomson) (in a
physical context) assumed that a set of decreasing solutions for, e.g., how soap attaching to a wire
rim would assume a lower-bound position of least energy with suitable regularity (Riemann
called this assumption the Dirichlet principle). But Karl Weierstrass showed some counterexamples, provoking a famous crisis in the mathematics of the time. And Cantor's work arose
directly out of studying the accumulation points of sequences like this. The notion of accumulation points of sequences itself is clearly set theoretic in character, and the existence conditions
for differential equations are entangled with set-theoretic questions (and also with the general
question of measure). Wittgenstein may have been misled by G. H. Hardy and J. E. Little-
wood's view that the questions of pure mathematics they worked on are "useless" (and they
prided themselves on this uselessness). But they turned out to be very wrong.

It is even easier to show that the full importance of the investigations Wittgenstein pooh-poohed has become even clearer since his death. For example, in I, appendix 3, just before the II that I have been discussing, Wittgenstein writes:

> 19. You say: ". . . so P [the Gödelian proposition] is true and unprovable." That presumably means: "Therefore P." That is all right with me—but for what purpose do you write down this "assertion"? (It is as if someone has extracted from certain principles about natural forms and architectural style the idea that on Mount Everest, where no one can live, there belongs a châlet in the Baroque style.) And how could you make the truth of your assertion plausible to me, since you can make no use of it except to do these bits of legerdemain?

This looks like an argument against taking Gödel's work seriously on the ground that it has no use except to do bits of "legerdemain." In fact, *precisely* the argument that Gödel used to prove his theorems immediately generalizes to yield the major fixed-point theorems of recursion, theorems that have numerous applications in the whole domain of today's computer science. Moreover, Church's theorem is easy to prove once one has appreciated the techniques Gödel invented;[15] and my own work in logic and number theory (with Martin Davis, Julia Robinson, and Yuri Matyasevich) involved—as is well known—showing that the same technique can be extended to show that there is no decision method for Diophantine equations (Hilbert's tenth problem).[16] The very arguments in which Wittgenstein saw only useless "legerdemain" are in fact the foundation of a whole new science.

In the rest of this paper, however, I shall focus on just one question: is the assertion that there is a set of all real numbers simply a useless and misleading picture? This claim is, after all, one that Wittgenstein clearly makes in II (more precisely, he asserts that there is no set of all irrational numbers, and he also puts "real number" in quotation marks in almost all the passages I quoted above). To put it another way, is the idea that there is a set of all real numbers just a piece of useless "legerdemain"?

15. Church's theorem is that there is no decision method for quantification theory.

16. Martin Davis, Hilary Putnam, and Julia Robinson, "The Decision Problem for Exponential Diophantine Equations," *Annals of Mathematics* 74, no. 3 (November 1961): 425–436; Yuri Matyasevich, "Enumerable Sets Are Diophantine," *Soviet Mathematics Doklady* 11, no. 2, (1970): 354–357.

In the present section I will move away from Wittgenstein's text and try to explain from my own "scientific realist" point of view why it would be very wrong to say this.[17]

So far I have said that Wittgenstein clearly rejects set theory, including the idea that there is such a thing as a set of all real numbers and the theory of infinite cardinals. I could have added that in *Remarks on the Foundations of Mathematics,* V, he reinterprets the theory of Dedekind cuts (the heart of the theory of functions of a real variable) in the same way: he says that "the idea of a 'cut' . . . is a dangerous illustration," and the thought that we understand the idea of an arbitrary Dedekind cut (i.e., an arbitrary real number) is called "a frightfully confusing picture." He treats it with the same scorn that he treated the notion of a "set of all irrational numbers":

> 37. The misleading thing about Dedekind's conception is the idea that the numbers are there spread out on the number line. They may be known or not; that does not matter. And in this way all that one needs to do is cut or divide into classes and one has dealt with them all.

Now, it is a fact about mathematical physics, and one of which Wittgenstein was surely aware, that ever since Descartes' invention of analytic geometry, which is the beginning of mathematical physics as we now know it, the notion of spatial point—and today the notion of a space-time point—is dependent on the notion of a real number. That is, in physics we fix the logic of talk about points precisely by making the Cartesian assumption that every point in space can be associated with a triple of real numbers. Thus the consequences for our understanding of physics of the view that we do not have a determinate notion of a real number are immediate and large. If Wittgenstein was right, then we lack a determinate notion of a point in space, as that notion is used in mathematical physics. In addition to saying, "There is no 'set of irrational numbers,'" Wittgenstein might as well have said, "There is no 'set of all points in space.'" To understand physics as a scientific realist does—that is, to understand it *without philosophical "reinterpretation"*—is dangerous naïveté on such a view.

---

17. By "scientific realism" I mean just what I meant in "Introduction: Science as Approximation to Truth," in Hilary Putnam, *Philosophical Papers,* vol. 1, *Mathematics, Matter and Method* (Cambridge: Cambridge University Press, 1975), vii–xiv. In particular, a scientific realist (in my sense) counts mathematics as a central part of science, and she rejects the idea that any significant part of science needs justification from *philosophy.*

In contemporary physics the notion of a particle has lost the fundamental role that it had in atomistic physics. That role has been taken over by the notion of a field (or, in quantum mechanics, by the notion of a quantized field). But a field—for example, the electromagnetic field or the gravitational field—is something that has a magnitude that can be measured by a real number (or by an *n*-tuple of real numbers) *at each point in space* (or each point in a certain region of space). If the notion of "any point in space" is indeterminate and the notion of a real number is indeterminate,[18] then the notion of a field must suffer a corresponding indeterminacy.

Here it might be argued that the idea that there are infinitely many points in space is "only an idealization." "Given that we use just the theory of the real numbers that Wittgenstein is criticizing to do mathematical physics," it might be said, "it is not surprising that we use just that idealization. But we must not be misled by what is only a picture."

To this I would reply that although it may, of course, *turn out,* as a matter of physical fact, that space is discrete, or that space-time is discrete, or that quantum mechanics can be done with discrete mathematics—perhaps because there are minimal volumes of space-time that can carry information, as postulated by some (speculative) theories of quantum gravity—these are clearly questions to be decided by *empirical science.* The best physical theories we have today, and the best physical theories we have had ever since the time of Newton and Descartes, are ones that postulate that space and time are continua in precisely the sense Wittgenstein finds unclear. If that view has to be abandoned, it should be for physical reasons, not for "Wittgensteinian" ones. The choice is a choice among *meaningful* physical pictures, at least by any ordinary scientific standard of meaningfulness. The fact that Wittgenstein

---

18. It is true that Wittgenstein does not use the word "indeterminate," but he does say (II, §30) that we have not "determined" whether the result of Cantor's diagonalization is an "expansion." And his denial that there is a "set of irrational numbers" is based on the idea (§10) that we have to "stipulate" whether an expansion arrived at by the diagonal process applied to a series ("system") of real numbers is also a "real number." It is clear that he does not think that just this one "stipulation" is needed, and *then* we can speak of a "set of irrational numbers"; what he thinks is that there is no surveyable totality of "developments," and *hence* the notion of a real number may need an indefinite number of future "stipulations." (His view that the intentional notion of a development is prior to the extensional notion of a cut also plays a role here, as do his discussions with Ramsey over whether logic is to be understood intensionally or extensionally, but these matters must be left for another occasion.)

questions continuum mathematics (which is to say, virtually *everything* that mathematicians know as twentieth- and now twenty-first-century mathematics) on *philosophical* grounds, as he does in these parts of *Remarks on the Foundations of Mathematics,* shows, to my mind, that Wittgenstein was led fundamentally astray. (Indeed, I suspect that if Wittgenstein had written something called *Remarks on the Foundations of Physics,* it would, like *Remarks on the Foundations of Mathematics,* contain much that is fascinating, but that it would also, in the end, amount to an attack on much that is essential to the science.)

✌ Why Wittgenstein Went Astray

In the Appendix to the original version of my essay "Was Wittgenstein *Really* an Antirealist about Mathematics?" I wrote:[19]

> In the preceding essay, I said that I do not believe that Wittgenstein's *Remarks on the Foundations of Mathematics* spring from an antirealist philosophy. That doesn't mean that there are no problems with Wittgenstein's remarks, and I understand perfectly well why some of them would seem to invite an antirealist reading. In particular, it does seem that Wittgenstein thought, when he wrote the material that the editors have collected under this title, that a mathematical proposition cannot be true unless we can decide that it is true on the basis of a proof or calculation of some kind.[20] Not only is this quite explicit in his 1937 remarks about the Gödel theorem,[21] but the view appears as late as the 1944 remark that even God's *omniscience* cannot decide whether people would have reached "777" in the decimal

---

19. Hilary Putnam, "Was Wittgenstein Really an Antirealist about Mathematics?" in Timothy G. McCarthy and Sean Stidd, eds., *Wittgenstein in America* (Oxford: Clarendon Press, 2001), 140–194. The article is Chapter 23 in this volume; the content of its Appendix appears here in the opening of Chapter 25 ("Wittgenstein, Realism, and Mathematics").

20. But see note 16 of Chapter 25 of this volume, in which I contrast what Wittgenstein published on this question—§516 of the *Investigations*—with the unpublished material collected as *Remarks on the Foundations of Mathematics.*

21. What I had in mind here was *Remarks on the Foundations of Mathematics,* I, appendix III, §§6–8. However, now [2011] I am no longer convinced that these remarks should be read in this way. The equation of truth in Russell's system with provability in Russell's system does not tell us much about Wittgenstein's attitude toward truth in mathematics; Russell's system is full of set theory, which Wittgenstein did not think was mathematics.

expansion of π "after the end of the world."[22] How can such a view *not* spring from antirealism?

And I answered my own question thus:

> What I want to say, in brief, is that the inadequacies that I find in *some* (by no means all) of Wittgenstein's *Remarks on the Foundations of Mathematics* represent a peculiar combination of genuine insight with an inadequate knowledge of actual mathematical practice and of sciences which depend on mathematical practice, in particular mathematical physics.

This is still my diagnosis for now. But there is something more we can learn about Wittgenstein as a philosopher if we look, in addition, at the "Lectures on Aesthetics."[23]

What I have in mind is this. Very often both the critics and the admirers of the later Wittgenstein share the view that Wittgenstein is concerned only to clear up confusions (or better, to begin to show us how to clear up confusions, or to show us how to begin to clear up confusions) in *philosophy*. But in the "Lectures on Aesthetics," as well as in the material I have discussed, Wittgenstein makes it quite clear that he is interested—passionately interested—in clearing up confusions in science—or in what we are inclined to think about science. Unfortunately, as the preceding discussion illustrates, much of what Wittgenstein mistakenly regards as just misleading pictures is actually good and important science. Nevertheless, the fact that Wittgenstein had these larger ambitions for his philosophy should lay to rest the idea that he is just an "end-of-philosophy" philosopher. The philosophical investigation of conceptual problems connected to science (and not only to science[24]) is enormously

---

22. Wittgenstein, *Remarks on the Foundations of Mathematics,* V, §34.

23. These lectures, delivered in 1938, are collected in Ludwig Wittgenstein, *Lectures and Conversations on Aesthetics, Psychology, and Religious Belief,* ed. Cyril Barrett (Oxford: Blackwell, 1966).

24. Wittgenstein's famous remark to Norman Malcolm, criticizing the latter for talking about "national character," indicates that our everyday life, and the way we talk about it, is something he wanted to teach us to criticize as well. Norman Malcolm, *Ludwig Wittgenstein, a Memoir* (Oxford: Oxford University Press, 1958), 39. Malcolm quotes a letter from Wittgenstein in which Wittgenstein writes, "You & I were walking along the river towards the railway bridge and we had a heated discussion in which you made a remark about 'national character' that shocked me by its primitiveness. I then thought: what is the use of studying philosophy if all that it does for you is enable you to talk with some plausibility about some abstruse ques-

important. But I believe that Wittgenstein failed to see that it requires a much better knowledge of science, and much more *respect* for science, than he possessed.[25] Here, for once, the philosopher who famously said "Take your time" failed to take his time.

---

tions of logic, etc., & if it does not improve your thinking about the important questions of everyday life?"

25. In *Lectures and Conversations on Aesthetics, Psychology, and Religious Belief,* §36 (p. 27), Wittgenstein's auditors record him as saying: "Jeans has written a book called *The Mysterious Universe* and I loathe it and call it misleading. Take the title. This alone I would call misleading. . . . I might say the title *The Mysterious Universe* includes a kind of idol worship, the idol being Science and the Scientist." Apparently Wittgenstein so feared science worship—which I agree is a bad thing—that he felt that he had to debunk the idea that the universe is mysterious. But it is! Immediately after this remark he adverts again to Cantor's proofs, which he had discussed with Ursell, and he says, "It has no charm for me. I loathe it" (the same verb he used in connection with *The Mysterious Universe*). He goes on to say: "38. Cantor wrote how marvelous it was that the mathematician could in his imagination transcend all limits. 39. I would do my utmost to show it is this charm that makes one do it. Being Mathematics or Physics it looks incontrovertible and this gives it a still greater charm. If we explain the surroundings of the expression we see that the thing could have been expressed in an entirely different way. I can put it in a way in which it will lose its charm for a great number of people and certainly will lose its charm for me." These lectures are from the same period as *Remarks on the Foundations of Mathematics,* II, and these remarks illustrate how important attacking Cantorian set theory was for Wittgenstein at that time (and later, apparently, if he really planned to add the material in II as an appendix to *Philosophical Investigations*).

27

Wittgenstein's "Notorious" Paragraph about the Gödel Theorem: Recent Discussions

How would it be if *Principia Mathematica* (hereafter *PM*) turned out to be ω-inconsistent?[1] In our brief "Note on Wittgenstein's 'Notorious Paragraph' about the Gödel Theorem" (2000)[2] we showed that in 1937, when he wrote down the most "notorious" of his remarks about Gödel, Wittgenstein may well have been pondering this question.[3] Viewed in this light, we argued, his remarks may be seen to contain the germ of a significant philosophical insight about Gödel's theorem, rather than a hopeless effort to refute it. The

---

This paper is co-authored with Juliet Floyd.

1. A formal system L is ω-inconsistent if there exists some well-formed formula $P(v)$, expressing a predicate of natural numbers and with no free variable other than v, such that $(\exists v)$ Pv is provable in L, but so are all the formulas $\sim P(\underline{0})$, $\sim P(\underline{1})$, $\sim P(\underline{2})$, . . . (where $\underline{0}, \underline{1}, \underline{2}, . . .$ are the formal expressions for the natural numbers in L). L is ω-consistent if there is no such well-formed formula.

2. Juliet Floyd and Hilary Putnam, "A Note on Wittgenstein's 'Notorious Paragraph' about the Gödel Theorem," *Journal of Philosophy* 97, no. 11 (November 2000): 624–632.

3. The "notorious" remarks, written in the fall of 1937, are published in Ludwig Wittgenstein, *Remarks on the Foundations of Mathematics,* rev. ed., ed. G. H. von Wright, R. Rhees, and G. E. M. Anscombe (Oxford: Blackwell, 1978), I, appendix III, §8.

purpose of our "Note" was to characterize this insight, detaching it from the disputed question whether Wittgenstein fully understood the Gödel theorem.

The insight, as we construed it, asks us to appreciate the philosophical naïveté and/or unclarity[4] involved in taking the following claim to be a straightforward truth:

Claim:

Gödel's theorem *proves* that

1. there is a well-defined notion of "mathematical truth" applicable to every formula of *Principia Mathematica*

and that

2. if *Principia Mathematica* is consistent, then some "mathematical truths" in the sense of (1) above are undecidable in *PM*.

This insight had never been specifically associated with Wittgenstein's "notorious" remarks and does not leap to the eye on their surface. Moreover, there are rival interpretations of the remarks that commit their authors to the view that Wittgenstein *could* not have been thinking along such lines, and/or to the view that if he was so thinking, then he was mistaken.

In what follows we explain why we chose to extract and emphasize this particular insight from the notorious remarks. Our treatment of Wittgenstein takes place within a rather complicated context of contemporary debate about the philosophical significance of Wittgenstein's writings on Gödel; another aim of this essay is to survey this terrain.

We do not claim that our reading exhausts the themes at stake in Wittgenstein's remarks about the Gödel theorem, nor that it offers a general approach to Wittgenstein's remarks on mathematics as a whole, which were, after all, written down tentatively for his own use and were not intended for publication. We do wish to defend the force and interest of our interpretation, however, on the ground that it does more credit not only to Wittgenstein's philosophical perspicacity but also to the interest of the paragraph itself. Certainly we think it important to stress that Wittgenstein's remarks, and the insight we see contained in them, are not at all an attempt to refute Gödel. Moreover, insofar as they may be seen to broach the insight we describe above, the remarks are of interest to the philosophy of mathematics today.

4. Juliet Floyd and I labeled this unclarity "metaphysical" in "Note on Wittgenstein's 'Notorious Paragraph' about the Gödel Theorem," 632.

First, we review the claims that were forwarded in our "Note." Second, we consider a rival interpretation of Wittgenstein's remarks offered by Mark Steiner in his essay "Wittgenstein as His Own Worst Enemy: The Case of Gödel's Theorem" (2001).[5] Steiner's essay, written partly in response to earlier work of Floyd's (1995),[6] was composed before he knew of our interpretation, so here we will be laying out explicitly where we take our differences with Steiner to lie and offering a response, although a full consideration of Steiner's interpretation of the notorious passage lies outside the scope of this essay. Steiner has said in conversation that our "Note" would not have led him to withdraw his criticism of Wittgenstein's remarks on the Gödel theorem, but we hope that what we write here may help begin sorting out the underlying issues. In our third and concluding section we consider the implications of our objections to Steiner's interpretation for some recent criticisms of our "Note" authored by Timothy Bays. Bays alleged that we have given a philosophical answer to the question "How would it be if PM turned out to be ω-inconsistent?" that is "almost certainly false" and is based on an argument that is "inadequate."[7] We defend our reading against his allegations here. Although Bays cites Steiner's essay and appears to be indebted to parts of Steiner's interpretation of Wittgenstein, he departs from Steiner in significant philosophical ways, as he has come to stress in subsequent remarks (as yet unpublished).[8] In particular, he forwards a view of the notion of truth that we find problematic, both in application to Wittgenstein and in application to the insight we extracted from his "notorious" remarks.

5. Mark Steiner, "Wittgenstein as His Own Worst Enemy: The Case of Gödel's Theorem," *Philosophia Mathematica* 9 (2001): 257–279.

6. Juliet Floyd, "On Saying What You Really Want to Say: Wittgenstein, Gödel and the Trisection of the Angle," in Jaakko Hintikka, ed., *From Dedekind to Gödel: The Foundations of Mathematics in the Early Twentieth Century,* (Dordrecht: Kluwer Academic Publishers, 1995), 373–426.

7. Timothy Bays, "On Floyd and Putnam on Wittgenstein on Gödel," *Journal of Philosophy* 101, no. 4 (April 2004): 197–210.

8. Bays has posted a manuscript replying to the earlier version of this essay on his website at http://www.nd.edu/~tbays/papers/index.html titled "Floyd, Putnam, Bays, Steiner, Wittgenstein, Gödel, Etc." So the debate is ongoing. Although there is much to say, we will not fully reply to this in-progress work here. We will, however, separate out his reply from Steiner's interpretation, for Bays has retracted the footnote that led us to think that he was relying on Steiner in the originally published version of this essay (see Bays, "On Floyd and Putnam on Wittgenstein on Gödel," 208 n. 27). See note 22 below.

Some may feel that the most we can hope to show is that other readers have been uncharitable to Wittgenstein, and that this would not by itself suffice to show that they are *wrong* about him. But we believe that, other things being equal, charitable interpretations should be preferred to uncharitable ones, especially when one is dealing with a great philosopher. This, of course, is not to say that one should always interpret a philosopher so that he or she comes out *right:* we are not Wittgensteinian "fundamentalists." We agree, for example, that Wittgenstein's rejectionist attitude to set theory was mistaken.[9] This, however, is a separate issue so far as the argument of our "Note" is concerned.

What Did We Actually Claim?

Here are the "notorious" remarks on Gödel, consisting of two paragraphs published subsequently as *Remarks on the Foundations of Mathematics,* I, appendix 3, §8:

> I imagine someone asking my advice; he says: "I have constructed a proposition (I will use '*P*' to designate it) in Russell's symbolism, and by means of certain definitions and transformations it can be so interpreted that it says: '*P* is not provable in Russell's system.' Must I not say that this proposition on the one hand is true, and on the other hand is unprovable? For suppose it were false; then it is true that it is provable. And that surely cannot be! And if it is proved, then it is proved that it is not provable. Thus it can only be true, but unprovable."
>
> Just as we ask, " 'Provable' in what system?," so we must also ask, " 'True' in what system?" "True in Russell's system" means, as was said, proved in Russell's system, and "false in Russell's system" means the opposite has been proved in Russell's system.—Now what does your "suppose it is false" mean? *In the Russell sense* it means, "suppose the opposite is proved in Russell's system"; *if that is your assumption* you will now presumably give up

9. Steiner discusses this rejection of set theory in "Wittgenstein as His Own Worst Enemy." I gloss the form it takes in *Remarks on the Foundations of Mathematics,* II, in Hilary Putnam, "Wittgenstein and the Real Numbers," in Alice Crary, ed., *Wittgenstein and the Moral Life: Essays in Honor of Cora Diamond* (Cambridge, Mass.: MIT Press, 2007), 235–250, reprinted as Chapter 26 in the present volume. Compare the closing paragraph of William Tait, "Wittgenstein and the 'Skeptical Paradoxes,'" *Journal of Philosophy* 83 (1986): 475–488; reprinted in Tait, *The Provenance of Pure Reason: Essays in the Philosophy of Mathematics and Its History* (New York: Oxford University Press, 2005), 198–211.

the interpretation that it is unprovable. And by "this interpretation" I understand the translation into this English sentence.—If you assume that the proposition is provable in Russell's system, that means it is true *in the Russell sense,* and the interpretation "*P* is not provable" again has to be given up. If you assume that the proposition is true in the Russell sense, *the same* thing follows. Further: if the proposition is supposed to be false in some other than the Russell sense, then it does not contradict this for it to be proved in Russell's system. (What is called "losing" in chess may constitute winning in another game.)

Inspired by these remarks, we argued as follows. Suppose that to our surprise we have discovered a proof of the negation of a Gödel sentence, "~*P*," in *Principia Mathematica.*[10] Suppose too that *PM* is consistent. Under these suppositions it is a well-known, uncontroversial consequence of Gödel's theorem that *PM* will be ω-inconsistent. As may be seen from inspection of the definition of "ω-inconsistent" (see note 1), this means that every model of *PM* must contain entities that are not natural numbers. In fact, in any such model *every* numerical predicate with an infinite extension will "overspill," that is, contain some elements that are not natural numbers. But then, because our original rigorization of the syntactic notions applied in the English sentence came through Gödel numbering with the natural numbers alone (as in Gödel's original 1931 paper),[11] our initial way of translating into English or German "*P*" (the formula of *PM* whose proposed translation was "*P* is not provable in *PM*") would, in this context, become more nuanced and complicated, would have to be "given up" in its original form, just as Wittgenstein observed.

Neither we nor Wittgenstein as we understand him were trying to argue that there are *no* circumstances under which we may legitimately translate "*P*" into the English sentence "'*P*' is not provable" (or say that *P* itself is "true

---

10. *P* has the form: ~(∃*x*)(NaturalNo.(*x*).Proof(*x*,t)), where "t" abbreviates a numerical expression whose value calculates out to be the Gödel number of *P* itself, "Proof" abbreviates a predicate that is supposed to define an effectively calculable relation which holds between two natural numbers *n* and *m* just in case *n* is the Gödel number of a proof whose last line is the formula with Gödel number *m*, and "NaturalNo.(*x*)" is the predicate of *PM* we interpret as "*x* is a natural number."

11. Kurt Gödel, *Collected Works,* vol. 1, *Publications, 1929–1936,* ed. Salomon Feferman, J. W. Dawson, S. C. Kleene, G. H. Moore, R. M Solovay, and John van Heijenoort, with the German original and the English translation on facing pages (New York: Oxford University Press, 1986), 145–195.

but unprovable").[12] Indeed, our interpretation prescinds altogether from talk about all proper or possible translations or interpretations of P.[13] Instead, we take Wittgenstein to have been exploring the context-dependent *sense* of the claim that "there are true but unprovable sentences of arithmetic" as it applies both to PM and to our natural language, and to have been doing so within a particular philosophical and historical situation. We also connect Wittgenstein's remark about "giving up" an interpretation of "P" with his (commendable) rejection of a notion of *interpretation* that in principle depends wholly on formalized mathematical language: as we wrote, Wittgenstein was "denying that a formal system *could* provide us with a standard of truth or clarity that is, in principle, inaccessible to a natural language."[14] Of course, the argument in our "Note" itself then turns on a partly informal understanding of the notion of a sentence's being "true in an interpretation" (or "satisfied in a model"). As all agree, in 1937 model theory was not yet an established branch of mathematics, and Wittgenstein may be forgiven for leaving its development out of his purview. Moreover, we believe that the later development of formal semantics and model theory fails to impugn the importance of (what we are calling) Wittgenstein's philosophical insight about the Gödel theorem.

Steiner's Interpretation of the Notorious Remarks

Steiner's writing is unfailingly interesting, and the essay in which he claimed that in the notorious remarks Wittgenstein "slips into trying to refute the [Gödel incompleteness] theorem" is certainly no exception.[15] However, we will offer reasons to doubt that Steiner offers a satisfactory interpretation of the notorious remarks.[16]

---

12. Despite what Bays says in "On Floyd and Putnam on Wittgenstein on Gödel," 198, 201, 202–203, 208.

13. Again, unlike Bays's interpretation, which we will discuss below.

14. Floyd and Putnam, "Note on Wittgenstein's 'Notorious Paragraph about the Gödel Theorem,'" 632.

15. Steiner, "Wittgenstein as His Own Worst Enemy," 261.

16. A full survey of the issues raised by Steiner's own rather sophisticated understanding of Wittgenstein on Gödel lies outside the scope of this reply, but some of them have been discussed in Juliet Floyd, "Prose versus Proof: Wittgenstein on Gödel, Tarski and Truth," *Philosophia Mathematica* 3, no. 9 (2001): 901–928.

According to Steiner's interpretation, Wittgenstein aspired in the notorious remarks to show that the claim that "there are true but unprovable sentences of arithmetic" is false and is nothing but a misguided philosophical interpretation of the theorem. It is the last two sentences of the second paragraph (starting with the words "Further: if the proposition is supposed to be false in some other than the Russell sense") that Steiner describes as "[Wittgenstein's] 'refutation' of Gödel's theorem."[17] As Steiner reads the section, up to the words "Thus it can only be true but unprovable," Wittgenstein is describing what he takes to be Gödel's proof, and the words "Further: if the proposition is supposed to be false in some other than the Russell sense, then it does not contradict this for it to be proved in Russell's system. (What is called 'losing' in chess may constitute winning in another game.)" are supposed to refute that proof. As Steiner summarizes his reading of these sentences:[18]

> The refutation goes (I abbreviate): there is no contradiction in a false but provable sentence—what is false is context (or "game") dependent. The very same words might sometimes express a truth and sometimes a falsehood. Thus Gödel's proof rests on an elementary mistake.

This does look at first blush like a *possible* reading of what is going on in the Gödel remarks, and it squares with a long history of reading the passage as evincing Wittgenstein's supposed view that "true in *Principia Mathematica*" and "provable in *Principia Mathematica*" amount under philosophical analysis to the same thing.[19] The purpose of our "Note" was, however, to broach and defend a different reading, one that did not presuppose this reduction of the notion of "truth" to that of "proof" or "game."

Steiner maintains that the argument he attributes to Wittgenstein in the notorious paragraph must be wrong. For, as he explains, as a result of Tarski's analysis of the notion of a sentence being *true in a formalized language* (or *satisfied in a structure*), we can give a proof in formal semantics of the Gödel theorem that demonstrates the existence of "true but unprovable" sentences of arithmetic. He writes:

> We can use Tarski's concept of truth to prove the Gödel theorem directly. It is a *mathematical theorem* (of set theory) that the Gödel sentence P is true in Tarski's sense if and only if not provable in PA [Peano arithmetic].

17. Steiner, "Wittgenstein as His Own Worst Enemy," 274.
18. Ibid., 261.
19. This is discussed in Floyd, "On Saying What You Really Want to Say," 376–377 n. 21.

(That is, Gödel constructs the sentence P to have this property.) Then if P is false, it is provable in PA, and so we have a false theorem of PA, which is impossible, because we have another *mathematical theorem* that all theorems of PA are true (in Tarski's sense).[20]

This reasoning implies that if the Gödel sentence "P" were false, it would be provable in Peano arithmetic (hereafter "PA"), and we would then have a false theorem of PA. But, Steiner argues, this cannot be, because we have another mathematical theorem that all theorems of PA are true (in Tarski's sense). In other words, we may safely assume that PA is sound, and this implies that there are true but unprovable sentences of arithmetic.[21]

There are several things to note about this interpretation of Wittgenstein's notorious remarks on Gödel. Most important, it avoids engaging with the issue of ω-inconsistency altogether—something neither Gödel's original paper nor Wittgenstein's remarks on that paper (as we read them) do. For it uses the standard contemporary exposition of the Gödel theorem that employs an improved version of Gödel's proof due to Rosser.[22] Gödel showed that if PM is consistent, then P is not provable, but he did not see how to show that if PM is consistent, then $\sim P$ is not provable. Gödel realized that the latter argument requires the stronger assumption that PM is ω-consistent, so his proof was not "symmetric" about P and $\sim P$. In 1936 (in a paper that, so far as we know, Wittgenstein never saw) Rosser showed that this asymmetry, and Gödel's stronger assumption of ω-consistency, are not necessary for a proof of the incompleteness of PM.[23] For we can construct a sentence of PM, call it "R," such that a proof of R in PM can be converted algorithmically into a proof of $\sim R$, and vice versa, assuming only the simple consistency of PM. Freely translated, R expresses "For all n, if n is the Gödel number of a proof of R, then there is a proof of $\sim R$ with a Gödel number smaller than n," or, even more freely, "If R is provable, then $\sim R$ is provable even more quickly."

It strikes us as important that R was not the sentence P that either Wittgenstein or our "Note" were discussing. Steiner's shift to a new context—one

---

20. Steiner, "Wittgenstein as His Own Worst Enemy," 267.

21. Compare Bays, "On Floyd and Putnam on Wittgenstein on Gödel," 200, for a similar argument; in his posted manuscript "Floyd, Putnam, Bays, Steiner, Wittgenstein, Gödel, Etc." he says that he did not intend to represent Wittgenstein's, Steiner's, or our argument here, but only an accessible version of the Gödel proof.

22. So did Steiner, "Wittgenstein as His Own Worst Enemy," 259, 262.

23. J. B. Rosser, "Extensions of Some Theorems of Gödel and Church," *Journal of Symbolic Logic* 1 (1936): 87–91.

that dispenses altogether with the sentence of *PM* Wittgenstein and Gödel were contemplating—leaves behind the whole issue of a sentence that purports to say of itself that it is true but unprovable, full stop. For Steiner's purposes, this does not really matter: his reading, whether correct or incorrect, applies irrespective of whether we are considering Rosser's *R* or Gödel's *P*. But because *P* was the sentence and ω-inconsistency the issue in which Gödel and Wittgenstein were, as we see it, interested, Steiner's reading fails to engage with the philosophical claim we take Wittgenstein to have been investigating in his remark. If one assumes (as Steiner does) that Wittgenstein would not have minded if one replaced the Gödel sentence *P* by the Rosser sentence *R,* then the perfectly correct observation that if the undecidable sentence were refuted, then *PM* would have no model in which the predicate that has been interpreted as "*x* is a natural number" possesses an extension that is isomorphic to the natural numbers would be trivial. For if *R* were refutable, then *PM* would be inconsistent and have no models at all, whereas if the Gödel sentence *P* were refuted, *PM* could still be consistent.[24]

Here is Steiner's pithy summary of Gödel's argument: "[Gödel's theorem] exhibited a 'computer program' which converts a proof of *P* into a proof of 'not-*P*' and *vice versa*."[25] In a note he adds, "It is worth repeating that the '*vice versa*' is due to Rosser, not Gödel."[26] But one should be careful here about Steiner's "vice versa." As Steiner himself repeatedly points out in his paper, this was not Gödel's argument. The formula *P* Gödel constructed is such that Gödel indeed showed us how to exhibit a "computer program" that converts a proof of it into a proof of "not-*P*," and that is how Gödel proved that if *PM* is consistent, then *P* is not provable. But Gödel did not prove "vice versa," that is, he did not exhibit a computer program that converts a proof of "not-*P*" into a proof of *P*. In fact, Warren Goldfarb has pointed out that if *PM* is ω-consistent, then it cannot be proved even in a system as rich as *PM* that there is such an algorithm.[27] (By contrast, Gödel's own proofs can, as is well

---

24. Bays agrees that the move to the context of the Rosser sentence would "seriously trivialize" our discussion and would be "quite unfair" to it; see "Floyd, Putnam, Bays, Steiner, Wittgenstein, Gödel, Etc.," 3.

25. Steiner, "Wittgenstein as His Own Worst Enemy," 262.

26. Ibid., 262, n. 19.

27. Here is Goldfarb's proof:

Suppose that it is a theorem of *PM* that Provable($\sim P$) $\Rightarrow$ Provable(*P*). Since *P* is provably equivalent to $\sim$Provable(P), it must also be a theorem of *PM* that Provable($\sim\sim$Provable(*P*))

known, be carried out in finitistic arithmetic.) To repeat: Rosser did indeed show how to construct a sentence (called "R") such that a proof of R can be converted into a proof of not-R and vice versa. Thus Rosser showed that if PM is consistent, then neither R (the "Rosser sentence") nor its negation is provable. But Gödel did not see how to show the existence of undecidable sentences in PM without a stronger hypothesis than simple consistency. (If he had, Wittgenstein's paragraph would have indeed been vulnerable to Steiner's interpretation.) Instead, Gödel used the hypothesis of ω-consistency, and this is what our "Note" on the Gödel paragraph turned on. The proof of the famous first incompleteness theorem (the one Wittgenstein was discussing in the notorious remarks) does not claim to show that if $\sim P$ ("not-P") is provable, then PM is inconsistent, but only that if $\sim P$ is provable, then PM is ω-inconsistent. (The other half of the proof, that if P is provable, then PM is inconsistent, does not need the notion of ω-consistency and does proceed as Steiner describes.)

We claimed that what interested Wittgenstein in the section we are discussing was: what would it mean if Gödel's undecidable proposition P were actually *refuted* in PM? What Wittgenstein observed, we believe, was that if this happened, then PM would have *no model in which the predicate that has been interpreted as* "x is a natural number" *possesses an extension* that *is isomorphic to the natural numbers.* (To use Gödel's term, PM would be ω-inconsistent.) And in that case, Wittgenstein claimed, one "will now presumably" give up the "translation" of P by the English sentence "P is not provable," that is, will rethink or revise it. To repeat: we did not suggest, and we see no reason to believe, that this was intended as a *refutation* of the Gödel theorem.

But is it credible that Wittgenstein was *that* sophisticated? The answer we gave is that we have testimony (and not from particularly sympathetic sources) that Wittgenstein thought about what are now called nonstandard models of the natural numbers and connected them with the Gödel theorem. In discussion with Alister Watson and Alan Turing in the summer of 1937

$\Rightarrow$ Provable(P), or equivalently, that Provable(Provable(P)) $\Rightarrow$ Provable(P). But Löb's theorem (which is itself finitistically provable and *a fortiori* provable in PM) states that for any formula F, if it is provable that Provable(F) $\Rightarrow F$, then F itself is provable; taking F to be "Provable(P)," we conclude that Provable(P) is provable in PM. But Provable(P) $\Rightarrow \sim$CON(PM) is provable in PM. (This is, of course, just Gödel's theorem.) So $\sim$CON(PM) is provable in PM, on the assumption that "vice versa" is provable. But then PM is either inconsistent or ω-inconsistent.

references were made to the issue of ω-inconsistency; in fact, Watson later credited Wittgenstein with his understanding of it.[28] And in an essay by Goodstein from 1957 we find this:

> Wittgenstein with remarkable insight said in the early thirties that Gödel's results showed that the notion of a finite cardinal could not be expressed in an axiomatic system and that formal number variables must necessarily take values other than natural numbers; a view which, following Skolem's 1934 publication, of which Wittgenstein was unaware, is now generally accepted.[29]

If this was Wittgenstein's interest in the Gödel paragraph, then it was absolutely necessary that he should consider the possibility that *PM* was *unsound* (because ω-inconsistency counts as unsoundness for a system that is supposed to formalize mathematics). By contrast, there is no known reason to think that Wittgenstein was interested in the possibility that Peano arithmetic is ω-inconsistent; Peano arithmetic is unquestionably sound mathematics, while *Principia Mathematica* was, we believe, "set theory" (or as bad as set theory) in Wittgenstein's eyes, and thus essentially a misleading construal of mathematics, if not metaphysics in formal dress.

---

28. In Floyd and Putnam, "Note on Wittgenstein's 'Notorious Paragraph' about the Gödel Theorem," we discussed Alister Watson's "Mathematics and Its Foundations," *Mind* 47 (1938): 440–451. Our paper makes clear that Watson discussed the relevant undecidability results and the foundations of mathematics with Wittgenstein and Turing in the summer of 1937, just before Wittgenstein, having traveled to Norway in September, wrote down the notorious remarks. The parallels between the structure of the notorious remarks and the Watson paper's presentation of Gödel are, we believe, no accident.

29. R. L. Goodstein, "Critical Notice of *Remarks on the Foundations of Mathematics*," *Mind* 66, no. 264 (1957): 549–553; quotation from p. 551. In another essay ("Wittgenstein's Philosophy of Mathematics," in Alice Ambrose and Morris Lazerowitz, eds., *Ludwig Wittgenstein: Philosophy and Language* [London: Allen and Unwin, 1972], 271–286), Goodstein says (279): "I [Goodstein] do not think Wittgenstein heard of Gödel's discovery before 1935; on hearing about it his immediate reaction, with I think truly remarkable insight, was to observe that it showed that the formalization of arithmetic with mathematical induction and the substitution of numerals for variables fails to capture the concept of natural number, and the variables must admit values which are not natural numbers. For if, in a system *A,* all the sentences G(n) with n a natural number are provable, but the universal sentence (n)G(n) is not, then there must be an interpretation of *A* in which n takes values other than natural numbers for which G(n) is not true (in fact in 1934, Th. Skolem had shown that this was the case, independently of Gödel's work)."

These issues are important because, as our reconstruction of his argument makes clear, Steiner also assumes that Wittgenstein would not have minded if one replaced "*PM*" in the Gödel paragraph by "PA" (Peano arithmetic).[30] Thus although what Steiner writes about PA is true, it is *not* the case that we have "another *mathematical theorem*" that all theorems of *PM* are true (in Tarski's sense) in any system that is not essentially stronger than *PM* itself— any system that is not a rich system of what we would all regard as "set theory." And because Wittgenstein's remark was about *PM* and not about PA, the applicability of Tarski's theory of truth to PA is irrelevant. Steiner's assumption is that the soundness of PA can be proved in whatever system we employ to formalize Tarski's theory of truth. But that is not the same as assuming that that system can prove the soundness of Russell's *Principia Mathematica*. If we are right in our understanding of what Wittgenstein has in mind in writing, "Suppose the opposite is proved in Russell's system," then his whole point was to ask whether we would hold on to our English interpretation of *P* as "*P* is not provable" if ~*P* were proved *and we therefore realized that PM was not sound.* There is no reason to think that Wittgenstein would have regarded PA in the same light as *PM*—and no reason to think that we, looking back all these decades later, should do so either.[31]

This is not merely a historical aside; it is a logical point. Steiner's criticism of the argument he attributes to Wittgenstein ultimately turns on invoking a semantic proof of the Gödel theorem that is a nonconstructive proof in set theory. Tarski himself did not actually give such a proof, but his definition of "satisfaction-in-a-model" does yield the means to express in set theory not only the notion of a sentence being *true in arithmetic,* but also the idea of a set being first-order *definable* in a formalized language of arithmetic.[32] Putting

---

30. The first note to Steiner's "Wittgenstein as His Own Worst Enemy" includes the sentence: "For the purpose of this essay, 'Russell's system' can be understood as first-order Peano arithmetic." Bays too offers this substitution, as we shall see below.

31. With this Steiner and Bays would presumably agree; see Bays, "On Floyd and Putnam on Wittgenstein on Gödel," 207, and Bays, "Floyd, Putnam, Bays, Steiner, Wittgenstein, Gödel, Etc."

32. For a first-order language L, a structure M interpreting L, a formula ψ of L with k free variables and $a_1, \ldots a_k$ elements of the domain M of M, "$M \mid = \psi[a_1, \ldots a_k]$" is taken to assert that the formula ψ is satisfied in M when the k free variables are assigned sequentially to $a_1, \ldots a_k$. Tarski's definition of satisfaction in a structure (or model), his analysis of "truth in an interpretation," amounts to a precise, set-theoretic definition of this notion. Applied to the particular structure of arithmetic that model theorists call "N" (to be characterized in the final

this together with Gödel's manner of defining the notion of "provable in arithmetic" (via Gödel numbering), we can then express the following argument in set theory:

> *Truth in arithmetic* is not definable in a formalized theory of arithmetic, but, as Gödel showed, *provability in arithmetic* is. Hence the two notions cannot be coextensive.

Gödel claimed to have grasped this (nonconstructive) argument already in 1930, but he did not publish it, partly because of the finitistic context of the Hilbert program, and partly out of worry that philosophers hostile to the notion of *truth* might object to his result.[33] What Gödel's 1931 paper (the one Wittgenstein saw) gives, quite self-consciously, is a different, finitistically acceptable argument for the incompleteness of arithmetic via an explicit construction of an undecidable sentence. As Gödel wrote to Menger after reading Wittgenstein's notorious remarks about his theorem, his incompleteness result "is a mathematical theorem within an absolutely uncontroversial part of mathematics (finitary number theory or combinatorics)."[34]

If we assume that arithmetic is bivalent[35] and look at the Gödel theorem not through the eyes of Rosser's improvement, but with Wittgenstein's point in mind, then we can say that what the Gödel theorem shows is *either* that (1) there exists a true but unprovable sentence of number theory in *PM or* that (2) number theory is inexpressible in *PM*. For if *PM* is consistent, and yet we had a proof that ~P, then Gödel's theorem tells us that *PM* is ω-inconsistent.

---

section of our essay), it provides an analysis of the notion of a sentence being "true in arithmetic." Now with L and M as above, a set $X \subseteq M$ is definable in L if there is a formula ψ in $n+1$ free variables and "parameters" $b_1, \ldots b_n$ such that $X = \{a \in M \mid |M| = \psi[a, b_1, \ldots b_n]\}$.

33. See Hao Wang, *Reflections on Kurt Gödel* (Cambridge, Mass.: MIT Press, 1987), 84ff.; cf. J. W. Dawson Jr., *Logical Dilemmas: The Life and Work of Kurt Gödel* (Wellesley, Mass.: A. K. Peters, 1997), chap. 4.

34. *Kurt Gödel Collected Works*, vol. 5, *Correspondence H–Z*, ed. Salomon Feferman, J. W. Dawson, S. C. Kleene, G. H. Moore, R. M Solovay, and John van Heijenoort (New York: Oxford University Press, 2003), 133.

35. The assumption that arithmetic is bivalent can itself be given two very different interpretations: it can mean either (1) that the schema $p \lor {\sim}p$ is accepted as part of the deductive apparatus of arithmetic, or (2) that every sentence of arithmetic is (in language introduced by Dummett, who repeatedly emphasizes the difference between these assumptions) that every sentence of arithmetic is "determinately true or false." See Michael Dummett, *The Logical Basis of Metaphysics* (Cambridge, Mass.: Harvard University Press, 1991), 74–81. Here we have in mind the stronger assumption.

And then there will be no way of defining in the model theorist's sense[36] the set of natural numbers *in PM*.

Moreover, Putnam has observed that our interpretation of the first part of the notorious paragraph may be naturally extended to account for Wittgenstein's final, puzzling remark that "what is called 'losing' in chess may constitute winning in another game." Many readers (Steiner included) have seen in this remark a rejection of the idea of true but unprovable propositions via the idea that truth holds only relative to this or that "game." But on our interpretation Wittgenstein's remark makes much better sense. For if *P* were "false" in what Wittgenstein calls "the Russell sense"—that is, disprovable in *PM* (in which case, as we have said, *PM* would be, if consistent, ω-inconsistent)—and if we imagine another system of set theory, *S*, that is strong enough to prove the consistency of *PM*, then *P will be a theorem of S*. This is a way of showing that what is called "losing" in one game may be called "winning" in another.

Formally put, the argument is very brief:

Assume that   (1) $PM \vdash \sim\!P$, and *PM* is consistent.
Assume an *S* such that   (2) $S \vdash \mathrm{Con}(PM)$.
Then by (1), (2), and Gödel, (3) $S \vdash \sim(PM \vdash P)$.

If we assume that *S* uses the same definitions of the natural numbers, primitive recursive predicates, and so on as *PM*, then (3) implies that

$$(4)\quad S \vdash P.$$

Here is a sense in which, to repeat, "losing" in *PM* implies "winning" in *S*.

Bays's Reply to Our Note

According to Bays, the philosophical upshot of our interpretation is unacceptable in two ways: (1) It ultimately urges on readers the abandonment of *N,* the standard model of arithmetic, because the "core" of our argument (according to Bays) is that "what a given formula 'expresses' depends on the model at which we interpret it,"[37] and we give insufficient weight to what Bays repeatedly calls the "canonicity" of *N*. (2) It forces us to deny the meaningfulness of what Bays takes to be clearly meaningful if not true, namely, the claim that

36. See note 31.

37. Bays, "On Floyd and Putnam on Wittgenstein on Gödel," 201–202; quotation from p. 204.

there are true but unprovable propositions of arithmetic. The sense of this claim's meaningfulness is explored at length by Bays, primarily through a series of counterfactuals: he argues that if Gödel's ~*P* were derived in PA (or *PM*), mathematicians would most likely hunt for new axioms in light of nonstandard results about a particular formalized theory, and they would certainly not abandon *N;* and this hypothetical "abandonment" he takes us to be committed to "urging."[38] But we never did discuss these counterfactuals, and in fact neither (1) nor (2) follows from our reading of Wittgenstein's notorious remarks.

In his paper Bays takes for granted certain views about what Wittgenstein is up to; although he has recently pulled back from endorsing these as readings of Wittgenstein, his apparent presumptions are widely enough shared that they are worth discussing in some detail.[39] (If Bays's own harsh criticisms of our arguments applied, they would presumably also impugn our argument as a reasonable interpretation of Wittgenstein.) First, Bays states that "Wittgenstein is criticizing a relatively common interpretation of Gödel's first theorem: that the theorem shows—or helps to show—that there are true but unprovable sentences of ordinary number theory."[40] What is relevant here, however, is the difference between our view of this critique and Mark Steiner's, which is the most sophisticated version yet published of the (received) view that Wittgenstein (perhaps unwittingly) is presenting a line of thought the conclusion of which denies something that is true, namely, "There are true but unprovable sentences of ordinary number theory." Like Steiner, Bays thinks that (1) Wittgenstein objects to the "common interpretation" of the Gödel theorem "partially because he is skeptical concerning the notion of 'truth' in play here ('True' in what system?)";[41] he adds that (2) Wittgenstein is also "opposed in principle to the derivation of 'philosophical' claims from 'mathematical' arguments."[42] Our interpretation, by contrast, takes Wittgenstein to have been attempting a constructive clarification of the idea of "'true'

---

38. Ibid., 201ff.

39. In his subsequent manuscript "Floyd, Putnam, Bays, Steiner, Wittgenstein, Gödel, Etc.," Bays asserts that he really did not wish to take any stand on interpreting Wittgenstein. Hence he is withdrawing the statements about Wittgenstein (in his "On Floyd and Putnam on Wittgenstein on Gödel"), as well as his apparent acceptance of Steiner's interpretation of the notorious remarks at p. 208 n. 27 of that article.

40. Bays, "On Floyd and Putnam on Wittgenstein on Gödel," 198.

41. Ibid.

42. Ibid.; Steiner has made no such claim.

in a system" by way of his understanding of the actual details of Gödel's 1931 proof; his "skepticism" concerns (as we see it) claims of a specific interlocutor (perhaps himself), not a general contextualism or skepticism about "Truth" (or, e.g., bivalence). We believe that no such *general* principle as (2) is to be found in Wittgenstein's later thought.

At issue, of course, is not merely Wittgenstein interpretation, but the appropriate response to certain philosophical claims about Gödelian incompleteness. Bays correctly sees that our reasoning does not entail that it is simply *false* to say that "Gödel's incompleteness theorem shows that there are true but unprovable sentences of number theory,"[43] but rather that it is *unclear* until we look at the details of the proof. When we do, the original translation of P by the English sentence "P is not provable" must be "given up." What is worthy of discussion is precisely what is meant by "giving up" the original translation. Wittgenstein makes it clear that he is concerned with the translation into German, that is, the translation into ordinary language. This, as we see it, is crucial for determining the aim and structure of Wittgenstein's philosophical *investigation* of the Gödel proof.

Like Steiner, Bays assumes that the substitution of PA for *PM* in reading Wittgenstein's notorious remarks will not affect our understanding of their philosophical significance. He writes:

> Wittgenstein focuses on number theory as formulated in "Russell's system"—that is, the system of *Principia Mathematica*. There is, however, nothing in his argument that depends on this particular choice of background logic. For expository convenience I will recast the argument in terms of ordinary, first-order Peano Arithmetic. Later . . . I discuss the possible philosophical significance of formulating the argument in "Russell's system."[44]

In our rendition of Wittgenstein's notorious paragraph, the force of his insight crucially depends on the fact that he is discussing *PM*. But when Bays offers what he calls a "sketch of the claims Floyd and Putnam want to argue against" and an account of the "core" of our argument, he presents these as a matter of imagining that PA, and not *PM,* might turn out to be unsound.[45]

---

43. Ibid., 203.
44. Ibid., 198 n. 4.
45. Cf. ibid., 200–201. In "Floyd, Putnam, Bays, Steiner, Wittgenstein, Gödel, Etc." he continues to say that his argument "follows Floyd and Putnam in supposing that we have

He admits that given the relative security of the axioms of PA, the former scenario is "extremely implausible" and "would require such deep revisions of present mathematics that it is virtually impossible to adjudicate questions concerning 'what we would/should do' in such circumstances."[46] With this we would agree. But Bays goes on to so adjudicate and states that our argument "surely" entails only one answer to the "implausible" counterfactual, something that is for Bays *"unimaginable,"* namely, that in this "implausible" scenario mathematicians would "adopt recognizably nonstandard models of arithmetic as canonical for interpreting the language of number theory"[47] (and thereby throw out PA [or *PM*]).

But we never discussed the "implausible" counterfactual at all, and it is difficult to see how our interpretation would entail any position with respect to it. But the consequences of Bays's shift in presenting what we assumed are significant for his reading of our "Note." For our interpretation, according to Bays, entails that it is impermissible for anyone to "step back to make semantic generalizations," or to "recognize from the *outside* that [e.g.] contradictions are unacceptable."[48] As Bays insists (presumably with us and/or Wittgenstein in mind), "We certainly have *some* ability to step back and engage in critical reflection on (purported) foundations."[49]

Of course we have such ability, and our reading of the notorious remarks, whether wrong or right about Wittgenstein, shows one instance of critical reflection at work. But Bays claims to find in our discussion the suggestion that were PA to turn out to be unsound, "only models which satisfy PA should count as 'admissible interpretations' for our language,"[50] and this leads him to assert (erroneously) that "the insistence that we limit ourselves to models of PA when we interpret arithmetic runs rather deep in Floyd and Putnam's paper."[51] In the end, then, Bays thinks that our reading of the notorious remarks

---

discovered a proof that PA is *not* sound" (8). But again, we never discussed PA and do not take Wittgenstein to have been doing so.

46. Bays, "On Floyd and Putnam on Wittgenstein on Gödel," 205.

47. Ibid., 204–205.

48. Ibid., 210 n. 32.

49. Ibid.

50. Ibid., 204.

51. Ibid., 204 n. 17. Referring to earlier work of Putnam's, he holds that Putnam, in particular, is committed to "a similar insistence on the priority of axioms over interpretations" (ibid., 204 n. 18). But we are not applying earlier arguments of Putnam's; see note 56 below.

must work against conceiving the standard model of arithmetic, N, Th(N)—and even model theory generally—as viable objects of study for mathematicians. He alleges that our "Note" leads ineluctably to "the abandonment of N that Floyd and Putnam urge upon us,"[52] and he suggests that this could even lead to a principled rejection of "the semantic analysis of formal systems," to a "crazy" position that is offering "principled reasons for permitting us to step back to make syntactic generalizations, while forbidding us to step back to make semantic generalizations."[53]

The reader will see no trace of these ideas in either our "Note" or the argument sketched above. So what has happened?

Although Bays says that he will make no interpretive claims about Wittgenstein,[54] he makes several suggestions about the notorious remarks and our interpretation of them that are, unfortunately, mistaken. Because these are often linked to Wittgenstein explicitly (without interpretive argument), they are worth examining in themselves, even if now Bays would not wish to ascribe any particular interpretation to Wittgenstein. The first is the suggestion that inspired by Wittgenstein, we wrote about Gödel in the grip of antirealism, verificationism, and/or formalism, and that we identified a priori the notions of "mathematically true" and "proved" (or perhaps, less stringently, "provable" or "provable in some formal system or other"). As Bays writes, "At best, Floyd and Putnam suggest, Gödel's paper gives rise to a proof-theoretic conception of truth. . . . Clearly, however, this is an interpretation that leaves no room for "true but unprovable" sentences of arithmetic."[55] But, contrary to what Bays says, the argument presented in our "Note" neither

52. Ibid., 204 n. 19.

53. Ibid., 210; cf. n. 31. There Bays attributes this view quite explicitly to us (or perhaps to Wittgenstein), asserting that "as a point *tu quo,* I would note that Floyd and Putnam themselves engaged in purely semantic reasoning about Russell's system." Of course we do so engage; this should have suggested to Bays that he had misinterpreted our position.

54. Ibid., 197.

55. Ibid., 202 n. 14. In "Floyd, Putnam, Bays, Steiner, Wittgenstein, Gödel, Etc." 202, n. 16, Bays expresses astonishment that we record this as his suggestion, but in the same manuscript he again links our interpretation to an attempt to place "pressure" on what he calls "naïve realism" (12; the whole final section of the manuscript is relevant). This may be because he links our "Note" to earlier work of Putnam's applying a model-theoretic argument to question realism (Putnam's work and Bays's previously published criticisms of it are referred to in Bays, "On Floyd and Putnam on Wittgenstein on Gödel," 204 n. 18, and in Bays, "Floyd, Putnam, Bays, Steiner, Wittgenstein, Gödel, Etc." at 209–210, n. 27). The argument of our

depends on nor was used to advocate such a philosophical point of view: as we have said, one of our main points was to offer a reading of the notorious paragraph that contains something more than a facile philosophical prejudice identifying the notions of "truth" and "proof."

Nor does our reading depend on the assumption, which Bays quite explicitly attributes to Wittgenstein at the opening of his paper, that one should be "opposed in principle to the derivation of 'philosophical' claims from 'mathematical' arguments."[56] On the contrary, our interpretation credits Wittgenstein with sensitivity to the subtle interplay between ordinary and formalized language in the context of philosophical disputes about the foundations of mathematics. What we were arguing is that neither formalism nor verificationism about mathematical truth in general is at issue in Wittgenstein's 1937 remarks on Gödel.[57] (It is worth nothing that although Steiner does take Wittgenstein to be identifying "true in *Principia Mathematica*" with "proved in (the game of playing with) *Principia Mathematica*" in the notorious paragraphs, he rejects, just as we do, the idea that Wittgenstein was a verificationist.)[58]

In closing, we must consider Bays's reasoning about the so-called standard model of arithmetic and the notion of "truth." Throughout his paper Bays assumes that the notions of "truth," "interpretation," and "natural number" are and ought to be determined from within model theory alone, with reference—indeed, what Bays calls "perfectly canonical" reference—to the set (or model) N.[59]

Bays never defines N, but let us do so now. N is a structure in the model theorist's sense, that is, a set conceived of as coming to us equipped with two distinguished elements, zero and one, and the functions of addition and

---

"Note" is not to be read, however, as an application of Putnam's earlier model-theoretic arguments against realism.

56. Bays, "On Floyd and Putnam on Wittgenstein on Gödel," 198.

57. Several reasons for doubting that the notorious remarks express verificationism about mathematical truth as a whole are canvassed in Floyd, "On Saying What You Really Want to Say." For the larger issue of verificationist tendencies in Wittgenstein's other later remarks, see Hilary Putnam, "Was Wittgenstein *Really* an Antirealist about Mathematics?" in Timothy G. McCarthy and Sean Stidd, eds., *Wittgenstein in America* (Oxford: Clarendon Press, 2001), 140–194; reprinted in this volume as Chapter 23.

58. Personal correspondence with Mark Steiner; cf. Mark Steiner, "Wittgenstein: Mathematics, Regularities and Rules," in Adam Morton and Stephen Stich, eds., *Benacerraf and His Critics* (Cambridge, Mass.: Blackwell Publishing Company, 1996), 190–212.

59. Bays, "On Floyd and Putnam on Wittgenstein on Gödel," 210.

multiplication. N is to be regarded as an interpretation for a corresponding first-order language L containing the usual apparatus of quantification theory, two binary function symbols, and two constant symbols. The full theory of N, $Th(N)$, sometimes called "true arithmetic" by model theorists, is defined as the set of sentences of L that are satisfied in N in the sense of Tarski; $Th(N)$ is thus defined recursively via Tarski's notion of "satisfaction in a structure."

We are given by this construction that every sentence of L will either be in $Th(N)$ or will not be. What this definition does not tell us, however, is how to determine, for any arbitrarily given particular sentence of L, whether that sentence is or is not an element of $Th(N)$. Indeed, by establishing that $Th(N)$ is in principle not recursively axiomatizable, Gödel showed that there is no way of effectively determining this. He thereby showed that $Th(N)$ can have no categorical axiomatization—that, in other words, it must have nonisomorphic ("nonstandard") models.[60]

Bays tries to eliminate the issue of ω-consistency by reiterating several times that N is a "core" or "perfectly canonical" "interpretation" of "mathematical truth" applicable to every formula of PM and our natural language in the same way, across a wide variety of counterfactual contexts.[61] But why Bays thinks that we should assume that the notion of "interpretation" is reducible to the model theorist's notion of N he does not say. Nor does he explain what he means by a "perfectly canonical" "interpretation."

For our part, we were careful to distinguish, as most philosophers do, between the notions of "interpretation," "model," and "translation." In terms of the notorious paragraphs under discussion, the relevant contrast is between *Übersetzung* (translation) and *Deutung* (interpretation, implying clarification), a distinction that Wittgenstein's German and our discussion of the notorious paragraph respect. Unfortunately Bays slides equivocally between them.[62]

---

60. Note that Gödel's proof applies to any first-order formalized language L^* sufficiently rich to express the usual arithmetical operations of addition, multiplication, and quantification on the natural numbers. Thus even if it can express more than the minimal language L, any such L^* will fail to recursively axiomatize a full theory of the structure consisting of the natural numbers and the functions of addition and multiplication on these numbers.

61. Bays, "On Floyd and Putnam on Wittgenstein on Gödel," 204, 206, 208, 210.

62. A running together of the notions of "model" and "interpretation" occurs when Bays mistakenly says that the "core" of our argument is the idea that "*all* 'admissible interpretations' [of arithmetic] take place on nonstandard models [because] . . . what a formula 'expresses' depends on the model at which we interpret it" ("On Floyd and Putnam on Wittgenstein on

On our view, whether a given sentence of *PM* does or does not have a model of a particular kind, subject to a particular assumption about *PM*'s being ω-consistent, affects but does not exhaust the issue of whether, why, and to what extent we can give a translation or interpretation in an informal sense of sentences of *PM* in English apart from that assumption. As we explicitly wrote in our "Note," a kind of translation (a "correlation" or "transformation") of sentences of *PM* into English might be possible even if it turned out that there were no models, and hence no admissible interpretations, of *PM* itself as a whole.[63] In short, the relations among the notions of "model," "interpretation," and "translation" are as complex as the relations of any of these notions to that of "meaning": to assess them, we need to be clear about the context within which we are using them. Wittgenstein's insight, as we understand it, is but a special case of this more general point.

To summarize: Bays assumes throughout his paper that his remarks have a direct bearing on philosophical problems about the notion of "truth." Yet we find it odd that Bays finds it reasonable to grant (1) that there exist philo-

---

Gödel," 202–203). What a formula "expresses" may depend on there being a model, or it may not, as we say ("Note on Wittgenstein's 'Notorious Paragraph' about the Gödel Theorem," 626); this just shows that we have *not* reduced the notion of interpretation (or of truth) to something *only* model relative. Note, for the same running together, Bays, "On Floyd and Putnam on Wittgenstein on Gödel," 204 n. 17: "The insistence that we limit ourselves to models of PA when we interpret arithmetic runs rather deep in Floyd and Putnam's paper. At one point they even suggest that, were we to find PA inconsistent, we should conclude that there are no admissible interpretations of arithmetic," (see Floyd and Putnam, "A Note on Wittgenstein's 'Notorious Paragraph' about the Gödel Theorem," 626). As is easily verified, we never spoke about all admissible interpretations of arithmetic (nor even about PA), but, following Wittgenstein, only about "admissible interpretations of *PM*" (Bays has simply misread and misquoted the pronoun "it" in the passage he cites). By "admissible interpretation" we understood an interpretation "fitting" at least one model of *PM* (cf. Floyd and Putnam, "Note on Wittgenstein's 'Notorious Paragraph' about the Gödel Theorem," 625–626). If *PM* (or any other formal theory) is simply inconsistent, there will be no models of it, and hence a fortiori there will be no admissible interpretations of it (in our sense), although there could be (in our sense) translations and improvements of it. The relevant contrast is not between the merely "syntactic" and that which "involves interpretation" in the model theorist's sense, as Bays seems to assume (cf. "On Floyd and Putnam on Wittgenstein on Gödel," 202); nor is there any "insistence" on prioritizing axioms over models (despite Bays, "On Floyd and Putnam on Wittgenstein on Gödel," 204).

63. Floyd and Putnam, "Note on Wittgenstein's 'Notorious Paragraph' about the Gödel Theorem," 626.

sophical problems about arithmetical truth, and also (2) that the existence of model theory will resolve these problems. In fact, Bays seems to think that an appeal to N proves not only that the Gödel sentence has a truth-value, but also that every sentence of arithmetic has a truth-value—that, in other words, the principle of bivalence holds in arithmetic.[64]

It is not that the authors of the "Note" are defenders of any philosophy on which bivalence fails in arithmetic.[65] But neither would we want to read intuitionists (Brouwer, Heyting) or nominalists, finitists, and so on out of the camp of philosophy. And we certainly would not claim that Brouwer, Goodman, and others have been *refuted* by Tarski, Abraham Robinson, and the other founders of model theory.

Does Bays think otherwise?

APPENDIX

My task is, not to talk about (e.g.) Gödel's proof, but to by-pass it.[66]

Ludwig Wittgenstein

Steiner and Wittgenstein on Mathematical Truth

Steiner describes Wittgenstein as "a philosopher who connects the notion of mathematical truth with mathematical proof."[67] Whether Wittgenstein

---

64. In his manuscript "Floyd, Putnam, Bays, Steiner, Wittgenstein, Gödel, Etc.," 209–210, n. 27, Bays says that he finds our distinctions among interpretation, translation, and model "quite misleading," and our emphasis on "more nuanced and complicated" aspects of meaning irrelevant to the main "disagreement" between us, which is, he speculates, the issue of (what he calls) "naïve realism" (see his section 4, "What's the Issue?"). No such form of realism was discussed in Bays's "On Floyd and Putnam on Wittgenstein on Gödel," but in any case our "Note" was not preoccupied with any such view (see note 56 above for references to articles where realism is discussed; compare note 52). A discussion of Bays's own ideas about "naïve realism" lies outside the scope of this essay.

65. See Hilary Putnam's Tarski Lectures, "Paradox Revisited I: Truth" (reprinted here with the title "Revisiting the Liar Paradox" as Chapter 10) and "Paradox Revisited II: Sets—A Case of All or None?" in Gila Sher and Richard Tieszen, eds., *Between Logic and Intuition: Essays in Honor of Charles Parsons* (New York: Cambridge University Press, 2000), 3–26.

66. Wittgenstein, *Remarks on the Foundations of Mathematics*, VII, §19.

67. Steiner, "Wittgenstein as His Own Worst Enemy," 260–261.

did think that mathematical truth is coextensive with mathematical provability is currently a hotly debated question. Our view is that *Philosophical
Investigations* and most of Wittgenstein's *Remarks on the Foundations of Mathematics* do not presuppose a generally antirealist philosophy of any kind. That
does not mean that there are no problems with Wittgenstein's remarks, and I
[HP] claimed that at least one of them does flirt with the position Steiner
ascribes to him.[68] We suggest, however, that in any case the strange-sounding
remark that "true in Russell's system" means, as was said, "proved in Russell's
system" does *not* represent Wittgenstein's view concerning *mathematical*
proofs—that is, proofs that are given by mathematicians and accepted as part
of "real" mathematics—but rather expresses his attitude toward *systems* of the
sort that Russell and Whitehead constructed. Given that *PM* is a system of
set theory (or can be thought of as such, as Wittgenstein's friend and colleague
Ramsey had pointed out), and that Wittgenstein (wrongly in both our view
and Steiner's) thought of or interpreted set theory as a metaphysical game,[69]
it is not necessary to read into this remark any more than a dismissal of *PM*
as such a "game"—a dismissal that does not imply that Wittgenstein thought
of the "motley" of activities that he regarded as "real" mathematics as a formal game or set of such games, or truth in mathematics as provability in some
"game."[70] In fact, Steiner himself points out that Wittgenstein "viewed set
theory as academic philosophy in sheep's clothing: a pseudoexplanatory
'theory' having no redeeming applications to anything other than set theory.
To assault set theory, therefore, was not to revise *mathematics*."[71] I suggest that
similarly, for Wittgenstein, to assault *Principia Mathematica* was not to revise
mathematics.

68. Putnam, "Was Wittgenstein *Really* an Antirealist about Mathematics?" appendix.

69. I am aware that Wittgenstein always (that is, from the *Notebooks* on) granted that,
treated as a "calculus," *Mengenlehre* is all right and must be judged as mathematics (personal
communication from Juliet Floyd). However, as I read it, that is only to say that as a mere
formal system, set theory is a legitimate object of mathematical study. Since—as Steiner himself points out—Wittgenstein regarded the Cantor diagonal argument as metaphysical, he
could hardly regard the putative "objects" of set theory as legitimate objects of mathematical
study in the way, say, the complex numbers are.

70. Even if Wittgenstein was unaware of Ramsey's identification of *PM* with a set theory,
there is every reason to think that he would have regarded the theory of "propositional functions of infinitely many types," which is what *PM* presents itself as being, as being at least as
"metaphysical" as set theory, if not more so.

71. Steiner, "Wittgenstein as His Own Worst Enemy," 270.

Steiner's Criticism of Wittgenstein's Remark That "What Is Called 'Losing' in Chess May Constitute Winning in Another Game"

If our interpretation of the Gödel paragraph is right, then Steiner's criticism of Wittgenstein's remark that "what is called 'losing' in chess may constitute winning in another game" is certainly misguided. Steiner writes, "What is fallacious here is that we have a *mathematical theorem* that 'losing' in Russell's system implies losing in Tarski's! So if 'false in some other than the Russell sense' includes the 'Tarski sense' Wittgenstein is simply wrong."[72] (Of course, Steiner does not think that Wittgenstein knew of Tarski's theory of truth; Steiner is simply arguing that that theory invalidates Wittgenstein's remark.) Steiner's claim that it is a "mathematical theorem" that "losing" in Russell's system implies losing in Tarski's is explained in the first few sentences of the same paragraph in his article,[73] previously quoted on pages 464–465 of this chapter.

But—as Steiner himself well knows—Peano arithmetic (which Wittgenstein would have had no reason not to regard as "mathematics") is not the same as *PM* ("Russell's system"), which, we believe, Wittgenstein (rightly or wrongly) did not regard as mathematics. And the argument in the paragraph just quoted depends on assuming that Wittgenstein would have accepted Steiner's equation of PA with *PM*. In "Wittgenstein as His Own Worst Enemy" Steiner writes: " 'Russell's system' from now on will be referred to as Peano Arithmetic."[74]

If the interpretation of the Gödel paragraph Juliet Floyd and I have defended is right, then the crucial clause is "Suppose the opposite is proved in Russell's system." That supposition makes no sense if we suppose that we are working in a "Tarskian" system *strong enough to prove the soundness of PM*. For on our interpretation the whole point of Wittgenstein's paragraph is to ask us whether we would hold on to the interpretation of *P* as "*P* is not provable" if "the opposite" (of *P*) were proved, and *we therefore realized that PM was not sound*. Steiner's whole paper assumes that of course *PM* is sound (because PA is, and Steiner chooses to identify the two), and therefore we do not have to think about what it would mean if ~*P* were provable in *PM*. But even at the level of surface grammar, that is the assumption that the Wittgenstein paragraph is about.

72. Ibid., 267.
73. Ibid.
74. Ibid., 265.

28

Wittgenstein: A Reappraisal

The papers on skepticism collected in Part V were written during a period when I was less critical of some of Wittgenstein's views (especially the *meta*philosophical views) than I am now. I do not mean by those words to suggest that I regard Wittgenstein's influence on my thinking as a bad thing. I still admire and find inspiration and instruction in those passages in which Wittgenstein calls our attention to the plurality of our language games and the plurality of the forms of life that are interwoven with those games, and in his rejection of the idea that only scientific language is really "first-class" language,[1] his rejection of the idea that rule following must be accounted for either physicalistically or Platonistically, and many other insights. And the idea that giving a "perspicuous representation" of the ways in which we employ our words in quotidian life can *assist* us to avoid the grip of misleading philosophical pictures is also right. But the idea that metaphysics as a whole,

---

1. Contrast Quine's claims about what does and does not belong to our "first-grade conceptual system." W. V. Quine, *Ontological Relativity and Other Essays* (New York: Columbia University Press, 1969), 24.

starting with Socrates,[2] is valueless and, indeed, a sort of illness,[3] explained by "grammar" going wrong and to be cured by a mysterious sort of therapy, is, I believe, quite wrong.[4]

All this being said, I do still think that there are good grounds for questioning (as I did in these papers) the "full intelligibility" of some of the classical skeptical scenarios. But I have come to see those grounds as having to do with the intellectual world we inhabit as inheritors of the Enlightenment, and not with the supposed fact that metaphysical sentences somehow lack a "meaning." In this brief essay I want to explain what I mean by this.

What I Said about "Not Making Sense" in These Papers

Here is a sentence I wrote in one of the essays collected in this volume (Chapter 31, "Strawson and Skepticism"): "If our conception of what an experience *is* indeed derives from our mastery of ordinary talk of persons' experiencing particular objects, properties of objects, interactions between objects, and so on, then such ideas as the idea that persons might *be* experiences (or immaterial 'spirits' plus experiences) do not make sense (except in the way in which a myth or a fairy tale makes sense)." I blush at this sentence.

The reason I blush *is not*, however, that I said that such ideas "do not make sense." A glance at the history of philosophy will show that just about every philosopher from Plato on has sometimes questioned whether some of the ideas of other philosophers "make sense."[5] Even the term "nonsense" is sometimes in

---

2. "Reading the Socratic dialogues one has the feeling: what a frightful waste of time! What's the point of these arguments that prove nothing and clarify nothing?" Ludwig Wittgenstein, *Culture and Value*, ed. G. H. von Wright (Oxford: Blackwell, 1980), 14e.

3. E.g., "In philosophizing we may not terminate a disease of thought. It must run its natural course, and slow cure is all important." Ludwig Wittgenstein, *Zettel* (Oxford: Blackwell, 1981), §382.

4. The prejudice against metaphysics seems to have gone back to Wittgenstein's student days and is perhaps partly explained by his early admiration for Boltzmann.

5. Mario De Caro pointed out to me a spectacular example of this from Schopenhauer: "What was senseless and without meaning at once took refuge in obscure exposition and language. Fichte was the first to grasp and make use of this privilege; Schelling at best equaled him in this, and a host of hungry scribblers without intellect or honesty soon surpassed them both. But the greatest effrontery in serving up sheer nonsense, in scrabbling together senseless and maddening webs of words, such as had previously been heard only in madhouses, finally appeared in Hegel . . . [who was] a commonplace, inane, loathsome, repulsive and ignorant

order as a term of criticism of views with which one finds fault; but what justifies such strong language is, in each case, a philosophical (or sometimes a
scientific or simply a commonsense) argument, not an appeal to a theory of
language, and certainly not an appeal to something that claims to be about
language (or "grammar") but "not a theory." And to say, as I did in "Strawson
and Skepticism," that talk of disembodied spirits (i.e., disembodied minds)
or of experiences apart from bodies does not make sense on the grounds that
that is not how we acquired our mastery of ordinary talk now seems to me a
relic from the weakest side of "ordinary-language philosophy." As I point out
in Chapter 24, we frequently and legitimately and necessarily "project" concepts into contexts very different from the ones in which "we acquired our
mastery" of those concepts.

This is not to say that I am unhappy with *everything* I said in "Strawson
and Skepticism." In that same paper I question the internal coherence of
several skeptical scenarios on what still seem to me sound *philosophical*
grounds (e.g., that the Humean idea that the world might consist of a *temporal sequence* of "loose and separate" events assumes that the concepts "earlier" and "later" make sense *even if the world does not have a causal structure,*
something that both Leibniz and Kant powerfully challenged). I still believe
that conceptual analysis is important in philosophy, but I long ago abandoned the idea that conceptual analysis is a discovery of "analytic" truths;
what it is, as the term implies, is a study of our concepts, meaning by that
not Platonic entities, but our system of (putative) knowledge, and particularly of what has a framework character in that system. The great philosophers (Leibniz and Kant in particular) who first pointed out the deep relations between time order and causality were not discovering analytic truths
or any other sort of *linguistic* facts, unless one decides to stipulate that facts
about which of our beliefs have a framework character and about the ways
in which the framework as a whole is structured are "linguistic" facts. But
that seems a very bad idea. (Strawson's term "descriptive metaphysics" is,
I think, happier.)

---

charlatan, who with unparalleled effrontery compiled a system of crazy nonsense that was
trumpeted abroad as immortal wisdom by his mercenary followers." Arthur Schopenhauer,
Die Welt als Wille und Vorstellung (Leipzig, 1819); English translation, *The World as Will and
Representation* (New York: Dover, 1966), 429.

The "Two Wittgensteins"

This issue very much connects with Wittgenstein interpretation because the question of what Wittgenstein meant (both in the *Tractatus* and in the later philosophy) by describing metaphysical statements as nonsense has become *the* major bone of contention among his admirers and interpreters.

One group of interpreters, led by two brilliant philosophers, James Conant and Cora Diamond, has inaugurated a whole family of interpretations, often referred to as the "new Wittgenstein" interpretations.[6] These interpreters find important *continuities* between Wittgenstein's *Tractatus* and his *Philosophical Investigations*. Now, the notion of nonsense—in fact, *two* notions of nonsense, the notion of *unsinnige Sätze* and the notion of *sinnlose Sätze*—clearly plays an important role in the *Tractatus*. Thus the "new Wittgensteinians" are led to stress the role of the notion of "nonsense" in the *Investigations* as well.[7] Many other interpreters favor an "orthodox" interpretation of *Philosophical Investigations,* sometimes referred to as the "Baker & Hacker" interpretation.[8] According to that interpretation (which finds much less continuity between the *Tractatus and Philosophical Investigations*), *Philosophical Investigations* is supposed to teach us that the "criteria" Wittgenstein speaks of are actually a framework of rules such that philosophers speak nonsense when (and because) they "violate" those rules.[9] The "new Wittgensteinians" deny that "criteria" are supposed to play that role.[10] Indeed, they deny that Wittgenstein thought that there was some *one* reason that philosophers often think that they are talking sense when their sentences are actually senseless, although

6. See Rupert Read and Alice Crary, eds., *The New Wittgenstein* (London: Routledge and Kegan Paul, 2000).

7. I also believe that although Wittgenstein's contempt for metaphysics finds occasional expression in *On Certainty* (Oxford: Blackwell, 1969), the great insights of that work are utterly independent of it. Both "orthodox" Wittgensteinians and "new Wittgensteinians" would strongly disagree.

8. I take the expression "Baker & Hacker," including the ampersand, from Stephen Mulhall, "The Givenness of Grammar: A reply to Steven Affeldt," *European Journal of Philosophy* 6, no. 1 (April 1998): 32–44. This is a fine example of an "orthodox" Wittgensteinian replying to a "new Wittgensteinian." Gordon Baker defected from the Baker & Hacker camp late in his life; see Baker, "Wittgenstein on Metaphysical/Everyday Use," *Philosophical Quarterly* 52, no. 208 (2002): 289–302.

9. I criticize this interpretation in Chapter 24.

10. Here they follow Stanley Cavell in *The Claim of Reason* (Oxford: Clarendon Press, 1979).

James Conant has suggested that this may happen because the philosopher hovers between two incompatible meanings without realizing that they are incompatible.[11]

Even if I am right, and the idea of "nonsense" is far less important in *Philosophical Investigations* than it was in the *Tractatus* (and still less important in *On Certainty*), the combatants in the Wittgenstein wars are right that Wittgenstein *did* think that what other philosophers—apparently, just about *all* other philosophers—wrote was nonsense, and that he meant the term "nonsense" quite seriously. Moreover, both sets of combatants give *linguistic* explanations of the term "nonsense," whether they be in terms of violating rules of language (the "Baker & Hacker" interpretation) or simply failing to give a determinate meaning to words (the "new Wittgenstein"). It is here, precisely where the two "camps" *agree*, that I now *disagree*.

How This Change in My Views Relates to My Papers on Skepticism

Several of the students in a seminar on skepticism that I gave in 2010 argued that the statement "Two plus two could turn out not to be four" is true, on the ground that we have been wrong in the past even about statements that seemed at the time to be necessary truths—a point I myself emphasized in defending my "Cavellian" interpretation of Wittgenstein in Chapter 23, by the way. My reply to these students *was not* that this sentence is literally "devoid of meaning," which would have been the response of a new Wittgensteinian, or that that it "violates" our criteria for using such and such words, à la Baker & Hacker. I *did,* however, say, and would still say, that it lacks "full intelligibility," because the students who said this were totally unable to say *how* it could "turn out" that two plus two is not four.[12]

Similarly, I would still say that there *is* a sense in which I cannot regard such putative "possibilities" as the possibility that some human beings are zombies[13] as fully intelligible, because I cannot understand *how* something

11. James Conant, "On Wittgenstein's Philosophy of Mathematics," *Proceedings of the Aristotelian Society,* Suppl. 70 (1996): 243–265.

12. See Gary Ebbs, "Realism and Ration Inquiry" in *Philosophical Topics* 20, 1 (1992): 1–33, and my reply, 347–408.

13. An argument of Jaegwon Kim's that I criticize in lecture 1 of "Mind and Body," my Royce Lectures, reprinted as part 2 of Hilary Putnam, *The Threefold Cord: Mind, Body, and World* (New York: Columbia University Press, 1999), turns on this possibility.

could have the same physical makeup as a normal human being (and the same behavior, with the same physical causes) and lack consciousness. To think that zombies are a possibility, I would have to assume that my entire view of the world is wrong in a way I am utterly unable to make cohere with anything I believe. But that is not the same thing as being "nonsense" either in the sense of the Baker & Hacker camp or literally failing to have a meaning à la the "new Wittgenstein."

Although I was certainly not clear on this when I gave my Royce Lectures, "Mind and Body,"[14] the problem with understanding "Some individuals are actually zombies," like the problem Wittgenstein discusses in *Lectures and Conversations on Aesthetics, Psychology, and Religious Belief* with understanding such a sentence as "There will be a Last Judgment, and the saved will be separated from the damned,"[15] is not a *linguistic* one; *it is not that the words literally do not have a meaning in that sentence*; nor is it the case that when Jaegwon Kim argued that if reductionism is false, then the statement that some human beings whose behavior and physical properties are perfectly normal lack mental properties must describe a metaphysical possibility, he failed to notice that the words lack a meaning.[16]

My refusal to believe that the sentences of philosophers (e.g., metaphysicians and skeptics) are devoid of content and/or are in violation of "criteria" for making sense is not based only on the obvious implausibility of such views;[17] the fact is that I do not find on reading Plato, or Hegel, or Kant, or

14. Reprinted as part 2 of Putnam, *Threefold Cord.*

15. Ludwig Wittgenstein, *Lectures and Conversations on Aesthetics, Psychology, and Religious Belief,* Cyril Barrett, ed. (Oxford: Blackwell, 1966), 55 (Wittgenstein does not actually give an example of a specific sentence about the Last Judgment, but he indicates that "separate" is one of the words he is thinking of as being used in this connection, and I have therefore constructed a sentence he might well have been thinking of).

16. See Jaegwon Kim, "The Myth of Nonreductive Materialism," in Kim, *Supervenience and Mind* (Cambridge: Cambridge University Press, 1993). What Kim actually argued was that if reductionism is false, and Cartesian dualism is untenable, then "mental properties make no causal difference" (269–270); it follows that if some human beings lacked those properties (were "zombies"), it would make no difference to their behavior, or to the causation of their behavior, an implication Kim accepts.

17. By the way, I rejected the idea of "deep" facts about language accessible to Wittgensteinian philosophers but not to linguists and anthropologists as early as Hilary Putnam, "Dreaming and 'Depth Grammar,'" in R. J. Butler, ed., *Analytical Philosophy,* 1st ser. (Oxford: Basil Blackwell, 1962), 211–235; reprinted in Putnam, *Philosophical Papers,* vol. 2, *Mind, Language and Reality* (Cambridge: Cambridge University Press, 1975), 304–324.

Spinoza, or Leibniz that I am reading *meaningless sentences*.[18] I cannot inhabit the intellectual world of these philosophers, but to suggest that what they wrote was "nonsense" is a hangover from the mistaken idea that we should "just say no" to metaphysics.

But Is This Giving Point, Set, and Match to the Skeptic?

In the essay collected as Chapter 30 in this volume, I write,

> If my purpose is to put my own intellectual home in order, then what I need is a perspicuous representation of our talk of "knowing" that shows how it avoids the skeptical conclusion, and that my *nonskeptical* self can find satisfactory and convincing. . . . What I have tried to provide . . . is an argument *that convinces me* that *the skeptic cannot provide a valid argument from premises I must accept to the conclusion that knowledge is impossible*. In the same way, Russell showed that (after we have carefully reconstructed our way of talking about sets) a skeptic—or whoever—cannot provide a valid argument from premises we must accept to the conclusion that mathematics is contradictory, and Tarski showed that (after we have carefully reconstructed our talk about truth) a skeptic—or whoever—cannot provide a valid argument from premises we must accept to the conclusion that talk about truth is contradictory.[19]

For similar reasons, the fact that we cannot provide an argument *from premises that the skeptic must accept* to the conclusion that skepticism is mistaken does not show that if we wish to be rational, we have to agree that the skeptic is right. Nor does it show that the skeptic's utterances are just as *reasonable* as our utterances when, for example, we refuse to agree with him that we "do not know" that "two plus two equals four" will not turn out to be false, or that we "do not know" that Massachusetts is not any longer a British colony, or that we "do not know" that we have hands and feet. If our ability to defend

---

18. In fact, most of Wittgenstein's own students were fine interpreters of classical texts; at times it seems that their favorite example of a philosopher who was allegedly talking nonsense was Quine.

19. Hilary Putnam, "Skepticism," in Marcelo Stamm, ed., *Philosophie in synthetischer Absicht* (Stuttgart: Klett-Cotta, 1998), 239–268; reprinted in this volume with the title "Skepticism and Occasion-Sensitive Semantics" as Chapter 30. In particular see the passage about skepticism as a paradox or "antinomy."

our ways of thinking by arguments that *we ourselves* find convincing is less than rationalist philosophers hoped for, perhaps that is what Quine meant by saying that "the Humean predicament is the human predicament,"[20] and what Cavell meant by saying that "the truth of skepticism, or what I might call the moral of skepicism," is that "the human creature's basis in the world as a whole, its relation to the world as such, is not that of knowing, anyway not what we think of as knowing."[21] But Wittgenstein was right to emphasize in *On Certainty* that our ways of using the verb "to know" and the adjective "certain" do not need a "foundation" to be perfectly useful.[22] I began this short essay by saying that I do not agree with much of Wittgenstein's metaphilosophy (especially the idea of "therapy"); but I love it when he writes, "And one might almost say that these foundation-walls are carried by the whole house."[23]

So Why Do I Still Speak of "Lacking Full Intelligibility"?

I have now given up the idea that the skeptic's "scenarios" are *nonsense* in some supposedly "grammatical" sense of "nonsense," but I did say above that my students' claim that it could turn out that two plus two is not four (just as, e.g., it turns out to be possible—according to general relativity—that two straight lines could be perpendicular to a third straight line and still intersect)[24] "lacks full intelligibility," because "the students who said this were totally unable to say *how* it could turn out that two plus two is not four," and I also said that "there *is* a sense in which I cannot understand such alleged metaphysical possibilities as the possibility that some human beings are zombies, because I cannot understand *how* something could have the same physical makeup as a normal human being (and the same behavior, with the same physical causes) and lack consciousness." Is this not the same as saying that these claims violate conditions for making sense? And are not such conditions what Wittgenstein

20. Quine, *Ontological Relativity and Other Essays*, 72.

21. Cavell, *Claim of Reason*, 241. For further discussion, see Chapter 32.

22. Wittgenstein, *On Certainty*.

23. Ibid., §248.

24. See Chapter 24; I first argued this in "It Ain't Necessarily So," *Journal of Philosophy*, 59, no. 22 (1962): 658–671; reprinted in Putnam, my *Philosophical Papers*, vol. 1, *Mathematics, Matter and Method* (Cambridge: Cambridge University Press, 1975), 237–249.

is supposed (on the Baker & Hacker interpretation, at least) to have meant by "grammar"?

To see how one can find something unintelligible, or of problematic intelligibility, without saying that it is "literally nonsense" (like the "new Wittgensteinians"), or that it violates rules that determine what does and does not make sense (like Baker & Hacker), let me quote at more length a passage from Wittgenstein's *Lectures and Conversations on Aesthetics, Psychology and Religious Belief* that I referred to above, and that I agree with. The students who took the notes quote Wittgenstein as saying,

> If you ask me whether or not I believe in a Judgement Day, in the sense in which religious people have belief in it, I wouldn't say "No. I don't believe there will be such a thing." It would seem to be utterly crazy to say such a thing. . . .
>
> In one sense, I understand all he says, the English words, "God," "separate," etc. I understand. I could say: "I don't believe in this," and this would be true, meaning I haven't got these thoughts or anything that hangs together with them. But not that I could contradict the thing.
>
> You might say: "Well, if you can't contradict him, that means you don't understand him. If you did understand him, then you might." That again is Greek to me. My normal technique of language leaves me. I don't know whether to say they understand one another or not.[25]

Everyone knows that Wittgenstein showed considerable respect for religious language. He is obviously unwilling to say that the believer's sentences about the Last Judgment are "nonsense" either literally (à la the new Wittgenstein) or for "orthodox Wittgensteinian" reasons. But he does admit that he cannot say that he *understands* them. They are not fully intelligible by someone who stands (intellectually and religiously) where Wittgenstein stood.

I believe that if Wittgenstein had had the same respect for metaphysics and, more broadly, for other kinds of philosophy than his own, he would have seen that we may find philosophical utterances that have nothing wrong about them *linguistically* less than fully intelligible, precisely because, to use his own words, we "haven't got these thoughts or anything that hangs together with them," and let it go at that without trying to show that those utterances are literally nonsense. I will close with one relevant example.

---

25. Ludwig Wittgenstein, *Lectures and Conversations on Aesthetics, Psychology and Religious Belief,* ed. Cyril Barrett, (Oxford: Blackwell, 1966), 55.

Souls, Zombies, and So On

I said above that I cannot regard such alleged "metaphysical possibilities" as the possibility that some human beings are zombies as fully intelligible, because I cannot understand *how* something could have the same physical makeup as a normal human being (and the same behavior, with the same physical causes) and lack consciousness. In *The Threefold Cord: Mind, Body, and World,*[26] I pointed out that for thousands of years both ordinary people and intellectuals regarded talk of "souls" and other "purely spiritual beings" as intelligible.[27] Thus I was forced to address this question: could it be that *everyone* was wrong about what is and what is not intelligible?

My way of dealing with the prima facie difficulty was to distinguish between *religious* uses of soul talk[28] (which, following Wittgenstein, I did not regard as nonsensical) and uses of soul and immaterial mind talk in metaphysics (which, I said, have not been given any "positive meaning").[29] This is precisely the double standard (religious language is meaningful, metaphysical language is not—or not in any serious sense of "meaningful") that I now find unwarranted and unfortunate in Wittgenstein.

Wittgenstein himself gives an interesting example of a possible use of the adjective "soulless," namely, to justify the exploitation of some natives, whom he imagines the exploiters (or "scientists" hired by those exploiters) classifying as "soulless."[30] (He also points out the inconsistencies in the story of the exploiters.) Discussing this passage, I wrote:[31] "Consider, for example, the case of Wittgenstein's 'soulless tribe.' What would the 'scientists' say if asked, 'Just what is this "soul" that the natives lack? What is the etiology of this condition of "soullessness"? And, by the way, how did you find out that they are "soulless"?' It would quickly become clear that what the 'scientists' are doing isn't 'describing a possible state of affairs' but something very different."

26. Putnam, *Threefold Cord,* 84–85.

27. And if such talk *is* intelligible, why not talk of zombies, i.e., humans *without* such immaterial souls?

28. Putnam, *Threefold Cord,* 94–98.

29. Ibid., 98.

30. Ludwig Wittgenstein, *Remarks on the Philosophy of Psychology* (Oxford: Blackwell, 1980), 96.

31. Putnam, *Threefold Cord,* 100.

These words were right, but they are very different from the attempts to prove that talk of souls and of the mind as something separable from the body is and always was without cognitive meaning, which is what the rest of my discussion in those pages argued.[32] The questions I imagined us putting to Wittgenstein's racist "scientists" were questions that arise, not because of the ordinary use of our words, but because we inhabit a scientific worldview that makes answers to them preconditions for taking talk of immaterial souls seriously.[33] The sense in which we cannot understand the nonsense that the "scientists" are imagined to be talking is that we "haven't got these thoughts or anything that hangs together with them."

32. Ibid., 93–100.

33. In this connection, Descartes is an example of someone (the last great thinker?) who could still try to inhabit (large parts of) the worldview of medieval philosophy *and* the worldview of the new science that he was in the process of creating. The tensions are everywhere evident in his theories; for example, nonhuman animals were automata whose behavior did not require an appeal to anything immaterial to explain, but human brains required help from the soul *(mens sive anima sive intellectus),* but the soul could act only by applying very small impulses to the right spot (the famous pineal gland).

PART FIVE

The Problems and Pathos of Skepticism

29

Skepticism, Stroud, and the
Contextuality of Knowledge

In my previous writings on the subject of skepticism[1] I have relied on what might be called a contextualist view of language in general and of the verb "to know" and its counterparts in other languages in particular. A contextualist view of language (the view that, as Charles Travis has brilliantly explained, lies at the heart of the views of language of John Austin, as well as of the later Wittgenstein[2]), does not, of course, claim that the *meanings* of sentences vary from context to context, or at least it does not claim that in *every* sense of that multiply ambiguous word "meaning," the meaning of a sentence that one understands changes whenever one finds oneself in a new context. In some sense it must be true that a speaker (as we say) "knows the

---

1. In particular, Hilary Putnam, "Strawson and Skepticism," in *The Philosophy of P. F. Strawson* (La Salle, Ill.: Open Court, 1998), 273–287; reprinted here as Chapter 31; and "Skepticism," in Marcelo Stamm, ed., *Philosophie in synthetischer Absicht* (Stuttgart: Klett-Cotta, 1998), 239–268; reprinted here with the title "Skepticism and Occasion-Sensitive Semantics" as Chapter 30.

2. See Charles Travis's book on Austin, *The True and the False* (Amsterdam: J. Benjamins, 1981), and Travis, *The Uses of Sense* (Oxford: Clarendon Press, 1989), as well as Travis, *Unshadowed Thought* (Cambridge, Mass.: Harvard University Press, 2000).

meaning" of each sentence that he or she is able to use *before* using it or understanding another speaker's use of it in a new context, and that this "knowledge of its meaning" plays an essential role in enabling the speaker to know what the sentence is being used to say in the context.

(Let me also say here that I do not think of meanings as either Platonic objects or as mental objects; in my view, talk of meanings is best thought of as a way of saying something about certain world-involving[3] competences that speakers possess. And corresponding to those competences, there are constraints on what can be done with sentences without, as we say, "violating" or at least "extending" or "altering" their meaning.)

What contextualism *does* deny is that the "meaning" of a sentence in this sense determines the *truth-evaluable content* of that sentence. The thesis of contextualism is that in general *the truth-evaluable content of sentences depends both on what they mean (what a competent speaker knows before encountering a particular context) and on the particular context, and not on meaning alone.*

The easiest way to explain what I just said is with the aid of examples. Here is one that I used in a recent book:[4] every competent speaker of English knows the meaning of the sentence "There is a lot of coffee on the table." But, *consistently with what it means,* it is possible to *understand* that sentence as saying:

1. There is a lot of (brewed) coffee on the table (e.g., in cups or mugs).

Sample context: "There is a lot of coffee on the table. Help yourself to a cup!"

2. There are dozens of bags full of coffee beans on a table standing near a place where a truck has come to get them and take them to a warehouse.

Sample context: "There is a lot of coffee on the table. Load it in the truck."

3. A lot of coffee has been spilled on a table.

Sample context: "There is a lot of coffee on the table. Please wipe it up."

---

3. My reasons for saying "world-involving" competences are, of course, the by-now-familiar reasons for "semantic externalism" that I laid out in "The Meaning of 'Meaning,'" in Keith Gunderson, ed., *Language, Mind, and Knowledge,* Minnesota Studies in the Philosophy of Science, vol. 7 (Minneapolis: University of Minnesota Press, 1975), 131–193; reprinted in Putnam, *Philosophical Papers,* vol. 2, *Mind, Language and Reality* (Cambridge: Cambridge University Press, 1975), 215–271, and in Putnam, *Representation and Reality* (Cambridge, Mass.: MIT Press, 1988), chap. 2.

4. Hilary Putnam, *The Threefold Cord: Mind, Body, and World* (New York: Columbia University Press, 1999).

Note that in one and the same context, the *truth-value,* the truth or falsity, of the sentence "There is a lot of coffee on the table" (or of the "content" that the speaker means to convey by uttering the sentence) will be quite different if the appropriate understanding of the sentence is as in the first example or as in the second example or as in the third example (and the number of possible nondeviant *understandings* of the sentence is much greater than three; in fact, it is literally endless).

For example, if a speaker intends the first understanding and a hearer thinks that the third understanding is meant, they will seriously misunderstand each other. But neither speaker nor hearer can be said *not to know the meaning* of the English sentence "There is a lot of coffee on the table."[5] Nor can the speaker be accused of misusing the sentence, or the hearer of understanding it in a way that would be a violation of (or extension of, or some other deviant relation to) its meaning. Thus there are at least three (in fact, as we just said, an endless number) of possible understandings of this sentence. And if the view Travis ascribes to Austin and Wittgenstein is right, this is typical of sentences in any natural language. I call these understandings "truth-evaluable contents" (this is my terminology, not Travis's) because in the contexts I (very roughly) described they are typically sufficiently precise to be evaluated as true or false. (Note that even a vague sentence—"He stood roughly there"—can often be evaluated as true or false given an appropriate context.) But it is also the case that these "contents" themselves admit of further specification, admit of different understandings in different contexts.[6]

---

5. I certainly know the meaning of the words "there," "coffee," "a lot," "of," "is," "on," "the," and "table." But that knowledge by itself does not determine the "truth-value" of the sentence "There is a lot of coffee on the table"; in fact, the sentence, simply as a sentence, does not have a truth-value apart from particular circumstances. Moreover, as I have explained, the truth-conditions of the sentence "There is a lot of coffee on the table" are highly occasion sensitive: depending on the circumstances, the sentence can be used to express different truth-evaluable contents. Responding to the coffee example, a philosopher of language of my acquaintance— one wedded to Grice's distinction between the standard meaning of an utterance and its conversational implicatures—suggested that the "standard meaning" of "there is a lot of coffee on the table" is that there are many (how many?) molecules of coffee on the table. But if that is right, the "standard" sense is a sense in which the words are never used. In Paul Grice, "Logic and Conversation," Part 1 of *Studies in the Ways of Words* (Cambridge, Mass.: Harvard University Press, 1989).

6. Charles Travis stresses this in "Mind Dependence," *Revue Internationale de Philosophie* 55, no. 4 (2001): 503 (the issue is dedicated to my philosophy).

I have just given a rough description of a kind of contextuality—the dependence of the truth-evaluable content of a sentence on features of the context of its use that are not determined simply by its meaning.[7] If this is indeed ubiquitous, it will follow, of course, that the truth-evaluable content of a sentence of the form "X knows that p" will depend on the particular context of utterance (or inscription) of the sentence. In the present essay I will suggest that one way of reading Barry Stroud's extremely influential book *The Significance of Philosophical Scepticism*[8] is as an attack on the idea that "know" is a context-dependent verb, and that Stroud's attack must, in the end, be rejected as misguided. My procedure will be, first, to summarize briefly some of the ways in which it has been suggested that "know" is context dependent, and to explain how this context dependence has been claimed to be relevant to the problem of skepticism; and, second, to explain how I read Stroud as attacking this whole approach to skepticism, not just in the few pages in which he briefly considers (and quickly dismisses) it, but in the first three chapters of his book, and, simultaneously, to criticize his attack.

The Context Dependence of "Know"

One of the ways (by no means the only way) in which "know" has been claimed to function as a context-dependent verb is the following: typically, it has been claimed, when one says, "I know that p," one is claiming that one is in a position to exclude a possible doubt concerning the truth of p.[9] This is not

---

7. Of course, certain sorts of context dependence are recognized by all semantic theorists: for example, the dependence of the truth-value of a sentence containing a pronoun on the (contextually determined) bearer of the pronoun, the dependence of the truth-value of a sentence containing a demonstrative on the (contextually determined) reference of the demonstrative, and the dependence of the truth-value of a sentence containing a finite verb on the time at which the sentence is uttered. But the thesis defended by Wittgenstein, Austin, Travis, myself, and others is more radical, as the "coffee" example illustrates: the very extension of common nouns and verbs is (we hold) context dependent.

8. Barry Stroud, *The Significance of Philosophical Scepticism* (Oxford: Clarendon Press, 1984).

9. I believe that, in fact, the so-called causal theory of knowledge introduced by Peter Unger and later made popular by Robert Nozick, according to which "X knows that p" is true even if X can in no way justify the claim that p just as long as X's belief that p was caused by a process that reliably produces true beliefs (details vary from causal theorist to causal theorist), is wrong if it is regarded as a description of the most typical uses of "know," but it does point

in itself a contextualist claim, but it becomes a contextualist claim if one adds that not every *conceivable* way in which *p* might be false is a "possible doubt"; a "possible doubt" must satisfy context-sensitive standards of *relevance*.[10]

The relevance to external-world skepticism is immediate because the contextualist will point out that it is only in exceptional contexts that "I might be dreaming" (this is the skeptical possibility introduced by Descartes in the *First Meditation* and strongly emphasized by Stroud) counts as a relevant possibility, and if the skeptics were to concede that this shows that in a normal context the mere conceivability that "my experiences might be a dream" does not *contradict* my claim to know whatever it is I claim to know, they would have to abandon the most famous arguments that they have hitherto used to show "that we can know nothing about the world around us."[11]

Stroud's Criticisms of Contextualism

Here is an example that may illustrate what is at issue in the dispute between Stroud and the "contextualists":

Suppose that my wife and I are on our way home, and she asks me, "Do you know if we have enough milk for breakfast?" I reply, "Yes, we do." (I was the one who put the milk bottle back in the refrigerator, and I remember that it was more than half full.)

Now imagine that a student who knows my family well and who is impressed by Stroud's arguments overhears this and immediately raises the following objection:

"I know," the student says, "that your son has a key to your house. Isn't it possible that he stopped by your house today in your absence?"

---

to the fact that in some contexts we may say that someone "knows" even though there is no question of her being in a position to rule out relevant doubts at all.

10. When Stroud first (rather obliquely) refers to this sort of position, his formulation is that "there must be some 'special reason' for thinking a certain possibility might obtain" (*Significance of Philosophical Scepticism*, 63). This seems to me to be already loading the dice against the contextualist. Normally we are not required to give reasons at all, let alone "special" reasons, for taking what are obviously relevant possibilities to be so. Stroud would perhaps reply that he did not say that we are required to give the reason that a given possibility is relevant in a given context; it is just that there has to be such a reason (one that a theorist could give, if not the ordinary speaker herself).

11. Stroud, *Significance of Philosophical Scepticism*, 1.

"But he would have been at work," I object.

"Couldn't he have taken a day off?" the student says.

"Well, he *could* have," I admit. "But I very much doubt that he did."

"Still, you admit that he could have," the student continues. "And couldn't he have been thirsty and drunk up most or all of the milk?"

"Of course he could have," I say, "*if* he was at my house. But there is no reason at all to think he was."

"Still, you don't *know* that he wasn't, do you?" (The student closes in for the "kill.")

"No, I don't," I admit.

The student will now conclude that I *do not* know that there is enough milk in the refrigerator for breakfast, and in fact that I never know any such thing (because this argument can be run on just about any day—even if I think that my son is away in California, "Couldn't he have come back from his trip a day early?" and so on). But what I will tell my student is that I do not regard the logical (indeed, the physical) possibility that my son did such an unusual thing as a "relevant possibility," and that I regard what I said to my wife as appropriate and indeed as *true*.

This is a case in which it is features of the context that the person who is said to know/not-to-know is in that determine which logical/physical possibilities are relevant, but in the general case it may be features of the context of someone who hears the knowledge claim, or even someone who later hears *of* the knowledge claim, that determine this, on a contextualist view—so that one speaker may be entitled to regard the utterance as true and another as false because they appropriately assign different truth-evaluable contents to the utterance. (Whether in such a case the persons are correctly said to *contradict* one another is itself a context-dependent matter, or so at least one contextualist, Charles Travis, would claim, I believe.)

That Stroud is inclined to disagree with all this may be fairly inferred from a passage in *The Significance of Philosophical Scepticism*.[12] Stroud imagines a case in which he has "just about the most favorable grounds one can have" for saying that John will be at the party. We imagine that John fails to come, nonetheless, because he was struck by a meteorite just as he stepped out of his front door. After pointing out that his failure to know in this case can be correctly described as "due to the falsity of what [he] claimed to know," and so a "necessary condition of knowledge was unfulfilled even though no one was

12. Ibid., 61–62.

in a position at the time appropriately or reasonably to criticize [his] claim on that basis." Stroud says, "Perhaps the same is true of other necessary conditions of knowledge."

To set the stage for his answer—that he would ("perhaps") have been wrong to claim that he knew that John would be at the party even if there had been no meteorite and John had come just as Stroud said he would—and for his further claim that the mere conceivability that John would be hit by a meteorite "perhaps" defeats Stroud's claim to know that John was coming to the party, Stroud first imagines an unreasonable host raising that very possibility:[13]

> Suppose that as soon as I had hung up the telephone from talking with John and had said that I knew he would be at the party, the boorish host had said, "But do you really know he'll be here? After all, how do you know he won't be struck down by a meteorite on the way over? You don't know he won't be." . . . Not only is this "challenge" . . . unfair and inappropriate. . . , it is difficult to understand why he even brings up such a consideration at this point and thinks it is a relevant criticism. His doing so would normally suggest that there have been a lot of meteorites hitting the earth lately in this general area, some of them rather big and capable of causing harm. If that were so, perhaps I should have thought of it and considered it—or at least if I didn't know about it my ignorance might threaten my claim to know John would be there. But in the absence of any such special reason, the "challenge" seems . . . outrageous.

Now Stroud is by no means a philosopher who despises the work of the later Wittgenstein; indeed, he is a serious scholar of that work. In particular, he by no means wants to ignore or treat as irrelevant the ordinary employment of words, and he attempts to do justice to the ways in which one might actually employ the sentences the "boorish" host is described as uttering. But he fails to note that the "boorish host" in his little story is not merely speaking outside the language games that we normally play with such a sentence as "I know John is coming to the party," but is speaking outside *any* definite language game—outside any linguistic practice that determines, given the context, criteria for "knowing" and "not knowing" that masters of the language can project from the context. Indeed, *unlike* all other philosophers influenced by Wittgenstein that I know of, Stroud does *not* think that the fact that the "boorish host" is speaking outside language games *shows* that the "challenge" is

13. Ibid., 61.

unintelligible; Stroud thinks that it can be assigned a truth-value, and that what the "boorish host" says may well be *true*. For he immediately goes on to write:

> My act of asserting that John would be at the party was made on just about the most favorable grounds that one can have for saying such things. It is no reflection on me or on my saying what I did that I had not ruled out or even thought of the meteorite possibility. But once the question is raised, however inappropriately, can it be said that I do know that that possibility will not obtain? It seems to me that it cannot. . . . I do not think it was true when I hung up the telephone that I knew that John would not be hit by a meteorite. So, again, part of what the host says is true. I did not know any such thing. But still I said I knew John would be at the party.

We should note carefully that Stroud does not say that the skeptical possibility raised by the "boorish host" *shows* that his claim to know that John was coming to the party was false. What he claims is that it *may* show this: that the contextual irrelevance of the meteorite possibility *may* only show that it is *inappropriate* to say that Stroud did not know, and not that it was *false* that he did not know. But, after many re-readings of *The Significance of Philosophical Scepticism,* it has become clear to me that the gravamen of the first three chapters is that this is, in fact, the most plausible line to take. I cannot present all the textual evidence for this claim in one lecture, but I shall try to make it plausible in what follows.

That it is at least *open* whether the "inappropriateness" of the boorish host's remark counts against its *truth* is something Stroud himself immediately claims:

> All I am saying at the moment is that it does not follow directly from the admitted outrageousness of [the boorish host's] introducing that possibility that my ruling out the meteorite possibility is simply *not* a condition of my knowing that John will be at the party. Its being a necessary condition of my knowledge is so far at least compatible with the host's remarks' being inappropriate or outrageous, just as John's being at the party is a necessary condition for knowing he will be there.[14]

In fact, however, Stroud's discussion of contextualism, even at this early point, is not as "neutral" as his "perhaps" and "all I am saying at the moment" might suggest. Let us briefly go back to some sentences I already quoted:

14. Ibid., 62.

Once the question is raised, however inappropriately, can it be said that I do know that that possibility will not obtain? It seems to me that it cannot. . . . I do not think it was true when I hung up the telephone that I knew that John would not be hit by a meteorite. So, again, part of what the host says is true. I did not know any such thing. But still I said I knew John would be at the party.

What Stroud is saying is that the skeptical possibility raised by the host *does* show that "Stroud did not know that John would not be hit by a meteorite" is true. All that is left open is whether the supposed fact that Stroud does "not know" that the meteorite possibility does not obtain *shows* that Stroud also does not know that John will be at the party. ("I want to be careful here. I want to emphasize that I am not saying that in this second case I do not know that John will be at the party because I do not know when hanging up the telephone that he will not be struck down by a meteorite on the way over."[15]) But the step (which Stroud evidently thinks is obvious) from "It is not true that I knew" to "I did not know" *already assumes* the falsity of contextualist (specifically, of Wittgensteinian) critiques of skeptical arguments. For what contextualists maintain—or, at least, what Wittgenstein teaches us[16]—is that in the sort of context Stroud describes, both "Stroud knows that John will be struck by a meteorite on his way to the party" and "Stroud does not know that John will be struck be a meteorite on the way to the party" presuppose some definite *question,* and if the question whether John will be struck by a meteorite is one that does not remotely arise, then we have no space of relevant possibilities to speak of as excluded or not excluded. *Here and now,* if you ask, "Hilary, do you know or not know that Jim will not be struck down by a meteorite tomorrow?" I would not say either "I know" or "I do not know"; I would say, "I do not know what you mean to be asking me." If it is "not true" that "I know that Jim will not be struck down by a meteorite tomorrow," that is because, uttered here and now, with no specified context other than the empty context of "It is a logical possibility," "I know that Jim will not be struck down by a meteorite tomorrow" is not *merely* something that is "not true"; it is something that *has not been given a determinate content in the context.* By assuming that the negation of the knowledge claim "Stroud

15. Stroud, *Significance of Philosophical Scepticism,* 62.
16. Here I am following Stanley Cavell's interpretation of Wittgenstein's *Philosophical Investigations* in *The Claim of Reason* (Oxford: Clarendon Press, 1979).

knows that John will not be struck down by a meteorite on his way to the party" is true, Stroud has assumed that that knowledge claim has a truth-evaluable content, and this is precisely what I think we should deny.

The point I am making might also be put in pragmatist rather than "contextualist" language. As I interpret Peirce's celebrated distinction between "real" and "philosophical" doubt, the point Peirce is making is that *doubt must be capable of justification just as much as belief.* Both pragmatists and "Wittgensteinians" think that *not all questions of the form "Does X know that p?" make sense.* If we cannot tell from the context what a justified doubt that *p* might be, then no definite question has been asked.

Nonclaim Contexts

It is obvious that Austin's arguments in *Sense and Sensibilia* and in "Other Minds" are of the contextualist variety.[17] In *The Claim of Reason* Stanley Cavell makes a far more nuanced employment of contextualist considerations that criticizes Austin for missing important distinctions and also offers a reading of Wittgenstein's *Philosophical Investigations* that puts skepticism at the heart of Wittgenstein's concerns. Here I have room to mention just one of Cavell's notions, the notion of a "nonclaim context."

Cavell's point, whose connection with the preceding will be apparent, is most easily presented with the aid of an example. I am sitting alone in my office writing these words on a computer. If someone were to telephone and ask me, "What are you doing now?" I might, in fact, mention that I am working on the computer. But if the person were to ask, "How do you know?" I would be nonplussed. I am unquestionably in a position to *say* what I am doing: I am working on the computer. But that does not mean that I can intelligibly say that "I know I am working on the computer" when the question does not even arise. In Cavell's terms, "I am working on the computer" is not a *claim,* and defending it or justifying it is a demand that does not arise any more than the question "How do you know?" arises.

If the sort of treatment of skeptical doubt that I have described is sound, then the skeptic is wrong to say, "You do not know . . ." (that you are not dreaming, that there is a computer in front of you, that your name is . . .).

17. J. L. Austin, *Sense and Sensibilia*, ed. G. J. Warnock (Oxford: Oxford University Press, 1961); "Other Minds" in *Philosophical Papers*, ed. J. O. Urmson and G. J. Warnock (Oxford: Oxford University Press, 1964) 76–116.

But does not finding out that the skeptic is wrong for *these* reasons not leave us, in the end, with a feeling of disappointment? Not disappointment with the reasons, but disappointment with *human knowledge as such?* This is the large (and strikingly original) question that Cavell presses with amazing depth in *The Claim of Reason.* To discuss this question here would take me far from my topic, which is Stroud's *Significance.* But I cannot write of Cavell's work, even in passing, without mentioning it. To see why such disappointment can arise at all, it perhaps suffices to reflect that it is just the sorts of things that we are entitled to say without justification, the things we rely on without "knowledge," that we appeal to when there *is* a question of justification. "Our relation to the world as a whole . . . is not one of knowledge,"[18] Cavell tells us. (He calls this "the truth in skepticism."[19]) The language game rests on *trust,* Wittgenstein tells us.[20] The notion of resting the language game on knowledge, or on certainty, or on demonstration is ultimately unintelligible: there is no coherent possibility that we have discovered not to obtain. But one can be disappointed when something turns out to be meaningless, as well as when it turns out to be false. And perhaps Cavell's deepest insight is that such disappointment is not always childish, or something to be gotten over; it may be that *not* feeling it (or pretending—to oneself, at that—not to feel it) is one form that failure to grow up can take. But now I must return to Stroud.

The Airplane Spotters

A few pages after discussing the question whether he (that is to say, anyone) can know that John will come to the party (that is, just about any empirical fact), Stroud writes:

> One way to bring out what I think is the skeptical philosopher's conception of everyday life in relation to his epistemological project is to consider in some detail the following story adapted from an example of Thompson Clarke's.[21] Suppose that in wartime people must be trained to identify

18. Cavell, *The Claim of Reason,* 45.

19. Ibid., 47.

20. Ludwig Wittgenstein, *On Certainty* (Oxford: Blackwell, 1969): "§508: What can I rely on? §509: I really want to say that a language game is only possible if one trusts something. (I did not say 'can trust something.')"

21. Thompson Clarke, "The Legacy of Skepticism," *Journal of Philosophy* 64, no. 20 (1972): 759–764.

aircraft and they are given a quick, uncomplicated course on the distinguishing features of various planes and how to recognize them. They learn
from their manuals, for example, that if a plane has features x, y, and w it
is an E, and if it has x, y, and z it is an F. . . .

Suppose there are in fact some other airplanes, Gs say, which also have
features x, y, z. The trainees were never told about them because it would
have made the recognition too difficult; it is almost impossible to tell an F
from a G from the ground. The policy of simplifying the whole operation
by not mentioning Gs in the training manual might be justified by the fact
that there are not many of them, or that they are only reconnaissance
planes, or that in some other way they are not as directly dangerous as Fs;
it does not matter as much whether they fly over our territory.

When we are given this additional information I think we immediately
see that even the most careful airplane-spotter does not know that a plane
he sees is an F even though he knows that it has x, y, z. For all he knows, it
might be a G.[22]

Two paragraphs later Stroud draws his conclusion:

I think the skeptical philosopher sees our position in everyday life as
analogous to that of the airplane-spotters. There might be very good reasons [But can a "skeptical philosopher" grant that there are such things as
"good reasons"?] why we do not normally eliminate or even consider count-
less possibilities which nevertheless strictly speaking must be known not to
obtain if we are to know the sorts of things we claim to know. We therefore
cannot conclude simply from our having carefully and conscientiously followed the standards and procedures of everyday life that we thereby know
the things we ordinarily claim to know.[23]

Further:

The point is worth stressing. Many people [e.g., contextualists] are apparently disposed to think that if the philosopher holds that a certain condition must be met in order to know something, and we do not insist on that
condition's being met in everyday life, then the philosopher simply *must* be
imposing new or higher standards on knowledge or changing the meaning
of the word "know" or some other word. But if our position in everyday
life is like that of the airplane-spotters, that is not so.[24]

22. Stroud, *Significance of Philosophical Scepticism*, 67–68.
23. Ibid., 69.
24. Ibid., 69–70.

Although this seems clear at first reading, upon reflection it is harder and harder to see what Stroud is claiming. Although Stroud's official position is only that the contextualist has not proved his case, that the alternative suggested by the airplane-spotter analogy is a possibility—that is, that context is relevant only in fixing what it is practically useful "for very good reasons" to say we know, to treat as appropriate to say we know, and the like, but not in fixing what it is *true* to say we know—he so insistently presents this alternative that one can hardly avoid the conclusion that he simply thinks that the positions I have lumped together as "contextualism" are *wrong* (and this is also my impression from a recent, unfortunately brief, conversation with him). At the same time, he refrains from saying flat out that the skeptic is *right*. Whether the skeptic is right is supposed to be a deep and presently unsolved philosophical problem. But if contextualism is wrong to *this* extent: that the possibility that Descartes is dreaming is enough to show that Descartes does not know that there is a fireplace or a room or a dressing gown in his vicinity, and the fact (as we have seen, Stroud takes it to be such) that Stroud does not know that John will not be struck down by a meteorite is enough to show that he does not know that John will come to the party (right after talking to John on the telephone), and so on, then why does Stroud not just conclude that skepticism is true? In many ways *The Significance of Philosophical Scepticism*—or at least its opening chapters—makes more sense as a demonstration of the *truth* of skepticism rather than (as it sees itself as being) a demonstration that there is a "deep problem." If I jump out of a fourteenth-story window, there is not a "deep problem" whether I will fall; yet Stroud has jumped out of the skeptical window but seems to be unsure about the fate he faces.

Indeed, the airplane-spotter analogy itself seems to be problematic in more than one way. Presumably the idea is not that the people who wrote the manual and omitted the information that "there are in fact some other airplanes, Gs say, which also have features *x, y, z*" definitely *knew* that this was the case; for in that case, how could this be an analogy that a "skeptical philosopher" might use? (A skeptical philosopher would say that the writers of the manual might have been dreaming or hallucinating.) Rather, the idea must be that the relation in which the airplane spotters' use of "know" stands to the manual writers' use of "know" is *analogous* to the relation in which our "everyday" use of "know" stands to . . . what? Hardly the skeptics' use of "know." For it was essential that the manual writers are supposed to know more than the airplane spotters, but according to the skeptics' use of "know" (which may be the literally correct use, if I understand Stroud's worry), the skeptic, like everyone else, knows nothing. (This unclarity in the nature of the analogy is connected to the casual use of expressions like "very good reason" by Stroud in explaining what is supposed to be the skeptic's view.)

Finally, look at the following sentence closely:

> The policy of simplifying the whole operation by not mentioning Gs in the training manual might be justified by the fact that there are not many of them, or that they are only reconnaissance planes, or that in some other way they are not as directly dangerous as Fs; it does not matter as much whether they fly over our territory.

These are, in fact, very different cases. If, on the one hand, Gs are extremely rare, then it could well be that even someone as knowledgeable as a manual writer would feel justified in saying that he "knew" that what he had spotted was an F on the basis of features x, y, and z—just as we normally feel justified in ignoring the rare event of someone being struck down by a meteorite. Of course, Stroud can explain this away as a case of saying what it is convenient but not true to say, but because the whole point of the airplane-spotter story is that it is supposed to support this very move by being a clear case of something's being convenient to say but not true, this would make the entire argument circular. On the other hand, if Gs are common, but the manual writers ignored them because they cause little harm, then we would very likely say (on learning this) that the airplane spotter did not know that a given plane he identified as an F really was an F (given the low reliability of the method he relied on); but this reliability consideration can also point in an antiskeptical direction (the direction taken by "causal theories of knowledge"). So it is natural to ask why Stroud treats these very different cases as equivalent. It seems that for Stroud the only relevant feature of the cases is that they show that there is a *possibility* that something that meets the criteria the airplane spotter is relying on (namely, the features x, y, and z) does not have the relevant property (namely, being an F); and Stroud must be assuming that any reader who sees this will agree that the spotter does not *know* that it has the property. Stroud expects the reader to think that it is *irrelevant* whether the possibility is one that is frequently or very infrequently realized. Only the status of possibility is relevant. But the assumption that *bare* possibility is enough to defeat a knowledge claim is precisely the one on which the skeptic's argument turns.

Stroud Contra Malcolm

The principle defender of a contextualist view whom Stroud directly rebuts is Norman Malcolm.[25] From my point of view, Malcolm's contextualism suffers from his unwise insistence that there is such a thing as *the* proper use of the word "know." In the spirit, and under the inspiration, of Stanley Cavell's work,[26] I have argued elsewhere that in general we should not expect to capture the cases in which competent speakers will find that we have used a word in a way that is a "natural projection" of uses previously accepted as perfectly in order by a closed set of "rules." This insistence on the openness of our possible projections certainly seems more in keeping with the spirit of contextualism. (Think how hard it would be to capture all the present and possible future uses of "There is a lot of coffee on the table.")

In any case, in "Defending Common Sense"[27] Malcolm claimed that for the proper use of the word "know" there must be some question at issue or some doubt to be removed, the person saying that he knows something must be able to give some reason for his assertion, and there must be some investigation that, if it were carried out, would settle the question. Because Moore's celebrated "proof of the external world"[28] uses "know" in a way that violates these criteria, Malcolm concluded that he had exposed Moore's argument as fallacious (as turning on a violation of the correct use of "know").

Stroud contends[29]: (1) against Malcolm that there are uses of "know" that are *not* misuses that violate all three of Malcolm's criteria; but (2) against Moore that such uses may not be relevant to "the philosophical problem of the external world." Stroud's example is borrowed from Thompson Clarke:[30]

> To see that it is possible to use Moore's very words in contexts that appear nowhere in Malcolm's list of correct uses of "know," we can recall Thompson

---

25. Ibid., 88–114.

26. I discuss this work in Chapter 24 of this volume.

27. Norman Malcolm, "Defending Common Sense," *Philosophical Review* 58, no. 3 (1949): 201–220.

28. G. E. Moore, "Proof of an External World," *Proceedings of the British Academy* 25 (1939): 273–300; reprinted in Moore, *Philosophical Papers* (London: Macmillan, 1959), 126–148; see 127.

29. Stroud, *Significance of Philosophical Scepticism*, 100.

30. Clarke, "Legacy of Skepticism," 754–769.

Clarke's example of the [made-up] physiologist lecturing on mental abnormalities. Near the beginning of his lecture he might say:

> Each of us who is normal knows that he is now awake, not dreaming or hallucinating, that there is a real public world outside his mind which he is now perceiving, that in this world there [are] three-dimensional animate and inanimate bodies of many shapes and sizes. . . . In contrast, individuals suffering from certain mental abnormalities each believes that what we know to be the real world is his imaginative creation.

Here the lecturer uses the same words often used by philosophers who make or question general statements about the world and our knowledge of it. When he says that each of us knows that there is a public world of three-dimensional bodies, he is stating what can only be regarded as straightforward empirical fact. Most of us do know the things he mentions and those with the abnormalities he has in mind do not know. That is a real difference that can be observed and ascertained.[31]

As I mentioned, Stroud does *not* think that the "physiologist's" use of "know" is philosophically relevant. "I think we do not regard the lecturer in this context as having settled affirmatively the philosophical problem of our knowledge of the external world."[32] Presumably, then, the "physiologist's" use is analogous to the "airplane spotter's" use in some way. But equally clearly, Stroud does think that having shown to his own satisfaction that Malcolm has not listed *all* the "proper" uses of "know,"[33] he can take it that there *is* a proper use that is the one involved in the "philosophical problem of our knowledge of the external world." But what is that use?

Stroud is extremely coy on this crucial point. Consider what he writes in a lengthy note concerning Peter Unger's defense of skepticism. After explaining with the aid of the "boorish-host" and the "airplane-spotter" examples that it is possible that the facts about our everyday use reveal only appropriateness conditions for our utterances and do not tell us what their *truth-conditions* might be, Stroud writes that (if this possibility obtains) "one would then be in a strong position to defend the skeptical conclusions against any

31. Stroud, *Significance of Philosophical Scepticism,* 100–101.
32. Ibid., 101.
33. Actually, this "physiologist" does not talk like any neural scientist I have ever heard. One of the peculiarities of his odd little speech is the claim that "each of us who is normal" knows that he is now awake. Aren't any normal people now sleeping?

objection to the effect that it distorts the meaning of the very words in which it is expressed since it conflicts with obvious facts about how those words are ordinarily used. The evidence from usage would not support that conclusion about meaning on the conception of the relation between meaning and use that I have tried to identify."[34] His remarks concerning Unger's work are contained in a note to the last sentence and read in part as follows:

> Peter Unger rightly insists on the importance of this distinction in the defense of skepticism. He identifies a class of terms he calls "absolute terms" (like "flat" and "empty") which are appropriately applied on many occasions even though they are never literally true of any of the things to which they are applied. . . . For Unger the same holds for "certain," and since knowledge implies certainty, for "know" as well. . . . I do not agree that Unger can *establish* skepticism on the basis of his theory of "absolute terms" alone. I think his argument to show that no one is ever undogmatically certain (and hence doesn't know) anything about the world makes essential use of a step that is equivalent in force to Descartes' requirement that we must know we are not dreaming if we are to know anything about the world around us. Without that requirement, the "absoluteness" of "certain" and "know" will not yield the skeptical conclusion. And I have tried to show here that with that requirement, we have all we need to generate the skeptical conclusion, so the doctrine of "absolute terms" is not needed.[35]

Now, Stroud believes that if Descartes is right about "that requirement," then we do not even need to employ the empirical fact that people dream to argue for external-world skepticism. Early on, for example, when Stroud summarizes Descartes' argument in the *First Meditation,* he writes:

> When he first introduces the possibility that he might be dreaming Descartes seems to be relying on some knowledge about how things are or were in the world around him. . . . He seems to be relying on some knowledge to the effect that he has actually dreamt in the past and that he remembers having been "deceived" by his dreams. That is more than he actually needs for his reflections about knowledge to have the force he thinks they have. He does not need to support his judgment that he has actually dreamt in the past. The only thought he needs is that it is now *possible* for

---

34. Stroud, *The Significance of Philosophical Scepticism,* 75.
35. Ibid., 75, n. 12.

him to be dreaming that he is sitting by the fire, and that if that possibility were realized he would not know that he is sitting by the fire.[36]

So the threat of skepticism turns out to be the threat that the philosophically relevant use of "know"—or rather, given the way Stroud distinguishes between meaning and use, the philosophically relevant *meaning* of "know"— might be such that *the mere logical possibility of dreams establishes that we do not and could not have any knowledge of "external" objects.* Such a "meaning" *would* make "know" behave the way Unger thought "flat" behaves—that is, as a term that never applies (at least to any empirical claims about "external" objects). It *would* make "know" an "absolute" term in Unger's sense. Like Unger's semantics for "flat," it would violate the principle of charity to the maximum possible extent.[37] But the question that neither Unger nor Stroud faces is this: why should the language contain a term for "knowing" in a sense in which it is an *obvious* logical truth that we have no empirical knowledge?

Stroud would, of course, reply that he does not say that it is an *obvious* logical truth. After all, Descartes's argument depends on more than the premise that I do not know that I am not dreaming (which Stroud appears to accept, just as he accepts that he does not know that John will not be struck down by a meteorite). It depends, as Stroud reminds us in the note I just quoted, on the further premise that to know that I am in front of the fire, or whatever, I need to know that I am not dreaming (to know that John will come to the party, I need to know that he will not be struck down by a meteorite). And Stroud claims that it is a deep problem whether this is right or wrong.

With all my respect for Barry Stroud's scholarship, intelligence, intellectual integrity, and unbelievable willingness to follow this investigation to the very end, I have to say that at the end of the day I feel, in reading *The Signifi-*

---

36. Ibid., 17.

37. Unger defended the context insensitivity of such "absolute" terms as "flat" in chapter 2 of his book *Ignorance: A Case for Scepticism* (Oxford: Oxford University Press, 1975). David Lewis, in response, advocates a context-sensitive approach in his "Scorekeeping in a Language Game," *Journal of Philosophical Logic* 8 (1979): 339–359; reprinted in Lewis, *Philosophical Papers,* vol. 1 (Oxford: Oxford University Press, 1993), 233–249. Incidentally, Unger no longer endorses the view that "flat" context insensitively expresses a property only ideal geometric surfaces possess. In his book *Philosophical Relativity* (Oxford: Oxford University Press, 1984) he hypothesizes that there is no fact of the matter about whether his earlier view or Lewis's view is correct. (I thank Steven Gross for these references.)

cance of Philosophical Scepticism, as if I am in Jabberwock Forest. Stroud has separated "meaning" and "application" (on the basis of what seems to me an overestimation of the value of Grice's arguments);[38] then he asks whether the "meaning" of "know" is such that the skeptical "requirement" holds, and apparently nothing but a conclusive demonstration will satisfy him one way or the other; and finally, having failed to specify any procedures (let alone a justification) for determining this application-transcendent "meaning," he finds that he is unable to say whether the skeptic is right or wrong.

---

38. See the papers collected in Paul Grice, *Studies in the Way of Words* (Cambridge, Mass.: Harvard University Press, 1989) especially "Logic and Conversation," 22–40. For a criticism of Grice's arguments, see Charles Travis, "Annals of Analysis," *Mind* 100, no. 398 (April 1991): 237–263; and Travis, *Uses of Sense.*

30

Skepticism and Occasion-Sensitive Semantics

What originally provoked these reflections is a certain movement away from Kant and back in the direction of Hume that is visible in Strawson's later philosophical writing. Like Hume, whom he explicitly cites as his model, Strawson argues that such beliefs as our belief in the uniformity of nature and the existence of the external world are not based on reasoning and can neither be undermined nor defended by it. This has led me to ask (a question Strawson has himself considered in a more Kantian period) whether the nonexistence of the external world is really a coherent idea. In Chapter 31 I will offer reasons for thinking that the answer is "no, it is not," or at least that we have not been so far enabled (by the skeptic or, for that matter, by his familiar opponent, the traditional epistemologist) to give it a coherent sense.

In the first section of the present paper I consider what sort of failure to achieve intelligibility is involved here, and I argue that intelligibility is, in general, not a matter of words and syntax alone, but a matter of the circumstances under which a sentence is uttered, and that the failure of such a

For my present position on skepticism, which differs from the position defended in this chapter on certain issues, see Chapter 32.

sentence as "We are immaterial beings undergoing the illusion of being in a material world" to achieve the kind of sense it would have to have to represent a challenge to our worldview is analogous to the failure of a sentence in a fairy tale to achieve the status of a claim whose truth-value could be seriously discussed. This emphasis on *context sensitivity* leads, in the second section, to a discussion of what I will call the "occasion sensitivity" of the verb "know." The third section discusses the idea of the skeptic as "challenging our conceptual scheme as a whole," and the last sections distinguish my approach from an approach (due to Michael Williams) that also stresses the context sensitivity of claims to know, but does so in a way that, in my opinion, gives the skeptic far too much.

Because these reflections belong, in large part, to what is called "philosophy of language," they run the danger of giving the impression that (I hold the view that) skepticism has its powerful appeal simply because we make certain mistakes about the way language works. This would indeed be absurd. Rather, my view is that the appeal both of skepticism *and* of a wrong way of viewing our language rests on much deeper facts about us—such as the fact that we humans oscillate between illusions of omnipotence and illusions of impotence with respect to many of our powers (emotional, intentional, and cognitive alike). But to develop that thought must be a task for another occasion.[1]

Austin on Meaning

In chapter 10 of *Sense and Sensibilia* Austin remarks, "It seems to be fairly generally realized that if you just take a bunch of sentences . . . impeccably formulated in some language or other, there can be no question of sorting them out into those that are true and those that are false; for (leaving out of account so-called 'analytic' sentences) the question of truth and falsity does not turn only on what a sentence *is,* nor yet on what it *means,* but on, speaking very broadly, the circumstances in which it is uttered."[2] "Knowing the

---

1. A related diagnosis is developed brilliantly in part 4 of Stanley Cavell's seminal *The Claim of Reason* (Oxford: Oxford University Press, 1979).

2. John Austin, *Sense and Sensibilia*, ed. G. J. Warnock (Oxford: Oxford University Press, 1961), 110–111. The "general realization" of which Austin speaks seems, in the meanwhile, to have been largely neutralized by the pretense—for that is what it is—that the dependence of the truth-conditions of a sentence (or better, of what is *said* when a sentence is used on a particular occasion) can be simply—indeed, mechanically—specified by just giving the

meaning" of the words in a sentence, in the sense of knowing how the words are used in a sufficient number of typical situations, is not knowing a list of all the circumstances that might affect exactly what the sentence is used to say, and of the ways in which the words are to be used in those circumstances, nor does it require possession of an algorithm for calculating what the sentence is used to say in an arbitrary circumstance—that is something that has to be determined with the aid of what Kant called "good judgment."[3] In addition, this sort of occasion sensitivity is not something that can or should be handled by postulating a host of different "meanings" for our words, one corresponding to each difference in the circumstances that affects the truth-conditions. This is something that Austin argues at some length (in some special cases) in chapter 10 of *Sense and Sensibilia*.

I will give an example from Charles Travis that may help indicate what I am thinking about.[4] I have an ornamental tree in my garden with bronze-colored leaves. Suppose that a prankster paints the leaves green. Depending on who says it and to whom and why, the sentence "The tree has green leaves," said with my tree in mind, may be *true, false,* or *indeterminate!*

The idea of occasion sensitivity that Austin referred to in the late 1950s is directly relevant to questions of skepticism and to the intelligibility of both skeptical challenges and various responses to those challenges. One can read Stanley Cavell's book *The Claim of Reason* as an exercise in showing how Austinian and Wittgensteinian considerations of occasion sensitivity can be brought to bear on philosophical skepticism. Cavell brings out how a skeptical claim to doubt a "best case" of knowledge and a counterskeptical claim to "know," such as G. E. Moore's claim to know that "this is a hand," fail, in the end, to be intelligible as *claims.*[5]

---

values of a few "indices," which, it is claimed, describe the relevant features of each of the circumstances in which one and the same sentence might be uttered. This claim is an instance of the prevalence of science fiction, often disguised under the name "cognitive science," in present-day philosophy and linguistics; but it is not necessary to go into that issue here. In his important book *The Uses of Sense* (Oxford: Oxford University Press, 1989), Charles Travis goes into it in detail.

3. Immanuel Kant, *The Critique of Pure Reason* [1781], ed. Paul Guyer and Allen W. Wood (Cambridge: Cambridge University Press, 1998), A133–135/B172–174.

4. Travis speaks of "occasion variability" and "S[peaking]-sensitivity," where I use "occasion sensitivity." I have changed his butter example to a coffee example.

5. A radical—but I have come to believe correct—interpretation/application/extension of Cavell's and Travis's points due to James Conant holds that even when Moore says "This is

Before applying the idea directly to skeptical scenarios, let me mention another example. We all "understand" the sentence "At the stroke of midnight, the coach turned into a pumpkin" in the sense of being able to enjoy *Cinderella*. But it does not follow that we should understand what was being said if—without providing any relevant context to indicate what he or she is doing—someone, obviously not engaged in telling a fairy tale, and with apparent seriousness, said "It has really happened that a coach turned into a pumpkin."[6] (And it would not be enough for the speaker to add, "I mean you to take this as a philosophical assertion.")

In the case of the tree on my property, someone who knows that my tree has (by nature) bronze-colored leaves might well be confused if I said, without explaining that a prankster had colored the leaves green, "The tree now has green leaves." But we would be much more confused if someone appeared to claim in all seriousness that a coach has turned into a pumpkin. Are the atoms of the coach supposed to have rearranged themselves? But the coach does not consist of the same *elements* as a pumpkin. And the mass of the coach is much greater than the mass of a pumpkin. There are, of course, possible occasions of use on which we could understand the claim (e.g., the "coach" was a "prop," made to be used in puppet shows, and it is part of its clever construction that it can "turn into a pumpkin"—that is, a pumpkin "prop"). But just saying "I mean it really happened" or "I am speaking in the context of

---

a hand" (i.e., *before* he introduces the word "know"), he has already failed to make a determinate statement. This seems extremely implausible at first, because we can think of many statements those words could be used to make (e.g., a teacher of English as a second language might use those words, and we would know exactly what she was saying—and she would have to be very cruel to say "This is a hand" if she were raising anything other than a hand). But Moore was not making the statement that the English teacher was making, nor was he clearly making any other statement. What makes it hard to see this, I think, besides our adherence to what Travis calls "classic semantics," is that we do think that we know what Moore is saying—in fact, we naturally understand him as saying that "this is a hand" is something he knows for certain (even before he introduces the word "know"). Whereas normally the intelligibility of "I know that p" presupposes the prior understanding of what p says in the context, in *this* context p itself is understood via the (illusion of) understanding "I know that p." But if "I know that p" has *no* determinate meaning, than neither does p, in Moore's case. One cannot give "This is a hand" a determinate use by just *staring* at one's hand.

6. An example of a relevant context for a related sentence (suggested by Paul Franks): in certain imaginable circumstances, someone very insightful might say to me, "Your coach just turned into a pumpkin," and his remark might both be true and not adequately expressible in any other way.

philosophy" does nothing to make the alleged claim "rationally meaningful." (Recall the early modern dispute about the meaningfulness of the doctrine of transubstantiation. The Christian claim that the Communion wine "is" Jesus's blood is intelligible from the point of view of a committed Christian [and problematic from any other point of view];[7] but is the philosophical interpretation of that religious utterance, which holds that the Communion wine has all the essential properties of blood but lacks the chemical composition of blood, supposed to be really intelligible, even from that point of view?) The point is not that we could not find an occasion of use for the expression "Some coaches really do turn into pumpkins," but rather that here and now we cannot understand *what* those words are to be taken as asserting.

But if we cannot understand "Some coaches really do turn into pumpkins," how much less can we say that we understand the claim that a person— indeed, oneself—might be a "disembodied spirit"? At least we know what *pumpkins* are. But do we have any account at all of what disembodied spirits are? (Even the medieval Schoolmen, or many of them, balked at the idea that the souls of the departed have literally *no* matter of any kind at all.)[8] Talk of disembodied spirits, ghosts, and the like does obviously have the kind of intelligibility appropriate to myth; but does that make such talk intelligible if it is intended as factual description?

One might, as a desperate last resort, try explaining the notion of one's being a disembodied spirit by a *via negativa:* "As a skeptical hypothesis, that you are a disembodied spirit need only mean that you have exactly the experiences you seem to yourself to have, but you do not actually have a body." But recall, the "possibility" that we do not actually have bodies was supposed to be *made intelligible* by pointing out that we "might be disembodied spirits." Now we have come full circle. We would have to know just *how* we might actually not have bodies in order to know what it would be to be a "disembodied

---

7. Referring to a religious person speaking about the Judgment Day, Wittgenstein says, "In one sense I understand all that he says—the English words 'God,' 'separate,' etc. I understand. I could say, 'I don't believe in this,' and that would be true, meaning I haven't got these thoughts or anything that hangs together with them. But not that I could contradict the thing." Ludwig Wittgenstein, *Lectures on Aesthetics, Psychology, and Religious Belief,* ed. Cyril Barrett (Berkeley: University of California Press, 1966), 55. A little later he says that he does not know "whether I understand him or not. I've read the same things as he's read. In a most important sense, I know what he means" (58).

8. Cf. Caroline Walker Bynum, *The Resurrection of the Body in Western Christianity, 200–1336* (New York: Columbia University Press, 1995).

spirit," and we would have to know what a disembodied spirit was in order to understand the possibility that we might not really have bodies.[9]

Hume thought that he could explain how we might not really have material bodies *without* invoking the Berkeleyan notion of a "spirit."[10] According to Hume, we are just collections of impressions and ideas. But although Hume never faces it, this would seem to imply the scarcely intelligible possibility that there might be *experiences without any experiencer at all* (a problem that haunts the sense-datum tradition[11]); for if sense data (ideas and impressions) are the atoms out of which "selves" are built, it would seem to be a purely contingent truth, if it is a truth, that all these atoms belong to the sorts of bundles that constitute selves. Why should there not be stacks of them lying around that do not belong to any self? And if *that* is possible, why should there not be grins without grinners? To raise the question of the intelligibility of these skeptical hypotheses is by no means to commit oneself to verificationism, as Stroud (and, implicitly, Strawson) suggest. Very strange ways of using words like "idea" and "impression" (or, alternatively, "experience") occur here; and the context of their use hardly clears up our difficulties. To say, "We do know the relevant context; it is the context of Hume's *Enquiry*," for example, is not to tell us *how* a person could be a bundle of experiences; the *Enquiry* simply takes that as already clear for reasons that have to do with Hume's failure to see the weakness of some of his own key arguments.

Denying, as I do, that Berkeley or Hume or any of their successors have provided us with a "rationally meaningful" scenario to discuss—denying that they have so much as made sense of the claim that we are nothing but "spirits and their ideas," or nothing but "ideas and impressions"[12]—does indeed involve rejecting the idea that traditional philosophical claims are automatically to be allowed to be clear and to possess truth-value, without looking

---

9. It is well to recall in this connection Austin's famous observation that "real" is a word whose sense in a context depends on what sort of "nonreal" *X* is relevant in the context. "Is that a real duck?" as opposed to a decoy? Or as opposed to a goose? "Are our bodies real?" Well, they certainly are not *fakes*.

10. See Chapter 31 in this volume for a fuller version of this criticism of Hume.

11. As is amply evidenced in Moore's replies to his critics in the volume on him in *The Philosophy of G. E. Moore: The Library of Living Philosophers*, vol. IV, ed. Paul A. Schilpp (Petersborough: Open Court, 1942).

12. The idea of an "immaterial spirit" has a kind of intelligibility in a ghost story (an intelligibility that becomes more questionable the more seriously the ghost story is taken), but does it acquire a determinate sense *just* from Berkeley's or Hume's philosophy?

carefully at the different circumstances in which the key terms are nonphilo-sophically used.[13] But rejecting that idea is putting the burden of proof back where it belongs, on the prima facie unintelligible claims, and that burden can be shifted without appealing to some controversial philosophical procedure for classifying all claims into those that are meaningful and those that are not.

The Occasion Sensitivity of Knowledge Claims

I turn now to an exploration of some features of our ordinary use of "know." The phenomenon of the occasion sensitivity of truth-claims in general that Austin remarked on is especially important in connection with knowledge claims. The reason for this is that as we actually use the word "know," a state-ment of the form "X knows that so-and-so" does not claim that either X or the speaker (who may be a third person) is in a position to defuse *every* conceiv-able "doubt" (or "possibility of error").

Here is a relevant example:

1. I know that the color of my house is "off-white."
2. I know the (obvious) implication: if some pranksters painted my house blue this morning, then the color of my house is now blue and not off-white.
3. But I do not *know* that some pranksters did not paint my house blue this morning (I have been at the office all day)—and, by the way, I also do not "not know" that some prankster did not paint my house blue this morning—in the absence of some legitimate reason for doubt, one can only say that the question does not arise.[14]

13. I believe that the resistance to acknowledging the pervasiveness (and the depth) of the phenomenon of occasion sensitivity is related to the resistance that I have found that people have to accepting that if we were "brains in a vat," we would not be able to think or say that we were, and even more to the resistance to granting the next step in the argument I offered in *Reason, Truth and History* (Cambridge: Cambridge University Press, 1981), chap. 1: if we could not think or say that we were brains in a vat if we were brains in a vat, then it follows that we are not brains in a vat. We are, I think, deeply attracted to the idea (call it "semantic omnipo-tence") that we can refer to and think about *anything whatsoever,* independently of our envi-ronment and the nature of our embeddedness in the world.

14. At one time such examples convinced me that knowledge is not "closed under known implication." But, as Paul Franks has pointed out to me, that way of putting it already accepts

Moral: when we say "I know," what we mean is that we can rule out certain relevant "doubts," but *not* all possible "doubts." Which "doubts" depends on the circumstances; there is no rule that we can give once and for all that covers all the occasions on which we might use the word "know." We have to use good judgment and assume that our conversational partners are attuned to us,[15] that they will share our implicit knowledge of what that comes to in the circumstances.

There is an intimate connection between this occasion sensitivity of "I know" and Stanley Cavell's point that Moore's "I know that this is a hand" (said while holding up one of his hands) does not succeed in making a claim. Normally we can tell which possibilities a speaker is claiming to be able to rule out and which he is not claiming to be able to rule out because we know the *point* of his assertion. If my hearers know that the point of my assertion is to tell them what color house I have, I do not have to guarantee (what is anyway beyond my power to guarantee) that there has not been a prank in the last few hours, unless, of course, there has been a rash of such pranks lately, and the questioner really is interested in the color of the house at *this moment*. And this observation is not merely a piece of "pragmatics" as opposed to "semantics"; what determines *what* is being said, in such a case, is our understanding of *why* one would say it in such circumstances.

Contrast this familiar use of "know" with G. E. Moore's. The point of Moore's utterance was what we call "philosophical"—that is, to establish that he "really knows" that there are material objects—and not practical in the way in which my saying that I know that I will be in Emerson Hall a week from Wednesday serves a practical purpose. But there is no standard of "really knowing" independent of an ordinary context for using the word(s) "know" (and "really"). One might understand the statement that one "really knows" that something is a hand if one were to encounter an ordinary context that gave sense to the question—a context that also gave sense to "not knowing" that it is a hand (for example, there could be doubt whether one could identify a hand, even one's own, if there were a real possibility that one had been brain damaged)—but to establish in such a context that one does really know that it is a hand would not answer the skeptic's doubt of the

---

the picture that there is some determinate set of things that I can be said to "know" independent of what the context is in which the question is asked.

15. The metaphor of attunement is Cavell's; cf. *Claim of Reason*.

external world; it would only establish (on the basis of a fund of background knowledge) that one is not brain damaged (or whatever the problem might be), and this would lack the requisite generality to address the skeptical challenge. On the other hand, to use "know" with no specifiable point other than to establish that one "really knows" that there are material objects is to use it in such a way that it has come loose from the criteria that allow it to be recognizable as an application of our concept of knowledge.[16] And when "know" is used without a criterion, we can just as well say (with the skeptic) that Moore *does not* know as that he does know. Moore wants to exploit the appropriateness of saying "I know" in a host of ordinary contexts without recognizing the *occasion sensitivity* that is part and parcel of the use of "know" in those ordinary contexts.

Ordinary and "Philosophical" Doubt

I have said that the point of saying "I know" is to rule out certain "possibilities" that what one claims to know is not the case, and that *which* possibilities these are depends on the *ordinary* context in which the knowledge claim is made.

"Ordinary context" here is meant to contrast with "philosophical context,"[17] not, for example, with technical, or scientific, or learned context. I do not suppose that such a contrast is one that can be drawn once and for all, independently of what particular context is being characterized as "philosophical" and what the point of so characterizing it may be. Instead, let me illustrate the notion of a "philosophical context" by saying briefly how both the skeptical use of "know" and Moore's use of "know" go wrong, wrong in ways that are characteristic of philosophical discussions of knowledge.

These two ways of going wrong correspond to opposing ways of attempting to transcend the context sensitivity of ordinary knowledge claims: both Moore and the skeptic attempt to use "know" in such a way that everything that ordinarily determines a human context—what our interests and objectives

16. Let me say that I follow Cavell (in *Claim of Reason*) in not taking "criteria" to be necessary and sufficient conditions for being an *X*, but rather for being the "grammatical" conditions that must be satisfied for it to be the case that what one is talking about / identifying / perhaps misidentifying is *X*'s.

17. I am indebted to Cavell for suggesting that this is how Wittgenstein's notion of the ordinary is to be understood.

are, what our background knowledge is, what our loyalties are, what our communal affiliations and disaffiliations are, what our relations to the person(s) we are speaking to are—become *irrelevant*. That attempt at an impossible transcendence is what I take to characterize a "philosophical context" for the purposes of this discussion.

Notoriously, skeptics will not allow one to ignore any "possibility" as too far fetched; they also will not allow one to appeal to background knowledge in one's reply (if one does, they will simply point to conceivable—or allegedly conceivable—"possibilities" of error with the background knowledge). But this makes the skeptics' use of "know" one in which it is virtually *tautological* that one does not "know" anything. The skeptic often conceals this (as does Descartes, for example, in his skeptical moment) by beginning as if all he were doing was insisting on "tightening up" our standards and introducing some praiseworthy "methodological rigor" into our attributions of knowledge. But when the skeptic is done, what he says seems to have no methodological implications whatsoever. This fact is reflected in Stroud and others saying that they are interested in a "special philosophical question." It is a live question whether this "philosophical" use of know has not just produced a homonym of our word if it does not feed back into any conclusions about when we should attribute knowledge in the ordinary use.

The skeptic's use of "know" is one example of a "philosophical" use; but what of Moore's? Moore's claim, as I just said, trades on the use of "know" in ordinary contexts (his "I know that this is a hand" is a parody of such ordinary knowledge claims as "I know that this is such and such a sort of object" and "I know that this is a tool for such and such a purpose," uttered in a context in which "what it is" is in genuine doubt); but precisely because it is not responsible to any ordinary context, what *licenses* Moore to make his claim is just as unclear as is the skeptical "possibility" that it is all a dream or a pervasive hallucination. We might say that the skeptic requires *absurd* criteria for knowledge, while Moore tries to make a knowledge claim *without* criteria.

But why are we tempted to suppose that such a use of "know" as the skeptic employs is part of the language game? Perhaps the deepest reason is the tendency to oscillate between illusions of omnipotence and illusions of impotence that I mentioned at the outset of this essay, or, to put it in a way that is only apparently different, the tendency to think that anything (any "knowledge") that is not omnipotent is impotent.

A more superficial—but nonetheless very plausible—reason was once suggested by Max Black.[18] Black pointed out that we sometimes speak of using the word "know" in a "stricter" or "less strict" sense. And he suggested that this may give rise to the idea that the skeptic's "sense" is *the absolutely strict sense.* Perhaps that is why it seems that the skeptic's use of "know" is the proper one; why it seems that it is we who normally speak "loosely"—we who are being too "sloppy" to see the correct point the skeptic is making: that "strictly speaking" we do not know anything at all.

To help us resist this temptation, Black pointed out that at one time mathematicians thought that it made sense to speak of the "sum" of any infinite series at all, without reference to whether the series of partial sums converges or diverges. (A celebrated early nineteenth-century mathematician suggested that the divergent series $1-1+1-1+1-\ldots$ has the sum $\frac{1}{2}$.) Just as it does not follow from the fact that the series $1-1+1-1+1-\ldots$ can be calculated to any finite number of terms that there is such a thing as the "sum" of the whole infinite series, or such a thing as the "limit" of the partial sums, so from the fact that one can use "know" in increasingly strict senses (in certain contexts, anyway) it does not follow that there must be such a thing as "the absolutely strict sense." The limit of a series of stricter senses of "I know" may not be a sense of "I know" at all. And, indeed, the skeptic's "sense" of "I know" is one that cannot do the work that knowledge claims do.

In distinguishing in this way between ordinary and philosophical uses of "I know," I am in the ballpark of Peirce's celebrated distinction between "real" and "philosophical" doubt. Unfortunately, Peirce drew that distinction in psychological terms; and this has allowed some philosophers (including Michael Williams) to say that there is no difference, or only a verbal difference, between Peirce's rejection of skeptical doubt and Hume's observation that such doubt cannot be sustained outside the study. But Peirce's observation need not be understood in this way. The real point that Peirce—and after him James and Dewey—were making, I believe, is a genuinely logical point: that *doubt requires justification as well as belief.* And pointing to remote possibilities of error, mere logical possibilities, or even far-fetched empirical possibilities does not in itself justify doubt. That the language game is not played the way the skeptic plays it is a fact about the *structure* of our concepts. It

18. I have been unable to locate the place in which (I remember) Max Black makes this comparison.

is not a mere contingent fact about our psychology that we do not use "know" in such a way that the only knowledge statement we can make is the one negative statement that we do not know anything; for such a use of "know" would defeat the purposes for which we have the concept, and *that* is as much a logical fact as is the fact that if we have one contradiction in a logical calculus (whose rules include the standard rules of two-valued logic), we can derive every statement as a theorem in the calculus, and then the calculus would be useless (if it were intended as a formalization of some branch of discourse).

Skepticism as a Challenge to the Whole Language Game

A number of philosophers will find all this argumentation futile; indeed, they will find any attempt to show the incoherence of skeptical doubt futile. (My worry in this essay has been that Strawson seems to agree with these philosophers, at least in his Humean mood.) Although their arguments against the enterprise vary, in the case of Barry Stroud's article cited by Strawson,[19] a central reason seems to be that such an attempt necessarily proceeds from *within* our conceptual system itself. But the skeptic, they say, is offering a *challenge to our ordinary conceptual system as a whole.* They believe that the sort of criticism of the skeptic's argument we have been engaged in necessarily begs the whole question.

To this, there are two replies I would offer. First, it need not be—and it had better not be—our purpose to give a reply to the skeptic that *the skeptic must accept.* To think of the target as the skeptic, or even "the skeptic inside me" (this is a suggestion of Michael Williams), is to engage in a losing enterprise. The skeptic will go on asking "How do you know that?" ad infinitum. The challenge of skepticism, insofar as it is an *intellectual* challenge at all (and, as Cavell has discussed at length in part 4 of *The Claim of Reason,* for the skeptic himself it is something much deeper), lies in the fact that the skeptic threatens our conceptual system from *inside.* The reason skepticism is of genuine intellectual interest—interest to the *nonskeptic*—is not unlike the reason that the logical paradoxes are of genuine intellectual interest: paradoxes force us to rethink and reformulate our commitments. But if the reason I undertake

19. Barry Stroud, "The Significance of Scepticism" in P. Bieri, R. P. Horstmann, L. Krüger, eds., *Transcendental Arguments and Science* (Dordrecht: Reidel, 1979) 277–297.

to show that the skeptical arguments need not be accepted is, at least in part, like the reason I undertake to avoid logical contradictions in pure mathematics (e.g., the Russell paradox), or to find a way to talk about truth without such logical contradictions as the liar paradox); if my purpose is to put my own intellectual home in order, then what I need is a perspicuous representation of our talk of "knowing" that shows how it avoids the skeptical conclusion, and that my *nonskeptical* self can find satisfactory and convincing. (Just as a solution to the logical paradoxes does not have to convince the skeptic, or even convince all philosophers—there can be alternative ways to avoid the paradoxes—so a solution to what we may call "the skeptical paradoxes" does not have to convince the skeptic, or even convince all philosophers—perhaps here too there may be alternative solutions.) It is not a good objection to a resolution to an antinomy that the argument to the antinomy seems "perfectly intelligible," and, indeed, proceeds from what seem to be "intuitively correct" premises, while the resolution draws on ideas (the theory of types in the case of the Russell paradox; the theory of levels of language in the case of the liar paradox; —and on much more complicated ideas than these as well, in the case of the follow-up discussions since Russell's and Tarski's) that are abstruse and to some extent controversial. That is the very nature of the resolution of antinomies. What I have tried to provide in this essay is an argument *that convinces me* that *the skeptic cannot provide a valid argument from premises I must accept to the conclusion that knowledge is impossible.* In the same way, Russell showed that (after we have carefully reconstructed our way of talking about sets) a skeptic—or whoever—cannot provide a valid argument from premises we must accept to the conclusion that mathematics is contradictory, and Tarski showed that (after we have carefully reconstructed our talk about truth) a skeptic—or whoever—cannot provide a valid argument from premises we must accept to the conclusion that talk about truth is contradictory.

Second, the notion of challenging the conceptual system as a whole is (as I think it was the merit of the Kantian tradition to have seen) an incoherent one. To challenge the conceptual system "as a whole" would require standing outside one's own concepts, and there is no place from which to do that. The idea that I have already criticized as incoherent, of a general problem of the existence of the external world, was a failed attempt to find just such a place. (Not surprisingly, it depended on the Humean-cum-Berkeleyan view of experiences—"ideas and impressions"—as freestanding objects.)

To be sure, the skeptic may, and often does, try to find a place "inside" our concepts from which he can formulate *something like* the traditional problems. The brain-in-a-vat scenario that I discussed in *Reason, Truth and History* is just such an attempt. But incoherence is not the only problem to threaten the skeptic's attempt to criticize our conceptual scheme "as a whole." Let us think of what goes on in brain-in-a-vat scenarios for a moment. Even if we bracket the question how, if we are indeed brains in a vat, we are supposed to be able to so much as refer to real vats, real objects, or a real computer (which we have to do to so much as formulate the skeptical possibility), the fact is that such scenarios do not even attempt to describe a possible world in which there are no external objects at all. Not only are there external objects in a brain-in-a-vat scenario (although the brains in the vat are supposed to be mistaken about what external objects are), but they are supposed to obey just the standard physical laws. So the brain-in-a-vat scenario does not actually contradict either our "belief in an external world" *or* our "belief in the uniformity of nature." And it even concedes that we have the general character of the external world right and the general character of the physical laws right. It is the details we have wrong. But precisely because the scenario assumes just our standard picture of the physical world at the level of natural law, it is appropriate for us to point out to the skeptic that the world picture that he needs to give content to his talk of computers, brains, electrical impulses, chemicals needed to keep the brains alive, and so on is simply stolen from *our* world picture, and our world picture does not imply that there is any serious likelihood of the existence of a computer large enough and powerful enough to simulate a "real world" so well as to fool billions of different human brains.[20] The skeptic can indeed say that the brain-in-a-vat scenario is meant only as an *illustration* of the possibility that the external world is an illusion; but then he is back at the problem of giving *that* "possibility" content. Of course, we must concede that the possibility of constructing such scenarios from within our own world picture does show that what our claims rest on is not "deductive certainty." But in the end, that is all the skeptic can get us to concede. And that is hardly news.

---

20. Recall our Peircean principle that doubt requires justification just as much as belief. If there is some version of the brain-in-a-vat scenario that overcomes the semantic objections that I raised in *Reason, Truth and History,* I do not think that a Peircean would say that "we know that we are not brains in a vat"; he would say, "The question does not genuinely arise."

How the Present Discussion Differs from Michael Williams's Claims Concerning the Contextuality of Knowledge Claims

In the remaining sections of this essay I will further explore the occasion sensitivity of knowledge claims and the nature of skepticism by contrasting the view I have been defending with some views defended by Michael Williams in *Unnatural Doubts*.[21]

Superficially, Williams's views on skepticism resemble the ideas about the "occasion sensitivity" of knowledge claims that I have been defending. Insightfully, he sees the skeptic and his traditional opponent as sharing a view he calls "epistemic realism": in effect, the view that the knowledge relation is a natural kind, like the parent relation, or like the metric relation in general relativity. For the epistemic realist, to say that one knows that *p* is always, no matter what the context, to say that one and the same relation *R* (the "natural kind" in question) obtains between the true proposition one claims to know and one's "evidence" (including memory and the "evidence of the senses"). This view implies that "knowing" cannot come to different things in different contexts; it implies that what reasons one has for claiming to "know," what doubts actually *arise* in the context, what is going to be *done* (and by whom) if one's knowledge claim is accepted, and the nature and seriousness of the possible consequences are all irrelevant to whether one can say "I know."

In opposition to epistemic realism, Williams proposes the thesis that whether or not one knows depends very much on these contextual factors, and particularly on what doubts actually are at issue in the context.[22] So far this resembles the claim I have attributed to various philosophers (J. L. Austin, Stanley Cavell, and Charles Travis, in particular) that knowledge claims are, by their very nature, occasion sensitive. But clearly Williams understands what he calls the "contextuality" of knowledge claims very differently than these authors. For he does *not* find skeptical arguments incoherent (although, significantly, he considers no skeptical arguments in detail, but only refers vaguely to the claim that external objects are an "illusion" and the like), and he finds Cavell's contention that Moore and the skeptic are alike trying to

---

21. Michael Williams, *Unnatural Doubts* (Princeton: Princeton University Press, 1998).
22. Ibid., section 8.

speak outside our criteria too recherché to be a successful rejoinder to the skeptic's allegedly clear and convincing arguments.

According to Williams, what I have called "philosophical contexts" are conceptually in perfect order; it is just that in those contexts, contexts in which such possibilities as the possibility that the whole external world might be a "dream"[23] or an "illusion" are raised, *every doubt becomes relevant,* and hence, according to Williams, the skeptic is *right* (in those contexts) to say that we have no knowledge of external things. However, Williams adds, that does not mean that the skeptic is right *tout court.* In an *ordinary* context, one in which we do not think about either the skeptic's *arguments* or the traditional philosopher's responses—or, for that matter, Williams's own responses—to those arguments, one in which philosophical considerations do not even arise, it is *right* to say that we *do know* that, for example, there are tables and chairs in the room. (In effect, Williams is offering a logical reconstruction of Hume's distinction between what we know in the "study" and what we know when we are not in the "study.")[24]

One way in which one can see the difference between Williams's notion of contextuality and the notion of occasion sensitivity that I have defended is to examine the way Williams treats the dispute about whether knowledge is "closed under known implication" (the very issue that first led me to think about the occasion sensitivity of knowledge claims). According to Williams, knowledge *is* closed under known implication. Thus, consistently with his position about the "contextual" truth of skepticism, Williams holds that in the sort of ordinary context just mentioned, in which I know, according to Williams, that there are tables and chairs, and that I see them, I also know (because it follows from the fact that I see real tables and chairs by an implication that is also known to me) that it is not the case that the external world

23. Of course, when this possibility is raised, the question of just how the dreamer (or his unconscious) is supposed to be able to dream a dream that is that complex and that coherent is never seriously discussed. In effect, the dreamer—or his brain—becomes the supercomputer in the brain-in-a-vat scenario.

24. Unfortunately, Williams is quite wrong to think that Hume claimed that in the "study" we know that the skeptic is right; as I have already pointed out, with respect to external (persisting) objects, the Hume of the "study" was an *idealist,* not a *skeptic.* The mistake about Hume is symptomatic of the larger error of supposing that so-called skeptical arguments are either simple or timeless, as if the Greek skeptics, Descartes, Hume, and our contemporaries who imagine scenarios involving brains in a vat and computers all deal with the same "skeptical arguments."

as a whole is an illusion. *But I only know it when I do not think about it;* as soon as I think about the question whether the external world is an illusion, the skeptical doubts become relevant, and then my knowledge is destroyed. Williams explicitly "buys" the counterintuitive consequence: there are true statements of the form "I know that *p*" that I can never know to be true at the time at which they are true. I can *know* that the external world is not an illusion—indeed, most of the time I do know this—but I can never truly *claim* that "I know that the external world is not an illusion."

I will argue that there are at least three things wrong with this account:

1. Williams is quite wrong about the concept of knowing.
2. The particular way in which he is wrong makes him an "epistemic realist" *contre lui.*
3. Williams misunderstands the nature of both skeptical and ordinary knowledge contexts.

I now take up each of these points in turn.

Williams Is Wrong about the Concept of Knowing

Even though I cannot say that "I know the external world is not an illusion" in an "ordinary" context, according to Williams (because to even think about the question would transform the context into a philosophical one), a third person—for example, Williams himself—*can* say that I knew this in that context. In fact, according to his theory, Williams can say of himself that *when* he was in an ordinary context, he knew (but did not know that he knew and could not claim that he knew) that the external world was not an illusion. The theory thus implies that the following utterances are perfectly acceptable:

1. "When I was counting the chairs to make sure there were enough for the party, I knew that the external world was not an illusion, but I do not know that now."
2. "In some contexts I know that I will come to Emerson Hall next Wednesday, and in other contexts I do not know that I will come to Emerson Hall next Wednesday."
3. "I would have known that I would come to Emerson Hall next Wednesday if you had not mentioned the fact that I might get killed in a traffic accident."

I submit that none of these utterances *is* a nondeviant English sentence. Our actual use of "know" has no place for such utterance forms as "I would have known that *p* if you had not mentioned the fact that *q*" or "I know that *p* when I do not think about *q*" or "I know that *p* in that context, but I do not know that *p* in this context." Williams is simply wrong about how the concept of knowledge functions. But how did it come about that, starting with promising observations about "epistemic realism" and "contextuality" as he did, he ended up with such mistaken (not to say bizarre) views about the concept of knowledge?

Has Williams Really Given Up Epistemic Realism?

A central difficulty is that Williams has tried to handle the occasion sensitivity of knowledge claims by assimilating it to the kind of simple relativity we find with such predicates as "large." There is nothing wrong with saying that someone is "large for a man but small compared with an elephant," or that something is "large for a piece of fabric but small for a wall covering." That knowledge claims exhibit *some* relativity is something no one would deny; on *any* account, whether I know that *p* or not depends on my *evidence,* my experiences, my memories, and my other beliefs. Thus it can be perfectly correct to say, "When I only had such and such evidence, I did not know that *p,* but now that I have such and such additional evidence, I do know that *p.*" Williams has tried to handle occasion sensitivity by simply adding such factors as *what doubts actually arise in the context* to the list of things (evidence, experiences, memories, other beliefs) on which knowledge depends. But *occasion sensitivity is quite a different phenomenon from dependence on evidence.* Indeed, the idea that *one* and the same knowledge relation is involved in all contexts (it is only that the *relata* are different in different contexts) is what Williams calls "epistemic realism,"[25] and as we have just seen, Williams does not really give this idea up; instead, he simply expands the list of permissible relata. In this sense, Williams is still an epistemic realist *contre lui.*

It is, of course, true that sometimes we have to take back or at least qualify a knowledge claim that we have made when new doubts become relevant. This can happen in two ways: the more drastic case arises when raising new doubts, doubts we had overlooked, actually causes us to stop believing that *p,* where *p* is whatever we claimed earlier to know. (So it is not, in this case, just

25. Williams, *Unnatural Doubts,* preface, xx.

that we will not say "I know that *p*"; we will not even say "I think that *p*.") In such a case we certainly would not say, "Well, in the former context I knew that *p*"; rather, we will say, "I thought I knew, but I did not, because I overlooked such and such a possibility." The less drastic case arises when we still believe that *p*, but the circumstances force us to qualify our claim to know. This could happen in a court of law, where one might admit the relevance of some possibility that in a less stringent context one would feel free to ignore, or, even outside a law court, it might happen that the possible damage that will result if one turns out to be mistaken is greater than one thought, and so one hedges one's claim in various ways. In the latter kind of case one might say, "I still believe that *p*, but I was speaking loosely when I said I *knew* that *p*," but again, one would not say, "I *did* know that *p*, but I do not know it in *this* context." The relevant features of the context determine whether one is licensed to make a knowledge claim at all; but they are not *relata*. (Indeed, strictly speaking, even the evidence is not a *relatum;* for although I may say that "I did not know when I only had such and such evidence, but I do know now that I have further evidence," I will not say, even when I still believe the *p* in question to be true, "When I had only such and such evidence, I did know that *p*, but now that I have additional evidence, I do not know that *p*"; for if the additional evidence causes me to retract my claim to know that *p*—perhaps by raising a "Gettier problem" with my earlier reasoning[26]—then I will say "I *thought* I knew" about my earlier epistemic condition, not "I did know, relative to that evidence.")

These are not, needless to say, mere linguistic oddities. Rather, they reflect fundamental features of our cognitive situation: that we are *fallible* (knowledge claims are defeasible), and that *we have the right to claim to know, in certain situations, at certain times, and for certain purposes.* If these two features of our cognitive situation have come in for so much emphasis from pragmatist writers, it is because pragmatism has from the beginning been simultaneously *fallibilistic* and *antiskeptical.* Although Williams also wishes to be fallibilistic (and to quarantine skepticism to philosophical contexts), the locutions he allows us, locutions like "I know it in this context, but I do not know it in that," cannot do the work of making knowledge claims, and without genuine knowledge claims, there is nothing to be fallibilistic about.

26. Edmund L. Gettier, "Is Justified True Belief Knowledge?," *Analysis* 23, no. 6 (1963): 121–123.

Williams Misunderstands the Nature of Both Skeptical and Ordinary Knowledge Contexts

This last remark about Williams's view may seem overly harsh. If I say in a given context, "I know that *p*," Williams may reply that I *have* made a genuine knowledge claim; I have not *merely* said, "Relative to this context, I know that *p*."

The problem, however, is that for Williams, what fixes or changes a context is merely what doubts are *thought of.* This makes knowledge contexts an essentially superficial matter. After all, the possibility of "illusion" may be mentioned in all sorts of contexts that are not technically philosophical! To revert to the law-court example: a hostile attorney may say to the witness, "Perhaps you only imagined it." If merely mentioning the possibility that what I claimed to see was only imagined or only an illusion, or that I was dreaming, sufficed to change the context to a philosophical one (and if, as Williams contends, the skeptic is right, as soon as the context becomes a philosophical one), then any attorney worth his salt could get anyone at all to admit that he does not know that *p*, no matter what the relevant *p* is. Williams's account loses what I described as the pragmatist insight—that raising a doubt is something that requires a *justification.*

For Williams, the Skeptic Is Making a Straightforward Conceptual Blunder, Misconstruing the Grammar of Our Language

For Williams, the skeptic's argument is perfectly correct; however, the skeptic has misconstrued the grammar of our language and thus has failed to see that from the fact that he (in his philosophical context) can truly say, "We do not know that there is an external world," it does not follow that in a different context one cannot perfectly well know (even if one cannot claim to know, because doing so would change the context!) that there is an external world. The skeptic is right (in the study), but, for conceptual reasons, it does not matter. But this account makes skepticism both less serious and more serious a matter than it is.

It makes skepticism less serious than it is because it misses what one may call the *pathos* of skepticism. Even if we can show within our conceptual system, or within a plausible rational reconstruction thereof, that doubting whether we "know that the external world is not an illusion" is incoherent, this will not satisfy the skeptic—in fact, it would not satisfy the skeptic *even if it were intellectually invulnerable in every respect.* Even if skeptics were to concede that they cannot coherently state what worries them without doing violence to the concepts of "knowing" and "doubting" that we actually have, they would still not feel that their worries have been laid to rest by that discovery; they would only come to feel that their worries are unreal if they were able comfortably to *inhabit* our conceptual system, were able to make those concepts of knowledge and doubt *their own*; and the skeptics' problem, as Stanley Cavell has emphasized, is precisely their *alienation* from our concepts (an alienation that has, indeed, its positive aspect[27]), and not just a conceptual worry occasioned by an argument. For the skeptic, skepticism is not just or even primarily a conceptual problem. And Williams makes skepticism more serious than it is by conceding so easily, almost frivolously, that the skeptic's arguments are correct.

27. See Hilary Putnam, foreword to Ted Cohen, Paul Guyer, and Hilary Putnam, eds., *Pursuits of Reason: Essays Presented to Stanley Cavell* (Lubbock: Texas Tech University Press, 1992), vii–xii.

31

Strawson and Skepticism

Peter Strawson is one of the great philosophers of the twentieth century. If I have long been an admirer of his work, it is because of the exemplary way in which, time and time again, he has advanced the state of philosophical discussion and opened new avenues for us to explore. He has done this in area after area (while always keeping in mind the interrelatedness of philosophical issues). I particularly value the fact that he opened the way to a reception of Kant's philosophy by analytic philosophers. In this essay, however, I explore a tension I find between Strawson's Kantian and his Humean sympathies.

Strawson's Humean Tendency

In *Skepticism and Naturalism: Some Varieties,* Peter Strawson takes a rather hard Humean line.[1] Belief in the external world or in the uniformity of nature is a matter of natural inclination, not received conviction, and we could not be brought to give such beliefs up by any reasons. Moreover, the skeptic's

---

1. P. F. Strawson, *Skepticism and Naturalism: Some Varieties* (New York: Columbia University Press, 1985).

doubts and the various philosophical attempts to answer the skeptic are alike pointless; beliefs that are not based on reasoning can be neither undermined nor defended by it.

Strawson's tendency to appeal to Hume is *not* a recent development. It is worth recalling a short debate between Strawson and Wesley Salmon as long ago as 1957–1958. In *Introduction to Logical Theory* Strawson had denied that induction can be given a general justification.[2] At the same time, he had written, "To call a belief reasonable or unreasonable is to apply inductive standards,"[3] and further, that "to ask whether it is reasonable to place reliance on inductive procedures is like asking whether it is reasonable to proportion the degree of one's convictions to the strength of the evidence. Doing this is what 'reasonable' means in such a context."[4] Criticizing all this in an article published in 1957, Salmon wrote, "If the foregoing theory [that induction cannot be given a general justification] is correct, empirical knowledge is, at bottom, a matter of convention. We choose, quite arbitrarily it would seem, some basic canons of induction; there is no possibility of justifying the choice."[5] He went on:

> The attempt to vindicate inductive methods by showing that they lead to reasonable belief is a failure. . . . If we regard beliefs as reasonable simply because they are arrived at inductively, we still have the problem of showing that reasonable beliefs are valuable. This is the problem of induction stated in new words. . . . It sounds very much as if the whole argument [that reasonable beliefs are, by definition, beliefs that are inductively supported] has the function of transferring to the word "inductive" all of the honorific connotations of the word "reasonable," quite apart from whether

2. P. F. Strawson, *Introduction to Logical Theory* (London: Methuen, 1952).

3. Ibid., 249.

4. Ibid., 257.

5. Wesley Salmon, "Should We Attempt to Justify Induction?" *Philosophical Studies* 8, no. 3 (April 1957): 33–48. The quotation is from p. 39. Salmon himself believes that it is possible to give a deductive "vindication" of induction, that is, a proof that induction must lead to successful prediction if any method does. For a refutation of this claim, see Hilary Putnam, "Reichenbach and the Limits of Vindication," reprinted in Putnam, *Words and Life*, ed. James Conant (Cambridge, Mass.: Harvard University Press, 1994), 131–148; the article originally appeared with the title "The Limits of Vindication" in Dag Prawitz, Brian Skyrms, and Dag Westerståhl, eds., *Proceedings of the Ninth International Congress of Logic, Methodology, and Philosophy of Science,* Uppsala, Sweden, August 7–14, 1991 (Amsterdam: Elsevier Science Publishers, 1994), 867–882.

induction is good for anything. The resulting justification of induction amounts to this: If you use inductive procedures you can call yourself "reasonable"—*and isn't that nice!*[6]

Strawson, of course, could not let this go by unanswered. Two issues later in the same journal, he retorted:

[Salmon] says that if [the view that induction cannot be given a general justification] is correct, then it must also be true that inductive beliefs are *conventional;* that empirical knowledge is, at bottom, a matter of *convention;* that it is a matter of our arbitrary *choice* that we recognize the basic canons of induction which we do recognize.

I find it mysterious that Mr. Salmon should think this. For he refers more than once, in the course of his paper, to Hume. Hume, I suppose, did not think that induction could be given a general justification. He did not, on this account, think that inductive beliefs were *conventional,* he pointed out that they were *natural.* He did not think that our "basic canons" were arbitrarily *chosen;* he saw that this was a matter in which, at the fundamental level of belief-formation, we had *no choice at all.* He would, no doubt, have agreed that our acceptance of the "basic canons" was not forced upon us by "cognitive considerations" (by Reason); for it is forced upon us by nature. . . .

If it is said that there is a problem of induction, and that Hume posed it, it must be added that he solved it.[7]

Puzzlingly, in *Individuals,* written about the same time (published in 1959), Strawson claims that Kantian insights are effective in exposing certain confusions involved in skeptical arguments: "[The skeptic's] doubts are unreal, not simply because they are logically irresoluble doubts, but because they amount to the rejection of the whole conceptual scheme within which alone such doubts make sense."[8] This asserts that apart from the conceptual scheme that the skeptic rejects, the skeptic's doubts do not so much as make sense. It follows, at the very least, that the skeptic is in the following position: if her argument is correct, then her argument makes no sense. (That would seem to be a *reductio* if I ever saw one.)

---

6. Salmon, "Should We Attempt to Justify Induction?" 42.

7. P. F. Strawson, "On Justifying Induction," *Philosophical Studies* 9, nos. 1–2 (January–February 1958): 20–21. The quotation is from pp. 20–21.

8. Peter Strawson, *Individuals* (London: Routledge, 1959), 35.

But in his later *Skepticism and Naturalism: Some Varieties* he returns to the themes of his short reply to Salmon. In particular, he argues again that the basic categories of our conceptual system are not chosen for reasons *because they are not chosen at all;* and he suggests that to so much as ask for reasons for them is accordingly a mistake.

"If It Is Said That There Is a Problem of Induction, and That Hume Posed It, It Must Be Added That He Solved It"

To begin with, the idea that Hume "solved" the problem of induction is decidedly problematic. My disagreement with this claim is not based on the textual point that Hume does not speak of "canons" of induction. In fact, it is not clear that there *are* "canons" in the sense in which Salmon's quarrel with Strawson concerned the issue of whether the (supposed) canons of induction can be justified.

The canons of induction Salmon writes about are supposed to describe a method by which the totality of our empirical knowledge (all of our knowledge that has predictive import) could in principle have been arrived at starting with nothing but observation (and, of course, deductive logic). Salmon believes (to this day) that there are canons of induction in this sense because he follows Reichenbach in supposing that the so-called straight rule of induction, namely, posit that the limit of the relative frequency will be close to the so-far-observed relative frequency, is such a canon (in fact, the only one.)[9]

Unfortunately for this view, all such "rules of induction," if unrestricted, lead immediately to inconsistencies. One example of such an inconsistency is Goodman's famous "grue" example.[10] A commonsense response to this difficulty might well be to say that we do not project the Goodmanian "regularity" "All emeralds are grue" because to do so would require us to predict that

9. Salmon accepts Hans Reichenbach's conception of the task of philosophy as "rational reconstruction." The sort of reconstruction he has in mind is illustrated by Reichenbach's *Experience and Prediction* (Chicago: University of Chicago Press, 1938).

10. In *Fact, Fiction, and Forecast* [1954] (Cambridge, Mass.: Harvard University Press, 1983), Nelson Goodman gives a celebrated argument to show that not all predicates are equally projectible. The argument depends on his definition of the strange predicate "grue." Goodman defines something as grue if it is either observed before a certain date and green or observed after that date and blue.

emeralds will *change color* at the stroke of midnight on December 31, 1999, and we have no reason to expect such a change.[11] To the expected Goodmanian response, "So what? To project the regularity we do project, namely, 'All emeralds are green,' requires us to predict that emeralds will change gruller at the stroke of midnight on December 31, 1999, and what reason do we have to expect *that* change?" We might answer that gruller changes are mere "Cambridge changes,"[12] while color changes are what we consider "real changes."

Can we save Reichenbach's "straight rule of induction," or any alternative such rule or system of rules, by requiring that the predicates we project be such that something can remain S and remain P (where "All Ss are Ps" is the induction we wish to make) without undergoing a "real change"? No, because many predicates we wish to project would be excluded by such a criterion. (Something cannot remain a mammal for any length of time, for example, without undergoing a variety of "real changes," including *aging*.) "But those are *normal* changes, whereas the kind of change that would be involved if all emeralds *were* 'grue'—namely, that every emerald in the world would change color from green to blue simultaneously at midnight, December 31, 1999—would be decidedly abnormal." But now we are appealing to a fund of background knowledge, not to some supposed "canon of induction" by which all that background knowledge was allegedly obtained.

Goodman's own solution is to seek canons of induction that are restricted to what he calls "projectible" predicates. Projectibility depends, for Goodman, simply on how often a predicate has been projected in the past (the technical criterion is quite complicated, but this is the basic idea). Yet new predicates,

---

11. This sentence was a slip on my part. What follows from Goodman's definition is that the emeralds that are first *observed* on or after midnight on December 31, 1999, will be blue rather than green, not that the already observed emeralds will change from green to blue. Similarly, instead of saying that projecting "all emeralds are green" requires us to predict that emeralds will *change gruller* at the stroke of midnight on December 31, 1999, I should have written that it requires us to predict that the emeralds that are first *observed* on or after midnight on December 31, 1999, will be a different gruller. [2011]

12. The term "Cambridge change" was introduced by Peter Geach, *God and the Soul* (London: Routledge and Kegan Paul, 1969). He defined (71–72) a Cambridge change ("since it keeps occurring in Cambridge philosophers of the great days, like Russell and McTaggart") as follows: "The thing called 'x' has changed if we have '$F(x)$ at time t' true and '$F(x)$ at time t_1' false, for some interpretations of 'F', 't', and 't_1'." He observed, "But this account is intuitively quite unsatisfactory. By this account, Socrates would change by coming to be shorter than Theaetetus. . . . The changes I have mentioned, we want to protest, are not 'real' changes."

with no history of prior projection, are frequently introduced by physical theories and immediately thereafter used in "inductions." Moreover, I remain unconvinced that (even if we accepted Goodman's criteria for projectibility) we *could* state any set of "canons" by which it would have been possible to arrive at the knowledge we do have. (Goodman's own canons presuppose a vast amount of knowledge about the history of prior projections, much of which it is not at all clear we actually have.) In sum, as long as no one has ever succeeded in exhibiting the supposed "canons of induction" on which our inductive practices are thought to depend, the problem of their "justification" must remain utterly unclear. If we cannot even *state* canons of induction, then the problem *cannot* be that we accept them "without choice," or that they represent "natural beliefs."

Philosophers often suppose that wherever there is a practice that can be correct or incorrect, there must be a set of "rules" or "standards" that could be explicitly stated. Otherwise, it is thought, "anything goes." But this is quite groundless. There are right and wrong ways of using a screwdriver, but there are not explicitly statable rules or standards for screwdriver use. Moreover, even where our practices do have rules, it is rare indeed that those rules are as precise as computer algorithms. Usually what we have is the sort of "rules" Kant had in mind when he observed that there cannot be such a thing as a rule for good judgment, because if there were, we would need good judgment to apply the rule.[13] Thus there is nothing contradictory in saying that we have practices—indeed, an innumerable number of practices—that could be described as procedures of "inductive inference" in the sense that they lead from observed phenomena to confirmed hypotheses about phenomena not yet observed, while denying that those practices are governed by "canons of induction" of the kind that Reichenbach (and Carnap) sought with their ventures in "inductive logic."

In an older sense of the term, there are some "canons of induction"; J. S. Mill's methods are an example. But those methods apply only to situations of

---

13. Immanuel Kant, *Critique of Judgment,* ed. Paul Guyer; trans. Paul Guyer and Eric Matthews (Cambridge: Cambridge University Press, 2000), Ak. 169. In Kant's language, an a priori standard of *good* judgment can only be "subjective," or, better put, transcendental logic cannot play the role of supplying "mother wit." Cf. Juliet Floyd, "Heautonomy: Kant on Reflective Judgment and Systematicity," in Herman Parret, ed., *Kants Ästhetik, Kant's Aesthetics, L'esthétique de Kant* (Berlin and New York: Walter de Gruyter, 1998), 192–218.

a special kind; in fact, they are rules for eliminating the factors that are not causally responsible for the phenomenon by carefully controlling all the potentially relevant factors. Thus they apply to what Reichenbach would have called "advanced knowledge"; they are not methods by which all our knowledge could have been arrived at to begin with.

What Strawson shares with some of his critics (not only Wesley Salmon, but also Michael Williams[14]) is a tendency to assume that the skeptical demand to "justify the canons of induction" represents a coherent problem, that is, to assume that it is clear just *what* we are unable to do or to justify by giving reasons. But it is in fact far from clear.

I said that the idea that Hume "solved" the problem of induction is decidedly problematic. Not only did he not discuss the "problem of justifying induction" that Salmon and Strawson debated, but his discussion of the problem he did discuss, the problem whether reason can tell us anything about what will happen in the future, depends on premises that, I will argue, Strawson should be the last philosopher in the world to accept.

Hume's "Loose and Separate" Existences

Hume did claim that reason is powerless to tell us what will happen in the future. Strawson's reply to Salmon asserts, as I have mentioned, that "if it is said that there is a problem of induction, and that Hume posed it, it must be added that he solved it." The double qualification in this sentence may well reflect an awareness that there is something problematic about talk of "a problem of induction," but the qualification does not save Strawson's claim. I shall now argue that, contrary to Strawson, if it is said that there is a problem of induction, and that Hume posed it, then it must be added that he posed it in a seriously confused way.

The problem, quite simply, is that Hume's argument depends on other views of Hume's that are utterly untenable. What Hume claims (this is the central lemma, so to speak, in Hume's demonstration that reason is powerless to tell us what will happen in the future) is that *events at different times are completely logically independent in the sense that what happens at one time has no logical or conceptual connection with what happens at any other time*. In Hume's own terminology, the "existences" at different times are "loose and separate."

14. Michael Williams, *Unnatural Doubts* (Oxford: Blackwell, 1991).

Hume defends this view (which, I shall argue, is itself untenable) with the aid of the following further views: (1) Reason, when pushed to its limits ("in the study," i.e., in the context of careful philosophical reflection, in which our natural belief in the existence of enduring and unobserved objects is temporarily suspended), inevitably leads to the demonstration—that is, to the *valid* demonstration—that the very notion of an enduring physical object as something that is capable of existing unperceived is incoherent. On this much, Hume agrees completely with Berkeley. What can coherently be thought to exist is only "ideas and impressions," and these do not logically or conceptually depend on the existence of material bodies or an objective causal order (indeed, as just said, the notion of "material bodies" that we think we have is incoherent). (2) Any sequence whatsoever of ideas and impressions represents a possible future course of the world; that is, it could be the case (regardless of what has existed in the present and past) that all that will exist in the future is that sequence of ideas and impressions occurring in that order.

Hume needs these premises because it would be enormously implausible to contend that any sequence of events whatsoever is logically possible if the "events" in question are described in our normal conceptual scheme—that is, in terms of kinds of material objects acting on and being acted on by one another in familiar sorts of ways. (This is a point that Strawson himself is extremely aware of when he wears his Kantian hat. After all, he did write that "[the skeptic's] doubts are unreal, not simply because they are logically irresoluble doubts, but because they amount to the rejection of the whole conceptual scheme within which alone such doubts make sense." What puzzles me is that he seems to forget all that when he puts on his Humean hat.) A material object, say, a cat or a rock or a table, is something that has certain familiar causal powers. To be sure, we cannot determine what those causal powers are simply by conceptual analysis; our ideas of them are subject to revision. But the idea that something might be a cat while having none of the familiar causal properties of cats has, as it stands, no intelligible sense. (I say "as it stands" because it can happen that we tell a story that provides a context in which it makes sense to say that something is a cat but is not, say, visible; but then the story makes sense of the statement by explaining how something or other interferes with the seeing of the cat; that is, the story introduces additional causal elements.) But the supposition that the world might, from now on, consist of "events" in a random order—at one moment there is a cat chasing a mouse, the next instant the cat is twelve feet tall and singing "Yankee

Doodle," the next instant there is a purple tidal wave sweeping over a field of flowers with heads like Charlie Chaplin . . . —implies that from now on, *there will be no significant regularities at all,* and, a fortiori, that there will be no causal connections. But if there are not going to be any causal connections, then there are not going to be any causal powers, or objects with causal powers, and, a fortiori, any cats, mice, waves, or flowers.

Incidentally, it is not only Strawson who overlooks the relevance of this point to Hume's argument. Reichenbach himself has brilliantly argued that all our empirical claims have predictive import.[15] But when he writes about induction, he simply assumes that we begin the enterprise of empirical inquiry in the following situation: we have accumulated knowledge—observational knowledge—that certain things have happened in the past, but no knowledge about the future, and our problem is to make justifiable "posits" about what will happen in the future. But the position Reichenbach imagines in his writing on induction, namely, that we have *no knowledge whatsoever about the future and have to justify our very first "posit,"* is totally incoherent. (Under ordinary circumstances we would not be willing to say, "I know there is a chair here," if we were not also willing to say, "If I go forward and touch it, I will feel a chair"; observational knowledge and predictive knowledge are not independent—"loose and separate"—in the way Reichenbach's discussion of induction presupposes.)

Of course, this criticism does not apply—or at least, does not apply at first blush—to Hume himself. Hume, as I said, is not talking about such events as a cat chasing a mouse. He is talking about sequences of *ideas and impressions,* that is, sense data considered as immaterial objects of which we are directly aware. But—as Kant very well saw—this does not rescue Hume's argument. For even if we assume that we know what we are talking about when we talk of *immaterial* "ideas and impressions"—a very big "if" indeed[16]—the notion of a *future* is inextricably connected with the notions of space, time, and causality. This is not just a fact about Newton's physics, or Einstein's, but about

15. See Hans Reichenbach, "Are Phenomenal Reports Absolutely Certain?" *Philosophical Review* 61, no. 2 (April 1952): 147–159; and Hilary Putnam, "Reichenbach and the Myth of the Given," in Putnam, *Words and Life,* 115–130.

16. For a criticism of sense-datum ontology and epistemology, see Hilary Putnam, "The Dewey Lectures 1994: Sense, Nonsense, and the Senses; An Inquiry into the Powers of the Human Mind," *Journal of Philosophy* 91, no. 2 (September 1994): 445–517.

our ordinary conceptual scheme as well. Imagine, thus (assuming the existence of Humean or Berkeleyan immaterial "ideas and impressions" for the sake of argument), that there is a (more or less instantaneous) world-state, call it "*A*," consisting of a sense impression as of a cat chasing a mouse, a world-state, call it "*B*," consisting of a sense impression as of a twelve-foot cat singing "Yankee Doodle," and a world-state, call it "*C*," consisting of a sense impression as of a purple tidal wave sweeping over a field of flowers with heads like Charlie Chaplin. What sense does it have to say that these are *states of one and the same world,* let alone to speak of them as temporally ordered, if there are no causal connections of any kind among them?[17] Hume's argument depends on our thinking of the concepts "experiences *A* and *B* [think of experiences at different times here] lie in one and the same phenomenal world" and "*A is earlier than B*" as themselves *presuppositionless.* But they certainly are not.

Of course, Hume was right if all we imagine him to have been arguing is that we do not have deductive certainty concerning what will happen in the future; but no philosopher—no ancient, medieval, or modern philosopher—ever imagined that we did. Hume was not out to refute a straw man. Hume wanted to show something truly surprising, namely, that knowing the nature of a current existent (when our notion of that nature is not allowed to be inflated by our tendency to project various ideas subjectively—the idea of causality, or the idea of enduring matter, for example—onto what is given) tells us nothing about what is likely to happen in the future. This amounts to claiming that the nature of the current existent, as it is in itself, is "loose and separate" from all dispositions, causal connections, or any other matter of fact. And that is an incoherent idea.

Why Do I Go into All of This?

The debate with Salmon was a long time ago, and Strawson's part in that debate was only two pages in length. It would hardly be worth discussing if

---

17. Even if one tries to reconstruct time order phenomenalistically, as Carnap did in *Der logische Aufbau der Welt,* by defining *A* to be earlier than *B* (in a solipsistic version of the world) if a memory of *A* coincides with *B*, we are committed to there being a lot more in the world than an arbitrary sequence of sense impressions if there is to be time order. For one thing, there have to be all those memories; and—although Carnap chooses to ignore this—a lot has to be in place before it makes sense to speak of an experience as a "memory."

Strawson had not returned to something like his earlier position in *Skepticism and Naturalism: Some Varieties.* There he argues that it is a mistake even to raise the question of justifying such beliefs as our belief in "the uniformity of nature" and our belief in the existence of "the external world" (so the skeptic and the philosopher who tries to "answer" the skeptic are alike victims of some sort of confusion). The reason, according to Strawson, is, as before, that these are *natural* beliefs; we do not hold them for *reasons;* if we did, then we would have had to be able at some time to seriously consider *not* having them. But we have no *choice* here. And where we have no choice, it is "idle" or "vain" to speak of reasons for and against. (It is clear from such statements that Strawson thinks that the question of the *truth* of skepticism does not genuinely arise, but it is not clear what this is supposed to mean. It is not clear whether the supposed fact that the question is "idle" implies that it *does not make sense,* or whether Strawson is conceding that it makes sense.) Michael Williams has also (very legitimately) questioned this last step, asking whether Strawson's point is supposed to be a logical one or only a psychological one;[18] however, my problem begins earlier. The very idea that there is a problem of showing that it is not an illusion that nature is "uniform," that is, that it is governed by causal laws and things exhibiting a variety of causal dispositions, assumes that we can make coherent sense of a world in which there are events and a time order but no causal laws or types of objects with determinate causal powers and dispositions. But this is just the Humean idea that the world might consist of a series of "loose and separate" events all over again.

Additional Remarks on "the Uniformity of Nature"

To say that there is no genuine problem of "the uniformity of nature"—and that is what I am saying—is not to deny that we can coherently imagine that *many* of the regularities we depend on in science or in daily life will someday break down. But suppose that we imagine a bizarre science-fiction future— for example, that one morning we wake up and find that although there is a strange uniform illumination permitting us to see one another, there is no sun in the sky. (So the famous induction to the conclusion "the sun will rise tomorrow" will break down.) Imagine also, if you like, that any of the events that one reads about in "fantasy" novels takes place—wizards cast spells

18. Williams, *Unnatural Doubts,* 13–15.

successfully, horses speak, and so on. Still, the very fact that we imagine that we can identify ourselves, see other people and many familiar sorts of objects, and identify and reidentify them—not to mention performing such familiar acts as putting on one's clothes, eating a meal, and walking down the street— means that we have conceived of a world where a great many familiar regularities (indeed, most of the ones that have mattered to the daily life of human beings through the centuries) continue to hold.

Moreover, if the problem is that particular regularities on which we rely can be conceived of as breaking down, *that* problem arises even if we suppose that at a fundamental level nature is much as contemporary science conceives it to be. Is the problem that "the sun might not rise tomorrow"? Well, it could conceivably go nova. Or a space-time singularity could capture Terra and take it far away from our sun. Indeed—although this would take an alteration in our current physical theory—it could be that space-time has an end "in the future direction": that for some time *t,* there are no events at all after *t,* and there is not even such a thing as "time after *t.*" But a space-time that is finite in both the past and future directions is still not one in which "the uniformity of nature" fails. I do not deny that the conceivability of such scenarios can be and has been used to generate skeptical difficulties. But these are *particular* scenarios, particular future "possibilities" (as the skeptic terms them), that we are unable to rule out *deductively.* What to make of that is a question that has often been discussed, and I will discuss it again shortly. But the difficulty is *not* that there is *one* overriding "possibility," namely, that "nature might not be uniform," such that if we could only rule *that one* out, then our inductive house would be in order. (Indeed, the uselessness of any supposed "principle of the uniformity of nature" from the point of view of providing a support for the particular inductive inferences we make has often been pointed out.)[19] Strawson's picture in *Skepticism and Naturalism: Some Varieties* is a familiar one, in which there are, as it were, certain "presuppositions of empirical knowledge" à la Bertrand Russell—the uniformity of nature, the existence of the external world, and the like—and what we have to discuss is whether *these* can be justified by means of reasons or whether, instead, they are "natural beliefs." But that is not a picture we should accept, nor (in view of his insightful remark in *Individuals* that "[the skeptic's] doubts are unreal, not simply

19. For an especially clear discussion, see Ernest Nagel, *The Structure of Science* (Indianapolis: Hackett, 1979), 317–318.

because they are logically irresoluble doubts, but because they amount to the rejection of the whole conceptual scheme within which alone such doubts make sense") is it a picture Strawson himself should accept.

The "External World"

What of that other old chestnut, the "existence of the external world"? Here too it is extremely questionable whether there is a coherent alternative, that is, whether it is coherent to think that there is something that would properly be described as discovering that the existence of "external objects" is illusory. Of course, for eighteenth-century philosophers it seemed to make perfect sense. Sense impressions were conceived of by rationalists and empiricists alike as immaterial; it was the coherence of the idea of *matter* that was thought (by certain empiricists) to be problematic. But today, not only do Berkeley's and Hume's difficulties with the conceivability of "unperceived matter" seem wrongheaded,[20] but in addition, the very idea of "sense data"[21] (Hume's "impressions and ideas"), thought of as immaterial particulars that alone we directly observe, has been repeatedly attacked by some of the best philosophical minds of the century. Without attempting to review a vast literature (and without attempting to deal with all the confusions and unwarranted assumptions that hover around the notion of "experience" in the philosophical literature like a swarm of hornets), let me make just three points here.

First, the very idea that experiencing something is being related to an immaterial particular or universal[22] ignores an obvious possibility (sometimes called the "adverbial" view of perception), namely, the possibility that experiencing something is being in a certain state or condition, in short, *having a certain sort of predicate apply to one,* and not being in a mysterious relation R to a mysterious sort of entity.[23] The adverbial view has the advantage of being

---

20. Cf. the discussion of Berkeley's difficulties with the notion in Hilary Putnam, "Language and Philosophy," reprinted in Putnam, *Philosophical Papers,* vol. 2, *Mind, Language and Reality* (Cambridge: Cambridge University Press, 1975), particularly 15–16.

21. Many terms have been used in the course of four hundred years of "modern" epistemology.

22. At one time Russell thought that sense data were universals.

23. In logical notation "Helen experiences (or seems to herself to be experiencing) the presence of a blue sofa" trivially has the logical form $P(\text{Helen})$, but we do not have to agree with sense-datum theorists that P has the "analysis" $P(x) = (\exists y)(xRy \wedge y$ is a blue-sofa sense

nicely neutral on a number of contested metaphysical issues.[24] Although, when I was young, one still found some philosophers who thought that the relational analysis of experience (i.e., the analysis that says that having a particular experience is being in a relation R—a relation of "directly experiencing"—to an immaterial something [a something that is not itself a state of the experiencer, although it may be dependent on his mind for its existence]) was *obviously* correct, indeed, thought that it was not an "analysis" at all, but just a report of the plain facts of (sufficiently sophisticated) introspection, today I know of no philosopher who thinks this. Now, I believe, all philosophers of perception recognize that all these "accounts" are compatible with the "subjective quality" of experience itself. Indeed, at one point in *Philosophical Investigations,* Wittgenstein seems to suggest that such accounts mistake what is just a new way of talking for the discovery of a novel feature of reality.[25]

Second, the idea that the experiencing subject could herself be just a "collection of ideas and impressions" (this is Hume's account of the self) is enormously problematic. Not only did Hume himself, famously, see deep difficulties with his own account, but the fact is that we learn what perception is (as Strawson himself has more than once pointed out) in the course of talking about transactions between persons and a material world that they sometimes perceive rightly and sometimes wrongly. Indeed, in the *Metaphysical Foundation of Natural Science* Kant suggests that even our idea of "color" involves notions of body, of surfaces of bodies, and of laws ("rules") relating our per-

---

datum). Note that an adverbialist need not, as is often claimed by critics, analyze all the aspects of an experience as further properties P'', P''', . . . of the subject; if I see a chair with certain features, the various qualifications of my experience can be analyzed as *properties of a property of me,* not simply as further properties of me. I believe that, in fact, all the "formal" objections to the adverbial analysis are easy to meet.

24. These issues include the tenability of traditional mind-brain reductionism (Reichenbach thought that having a certain sort of experience is being in a particular brain state; Sellars opposed this sort of reductionism on the ground that it was being in a kind of state that is sui generis); the question of functionalism (applied to experiences, this is the philosophical theory that having a certain sort of experience is being—or one's brain being—in a particular *computational state;* I proposed this view in a series of papers starting in 1960, but I reject it now, for reasons given in Hilary Putnam, *Representation and Reality* [Cambridge, Mass.: MIT Press, 1988]); and also the question whether we should think of the "qualitative identity" that (it is claimed) sometimes obtains between "veridical" and "nonveridical" perceptions as due to the presence of something that is literally "numerically identical."

25. Ludwig Wittgenstein, *Philosophical Investigations* (Oxford: Blackwell, 1953), §400.

ceptions to what goes on with the surfaces of bodies.[26] The idea of taking away all the bodies (including the body of the experiencer) and leaving the experiences makes no more sense than the idea of taking away the Cheshire cat and leaving the grin.

Third, our whole experience of persons, including ourselves, is as *embodied* beings. The Berkeleyan notion of a "spirit" is, as the etymology suggests,[27] a survival of a time when the human mind was thought to consist of *wind*. Why should we grant that this is an intelligible notion?

If we take these points into account, then the natural upshot may well be a view of perception like the one that Strawson adopted in *Individuals*. The main elements of such a view are the following: (1) We understand persons as having special properties (Strawson's "P-predicates"), including experiential properties (as in the "adverbial view"), properties that figure in explanations of, in particular, the *reasons* for our attitudes and doings. (2) Because the analysis of conduct from the point of view of its intentional meaning and its justification or lack of justification in terms of reasons is a different enterprise from the enterprises of physics and chemistry, the explanations we give using "P-predicates" employ a different sense of "because" than do physical (or chemical, and so on) explanations of our bodily motions, and because the "becauses" are different, there is not, in principle, any *conflict* between the physical and the psychological perspectives on our behavior. (3) We need not be driven to a metaphysical dualism by the fact that P-predicates are not reducible to physical (chemical, electrochemical, and so on) predicates in any sense of "reducible" we presently possess. (4) Being a possessor of P-predicates is not the same thing as being an immaterial "spirit" or "soul" in a mysterious relation to a body.

But is it not the case that on such a view claims that "nothing exists except spirits and their ideas [experiences]" (this was Berkeley's claim) and "nothing exists except ideas and impressions" (this was Hume's view, at least "in the study") rest on a misunderstanding of the necessary preconditions for ascribing predicates at all? If our conception of what an experience *is* indeed derives from our mastery of ordinary talk of persons' experiencing particular objects,

26. Immanuel Kant, *Metaphysical Foundations of Natural Science* [1786] (Cambridge: Cambridge University Press, 2004).

27. This etymology is not only in Indo-European languages; the Hebrew word for "spirit," *ruach,* originally means "wind."

properties of objects, interactions between objects, and so on, then such ideas as the idea that persons might *be* experiences (or immaterial "spirits" plus experiences) do not make sense (except in the way in which a myth or a fairy tale makes sense).

I have expressed surprise at the fact that Strawson does not see his Humean tendencies and his Kantian tendencies as in any way in conflict. In *Skepticism and Naturalism: Some Varieties* he reconciles them in the following way: Kantian arguments show us how our concepts hang together, but they do not speak to the skeptic's challenge to justify our conceptual scheme as a whole; Strawson's Humean contention that the question of justification cannot even be raised, because reasons are not and cannot be in question here, appears to speak to that challenge. But the contention employs the wrong "cannot." Strawson's contention speaks to the issue of pointlessness, given the way we are "hard-wired," not to the issue of intelligibility. Surely the skeptic's challenge to justify "the uniformity of nature" and "the existence of the external world" presupposes that these can be *coherently* doubted. And that is just what the Kantian arguments call into question. How our concepts hang together has everything to do with whether there *is* an intelligible skeptical challenge.

But *Are* the Skeptical Scenarios "Rationally Meaningful?"

In *Skepticism and Naturalism: Some Varieties* Strawson does briefly discuss the issue of the intelligibility of the skeptical challenge, or, rather, he briefly discusses Stroud's discussion of that issue.[28] The challenge to the intelligibility of the skeptical challenge that Strawson considers is Carnap's,[29] and Stroud, in Strawson's summary, finds that "Carnap does not altogether miss the point [of the skeptical challenge], but seeks to smother or extinguish it by what Stroud finds an equally unacceptable verificationist dogmatism. It is all very well, Stroud says, to declare the philosophical question to be meaningless;

---

28. The article Strawson discusses is Barry Stroud, "The Significance of Skepticism," in Peter Bieri, R. P. Horstmann, and Lorenz Krüger, eds., *Transcendental Arguments and Science* (Dordrecht: Reidel, 1979), 277–297.

29. Rudolf Carnap, "Empiricism, Semantics and Ontology," *Revue Internationale de Philosophie* 4 (1950): 20–40; reprinted in Leonard Linsky, ed., *Semantics and the Philosophy of Language* (Urbana-Champaign: University of Illinois Press, 1952), 208–228.

but it does not *seem* to be meaningless. The skeptical challenge, the skeptical question, *seem* to be intelligible. We should at least need more argument to be convinced that they were not."[30] And Strawson adds: "Many philosophers would agree with Stroud, as against Carnap, on this point; and would indeed go further and contend . . . that the skeptical challenge is perfectly intelligible, rationally meaningful." Strawson then proceeds to discuss the prospects for answering the skeptical challenge by rational argument, leaving unchallenged the view of "many philosophers" that "the skeptical challenge is perfectly intelligible, rationally meaningful." (Contrast Strawson's own statement in *Individuals* that I quoted earlier, that the skeptic's doubts "are unreal, not simply because they are logically irresoluble doubts, but because they amount to the rejection of the whole conceptual scheme within which alone such doubts make sense.")

Stroud (and, if he indeed agrees with him on this point, Strawson) is, of course, right that the way to meet the skeptical challenge *is not* to resort to "verificationist dogmatism." But, as I have been explaining, there are grounds for doubting whether the traditional skeptical scenarios in connection with "the existence of the external world" really are meaningful, and none of the ones I have rehearsed turns on a prior commitment to a philosophical theory (such as verificationism) that is supposed to yield a general method for assessing the meaningfulness of an arbitrary statement.

30. Strawson, *Skepticism and Naturalism*, 7.

32

Philosophy as the Education of Grownups: Stanley Cavell and Skepticism

At the close of the section titled "Natural and Conventional" in *The Claim of Reason,* Stanley Cavell wrote the following remarkable words:

> In philosophizing, I have to bring my own language and life into imagination. What I require is a convening of my culture's criteria, in order to confront them with my words and life as I pursue them and as I may imagine them; and at the same time to confront my words and my life as I pursue them with the life my culture's words may imagine for me: to confront the culture with itself along the lines in which it meets in me.
>
> This seems to me a task that warrants the name of philosophy. It is also the description of something we might call education. In the face of the questions posed in Augustine, Luther, Rousseau, Thoreau. . . , we are children; we do not know how to go on with them, what ground we may occupy. In this light, philosophy becomes the education of grownups.[1]

The last two sections of this chapter have been substantially revised since it was published in Alice Crary and Sanford Shieh, eds., *Reading Cavell* (London: Routledge, 2006). There I described the idea of treating skepticism as one treats antinomies in logic as "superficial," which I believe was wrong. In the present version I describe it as insufficient by itself, which I think is right [comment added January 7, 2010].

1. Stanley Cavell, *The Claim of Reason* (Oxford: Oxford University Press, 1979), 125.

The uniqueness of Stanley Cavell extends not just to the *way* he speaks, but also to *what* he has to say to us—indeed, the idea that one can "factor" what Cavell says into a *way* of speaking, a "style," and a substance independent of the way it is said and the way it is listened to is as wrong as the idea (which Cavell has brilliantly criticized[2]) that one can do this with Wittgenstein's writing. The purpose of this paper will be to bring out the way in which Stanley Cavell's reflections on *skepticism* are unique, and the way in which they exemplify the idea of philosophy as the education of grownups. But this will require a good deal of stage setting.

Skepticism

If Cavell's interest in skepticism is well known, it is also frequently misunderstood, which is unfortunate because properly understood, it is the key to understanding his Wittgenstein interpretation—or, rather, a proper understanding of each of these is essential to a proper understanding of the other.

Here is an example of what I have in mind. Although leading Wittgenstein interpreters on both sides of the divide that separates Stanley Cavell and Peter Hacker[3] have long pointed out that Wittgenstein is not a behaviorist, the idea that he must be remains widespread. But the textual evidence for the behaviorist reading is, to put it mildly, extremely slim. Why, then, does it continue to appeal?

The answer, I think, has to do with the following question, which I often hear from students: "If Wittgenstein was not a behaviorist, what *was* his answer to skepticism about other minds?"

Although most "orthodox" (non-Cavellian) interpreters no longer read Wittgenstein as a behaviorist, they do read him as having *shown* that skepticism about other minds cannot even be coherently expressed, and, in that sense, as a philosopher—indeed, as *the* philosopher—who has refuted skepticism (and not only about other minds). But to understand either *The Claim of Reason* or Cavell's other writings on the topic of skepticism, the first thing

2. Stanley Cavell, "Epilogue: The *Investigation*'s Everyday Aesthetics of Itself," in Stephen Mulhall, ed., *The Cavell Reader* (Oxford: Blackwell, 1996), 369–389.

3. I discuss this divide in Hilary Putnam, "Rules, Attunement, and 'Applying Words to the World': The Struggle to Understand Wittgenstein's Vision of Language," in Ludwig Nagl and Chantal Mouffe, eds., *Pragmatism or Deconstruction* (New York: Peter Lang, 2001); reprinted here as Chapter 24.

one needs to be clear about is that Cavell's Wittgenstein is not out to "refute skepticism." Accept this, and what Cavell has to say about "criteria," about the philosopher's context being a "nonclaim context," and about other subtle issues connected with his reading of *Philosophical Investigations* will fall into place. Miss it, and all this will be inscrutable.

But the idea that Wittgenstein is *not* "refuting" skepticism can also be misunderstood. When Cavell denies that Wittgenstein is out to "refute" skepticism by showing that the skeptic has committed a conceptual blunder—when Cavell even speaks of "the truth in skepticism"[4]—he is not *agreeing* with those philosophers who would say that the skeptic (who, after all, speaks outside the normal language games in which the word "know" figures) makes perfect sense.[5] Cavell is not saying that skepticism is intelligible and (wholly or partly) right. Rather, he is making two points. First, the *concern* of *Philosophical Investigations*—and, importantly for what I am going to say here, Cavell's own purpose in *The Claim of Reason*—is not to do anything that could be called "refuting skepticism." To read either *Philosophical Investigations* or *The Claim of Reason* as if it were a longer and more careful attempt to say what John Austin said in "Other Minds" is to look at Wittgenstein's work and Cavell's work in the wrong way.

Second, if the skeptic's utterances do not clearly fail to make sense, neither do they clearly make sense. In the end (but *The Claim of Reason* thinks that this is something still to be shown in detail[6]) there is something we ought to find nonsensical about the skeptic's claim that we "do not know," something we are not in attunement with. (Still, Cavell wants us to ask, "Who is the 'we' here?") But the questions this raises for Cavell (and, he argues, for Wittgenstein) are: What is the *source* of the skeptic's being out of attunement? And why are we all moved, at certain moments, to be "the skeptic" ourselves?

That his failure to make what *we* call "sense"—or rather, to make "full" sense—is, at bottom, a failure of attunement is something the skeptic will

---

4. E.g., "The bond [between the teachings of Wittgenstein and that of Heidegger] is one, in particular, that implies a shared view of what I have called the truth of skepticism, or what I might call the moral of skepticism, namely, that the human creature's basis in the word as a whole, its relation to the world as such, is not that of knowing, anyway not what we think of as knowing." Cavell, *Claim of Reason*, 241.

5. Barry Stroud, in *The Significance of Philosophical Scepticism* (Oxford: Clarendon Press, 1984), is such a philosopher.

6. Cf. Cavell, *Claim of Reason*, 220: "This is no more than a schema for a potential overthrowing or undercutting of skepticism."

regard as a *confirmation* of skepticism. If Cavell and Wittgenstein, read through Cavell's eyes, increase our understanding of skepticism, they may well increase our unease, our "conflicts," in the face of its presence, inside ourselves as well as outside. This is all that I mean by remarking on the "uniqueness" of these philosophers; whereas English-speaking philosophers who write about skepticism generally think of themselves as "epistemologists" and are concerned to "answer" the skeptic's arguments (or to defend them, or to find new arguments for skepticism if they think the skeptic is right[7]), Cavell and Wittgenstein, as Cavell reads him, are concerned to make us see something that troubles the skeptic, something that can and should give us a sense of "vertigo" at certain times, without causing us either to become skeptics or to find illusory comfort in an overintellectualized response. If I find this even more in Cavell than I do in Wittgenstein, it is because it is, after all, Cavell who uses the disturbing expression "the truth in skepticism."

But it is not just a turn of phrase that is new. To see a "truth in skepticism" without seeing some "truths of skepticism" marks, I believe, a decisive shift in the treatment of epistemological questions, and it was the great achievement of *The Claim of Reason* to highlight that shift.

Is It Sufficient to Think of Skepticism as an "Antinomy"?

Thinking about the uniqueness of Stanley Cavell's approach to skepticism had an effect on me that I did not anticipate: it caused me to rethink something I wrote about skepticism a few years ago. I want to share with you both what I wrote then and why I find that when I think about what I wrote in the light of Cavell's insights, I find it insufficient.

Here is the passage I recalled:

> The reason skepticism is of genuine intellectual interest—interest to the *nonskeptic*—is not unlike the reason that the logical paradoxes are of genuine intellectual interest: paradoxes force us to rethink and reformulate our commitments. But if the reason I undertake to show that the skeptical arguments need not be accepted is, at least in part, like the reason I undertake to avoid logical contradictions in pure mathematics (e.g., the Russell Paradox), or to find a way to talk about truth without such logical contradictions

7. As Peter Unger did in *Ignorance: A Case for Scepticism* (Oxford: Oxford University Press, 1975). He takes a different position in *Philosophical Relativity* (Oxford: Oxford University Press, 1984).

as the Liar Paradox; if my purpose is to put my own intellectual home in order, then what I need is a perspicuous representation of our talk of "knowing" that shows how it avoids the skeptical conclusion, and that my *nonskeptical* self can find satisfactory and convincing. (Just as a solution to the logical paradoxes does not have to convince the skeptic, or even convince all philosophers—there can be alternative ways to avoid the paradoxes— so a solution to what we may call "the skeptical paradoxes" does not have to convince the skeptic, or even convince all philosophers—perhaps here too there may be alternative solutions.) It is not a good objection to a resolution to an antinomy that the argument to the antinomy seems "perfectly intelligible," and, indeed, proceeds from what seem to be "intuitively correct" premises, while the resolution draws on ideas (the theory of types in the case of the Russell Paradox; the theory of levels of language in the case of the Liar Paradox—and on much more complicated ideas than these as well, in the case of the follow-up discussions since Russell's and Tarski's) that are abstruse and to some extent controversial. That is the very nature of the resolution of antinomies. What I have tried to provide in this essay is an argument *that convinces me* that *the skeptic cannot provide a valid argument from premises I must accept to the conclusion that knowledge is impossible.* In the same way, Russell showed that (after we have carefully reconstructed our way of talking about sets) a skeptic—or whoever—cannot provide a valid argument from premises we must accept to the conclusion that mathematics is contradictory, and Tarski showed that (after we have carefully reconstructed our talk about truth) a skeptic—or whoever— cannot provide a valid argument from premises we must accept to the conclusion that talk about truth is contradictory.[8]

Here I compared skepticism to an "antinomy" in the sense that word has acquired since Russell and Tarski. And I now think that although what I wrote is correct, it is insufficient.

---

8. Hilary Putnam, "Skepticism," in Marcelo Stamm, ed., *Philosophie in synthetischer Absicht* (Stuttgart: Klett-Cotta, 1998), 239–268; reprinted here with the title "Skepticism and Occasion-Sensitive Semantics" as Chapter 30. David Macarthur objects (in a private communication), "You don't offer a rational reconstruction but an attempt to explain the conditions of sense-making and how we are led to forget or overlook those conditions. This activity of 'trying to make sense' seems different from Tarksi on truth, etc." It is true that I am more interested in describing our practice than in "reconstructing" it, but I have always believed that philosophically relevant descriptions of our practice always involve some element of "rational reconstruction."

Skepticism as an Antinomy

It is no accident that I chose the Russell paradox and Tarski's formalization and treatment of the ancient liar paradox.[9] These are the two great examples that gave the notion of an "antinomy" (or "logical paradox") the sense it has in modern logic.

I do not suppose that anyone did, but it would have been perfectly possible for someone in the Victorian era with a penchant for logical paradoxes (Lewis Carroll, for example) to write the following sentence on a piece of paper:

(1) The sentence (1) is not true.

Lewis Carroll would then (let us imagine) have gone on to point out that if (1) is true, then it is not true. So, by reductio ad absurdum, (1) is not true. But we have just proved (1), so (1) must be true after all. So (1) is both true and not true, which is an absurdity.

Before one has any special reason to spend time worrying about this curiosity, a plausible response would be to be amused and then to forget all about it. After all, one might think, when in our ordinary use of the word "true" do we ever encounter puzzles like this? We do not need to be able to say "what is wrong" with the liar sentence to go on using the word perfectly successfully. (And the Russell paradox cannot even be explained without first explaining the Cantorian notion of a "set," as something that can itself have "sets" as elements.) But once reasons arise for worrying about the liar sentence—and there *are* good reasons for worrying about it *in certain mathematical and philosophical contexts*—what one naturally does is to write down the assumptions that generate the contradiction. There are a number of ways of doing this— for example, one may highlight the role of what we now know as Tarski's *T*-schema,

"*P*" is true iff *P*,

and, in the present case, the particular instance ("*T*-sentence")

"The sentence (1) is not true" is true iff the sentence (1) is not true

9. For a detailed examination of the liar paradox, see Hilary Putnam's Tarski Lecture, "Paradox Revisited I: Truth," in Gila Sher and Richard Tieszen, eds., *Between Logic and Intuition: Essays in Honor of Charles Parsons* (Cambridge: Cambridge University Press, 2000), 3–15; reprinted here, with the title "Revisiting the Liar Paradox," as Chapter 10.

by taking it as the relevant assumption.[10] In more sophisticated treatments one may make explicit the role of the assumption that "true" can be thought of as a single predicate rather than a hierarchy of predicates (as in Tarski's solution to the liar paradox[11]), or the assumption that sentences are true and not "statements" or "propositions," or the role of the assumption that there is just *one* relevant interpretation of the sentence (1),[12] or the assumption that there is *at most one* relevant interpretation of the sentence (1). In still more complicated treatments one may challenge the idea that there is even such a thing as a *totality* of all relevant interpretations.[13] And one may show how giving up one or more of these assumptions enables one to use the word "true" without risk of formal contradiction.

What I was pointing out when I wrote the passage I now regard as insufficient was that one can similarly turn a standard skeptical argument—say, Descartes' in the *First Meditation*—into a formal proof, not, to be sure, of a contradiction, but of "I do not know that *p*," where *p* can be *any* proposition "about the external world." In a sense, this is what Thompson Clarke's famous essay "The Legacy of Skepticism" did.[14] In *The Significance of Philosophical Scepticism,* a book strongly influenced by Clarke's essay, Barry Stroud offers the following example of a skeptical argument (which I state in my own words): suppose that I am at a party, and someone asks whether John is coming. I reply that he is. Asked by my skeptical host, "How do you know?" I reply, "I just spoke to him on the phone, and he told me he is on the way." My host points out (or, better, claims) that "it is possible that John will be struck by a meteorite on the way." Although Stanley Cavell and my beloved friend Rogers Albritton always regarded this as a highly suspicious use of the word "possible," Stroud accepts that

10. Because the sentence (1) = "The sentence (1) is not true," this implies "The sentence (1) is not true" is true iff "The sentence (1) is not true" is not true—which is a contradiction.

11. Alfred Tarski, "Der Wahrheitsbegriff in den formalisierten Sprachen," *Studia Philosophica* 1 (1935): 261–405; translated as "The Concept of Truth in Formalized Languages," in Alfred Tarski, *Logic, Semantics, Metamathematics: Papers from 1923 to 1938,* ed., J. Corcoran, trans. J. H. Woodger (Indianapolis: Hackett, 1983), 152–278.

12. As proposed by Charles Parsons, "The Liar Paradox," *Journal of Philosophical Logic* 3 (1974): 381–412; reprinted in Parsons, *Mathematics in Philosophy* (Ithaca, N.Y.: Cornell University Press, 1983), 221–267.

13. See Putnam, "Revisiting the Liar Paradox."

14. Thompson Clarke, "The Legacy of Skepticism," *Journal of Philosophy* 69, no. 20 (1972): 754–769.

1. it is possible that John will be struck by a meteorite and prevented from coming to the party; and hence
2. I do not know that John will not be struck by a meteorite and prevented from coming to the party. But
3. If John is struck by a meteorite and prevented from coming to the party, then John will not come to the party.

So, if (and, surprisingly, this is the only premise that Stroud thinks is problematic in this reasoning)

4. *X* knows that *p,* and *X* knows that *p* implies *q,* together imply that X knows that *q,*

it obviously follows that if I know that John is coming to the party, then I know that he will not be struck by a meteorite and prevented from coming, and because we have agreed that I do not know any such thing, it follows I *do not* know that John is coming to the party. My knowledge claim is defeated.[15]

Stroud's analysis of Descartes' dream argument is very similar. Stroud thinks that it is undeniable that it is possible that D is somewhere else and only dreaming that D is sitting in front of the fire, and clearly that statement implies that D is not sitting in front of the fire. And D knows the implication. So, if (4) ("closure of knowledge under known implication") holds, then Descartes is correct in his most skeptical moment. He does not *know* that he is sitting in front of the fire.

Interestingly, Stroud does not conclude that (4) must be given up; he thinks that it is a mysterious and deep problem whether it can be.

Here we have skepticism about "knowledge of future events" and skepticism about "the external world" nicely laid out as an antinomy (although Stroud is not sure that it is an antinomy, and not a sound argument for the truth of skepticism). And I do indeed think that it is important to see at what point in these proceedings we have departed from the concept of knowledge we actually have—the concept we are normally "attuned" in employing—although I would start at a very different place than Stroud does; I would start, in fact,

15. I critically examine Stroud's argument in Hilary Putnam, "Skepticism, Stroud and the Contextuality of Knowledge," in *Philosophical Explorations* 4, no. 1 (2001): 2–16; reprinted here as Chapter 29.

by questioning Stroud's Gricean assumption that "know" and "possible" have context-independent truth-conditions. And this—seeking for what Wittgenstein would have called a "perspicuous" overview of our uses of "know" and "possible"—is essential if skepticism is to be answered *at a purely intellectual level*. And skepticisim is, among other things, an intellectual problem.

But to treat the skeptic's argument as *only* an intellectual puzzle, an "antinomy" to be removed, as I suggested we do, is to *trivialize* skepticism. "Look," I was in effect saying in the passage I quoted earlier, "Here are some premises from which I can deduce 'I do not know that *p*' for any *p* about 'the external world.' Let us try to find a 'neat' way of giving up one of the premises, so we do not 'get' that conclusion." Surely this was a shallow response to a deep issue, for skepticism is much more than an intellectual problem.

The Depth of Skepticism

What I want to bring out now is the way in which Stanley Cavell has taught me to *see* skepticism as a deep issue, thus illustrating the education of at least one "grownup." In a sense this means that I want to explore the internal relation between part 4 of *The Claim of Reason* and parts 1, 2, and 3. In those parts Cavell was concerned to do justice to the very complex ways we speak of "knowing" and to understand how it can be that "masters of the language" (like Barry Stroud) find skeptical arguments (all but) irresistible. It cannot be the case that Austin was right to adopt the schoolmasterly tone he did, to treat the skeptic as someone who is just being childishly unreasonable.[16] The long and subtle discussion in the first three parts of *The Claim of Reason* includes, among many other things, the presentation of what Cavell calls "the projective vision of language," the vision of language as everywhere dependent on human attunements, attunements in projecting words into novel contexts. This is something I have written about elsewhere, and I will not repeat that discussion here.[17] But one feature I do have to mention is that there are also attunements in what we do not say, do not ask, do not question. That

16. This is what, in chapter 3 of *The Claim of Reason*, Cavell sees Austin as doing in "Other Minds" (*Proceedings of the Aristotelian Society*, suppl. 20 [1946]: 148–187; reprinted in J. L. Austin, *Philosophical Papers*, ed. J. O. Urmson and G. J. Warnock, 2nd ed. [Oxford: Oxford University Press, 1970], 44–84).

17. Putnam, "Rules, Attunement, and 'Applying Words to the World.'"; reprinted as Chapter 24 in this volume.

we do not so much as raise the question "Do I know that I am sitting here by the fire" when we are comfortably sitting by a fire—or that only philosophers raise it (or pretend to raise it) is no trivial fact, on the one hand, nor is it just evidence of the lack of philosophical insight on the part of the mass of mankind, on the other. That there is no clear job for that question to do in those circumstances is a fact that, if we reflect on it, can guide us in perceiving at least some of the jobs that talk of "knowing" does do. That epistemologists continue to take it completely for granted that "I know" that I am sitting in a chair when I am, that "I know" that I am eating a sandwich when I am, and so on, and do not so much as consider the Cavellian claim that my relation to the chair, the sandwich, or whatever is not one of knowing is something I find extremely disappointing.

But as long as we stick to examples involving chairs and fireplaces and sandwiches, it might still seem that skepticism is a purely "intellectual" affair, even if it is one that raises significant questions concerning the most perspicuous way to view human life with and within language. It is when Cavell shows us that skepticism about other minds is not a purely intellectual matter, in the way in which skepticism about tables and chairs seems to be, that the true depth of the problem of skepticism really comes into view.

The stage is set for this by early examples in *The Claim of Reason,* for example, the following passage:

> And then perhaps the still, small voice: Is it one? [Is the man having a toothache?] Is he having one? Naturally I do not say that doubt cannot insinuate itself here. In particular I do not say that if it does I can turn it aside by saying "But that is what is *called* having a toothache." That abjectly begs the question—if there is a question. But what is the doubt now? That he is actually suffering. But in the face of that doubt, *in the presence of full criteria,* it is desperate to continue: "I'm justified in saying, I'm almost certain". My feeling is: There is nothing any longer to be almost certain about. I've gone the route of certainty. Certainty itself hasn't taken me far enough. And to say now, "But that is what we call having a toothache" would be mere babbling in the grip of my condition. The only thing that could conceivably have been called "his having a toothache"—his actual horror itself—has dropped out, been withdrawn beyond my reach.—Was it always beyond me? Or is my condition to be understood in some other way? (What is my condition? Is it doubt? It is in any case expressed here by speechlessness.)[18]

---

18. Cavell, *Claim of Reason,* 69–70.

When we reflect on this passage, we of course see how strange it would be to ask, "Do I know that so-and-so is in agony?" when he obviously is. But it is also strange to ask, "Do I know that I am standing on a floor?" when I obviously am. A difference between the two cases—a big one—is that the latter question does not hurt the floor. It is not a mark of a "lack of respect for floors" or "an inhuman attitude to floors." But to really stop and ask the skeptical question when someone is in agony (in what Cavell called "the horror" of toothache), to ask it about the agony, would manifest a failure of humanity. It would be a refusal to acknowledge the other as a person; indeed, as Cavell describes the phenomenology, the other does not so much as exist for me as a person if I am in that state of nonacknowledgment. His agony has "been withdrawn beyond my reach." And what is not yet said at this stage in *The Claim of Reason,* but what we already begin to perceive, is that even if we do not go to the extreme of meditating about "the problem of other minds" in the presence of someone in agony, we all, at moments, fail to acknowledge the suffering of others. Nonacknowledgment is a real human possibility—an ever-present one, in fact.

In the closing pages of part 4 of *The Claim of Reason* this is illustrated, not by the example of a moral monster, a Himmler, say, a figure of "radical evil," but by "a study of Othello."[19] Othello is, in a sense, a genuine skeptic about one other mind—Desdemona's. His problem—and this is the horror of his situation—is not that he lacks "evidence" of Desdemona's faithfulness. It is that no evidence is good enough. And even to imagine one of the minor ordinary-language philosophers of the 1950s and 1960s saying to Othello, "This is what we *call* conclusive evidence of Desdemona's faithfulness" (or "This is what we *call* having an unreasonable doubt") makes one want to vomit. That is why *The Claim of Reason* may be described as a defense of Wittgenstein against "ordinary-language philosophy," not a defense of Wittgenstein as an ordinary-language philosopher.

My imaginary ordinary-language philosopher might have put the emphasis in a different place, and thereby have said something more interesting, by saying, "This is what *we* call conclusive evidence"—but then it would have been evident that Othello is no part of that "we." But it is not that Othello is part of a different "we" either. Othello resembles the philosophical skeptic in that he simultaneously treats "Do I know that *p?*" (that Desdemona is faithful) as a serious question, an all-important question, and makes it impossible for it

19. Ibid., 483–496.

to receive an affirmative answer; he makes it so that nothing could count as showing to his satisfaction that p. And what good could it do to "show" Othello that this is "logically absurd"?

Our fundamental relation to the world, Cavell teaches us, is not one of knowledge but one of acknowledgment. But for Othello there is no acknowledgment; and the cry, "I do not know if she is faithful," however "inappropriate" it may be when all the criteria for knowing have been suspended, is the natural, the inevitable, expression of despair in such a situation. It is not just an intellectual error.

One might grant all this, however, and see Othello as a tragic example of a pathology, a sort of spiritual cancer, from which (one might comfort oneself by supposing) most of us fortunately do not suffer. But this sort of comfort is something that Cavell wants us to repudiate. What is important—and this connects with many of Cavell's other interests, in Emersonian perfectionism, for example, and likewise with the ways in which Cavell takes films and operas as subjects of philosophical reflection—is that if Othello's is a pathology, it is an exaggerated form of, so to speak, a normal pathology. The point of saying this is not simply to awaken us to the suffering of others, as Levinas does (although I have heard Cavell speak about Levinas with deep respect); it is also to get us to see that an idea of being totally free of skepticism, in this deep sense, is itself a form of skepticism.[20]

To explain this last remark, I have to touch on some of the most profound and difficult parts of Cavell's philosophy.

Conclusion

One of the pitfalls into which many Wittgenstein interpreters have fallen is the pitfall of reading the private-language argument as some sort of "transcendental argument," say, a transcendental argument to the effect that a private language is impossible. James Conant, himself a student of Stanley Cavell, has beautifully explained why and how this is a misreading.[21] Cavell, needless

---

20. I elaborate on this idea in Hilary Putnam, foreword to Ted Cohen, Paul Guyer, and Hilary Putnam, eds., *Pursuits of Reason: Essays Presented to Stanley Cavell* (Lubbock: Texas Tech University Press, 1992), vii–xii.

21. James Conant, "The Earlier, the Later, and the Latest Wittgenstein"; published as "Le premier, le second et le dernier Wittgenstein," in Jacques Bouveresse, Sandra Laugier, and Jean-Jacques Rosat, eds., *Wittgenstein, dernières pensées* (Marseilles: Agone, 2002), 49–88.

to say, avoids this pitfall. But what is "shown," not by some controversial set of paragraphs in *Philosophical Investigations,* but by the whole "projective vision of language" that Cavell finds in that work, is that language speaking in a normative sense, the fully human use of language, necessarily involves both individual responsibility and community. Speaking without rising to the level of full responsibility for one's projections of words into situations—the sort of speaking that Emerson's great essay on self-reliance taught us to call "conformist"—is parroting; but "speaking" by means of projections that do not and cannot make sense to a human community is "acting out," an imitation of human language just as much as parroting is. And one of the deepest suggestions that I find in part 4 of *The Claim of Reason* is that skepticism universalized, skepticism that refuses to acknowledge any human community, is, to the extent that it is possible, a posture that negates not only its own intelligibility but also the very existence of a speaking and thinking subject, negates the skeptic's own existence and the world's.

The refusal to acknowledge a human community need not take a form that is "skeptical" on the surface, however. Every ideological posture that purports to free us from our human limitations, whether by means of political "revolution" or magical "therapy" or absolute religious "enlightenment," in the end replaces the real imperfect and limited acknowledgment of the other of which we are capable by a fantasy. That is why I said that the idea of being totally free of skepticism is, in the end, a form of skepticism in Cavell's sense; it is something that "repudiates or undercuts the validity of our criteria, our attunement with one another."[22]

22. Cavell, *Claim of Reason,* 46. A topic on which I want to suggest that we might want to reflect in the future is that—if these last remarks are at all on the right track—there is a way of seeing "skepticism about tables and chairs" as also a "deep" matter. It cannot, after all, be an accident that the idealist tradition in philosophy has always, in one way or another, passed through a skeptical moment, and that "external-world skepticism" has regularly been linked to something called "solipsism." (Neurath once remarked that "it is hard to say what the difference between 'methodological solipsism' and real solipsism is." Otto Neurath, "Physikalismus" *Scientia* 50, [1931]: 297–303; translated as "Physicalism" in *Philosophical Papers,* ed. R. S. Cohen and M. Neurath [Dordrecht: Reidel, 1983], 52–57.) But "solipsism" is just the attempt—or the feigned attempt—to universalize the nonacknowledgment of the other that Cavell speaks of.

PART SIX

Experience and Mind

33

The Depths and Shallows
of Experience

No one who has the temerity to address such broad themes as "science, religion, and the human experience," as I was originally invited to do, can hope to hide behind an academic facade of "professional expertise." To be sure, there are issues here that can benefit from being treated with scientific or philosophical sophistication, I believe. But the big issues: to believe in God or not to believe in God; to engage in such religious practices as prayer, attending services, and studying religious texts or not to do so (I am *not* equating this with the issue of believing or not believing in God, by the way); to look for "proof" of God's existence if one is religious (or thinking of being religious) or to regard such a quest as misguided; to be "pluralistic" in one's approach to religion or to regard one religion as "truer" than all the rest—these are deeply personal choices, choices of who to be, not just what to do or what to believe. I do not believe that philosophical or scientific discussion can provide compelling reasons for making them one way rather than another, although it can help us make whichever choices we make more reflectively.[1]

---

1. Avi Sagi once told me that in a still-unpublished fragment of—I think it was a diary—of Kierkegaard's, he found the words "Leap of faith—yes, but only after reflection."

I did say, however, that there are aspects of these issues that a philosophically sophisticated discussion (as well as a scientifically sophisticated discussion) can illuminate. The intentionally broad phrase "the human experience" raises the issue of what is meant by "experience" in the context of discussions of science and religion (as well as, perhaps, the issue of what it means to be "human" in our age, or in any age). In this essay it is the question of how we should understand "experience" that I shall address.

Both in life and in philosophical reflection, experience is sometimes seen as intrinsically shallow, as mere surface, and sometimes as deep. I want particularly to investigate the origins of our Western notion of experience in Cartesian and post-Cartesian philosophy and to explore the relevance of the long-standing philosophical disputes about experience for the broad themes of "science, religion, and the human experience."

Depths and Shallows

When I discuss "religious experience" in what follows, I will not mean experience that purports to be of supernatural beings, or of "revelation" conceived of on the model of having words dictated to one by a divine being. (One can find a very different model—the model of revelation as the ongoing connection between the individual and God—in the writing of Franz Rosenzweig.)[2] Rather, I will have in mind the way in which a religious person may at any time experience something or some event, whether it be an obviously significant one, say, the birth of a child or the sort of deep crisis in one's life that William James describes in *The Varieties of Religious Experience,* or a superficially "ordinary" one, as full of religious significance. Speaking for myself, I cannot imagine being religious in any sense, theistic or nontheistic, unless one has had and cherished moments of religious experience in this latter sense. But the concept of "experience" that we have operated with from Descartes and Hume to today's cognitive scientists has a troubled history, and it will repay us, I believe, to reflect on that history.

What I will be discussing for the most part will not be what I just called "religious experience." Rather, I am going to spend some effort trying to

2. See, for example, Franz Rosenzweig, "Revelation as the Ever-Renewed Birth of the Soul," part 2, book 2, of *Der Stern der Erlösung* (Frankfurt am Main: Kauffmann, 1921); English translation, *The Star of Redemption* (New York: Holt, Rinehart, and Winston, 1971).

explain why so many people have (and from where they got) a concept of experience that leaves literally no room for *depth,* a concept of experience as, so to speak, all psychological *surface,* one traditionally summed up in the concept of experiences as "sensations," and after that I shall try to explain why that concept is wrong, drawing especially on Kant's profound analysis of experience.

We all know that the philosophers of the seventeenth and eighteenth centuries are classified by the standard texts as "empiricists" and "rationalists." Although the classification is in many ways a procrustean bed, it certainly captures a broad divide between, say, the British philosophers Locke, Berkeley, and Hume and the continental philosophers Descartes, Spinoza, and Leibniz, and although the pattern of disagreements is by no means as tidy as the labels "empiricism" and "rationalism" suggest, it is certainly true that we find very different conceptions of experience in the two groups, and especially in Hume and Leibniz. (What is not often remarked is that Hume, the empiricist who makes experience—under the name "impressions and ideas"—the be-all and end-all of his philosophy, and who prides himself on being a sort of Newton of psychology,[3] is, in fact, far less subtle in his description of experience than Leibniz.)[4] Be that as it may, the line that has come to be recognized is between conceptions of experience that go back to Hume and conceptions that go back to Kant (who hoped, of course, to sublate the categories "empiricism" and "rationalism").[5] I shall briefly sketch these two conceptions because they epitomize the idea of experience as shallow and the idea of experience as deep.

Hume and the Shallow Conception

For Hume, the very paradigm of an "impression" (and the other sort of experience, "ideas," was identified by him with "faint copies" of impressions) is a visual image. Descartes and Berkeley had both tried to read the nature of visual impressions directly from the newly investigated nature of *retinal*

3. Norman Kemp Smith, *The Philosophy of David Hume* (London: Macmillan, 1949), 12–13.

4. For example, it is Leibniz and not Hume who sees that there is no sharp line between "conscious" and "unconscious" experience, and who is aware of the ways in which experience and cognition shade into each other.

5. For a discussion of these two conceptions with major implications for contemporary philosophy of mind, see John McDowell, *Mind and World* (Cambridge, Mass.: Harvard University Press, 1994).

images.[6] The result of this approach was a tendency to think of all "impressions" on the model of *pictures.* These impressions were not necessarily visual, of course—there were also tactile, olfactory, and other representations—but like pictures, these, and the "ideas" or faint copies that corresponded to them, were thought by Hume to refer only to what they *resembled.*[7] Content, on this resemblance semantics, is a rather primitive affair.[8] Hume rejected the very idea of a fact that cannot be *sensorily pictured.* The only other sort of "content" arises from "association"—especially the association of "passions" (feelings and emotions) with images.

Today there are very few, if any, old-fashioned empiricists in philosophy. But what survives of the older view is the very influential idea that experience (still identified by empiricists with sensory inputs) is "nonconceptual." Quine's idea that for philosophical purposes, experience talk could simply be replaced with talk of "surface irritations" (stimulation of the nerves on or near the surface of the body) in many ways foreshadowed this influential idea.

Kant and the Deep Conception

In Kant's writings one can find a response to the empiricist view of experience as consisting of sensory images, a response so deep that even today few philosophers who are not primarily Kant specialists have fully appreciated it (Strawson, Sellars, and more recently John McDowell and James Conant are among the happy exceptions). In this paper I cannot, of course, do justice to it, but I hope to point out at least some of the leading ideas of the Kantian conception. It is important, however, to realize that no one book of Kant contains all of it. From *The Critique of Pure Reason* to *Religion within the Bounds of Mere Reason,* Kant constantly broadens and deepens the presentation of his view, if not the view itself. The account in *The Critique of Pure Reason* is, nonetheless, the basis on which the deeper and broader reflections in Kant's subsequent writings depend.

6. Celia Wolf-Devine, *Descartes on Seeing: Epistemology and Visual Perception,* Journal of the History of Philosophy Monograph Series (Carbondale: Southern Illinois University Press, 1993).

7. David Hume, *A Treatise of Human Nature* [1739–1740], ed. L. A. Selby-Bigge (Oxford: Clarendon Press, 1976).

8. Elijah Milgram, "Hume on Practical Reasoning: *Treatise* 463–469," *Iyyun: The Jerusalem Philosophical Quarterly* 46 (1997): 235–265; Milgram, "Was Hume a Humean?" *Hume Studies* 21, no. 1 (1995): 75–93.

Hume, as we just saw, conceives of experiences on the model of pictures, and of their cognitive content as contained and communicated via (sensory) *resemblance*. Thus only sensory qualities are properly cognizable at all. If one accepts this, then many of Hume's other famous doctrines readily follow. For example, Hume's claim that we do not "observe" causal connection depends both on his limitation of what we can observe to sensory qualities and on his very narrow inventory of sensory qualities: because causal connection is not a sensory quality for Hume, it is evident to him that causal connection is never *observed*. On the other hand, although objective time hardly consists of sensory qualities either, Hume never worries about the question, "How and why are we able to think of impressions and 'ideas' as succeeding one another in an objective time?"

Kant did, however, worry about this question, and he concluded that our notions of objective time, causality, and lawful connection are *interdependent*. For example, our awareness of a boat's sailing down a river (coming, let us say, to a certain bridge) as *earlier* than the boat's sailing *beyond* the bridge even though we think of a building's back as existing at the same time as the front even if we look at the front *before* we look at the back is internally related to our beliefs that we could have chosen to experience the front before the back, but we could not, conditions being as they were, have chosen to experience the boat's sailing beyond the bridge *before* we experienced it approaching the bridge, and these beliefs are in turn related to the system of causal connections we accept.[9] The notion of time is inextricably connected with the notions of space and causality.[10]

My question, however, concerned how we *experience* things, and not how we *conceive* them. But—long before modern psychology—Kant questioned the coherence of such a dichotomy. We do not experience familiar objects and events—a cat's drinking milk, a tree swaying in the wind, someone's hammering a nail into a wall—as collections of "color-points" on a spatial grid. As William James put it, (in the case of a "presented and recognized material object") "sensations and apperceptive ideas fuse . . . so intimately that you can no more tell where one begins and the other ends, than you can tell, in those cunning circular panoramas that have lately been exhibited, where

9. Immanuel Kant, *The Critique of Pure Reason* (Riga, 1781/1787), trans. Norman Kent Smith (Houndmills: Macmillan, 1965), B237/A192.

10. For further discussion of this point, see Chapter 31 of this volume.

the real foreground and the painted canvas join together."[11] To employ Kant's language, in the sort of perception James described (or—an example Kant himself uses—in the case of experiencing something as a boat's sailing down a river), we have not mere unconceptualized sensations, whatever those might be,[12] but a *synthesis* of experiences and conceptual ideas, the ideas of space, time, and causation. This is something that the phenomenological school, beginning with Husserl, likewise emphasized: I see a building as something that *has* a back, Husserl pointed out, even when I do not see the back. Such perception is fallible, to be sure; but so is the perception that something is red or circular. And the retreat to "sense data" in the hope that *there* we can find something "incorrigible" has long been recognized to be a loser.

A second issue that plays a large role in *The Critique of Pure Reason,* and one that figures in contemporary attacks on what postmodernists consider to be the metaphysical illusion of the "ego," is the issue of right and wrong ways to think about what it means to be or have a self. (As Nicholas Boyle has observed, postmodernist doubts about whether there is such a thing as a self, or an "author," never stop the postmodernist from cashing a royalty check.[13]) Here again, paying more attention to Kant would help clear our heads.

For Kant, rational thought itself depends on the fact that I regard my thoughts, experiences, memories, and the like as *mine.* To illustrate Kant's point, imagine yourself going through a very simple form of reasoning, say, "Boiling water hurts if you stick your finger in it; this is boiling water; so it will hurt if I stick my finger in this." If the "time slice" of me that thought that "boiling water hurts if you stick your finger in it" was one person, person A, and the "time slice" that thought that the minor premise, "This is boiling

11. William James, *Essays in Radical Empiricism [1912],* (Cambridge, Mass.: Harvard University Press, 1976), 16.

12. This sentence reflected my views in 2002. At the present time [2011], Hilla Jacobson and I are working on a book on perception, and our view is that (1) it is important to distinguish between the phenomenal (or "qualitative") character of an experience (which is often what is meant by a "sensation") and an apperception of a feature or features of the world (both of which get referred to as "experiences" in the literature); (2) not all experiences are conceptualized (contrary to McDowell and to what I suggest here); (3) not all apperceptions have a phenomenal character (there are "amodal" apperceptions); and (4) it is apperceptions that are conceptualized; the phenomenon James called "fusion," in which the phenomenal character of experience "fuses" with the accompanying apperception, is not a feature of all experiences.

13. Nicholas Boyle, *Who Are We Now? Christian Humanism and the Global Market from Hegel to Heaney* (Notre Dame: University of Notre Dame Press, 1998).

water," was a different person, person B, and the person who thought that the "conclusion," "It will hurt if I stick my finger in this," was yet a third person, person C, then that conclusion was not warranted; indeed, the sequence of thoughts was not an argument at all, because the thoughts were thoughts of different thinkers, none of whom had any reason to be bound by what the others thought or had thought. We are *responsible* for what we have thought and done in the past, responsible *now*, intellectually and practically, and that is what makes us *thinkers*, rational agents in a world, at all. Kant, like Locke before him, can be seen as making the point that the thinking of my thoughts and actions at different times as *mine* does not depend on a metaphysical premise about "self-identical substances" and is nonetheless a form of conceptualization that we cannot opt out of when we are engaged in judgment in action.

As before, to say that Kant's point is valid for conceptualization but not for experience would be to miss the way in which experiences and concepts interpenetrate, the way in which they are "synthesized." When I reason (say, about the boiling water), I *experience* my successive thoughts as "mine." Hume is right in holding that this is not a sensory quality; there is no "impression" of "my-ownness"; and Kant would emphasize this just as much as Hume. But whereas Hume concludes that the self is an *illusion,* Kant sees that experience transcends Humean "impressions." Whereas for Hume experiences are sheer psychological surface, for Kant even the simplest perception links us to and interanimates such deep ideas as the ideas of time, space, causality, and the self. And this is something that Kant does not just claim, but that he argues in detail, and with incomparable brilliance. That experience is intrinsically *deep* is the heart of the Kantian conception. It is not something that was overthrown by the collapse of Kant's "synthetic a priori" and the metaphysics Kant tried to base on it.

Kant on Aesthetic Experience

I said above that Kant deepens the presentation of his views (and perhaps the views themselves) in successive books, and, I should add, not only in books. For example, a wonderful (and sadly neglected) discussion of what is right and wrong in mysticism may be found scattered in Kant's writing.[14] But nowhere is this more true than in *The Critique of the Power of Judgment.*

14. Immanuel Kant, *Religion within the Boundaries of Mere Reason and Other Writings* (Königsberg, 1793), trans. and ed. Allen Wood and George Di Giovanni (Cambridge:

I cannot, of course, even sketch the complex and rewarding aesthetic theory of that critique. Fortunately, that is not my goal. What I want to do is to extract one item from that complex discussion, although to do that, I will have to say a little about the ideas that surround it.[15] The "item" in question is the fascinating notion of an "indeterminate concept." When we experience a work of art, Kant tells us, we experience it as escaping capture by "determinate concepts," but we do perceive it as not being captured by, but evoking a kind of concept, an indeterminate concept, one that is deeply connected with what Kant calls "the free play of the faculties" (imagination and understanding—under the guidance of the former).

Here I do have to interpret the aesthetic theory I said I would not discuss to the extent of warning my readers against two common misinterpretations. The first, which I am indebted to Paul Guyer for pointing out, is the assumption that when Kant speaks of "pure" aesthetic experience, he is using "pure" as a value term. The reverse is the case; the art that Kant values and thinks that we should all value, Guyer has conclusively shown, is mixed, impure. "Pure aesthetic experience" in Kant's sense is concerned only with form; but to value, say, a painting that moves us on account of both its subject matter and its formal properties, or a novel or a poem, is to respond not only to the "purely aesthetic" features in this technical sense, but also to the *interplay* of description, valuation, and purely formal experience.[16] The second misunderstanding is that it is only "the concept of beauty" that Kant has in mind by the term "indeterminate concept."

To illustrate what I believe Kant actually had in mind, think of a painting by Vermeer (pick your favorite). It is not "indescribable"; a great deal about it

---

Cambridge University Press, 1998). For a fine discussion, see P. W. Franks, "Kant and Hegel on the Esotericism of Philosophy" (Ph.D. diss., Harvard University, 1993).

15. I thank Paul Guyer for discussions and for access to unpublished papers that enriched my understanding of Kant's aesthetics. I believe that the interpretation offered here is thoroughly consonant with Guyer's reading in his edition of Kant's *Critique of the Power of Judgment* (Cambridge: Cambridge University Press, 2000).

16. In "The Origins of Modern Aesthetics: 1711–1735," in Peter Kivy, ed., *The Blackwell Guide to Aesthetics* (Oxford: Blackwell, 2004), 15–44, Paul Guyer speaks of "the common caricature of Kant's purported reduction of aesthetic response, whether in the case of works of nature or works of art, to perceptual form apart from all content and significance," 39. Guyer points out that "when Kant turns to his explicit discussion of the fine arts—buried in the sections following the 'Analytic of the Sublime' and the 'Deduction of Pure Aesthetic Judgments' without the benefit of a heading of its own—it becomes clear that artistic imagination and aesthetic response can play freely with content as well as form," 39.

can be described. The notorious Vermeer forger Han van Meegeren could undoubtedly have given a precise (determinate) description of a great many features of this or of Vermeer paintings in general. But the description, although it might teach us a lot and even add to our appreciation of such a painting, would not answer the question: "Why is this painting so beautiful?" Indeed, as van Meegeren's rather unpleasant forgeries testify, a painting could satisfy this "determinate" description and *not be beautiful*. What Kant, interestingly, says about the discussion of works of art is not that it is impossible to describe what it is that strikes us as beautiful (which it would be if the only alternatives were to apply to them determinate concepts of the kind a van Meegeren—or an art historian—might offer, or to apply the single indeterminate concept "beautiful"). What he says is that the aesthetic ideas that are the content of works of artistic genius evoke so much thought that language cannot fully attain them or make them intelligible.[17] (He also says that we add to a [determinate] concept "a representation of the imagination that belongs to its presentation, but which . . . aesthetically enlarges the concept itself in an unbounded way.")[18] In short, certain concepts seek—and manage—less to *finish* a discussion or answer a determinate question than to further provoke both thought and imagination and to raise an "unbounded" number of further questions. And these are the concepts we need and have to use to talk meaningfully about art.

What connects the notion of an indeterminate concept with my topic of experience is that it is precisely in the context of discussing how we *perceive* works of art that Kant invokes it. Indeterminate concepts are not purely intellectual concepts; they require both a *sensible* subject matter and an active imagination to apply. That perception is fused with conceptual content is something we learned from *The Critique of Pure Reason;* that some of the perceptions we value most are fused with *indeterminate,* open-ended, conceptual

17. Kant characterizes the content of a work of artistic genius as "that representation of the imagination that occasions much thinking without it being possible for any determinate thought, i.e. *concept,* to be adequate to it, which, consequently, no language fully attains or can make intelligible." Kant, *Critique of the Power of Judgment,* 314.

18. Kant tells us that in a work of artistic genius "we add to a concept a representation of the imagination that belongs to its presentation, but which *by itself stimulates so much thinking that it can never be grasped in a determinate concept,* hence which aesthetically enlarges the concept itself in an unbounded way. . . . In this case the imagination is creative and sets the faculty of intellectual ideas (reason) into motion" (emphasis added). Kant, *Critique of the Power of Judgment,* §49, Ak. 317.

content, content in which imagination and understanding cooperate under the leadership of imagination, is something we learn from *The Critique of the Power of Judgment.*

The notion of an "indeterminate concept," understood in this way, naturally extends to moral notions. If Kant does not use it in the area of morals, it is because, I think, of a desire to keep morality rigorous and transparent. But morality, good morality, cannot always be rigorous and transparent, and a thinker who has seen that something like the notion of an "indeterminate concept" I just described applies also to the highest type of moral awareness is Iris Murdoch, even if she does not cite Kant or use his terminology. (Thus in her philosophical masterpiece, *The Sovereignty of Good,* she writes, "Moral tasks are characteristically endless not only because 'within', as it were, a given concept our efforts are imperfect, but also because as we move and as we look our concepts themselves are changing. . . . We do not simply, through being rational and knowing ordinary language words, 'know' the meaning of all necessary moral words. We may have to learn the meaning; and since we are human historical individuals the movement of understanding is onward into increasing privacy, in the direction of the ideal limit, and not backwards towards a genesis in the ruling of an impersonal public language.")[19]

Beyond both aesthetics, in the sense of the open-ended appreciation and discussion of works of art, and morality, in the sense of Murdoch's "loving attention" to the whole complexity of human beings and human moral life, it should be obvious, I think, that religious experiences both are guided by and spontaneously give rise to indeterminate concepts in a way analogous to the ways in which aesthetic and moral experiences do. And if we see religious, aesthetic, and moral experiences in this way, as I have been urging we should, we will avoid Hume's mistake of trying to analyze them as a chemist analyzes a compound, into so much of this factor ("ideas and impressions"), so much of that factor ("passions"), and so much of this other factor ("beliefs"). In the deepest human experiences, ways of perceiving things that are inseparable from those experiences but nonetheless conceptual, at least in the way indeterminate concepts are conceptual, fuse so intimately that you cannot tell where one begins and the other ends, to mimic William James's words quoted earlier.

Although the phenomenological school of philosophy that began with Husserl inherited and extended the Kantian insights I have been describing,

19. Iris Murdoch, *The Sovereignty of Good* (London: Routledge, 1970), 29.

the most influential twentieth-century phenomenologist, Heidegger, had a contemptuous attitude to science, which, for him, was merely an aspect of technological civilization (which he regarded as intrinsically evil). In Heidegger's writing, everything I have been saying about the depth of religious experience (including the experiences of "being" and of being "thrown" into the world and of finding a destiny that is one's "ownmost," which are Heidegger's versions of or substitutes for religious experience), as well as of artistic experience (especially the experience of poetry), and even of our everyday experiences with artifacts, is recognized and phenomenologically interpreted; but science is denigrated.

But by default, if we do not examine the impact of science on our ways of experiencing the world in a more sympathetic spirit than Heidegger was capable of, we are likely to fall back into the empiricist picture of science as consisting of deductive and inductive inferences from simple "sense data" (or Machian *Empfindungen*). To find a sustained critique of this way of thinking, we have to turn to the American pragmatists, and especially John Dewey. Extending the line of thought that William James had begun with his talk of apperceptive ideas and sensations as "fusing," Dewey saw that science endlessly and inventively creates new observation concepts, and that by so doing it institutes new *kinds* of data.[20] A scientist with a cloud chamber may now *observe* a proton colliding with a nucleus (without being able to answer the question "Exactly what visual sensations did you have when you observed it?" except by saying, "It looked like a proton colliding with a nucleus"), or *observe* a virus with the aid of an electron microscope, or *observe* a DNA sequence, and so on. And the impact of science on the conceptualization of experience is not confined to specialists; the way in which all of us experience the world was changed by Darwin, was changed by Freud (whether one thinks this or that claim of Freud's was well or ill founded) as the notion of the unconscious became part of our vocabulary, and is being changed today by computer science and the concepts and metaphors it adds to the language.

On the "metalevel," the level of the methodological appraisal of scientific theories, we also find something in science analogous to the indeterminate concepts involved in aesthetic judgment, indeterminate concepts that figure

---

20. John Dewey, *Logic: The Theory of Inquiry* [1938], in *The Later Works of John Dewey*, vol. 12, ed. Jo Ann Boydston (Carbondale: Southern Illinois University Press), 388–389; see also *Essays, The Study of Ethics*, vol. 4 (1929), 60–86, 87–111, 142–143.

in judgments that are internal to scientific inquiry itself: judgments of coherence, simplicity, plausibility, and the like. The similarity of such judgments to aesthetic judgments has, indeed, often been pointed out. Dirac was famous for saying that certain theories should be taken seriously because they were "beautiful," and Einstein talked of the "inner perfection" of a theory as an "indispensable criterion."[21]

But it is time to say something of the wider relevance of this picture of experience—the picture of experience as deep—for reflections on "science, religion, and the human experience."

Conceptuality and Skepticism

At first blush, recognizing that perception (and experience that purports to be perception or resembles perception) is always conceptualized may seem to make the problem of skepticism much worse, especially when religion is the issue. From Kant to John McDowell, philosophers who point out that experience is conceptualized have been told that they are problematizing our access to reality. Concepts can, after all, mislead as well as lead, conceal as well as reveal.

The fact that religious concepts are no longer intersubjectively shared within Western culture, and have not been for a long time, makes this more than a purely theoretical issue, as skepticism about the existence of houses and rocks happily has become.[22] Although no one can say that there are only so-and-so many possible answers a religious person can give to the atheist or to the religious skeptic, three main approaches are familiar to all of us.

The traditional approach, and the one that is still that of the Roman Catholic Church, is to continue (albeit with contemporary sophistication) the medieval attempt to "prove the existence of God" ("neo-Thomism"). This is not an approach I find possible for myself, at any rate, for the following reasons:

First, in order to understand talk about God, whether or not that talk takes the form of a "proof," one must be able to understand the concept "God." But

---

21. Paul Dirac, "The evolution of the Physicist's Picture of Nature," *Scientific American*, 208, no. 5 (1963): 45–63. Albert Einstein, "Autobiographical Notes," in P. A. Schilpp, ed., *Albert Einstein: Philosopher-Scientist* [1949] (LaSalle, Ill.: Open Court, 1970), 21–23.

22. For the ancient Greek skeptics, it is often pointed out, it was anything put a "purely theoretical issue," but that is another story, and one I need not tell here.

there are very different possible conceptions of what it is to understand the concept "God," in a way that has no analogue in the case of, say, mathematical proof. Second, even if one understands the concept "God," to accept any of the traditional proofs, one has to find a connection between that concept and the highly theoretical philosophical principles involved in those proofs, premises about conditioned and unconditioned existence and about what sorts of necessity there are. Some of the most profound religious thinkers of the last two hundred years (particularly the religious existentialists from Kierkegaard to Rosenzweig and Buber) have had no use at all for this sort of philosophizing; and I would be the last to say that they lacked the concept "God." What the traditional proofs of the existence of God in fact do is to *connect* the concerns of two different salvific enterprises: the enterprise of ancient and medieval philosophy,[23] which, after all, is the source of the materials for these proofs, and the enterprise of monotheistic religion. While it is certainly possible to have a deeply worthwhile religious attitude that combines these two elements—indeed, the effort to do so has contributed profoundly to Judaism, as well as to Christianity and Islam—it is also possible to have a deeply fulfilling religious attitude while keeping far away from metaphysics. Speaking for myself, I would say that although I do conceive of God as a "transcendent being," as a "necessary being," as an "unconditioned ground for the existence of everything that is contingent," I feel that insofar as I have any handle on these notions, I have a handle on them as *religious* notions, not as notions that are supported by an independent philosophical *theory* (certainly not by the theory of Aristotle's *Metaphysics*). For me, the "proofs" show conceptual connections of great depth and significance, but they are not a foundation for my religious belief. (In spite of Maimonides' prestige, they never played an important role in Judaism.) Nor are "proofs" the way in which I would try to bring someone else to religious belief of any kind.[24]

A second familiar response to religious skepticism is the dogmatists' "My religion is true and every other belief is wicked (especially atheism), or no better than witch doctoring (other religions)." (A friend remarked, "I understand

---

23. For the reasons for seeing "philosophie antique" (ancient and medieval philosophy) as a group of salvific enterprises, see Pierre Hadot, *Philosophy as a Way of Life* (Oxford: Blackwell, 1995).

24. I have taken this paragraph from Hilary Putnam, "Thoughts Addressed to an Analytical Thomist," *Monist* 80, no. 4 (1997): 487–499.

this is very popular among people philosophers don't talk to.")[25] Not only is this response a denial of the very raison d'être of philosophy itself, which John Dewey so well defined as "criticism of criticisms,"[26] but also, in a marvelous discussion of the psychology of "fanaticism" in *The Critique of the Power of Judgment,* Kant argues that this is, at bottom, not religion but a disease of religion.[27]

Part of Kant's point is that the "fanatic" (his term for what I just called "the dogmatist") treats religious beliefs as if they were as sure as ordinary perceptual beliefs. I remarked that skepticism about the existence of houses and rocks has happily become a purely theoretical issue. In practice, as Kant pointed out in *The Critique of Pure Reason,* perception of such objects is passive; we have no real choice about whether to believe that there is a house in front of us when we see one. Nor do we have to "take responsibility" for believing that there is a house there when we see one or walk into one. For the fanatic, it is as if he had as simply (and as unproblematically) seen God, or seen Jesus (or, in Kantian language, seen the unconditioned). Those who do not accept what is so obvious are wicked or stupid or both, or, in the best case, waiting for the fanatic to enlighten them. Such an attitude, Kant believes, misses the essence of true religious faith, which (for him) involves the recognition that what one believes is not simply forced on one passively. The uncertainty, the unprovability, of religious propositions is, Kant believed, a *good* thing; for if religious propositions could be proved, there would be nothing to take responsibility for. To put it in present-day language, the fanatic is unconsciously *fleeing responsibility.* I find that my perceptions are in accord with Kant's here: I find that both his psychology of fanaticism and the phenomenology of faith presupposed by that psychology are very deep.

A third approach to skepticism, often associated with existentialism, is to accept responsibility for believing what cannot be proved. I already mentioned the note Avi Sagi found in an unpublished bit of Kierkegaard's *Nachlass* that reads: "Leap of faith—yes, but only after reflection." In this approach the role of religious experience is not to *prove* something but to

---

25. The friend is Philip Devine.

26. John Dewey, *Experience and Nature* [1925] in *The Later Works of John Dewey,* vol. 1, ed. Jo Ann Boydston (Carbondale: Southern Illinois University Press, 1981), 298.

27. Eli Friedlander, "Kant and the Critique of False Sublimity," *Iyyun: The Jerusalem Philosophical Quarterly* 48 (1999): 69–93, is a beautiful analysis of Kant's discussion.

confront one with an existential choice, to make "believe or don't believe" a "live option," in William James's words.[28] A fine but deeply challenging account of this third option can be found in Wittgenstein's "Lectures on Religious Belief."[29] (Wittgenstein described himself as not a believer, "although I cannot help seeing every question from a religious point of view.")

Here is what Wittgenstein says:[30]

> These [religious] controversies look entirely different from normal controversies. Reasons look entirely different from normal reasons.
>
> They are, in a way, quite inconclusive.
>
> The point is that if there were evidence, this would in fact destroy the whole business.

Several paragraphs later, Wittgenstein discusses a "Father O'Hara"[31] who, he tells us, "is one of those people who make it a question of science." He continues:[32]

> Here we have [religious] people who treat this evidence in a different way. They base things on evidence which taken in one way would seem exceedingly flimsy. They base enormous things on this evidence. Am I to say they are unreasonable? I wouldn't call them unreasonable.
>
> I would say they are certainly not *reasonable,* that's obvious.
>
> "Unreasonable" implies, with everyone, rebuke.
>
> I want to say: they don't treat this as a matter of reasonability.

28. William James, *The Will to Believe and Other Essays in Popular Philosophy* [1897] (Cambridge, Mass.: Harvard University Press, 1979), 2.

29. Ludwig Wittgenstein, *Lectures and Conversations on Aesthetics, Psychology, and Religious Belief,* compiled from notes taken by Yorick Smythies, Rush Rhees, and James Taylor, ed. Cyril Barrett (Oxford: Blackwell, 1966), 53–72.

30. Ibid., 56.

31. Father O'Hara, we are told by the editors, wrote a contribution to a symposium on science and religion. James Conant and Cora Diamond have come up with the following information, which, as far as I know, has not been previously published: Wittgenstein came across Father O'Hara's piece by hearing it delivered as a talk on a BBC radio broadcast. The piece was part of a series of twelve broadcasts, including ones by Huxley, Haldane, Malinowski, and Eddington. The title of the series was "Science and Religion." The twelve talks were broadcast between September and December 1930. O'Hara's piece was subsequently published along with all the other pieces in the series in M. I. Pupin, ed., *Science and Religion: A Symposium* (London: Gerald Howe, 1931), 107–116. None of the individual pieces in the volume are titled. It is almost impossible to lay hands on a copy of the original 1931 volume, but fortunately, it was reprinted in 1969 by Books for Libraries Press in Freeport, New York.

32. Wittgenstein, *Lectures and Conversations,* 57–58.

Anyone who reads the Epistles will find it said: not only that it is not reasonable, but that it is folly.

Not only is it not reasonable, but it doesn't pretend to be.

What seems to me ludicrous about O'Hara is his making it appear to be *reasonable*.

The question these remarks of Wittgenstein's invite is the obvious one: is it ever *justified* to believe what is not "reasonable"? This is the question that William James dealt with in his celebrated essay, "The Will to Believe" (which he considered calling "The Right to Believe," which is what the essay actually defends). That often misrepresented and misinterpreted essay, it seems to me, gives exactly the right answer to this question, but it would take a much longer essay than this one to interpret and discuss it. I want, however, to make just one point about it, namely, that James emphasizes that saying that there is a right to believe is by no means to say that there is a right to be intolerant,[33] and that too seems to me exactly right.

Why Did I Focus on Experience, Then?

In view of what I just said, it will be clear that I did not focus on experience in this essay because I wish to argue that religious experience answers skeptical questions. But I did have a reason for focusing on it, just as Wittgenstein had a reason for focusing on the complexity of the phenomenon of religious belief. Wittgenstein began his lectures on religious belief by pointing out that believers and atheists regularly talk past each other. If you search the Web under "atheism," you will find a great deal of intelligent and painstaking proof that the Bible contains errors, that it is silly to think that every word of the Bible was literally dictated by God, and so on, but precious little recognition that most religious people are not fundamentalists, and many do not believe in the idea of divine *dictation* at all. It is as if atheists too were "fanatics" in Kant's sense; for them, too, their (negative) religious belief is, it seems, akin to a perceptual certainty, something that involves no *responsibility*. Wittgenstein, if I interpreted him correctly in "Wittgenstein on Religious belief,"[34]

33. William James, *The Will to Believe*, 33.

34. Hilary Putnam, "Wittgenstein on Religious Belief," in Leroy Rouner, ed., *On Community* (Notre Dame, Ind.: University of Notre Dame Press, 1991), 56–75; reprinted in Putnam, *Renewing Philosophy* (Cambridge, Mass.: Harvard University Press, 1992), 134–157.

did not want to make us believers (he was not religious himself), but he felt an enormous respect for the literature and the spirituality contained in religious traditions, and he wished to combat this sort of simplistic stereotyping. One way of overcoming the idea—and we need to overcome it—that it is simply *obvious* what having a religious faith consists in is to overcome the idea that it is simply obvious (or if not obvious, obviously irrelevant) what the words "religious experience" refer to. In this essay I have tried to suggest that what "experience" refers to is far more complicated a matter than we tend to think, and that understanding how deep experience can be is a necessary preliminary to any discussion of "science, religion, and human experience."

34

Aristotle's Mind and the
Contemporary Mind

Since the beginnings of philosophy as a subject there has been a struggle between two very different conceptions of perception. In ancient philosophy perception was thought of as supplying us with "appearances," and the form that the struggle in question took in that period concerned the nature of these appearances. Some classical philosophers conceived of appearances as *intermediaries* between us and things external to the soul. For the Democriteans, for example, appearances were affections of the senses (and ultimately of the soul, which was thought also to consist of atoms, albeit of a finer kind than bodies), and the particular sensory qualities were defined in terms of the kinds of atoms that caused them.[1] The Stoics thought of appearances as

[2011]: This lecture was written before I developed the criticisms of "disjunctivism" described in Chapter 36 or the optimistic attitude to the potential value of a nonreductive and noncomputational functionalism that I now have, and that I express in Chapter 3. In places it seems to represent the negative aspect of "Wittgensteinian" thought that I disavow in Chapter 28. I no longer believe that the notion of a mental representation has to be useless or nonsensical, although the idea that either meanings or concepts are *identical* with "mental representations" still seems to me confused. But I believe that the interpretation of Aristotle offered here is of interest in its own right, and that it is of value to see how relevant some of Aristotle's concerns still are.

"impressions" in the soul or "alterations" of the soul (which they also conceived of materialistically).[2] For Aristotle, on the interpretation I defend, in perception and thought the intellectual part of the soul is in *direct* contact with the properties of the things thought about. In modern jargon, the Democriteans and Stoics had a "representational" theory of perception, while the Aristotelians (according to me) were "direct realists." To be sure, there are difficulties with the Aristotelian view. Those difficulties arise from the essentialism involved in the Aristotelian notion of "form," and they become serious when we hold that the form of the object is in the mind when we *think* about the object, as well as when we perceive it.[3] It does seem that in the case of thought, the form Aristotle speaks of is what he regards as the "essence," and I find serious difficulties with the idea that we can think only about things whose "essence" we know. But Gisela Striker has suggested to me that in the case of perception it is unlikely that Aristotle has so demanding a notion of form in mind. It is plausible that the form that we receive in perception is simply the *sensible form*—the color or the shape or the texture or the sound or whatever.[4] If we read Aristotle in this way, it would seem that what he is saying is simply that in perception we are aware of sensible properties of external things—their shape, their color, and so on.

It may be, of course, that Aristotle also thought that his talk of the same thing (the form) being in two places at once (in the object perceived and in the mind) *explained* how our awareness of the sensible properties of external things is possible. If so, he was mistaken; taken as an explanation, such a metaphysics is unhelpful, to say the least. It is also possible, as indeed the

---

1. This is Democritus's view as reported by Theophrastus, *De sensibus,* ed. G. M. Stratton (London: Allen and Urwin, 1917; reprint, Chicago: University of Chicago Press, 1967), 60–61, 63–64. For a penetrating and philosophically sophisticated discussion of the state of our knowledge of Democritus's views and what those views are likely to have been, see Mi-Kyoung Lee, "Conflicting Appearances" (Ph.D. diss., Harvard University, 1996).

2. The Stoics standardly described a *phantasia* as an impression—as from a signet ring in wax—and Sextus Empiricus (*Adv. math.* 7.288ff.) reports a long debate among the Stoics about the appropriateness of this description, Chrysippus objecting that one could not imagine several conflicting "impressions" arising simultaneously, and proposing to speak of "alterations" in the soul instead. (Communication from Gisela Striker.)

3. Cf. Hilary Putnam, "Aristotle after Wittgenstein," in R. W. Shaples, ed., *Modern Thinkers and Ancient Thinkers* (Boulder, Colo.: Westview; London: UCL Press, 1993), 117–137; reprinted in Putnam, *Words and Life* (Cambridge, Mass.: Harvard University Press, 1994), 62–81.

4. I did not consider the possibility of this reading in "Aristotle after Wittgenstein," and if I had I would have been more charitable to the Aristotelian view.

rather strange prose of *De Anima* at this point suggests might be the case, that he is simply using a number of figures of speech to say that we really *are* aware of properties of external things (just as we today speak of something being "in one's mind"). But whatever Aristotle may have intended, and whether or not part of what he intended has to be rejected as confused, he *at least* believed that we *do* have an awareness of the sensible properties of "external" things, and that this is not to be interpreted as meaning that we merely have "images" or "representations" of those things before our minds (as the view that has been dominant ever since Descartes holds). And Aquinas followed Aristotle faithfully in all this.

I contend that this disagreement continues to the present day, in essence if not in the details of the positions, and that the confusion in speculation about the mind in much of contemporary cognitive science arises from a continuing allegiance to the picture of appearances—and, as we shall see, of "thoughts" as well—as intermediaries, and from a systematic neglect of the "Aristotelian" alternative.

Caston's Aristotle

That Aristotle was a direct realist is, however, challenged by Victor Caston in a boldly original study of Aristotle's views on intentionality.[5] Caston recognizes that Aristotle has long been read in this way, for example, by Aquinas, as well as in our day (in spite of important disagreements between us about the details) by Owens,[6] Burnyeat,[7] Sorabji[8] and myself.[9] Instead

---

5. Victor Caston, "Aristotle and the Problem of Intentionality," *Philosophy and Phenomenological Research* 58, no. 2 (1998): 249–298.

6. Joseph Owens, "The Aristotelian Soul as Cognitive of Sensibles, Intelligibles, and Self," in John R. Catan, ed., *Aristotle: The Collected Papers of Joseph Owens* (Albany: State University of New York Press, 1981), 81–98.

7. Miles Burnyeat, "Is an Aristotelian Philosophy of Mind Still Credible?" in Martha C. Nussbaum and Amelie Rorty, eds., *Essays on Aristotle's "De Anima"* (Oxford: Oxford University Press, 1992), 15–26.

8. Richard Sorabji, "From Aristotle to Brentano: The Development of the Concept of Intentionality," in Henry Blumenthal and Howard Robinson, eds., *Aristotle and the Later Tradition* (Oxford: Studies in Ancient Philosophy, suppl. vol., 1991), 227–259.

9. Putnam, "Aristotle after Wittgenstein." See also Hilary Putnam and Martha Nussbaum, "Changing Aristotle's Mind," in Nussbaum and Rorty, *Essays on Aristotle's "De Anima,"* 27–56; reprinted in Putnam, *Words and Life,* 22–61.

of the Aristotle who thought that the mind, in both thought and perception, is in contact with the forms themselves, Caston proposes an Aristotle who had a causal theory of reference and a causal theory of perception, as well as of memory and imagination. But although there is much to be learned from Caston's work, particularly on *phantasia*,[10] I think that Caston's arguments against the direct realist interpretation of Aristotle cannot be accepted.

Caston rightly points out that Aristotle's statement that "concerning sensation in general one must suppose that the sense is receptive of sensible forms without the matter, as when wax receives the device of the signet ring without the iron or gold, it takes up the golden or brazen device, but not insofar as it is gold or bronze"[11] is not, taken by itself, decisive. He cites three difficulties with reading this as an account of the intentionality of sensation: "(1) Aristotle takes the case of the wax and the signet ring to be a *genuine* case of receiving the form without the matter—this has occasionally been denied, but this reading goes hard against the text—even though the wax is not in any intentional state about the ring. (2) There is nothing in the phrase itself to suggest a notion of reference: how does 'received without matter' signal the notion of being about something? (3) The doctrine is still a causal doctrine . . . the doctrine cannot be extended to intentional states in general, many of which are not about any of their causal ancestors";[12] e.g., cases of thinking about something that does not exist, because there is no informed and enmattered object involved from which to abstract the form.

Following this, Caston develops his own extremely interesting account of how Aristotle does account for the possibility of thinking about things that do not exist, and especially for the possibility of error in memory and for perceptual illusion. I shall say a few words about the motivation behind Caston's account and reserve the detailed discussion of Caston's interpretation to the Appendix.

All three of Caston's difficulties assume that if Aristotle did make the claim that the form of the objects we perceive or even think about are both in

---

10. Cf. Victor Caston, "Why Aristotle Needs Imagination," *Phronesis* 41, no. 1 (1996): 20–55, which overlaps with Caston, "Aristotle and the Problem of Intentionality."

11. Aristotle *De an.* 2.12, 424a17–25.

12. Caston, "Aristotle and the Problem of Intentionality," *Philosophy and Phenomenological Research* 58, no. 2 (1998): 249–298, quotation from 256.

the objects and in the mind (but not enmattered in the mind), he must have intended that claim as a *metaphysical explanation* of the intentionality of thought and reference. I began this paper by saying that if that *was* Aristotle's intention, the explanation was unsuccessful. But, as I already remarked, a more charitable interpretation is possible, on which Aristotle was not describing a *mechanism to explain* intentionality (an "account of content" in the sense of our present-day philosophy of language), but simply committing himself to the commonsense view that in sense perception we are directly aware of properties of objects and not of representations, and that in thought we directly conceive of properties and things. The psyche does not merely manipulate "representations," conceived of as inner affections that are connected with those properties and things only by being some of their effects (as on Caston's interpretation).

When I say that on the commonsense view we are "directly" aware of the properties of external things, I do not mean, and I do not ascribe to Aristotle the view, that such perception is not *causally dependent* on bodily processes.[13] I mean that the *upshot* of a successful perception is a cognitive contact with those properties themselves, and not merely with some effects or representations. Similarly, when I say that in the case of thought the commonsense view is that we "directly" conceive of the properties of things, I do not mean, and I do not ascribe to Aristotle the view, that such thought is not *causally dependent* on bodily processes, or that it does not ever *involve* images or other representations.[14] I mean that the *upshot* of successful thinking is a cognitive contact with those properties themselves, and not simply with representations among whose efficient causes are the things with those properties.[15] Caston wants to read Aristotle as having a more "modern" view because he believes that the

13. But see Putnam and Nussbaum, "Changing Aristotle's Mind," and Burnyeat, "Is an Aristotelian Philosophy of Mind Still Credible?" for a debate concerning the nature of that dependence. Here, by the way, Caston sides with the Putnam-Nussbaum view.

14. In *De an.* 3.7, 431a15–20, Aristotle writes, "To the thinking soul images serve as if they were contents of perception. . . . That is why the soul never thinks without an image." Here and elsewhere I use the *The Complete Works of Aristotle: The Revised Oxford Translation*, ed. Jonathan Barnes, vol. 1 (Princeton, N.J.: Princeton University Press, 1984). That the images used in thought themselves depend on bodily processes is shown by the passages from *On Memory and Recollection* cited by Caston that I discuss below.

15. I discuss the significance of this distinction at length in Putnam, "The Dewey Lectures: Sense, Nonsense and the Senses; An Inquiry into the Powers of the Human Mind," *Journal of Philosophy* 91, no. 2 (September 1994): 445–517.

more modern view *works;* I believe that the modern view is a disaster, and that Aristotle was thinking in a better way.

But I must now leave the further discussion of Caston's interpretation to the Appendix I mentioned and turn to the "contemporary mind."

The Picture of the Mind as an Inner Theater

Although no modern representational theory would suppose that in perception what we are directly aware of are literally little images in the sense of physical likenesses, Wilfrid Sellars pointed out in a famous essay that the picture that underlies representational theories up to the present day is of an *inner theater* or *inner movie screen.*[16] On this conception, the "colors we see" are not really properties of the external objects but properties of the images on the inner movie screen; indeed, the objects we "directly see" are only objects in the inner theater or on the inner movie screen. It is this picture that direct realists have always been concerned to combat; indeed, as I have already indicated, "direct realism" is best thought of not as a "theory of perception" but as a denial of the necessity for and the explanatory value of the idea that what we are in immediate contact with in thought and perception is "internal representations."

The disagreement between the two conceptions that began in ancient philosophy continued in the Middle Ages. After Descartes, however, the situation changed drastically; direct realism virtually vanished from the scene until the early twentieth century, when William James and the American "new realists" led by Perry defended it (in a novel form). Then it disappeared again until Austin's vigorous defense of a commonsense direct realism in his *Sense and Sensibilia.* Although in the last few years both John McDowell and myself have defended varieties of direct realism, the dominant view in Anglo-American philosophy of mind today appears to be what we may call "Cartesianism-cum-materialism," that is, a combination of Descartes' conception of the mind as an inner theater with materialism.

I believe that the problems of philosophy that have become "traditional" since Descartes' time all rest on a mistaken conception of perception. On this "traditional" conception, what we are *cognitively* related to in perception is not

16. Wilfrid Sellars, *Empiricism and the Philosophy of Mind* [1956] (Cambridge, Mass.: Harvard University Press, 1997).

people and furniture and landscapes but *representations*. These "inner representations" are supposed to be related to the people and furniture and landscapes we ordinarily claim to see and touch and hear only as inner effects to external causes; and how they manage to determinately *represent* anything remains mysterious in spite of hundreds of valiant attempts by both "realists" and "antirealists" to clear up the "mystery." Although I will not repeat here the arguments I have offered elsewhere, I believe that it is only by giving up this picture of perception as mediated by a set of "representations" in an inner theater that we will ever be able to escape from the endless recycling of positions that do not work in the philosophy of mind (not to mention traditional epistemology and traditional metaphysics), a recycling that has been going on for at least four centuries.

The Picture of Thoughts as "Representations"

If the picture of the mind as an "inner theater" has had disastrous consequences for the philosophy of perception, it has had equally disastrous consequences for would-be scientific speculation about the nature of meaning and thought. For Aristotle, I claimed, thought, as well as perception, involves direct contact with the properties of the objects thought about. Descartes himself may have had a similar view, but with a more Platonic cast (the universals, for Descartes, are native to the mind itself, rather than requiring to be abstracted from experience of particulars). But once Descartes' notion of the mind as an inner theater was naturalized (e.g., by identifying the mind with the brain), the objects of thought were banished from the inner theater, and in their place we now find only "representations" whose connection to what they "represent" seems an insoluble puzzle. The result, I believe, is not scientific progress but a deeper and deeper sinking into confusion. But rather than leave this statement unsupported by any example, and thus as a mere dogmatic statement, I will illustrate what I mean by examining in detail one attempt to identify the contents of thought with scientifically describable "semantic representations," that of Noam Chomsky. However, the considerations I will bring to bear against Chomsky's particular version of this idea will bear against the whole idea that concepts or thoughts are *objects*, let alone *scientific* objects.

Chomsky's "Innateness Hypothesis"

Chomsky's view features the controversial claim that our linguistic skills are *innate*. This is not the feature of the view that I shall discuss here, although I shall begin by asking what exactly this is supposed to mean.

In Chomsky's own development it has meant different things at different times. Early on, the idea of innateness was that *the rules of "universal grammar" are represented in the brain*. Thus more than twenty years ago Chomsky was happy to write:

> With the progress of science, we may come to know something of the physical representation of the grammar and the language faculty—correspondingly, the cognitive state involved in language learning and the initial state in which there is a representation of UG [universal grammar] but of no specific grammar corresponding to UG.[17]

Chomsky later dropped the idea that there is a "representation of UG (universal grammar)" in the "initial state." He now agrees that it is not necessary to suppose that the rules of the postulated "universal grammar" are themselves recorded in the brain; it is only necessary to suppose that the brain is so composed that when it *functions* according to what he calls its "competence," the speaker uses language in accordance with those rules, and that the differences between the surface grammars of the various natural languages can be explained by supposing that certain parameters get "set" differently as a result of speakers' growing up in one linguistic environment or another. Thus in 1988 we find him writing:

> We may think of the language faculty as a complex and intricate network of some sort associated with a switch box consisting of an array of switches that can be set in one of two positions. Unless the switches are set one way or another, the system does not function. When they are set in one of the permissible ways, then the system functions in accordance with its nature, but differently, depending on how the switches are set. This fixed network is the system of principles of universal grammar, the switches are the parameters to be fixed by experience. The data presented to the child learning the language must be sufficient to set the switches one way or another. When these switches are set, the child has command of a particular language and

17. Noam Chomsky, *Reflections on Language* (New York: Pantheon, 1975).

knows the facts of the language: that a particular expression has a particular meaning and so forth.[18]

In current versions of the innateness hypothesis (e.g., the Managua Lectures, from which I just quoted) the emphasis is not on the *representation* of rules but on the hypothesis of what Chomsky refers to as a "language organ," or a system of language "modules." The idea is that just as there are certain organs, the hands, that are innately adapted to certain tasks (grasping, picking up, manipulating), so there are "mental organs" (brain organs), subpersonal processors in the brain, that have evolved to carry out various linguistic tasks, and that the structure of language (the "universal grammar") is just a reflection of the computational structure of these modules, the computational structure of the language organ.[19]

In the next section I will try to show that the claim that the "particular meaning" (narrow content or "semantic representation") of each word can be described by specifying the values of a finite list of parameters is deeply problematic. To do this, I turn to the relevance of Gödel's celebrated (second) incompleteness theorem.

A Gödelian Example and What It Shows

"The *Gödel theorem?*" I can imagine some of you thinking. "Of what possible relevance is that?" But actually the relevance is fairly immediate.

First, Chomsky's claim is a claim about the meanings of *all* words. So if it is correct, it applies also to the word "proof," or, as I shall say, "demonstration" (to remind you that I am discussing logical-mathematical proof and not empirical "proof," as when we speak of "proving that the suspect is guilty"). Presumably just about anyone can learn a tiny bit of geometry or a tiny bit of arithmetic and can learn to use the word "demonstration" in connection with the kinds of reasoning that go on here. But perhaps I am already moving too quickly.

Although Chomsky does not believe that there is such a thing as "general intelligence," he does believe that there is such a thing as "the science-forming

18. Noam Chomsky, *Language and the Problems of Knowledge: The Managua Lectures* (Cambridge, Mass.; MIT Press, 1988), 62–63.

19. I believe that this way of describing the innateness hypothesis entangles two issues that need to be separated: (1) the existence of linguistic universals and (2) the language-organ hypothesis. The failure to disentangle these issues has been detrimental to clarity about the foundations of linguistics.

faculty."[20] Perhaps he would say that although most words require only "the language organ" to learn, it requires the "science-forming faculty" (which is supposed to work more slowly and not to be as equally developed in all speakers as "the language organ") to learn such words as "demonstrated." However, Chomsky holds that the "science-forming faculty" *too* is "modularized," and for the same sorts of reasons that he accepts in the case of "the language organ" (such as the failure of behaviorism and of connectionism, and the existence of universal generalizations that cannot be otherwise explained). Because my argument is against the modularity of the "particular meaning" of "demonstrate," it does not matter for my purpose whether the supposed "module" is to be part of the "language organ" or part of "the science-forming faculty."

Quite simply, the argument is that if Chomsky is right, we could in principle describe our "competence" in the use of the word "demonstrate" *recursively,* and that would mean that we could *recursively enumerate* all the truths of, say, elementary number theory that the human mind can demonstrate. At this point we must be careful *not* to make the error that Roger Penrose makes in *Shadows of the Mind.*[21] It is tempting to argue that "but the consistency of this set of truths would then be one more arithmetical truth that we could demonstrate, contradicting Gödel's theorem." *That* argument is fallacious, as I have shown elsewhere.[22] However, we can argue along a different path.

Let us assume that Chomsky is right, and that the set of purely mathematical statements that a speaker who makes no performance errors can demonstrate is the output of a computationally describable "module." In other words, the set is recursively enumerable (this is technical jargon for "can in principle be written down by a computer that is allowed to continue making computations and writing down the results forever"). We shall also assume that any purely mathematical statement that a speaker who makes no performance errors can demonstrate is true. This certainly seems to be a part of our concepts of "error" and "demonstration." Then although the consistency of the recursively enumerable set of statements that we can *demonstrate to be true* is not itself one that we can *demonstrate* to be true, even with the aid of

20. Noam Chomsky, "Explaining Language Use," *Philosophical Topics* 30, no. 1 (1993): 205–231; see p. 218.

21. Cf. Hilary Putnam, "The Best of All Possible Brains?" review of Roger Penrose, *Shadows of the Mind* (Oxford: Oxford University Press, 1996) published in the *New York Times Book Review,* November 20, 1994, 7–12. 7. A slightly expanded version was reprinted (without the title) in *Bulletin of the American Mathematical Society* 32, no. 3 (1995): 370–373.

22. See Putnam, "Best of All Possible Brains?"

this assumption (because we have to use *empirical* knowledge—the description of the module in question and the values of the relevant parameters—to define the set precisely), still, if we can *know* the description of the set in question, the consistency statement is one we can empirically *justify*. But now we can run a Gödelian argument again, this time using the notion of "empirical justification" instead of the notion of "demonstration," and (if we can describe our competence in the use of the notion "justified given evidence *e*" *recursively,* as we should be able to do on Chomsky's picture of the mind), the result *will* be a real contradiction.[23]

The upshot of these Gödelian reflections is not, of course, what Roger Penrose hopes for—a proof that the *performance* of the brain as a whole cannot be simulated by a Turing machine. But it is a proof that if there is a recursive description of our *competence* in the use of such epistemic notions as "demonstrate" and "justify," *then it is beyond our powers*—our powers *in principle,* not just *de facto*—to *know* that description.

In recent publications I have argued that, at least at present, the functionalist program is *empty*—empty because the notion of a "computational description" of meaning presupposes that we have at least an idea of what a *possible form* for such a description might be. Without such a description, the notion of a "computational state" becomes a "we know not what." It cannot, of course, be ruled out *a priori* that in the future someone will come up with a definite proposal, with a computational formalism, and with at least a first step toward a genuine research program for describing propositional attitudes in such a formalism; but can anything at all be ruled out *a priori?* Do we have a metaphysical guarantee that in the future no one will give an intelligible sense to the claim that two and two are sometimes six? (Is the notion of such a guarantee even an intelligible one?) But the fact that someone may give a proposition *p* an intelligible sense no more implies that *p* now has an intelligible sense than the fact that I might someday learn to play the violin implies that I now know how to play the violin. Without a coherent body of embedding theory to which the claim "propositional attitudes are computational states of the organism-cum-environment" (or "two and two are sometimes six") belongs, the idea that we now *understand* the claims presupposes a Platonic picture of understanding as a completely free-standing ability, a mystery mental act.

23. For details, see Chapter 13, "The Gödel Theorem and Human Nature."

In the case of the proposal that "the science-forming faculty" itself is modularized, and that our scientific competence can be completely recursively described (Chomsky once confirmed to me in a conversation that this is what he believes, by the way), the situation is even *worse* than in the case of my former functionalist proposals. There, at least, there was no *contradiction* in the idea that we might someday give content to talk of the "computational description" of the states in question. If a proposal that we do not know how to "cash out"—do not know how to take even a *first step* to investigating (and this is what I claimed the proposal that propositional attitudes are computational states to be)—is a mere "we know not what," what are we to say about a proposal that we *know* could never be cashed out? The notion of a recursive description of our "science-forming faculty" is totally vacuous.

What my Gödelian argument shows is that there is at least one concept that we are able to acquire, but such that it is not the case that there is some describable "module" whose structure (whose innate "grammar," so to speak) completely captures our "competence" in the *use* of that notion. In short, either something is wrong with the modularity hypothesis, or something is wrong with the notion of "competence." But none of this goes against the thought that there may be significant universal generalizations to be made about how human speakers do and do not employ the notions of "demonstrations" and "justification."[24] But let us reflect on this example a little further.

Some Reflections on the
Competence/Performance Distinction

How might Chomsky evade the argument just given? What follow are some reflections triggered by this question.

1. Rejecting the *competence/performance* distinction and saying that what is at stake is the existence of a module responsible for our "performance," and not the description of our "competence," would short-circuit my Gödelian argument, but this is not a real option for Chomsky, or for modern linguistics itself, for that matter. From Chomsky's *Syntactic Structures* on, the arguments for the existence of linguistic universals have presupposed that *that* distinction is in place, as have the later arguments for modularity. To appreciate why

---

24. This illustrates the way in which inferences from the existence of linguistic universals to the existence of mental organs are apt to be fallacious.

this is so, it suffices to employ Wittgenstein's discussion of following as simple an arithmetical rule as "keep adding two to the number you are given to start with." If we were not allowed to prescind from performance errors—if the very *notion* of a "performance error" were not allowed—then we could not even say that what the speaker does when given the number "11" is to produce the sequence "13, 15, 17, . . . ," for that is *not* what the speaker "does" if every slip of the tongue, every response produced by fatigue, or every failure to remember the preceding digits correctly is taken to be a part of the "output" we are asked to account for. Indeed, no "module" smaller than the whole brain—or, better, the whole speaker plus the environment—can predict the *whole* of a speaker's output, warts and all. When we postulate a simple process—under an *idealized* description, according to which what the process does is to errorlessly perform the calculation $x = n + 2$ (this is what Wittgenstein calls using the "machine as symbolizing its action," as opposed to thinking about an actual machine as an engineer does[25])—as the process responsible for producing the series, we have *already* decided to discount any occasion on which the speaker says, for example, "$222634 + 2 = 22636$" as a "performance error." The whole structure of linguistic theory and cognitive science presupposes that the competence/performance distinction is in place.

2. A better line of argument against the relevance of Gödelian considerations would be to insist on the distinction between *linguistic* competence and *scientific* competence and to jettison the idea of a recursive characterization of the latter, but not of the former. Indeed, such a distinction is contained in Chomsky's own distinction between "the language organ" and "the science-forming faculty." At first blush this seems to do the trick. For we can then say that to know the *meaning* of "demonstrate," it is not necessary to be able to follow complex proofs (to possess "mathematical competence"). Someone who can follow only the very simplest arithmetical or logical or geometric arguments can easily show that he understands the *word* "demonstrate." If what is included in our grasp of the *literal meaning* of "demonstrate" is limited enough, no Gödelian argument can get started. For Gödelian arguments presuppose an understanding of metalogical reasoning (in addition to mathematical induction, rules of inference, and so forth). The Gödelian argument I referred to does indeed explode the claim that there is a recursively surveyable "science-forming faculty," but why should the linguist on the street or the

25. Wittgenstein, *Philosophical Investigations* (Oxford: Blackwell, 1953), §193.

psychologist on the street or the philosopher on the street follow Chomsky in *that* bit of metaphysical speculation? But now a different problem arises.

The very fact that I just suggested we exploit to avert the threat of Gödelian arguments (arguments that, however strange it may seem to introduce them into a lecture on the philosophy of mind, are nevertheless perfectly appropriate when the notion of "competence" becomes as utopian as it is in Chomsky's picture of the mind)—the fact that we require very little of speakers when it is just a question of *knowing the literal meaning* of the word "demonstrate" and not of exhibiting some idealized competence at demonstrating mathematical truths—cannot realistically be separated from such other by-now-familiar facts as the fact that no one *fixed* set of skills (linguistic skills and the nonlinguistic skills interwoven with them) is required before we can say that a speaker knows the meaning of "demonstrate" or, indeed, of any word. Moreover, even if a speaker exhibits a *lot* of skill at using the word, some further bit of "corpus" may reveal that he does not, after all, have the concept we thought he had. If a speaker one day refuses to call something a "demonstration" because the Dalai Lama does not approve of it, we may have to revise our whole account of the meaning of this lexical item in his idiolect. This is, of course, the notorious problem of "meaning holism."

To this, the stock response of antiholists[26] is to say that what is holistic is only the process of *finding out* what the meaning actually is; but, they say, meaning itself is not holistic.

Moreover, antiholists often accept a distinction between "wide content" and "narrow content." Only "narrow content" has to do with the language organ, they say; wide content depends on causal connections to the environment as well.[27] To describe "narrow content" is to describe one particular aspect of the language organ. But *what* aspect? If there is no one *fixed* competent use of the word—if it is a matter of *good judgment* whether a speaker uses a word "competently" or not[28]—then seeking a module, or whatever, that

---

26. Cf. Jerry Fodor and Ernest Lepore, *Holism: A Shoppers Guide* (Oxford: Blackwell, 1992).

27. This was Fodor's view until recently. In *The Elm and the Expert* (Cambridge, Mass.: MIT Press, 1994) he gives up the notion of narrow content, however.

28. Of course, Fodor would not agree that it is a matter of good judgment. On his view, it is a matter of whether the speaker is nomically connected to the right content; and contents, he now thinks, are extramental entities. In a way, his position seems to me now very close to Katz's explicit Platonism about meanings. See Jerrold J. Katz, *The Metaphysics of Meaning* (Cambridge, Mass.: IT Press, 1990).

accounts for *the* competent use is worse than a search for a "we know not what"; it is utter confusion.

3. This is a good occasion to rebut a common but absurd antiholist argument. It is often argued that if holism were true, every change in a speaker's beliefs would count as a change in the meaning of his words. But that is a fallacy. Real live holists, as opposed to straw men, maintain that the meaning of a word is not an object at all; the question should not be what meanings are but *what form the description of the reference and use of words had best take* for linguistic purposes. My own idea of a "meaning vector" in "The Meaning of 'Meaning'" was such a proposal.[29] Although holists have different answers to *that* question (Quine and Davidson, respectively, think that the answer is a recursive translation scheme for the whole language and a Tarski-style truth theory for the language), on any holist theory,[30] the *right description* is a function of the whole (actual and potential) corpus of utterances-in-contexts. But it does *not* follow that the meaning (the description) must change every time the corpus changes. That is like concluding from the fact that the least common multiple of a billion numbers is a function of *all* of them that it must change whenever any one of them changes. That would be a valid argument if *every function were one-to-one;* but meaning holists do not claim that correct meaning descriptions are *one-to-one* functions of the corpus.

4. If what I have been arguing is right, then such notions as "narrow content," "semantic component of universal grammar" (or of a particular grammar), and "mental representation of the concept" are one and all completely empty. It is high time we took seriously Wittgenstein's warning in the next-to-last paragraph of *Philosophical Investigations* that "the confusion and barrenness of psychology is not to be explained by calling it a 'young science'; its state is not comparable with that of physics, for instance, in its beginnings . . . for in psychology there are experimental methods and *conceptual confusion.*"[31]

29. Hilary Putnam, "The Meaning of Meaning," in *Mind, Language and Reality: Philosophical Papers,* vol. 2 (Cambridge: Cambridge University Press, 1975), 215–271.

30. Fodor tells me, "You aren't what I call a holist, you are an interpretivist." But then I doubt that there *are* any "holists."

31. Ludwig Wittgenstein, *Philosophical Investigations* (Oxford: Blackwell, 1953), Part 2, xiv.

A Brief Moral

The idea that there is a *scientific* problem of "the nature of the mind" presupposes the picture of the mind and its thoughts, or "contents," as *objects* (so that investigating the nature of thought is just like investigating the nature of water or of heat). I want to suggest that philosophy can expose this presupposition as a confusion. There are, to be sure, empirical facts about thought, as about everything else, and science can reasonably hope to discover new—and even some very surprising—empirical facts. But the idea that these facts (or some of them) *must* add up to something called "an account of the nature of mind" is an illusion—one of the dominant illusions of a certain part of contemporary science.

APPENDIX

A Discussion of Victor Caston's "Aristotle and the Problem of Intentionality"

I will begin with a few additional comments on Caston's three "difficulties." (1) The wax "genuinely" receives the form of the signet ring without the matter *of the signet ring.* And that form is *enmattered* in the wax—with a different matter. But reading *De anima,* one cannot suppose that when Aristotle speaks of the soul receiving the form of an object without the matter, he means that the form *becomes enmattered in the soul in a different way.* Indeed, as we shall see, this is something he explicitly denies. The analogy of the wax and the signet ring is only that—an analogy. (2) What makes the presence of a form in the soul a case of "being about something" is that it is *the thinking part of the soul* that form is in. Wax does not think. It is what the soul *does* when forms are in it that constitutes the activity of the soul as *thinking about those forms.* (In *this* sense, Aristotle was a "functionalist.")[32] (3) I agree with Caston that the way in which a sensation refers to a sensible quality "has no hope" of being extended to an account of error. In the case of the possibility of erroneous thought, Caston's interpretation is conventional, and indeed the text leaves little room for any other: it is when predication and combination of properties enter that thought may become erroneous. If I merely contemplate the property

---

32. See Putnam and Nussbaum, "Changing Aristotle's Mind."

"unicorn" without predicating it of anything, I am neither right nor wrong; but as soon as I predicate unicorn of something or think that the properties "horse" and "has a single horn" can be jointly predicated of something, then I open myself to the possibility of error. That it is not in error in supposing there to be such predicates as $P, Q, R \ldots$ does not mean that the soul cannot use them erroneously. In the case of other sorts of error—for example, false memories, illusions of all kinds, dreams, and hallucinations—Caston is right that another account is needed. A commonsense account would be that in these cases we have an experience that is similar to the experience we would have had if the memory or illusion had been a veridical memory or sense perception. In short, "experiencing an X" is, in some uses of the locution, *disjunctive:* it can refer to *either* perceiving an X (being in proper cognitive contact with an X) *or* seeming to perceive an X (in the foregoing sense). But it is important that we *do not have to postulate a "highest common factor" to the two disjuncts,* a "sense datum" that is "numerically identical" whether the perception is veridical or not. (Indeed, once we *do* postulate a "highest common factor," we are at once committed to the picture of the mind as an inner theater, in the familiar post-Cartesian fashion.)[33]

Caston's account (or rather his interpretation of Aristotle's account) has two elements: a causal theory of reference and an account of phantasmata (the products of the imagistic faculty Aristotle calls *phantasia*).[34] Sensations in (what I shall continue calling) Caston's account refer to sensible properties (color, shape, smell, sound, texture, and the like) by *being caused by them.* Phantasmata are bodily modifications that function as representations, and they refer by having the same relevant effects as certain sensations would have. Here is Caston's own statement of the causal theory he ascribes to Aristotle:[35]

> We can put this more formally. Aristotle is committed to something like the following *account of content for phantasmata:*[36]
>   (P) For any phantasma φ and time t, the total effect φ can

---

33. The term "highest common factor" was introduced in this connection by John McDowell, "Criteria, Defeasibility, Knowledge," *Proceedings of the British Academy* 68 (1982): 457–479.

34. Cf. Caston, "Why Aristotle Needs Imagination" and "Aristotle and the Problem of Intentionality."

35. Caston, "Why Aristotle Needs Imagination," 49–50.

36. Emphasis added.

produce at *t* on the central sense organ is the same as the total effect some sensory stimulation *s* would produce on the central sense organ were *s* to occur; and at *t*, φ is about whatever *s* would be about.[37] [Caston has already explained that a sensation is "about" its cause.]

The claim that phantasmata are bodily modifications is based, inter alia, on a reading of two fascinating passages in Aristotle's tiny treatise *On Memory and Recollection*.[38] Here is the first passage, in Caston's translation:

> For clearly one must understand the effect, the possession of which we call memory, to be like a kind of picture, occurring on account of sensation in both the soul and the part of the body which has [sensation] [sc. the soul].[39] For the change which occurs is imprinted like a kind of cast of the sensory stimulation, just as signatories do with their rings.[40]

According to Caston, Aristotle appeals to changes in the body that *represent* the objects of our mental states: a person stands in the appropriate relation to a phantasma that represents what is absent.

Caston's conclusion that the phantasma just is a "change in the body" may seem hasty, however. Aristotle, after all, speaks of a picture "occurring on account of sensation in both the soul and the part of the body which has [sensation]," and this says that the phantasma arises from sensation and that sensation in turn involves both the soul and the body; it does not identify the "picture" with the bodily change. More support is forthcoming from the second passage (*De mem. et remin.* 1, 453a16–28); Caston does not translate this, so I give the Revised Oxford Translation:

---

37. Gisela Striker has pointed out to me that one of Caston's key passages from *De motu animalium* (11, 703b18–20, cited in "Why Aristotle Needs Imagination," 24) is significantly misparaphrased. What Aristotle says is not, as Caston would have him, that "because they [*phantasia* and thought] have this ability [to affect the animal in the same way as the objects would], Aristotle reasons that they represent such objects." The sentence literally reads: "Thought and *phantasia* present [or bring to mind: *prospherousin*] the things that produce the affections because they present the forms [*eide*] of the things that produce [the affections]." So much for the one passage that allegedly mentions representation.

38. Caston, "Aristotle and the Problem of Intentionality," particularly 254–272.

39. Ross's translation takes the "it" in "which has it" to be "the bodily part where sense resides." Caston's translation agrees with Beare's in the Revised Oxford Translation, which has "in the soul and in the part of the body which is its seat."

40. Caston, "Aristotle and the Problem of Intentionality," 258. Aristotle *De mem. et remin.* 1, 450a27–32.

That the affection [recollection] is corporeal, i.e., that recollection is a searching for an image in a corporeal substrate is proved by . . . [here Aristotle cites certain phenomena[41] that he thinks can be explained by supposing that "he who recollects and hunts sets up a process in a material part, in which resides the affection"].

There is no doubt that here Aristotle is acting as a primitive cognitive scientist. He is speculating about the bodily phenomena on which memory and recollection are likely to depend, and he is suggesting that they involve phantasmata or (in the Revised Oxford Translation) "images" in and around the material part (possibly the heart, which most of the ancients took to be the central organ). Importantly, these images are not pictorial in the sense of representing sizes and colors of the objects or events they "depict" by means of sizes and colors—Caston cites convincing textual evidence that Aristotle rejects so crude a view.

Caston has provided convincing proof that Aristotle saw certain mental processes as supervenient on bodily processes, and that he was even willing to speculate in a strikingly modern way about what some of those processes might look like, for example, that they involve what we might call "representations." But what are we to make of Caston's claim that Aristotle had a causal theory of reference, and of his further claim that the formation of such affections in a corporeal substrate is all that Aristotle meant by his talk of forms being in the soul without their matter in perception and in thought? I offer three objections to these claims (three objections to counter Caston's "three difficulties").

1. That in the case of *thought* forms are present in the thinking part of the soul without matter is a doctrine that Aristotle propounds in *many* different ways, using different technical terms and different images. The evidence here does *not* rest on the use of the single phrase "receptive of sensible forms without the matter" or the single analogy of the wax and the signet ring. And Aristotle himself analogizes perception to thought in this respect.

---

41. Aristotle proposes the hypothesis that the reason we are disturbed when we fail to remember something, and that the disturbance continues even after we have abandoned the search, is that the search has stirred up the moisture (blood?) around "that part which is the center of sense perception," and "once the moisture has been set in motion it is not easily brought to rest, until the idea [which was sought for [more literally, "the object which was sought for"—HP] has again presented itself, and thus the movement has found a straight course." *The Complete Works of Aristotle De mem. et remin.*

2. Caston's claim that "sensation, for Aristotle, is clearly about its cause"[42] is inadequate. I shall argue that Aristotle did *not* have a simple "causal theory of reference" even in the case of sensation. But the claim that he did is essential to Caston's argument.

3. It is essential to Aristotle's whole epistemology that the thinking part of the soul can somehow extract universals from sensory experience.[43] The account that Caston has to give of how this is possible is, I think, completely unacceptable as *Aristotle.*

Re (1): Aristotle tells us that "the thinking part of the soul must therefore be, while impassible, capable of receiving the form [of an object]; that is, must be *potentially identical in character with its object without being the object*" (literally, "must be potentially such as this but not this"). And he adds, "The thinking part must be related to what is thinkable as what senses is to what is sensible."[44]

The phrase "potentially identical in character with its object" points to difficult doctrines, but I believe that at least this much is clear. For Aristotle, what we today call "properties" and what we call "concepts" are one and the same thing, regarded in two different ways. When we think of a predicate as qualifying an object, we now call it a "property," and when we think of a predicate as figuring in thought, we now call it a "concept"; but there is no such dualism in Aristotle. When I think of a flower as red, as it might be, or think of Callias as a man, the very same universal *red* or *man* is "in" the flower or the man and in my mind (in the thinking part of my soul), though not in the same way. But it is no part of Aristotle's doctrine that the universal is *enmattered* in the thinking part of my soul, and this is a point he makes in a number of different ways, one of which is to say "without being the object."

Similarly, Aristotle writes, "In the case of those [objects] which contain matter, each of the objects of thought is only potentially present [i.e., when thought is *actually* present, no matter is involved, so we get no gap between subject and object[45]]."[46] And in the case of immaterial objects (thought

42. Caston, "Why Aristotle Needs Imagination," 40.
43. E.g., Aristotle *Posterior Analytics* 100a and 100b.
44. Aristotle *De an.* 3.4, 429a14–18. Emphasis added.
45. This formulation was suggested by Rachel Barney.
46. Aristotle *De an.* 3.4, 430a6.

itself, or God, or essences) he writes, "In the case of objects which involve no matter, what thinks and what is thought are identical." [47]

Now, when Caston interprets Aristotle's statement (*De an.* 2.12, 424a17–25) that

> concerning sensation in general one must suppose that the sense is receptive of sensible forms without the matter, as when wax receives the device of the signet ring without the iron or gold, it takes up the golden or brazen device, but not insofar as it is gold or bronze,

he argues that although the image of the signet ring in the wax is literally an "image," it would be wrong in general to think that the phantasmata that are formed in the central organ are literally images. He writes:

> Aristotle goes on to offer a sketch of how we think of such things [Caston is here referring to Aristotle's account of how we can think of what does not exist], which turns on the presence of changes within the subject that *model* or *simulate* the objects thought about. Once again, he takes the subject to stand in relation, not to the objects thought about, but to "something distinct within the subject" which represents those objects, whether or not they exist. . . . He seems to have in mind an analog form of representation, whose parts represent the parts of the object, and whose content is decidedly nonpropositional. [48]

On this account, what is present in the thinking part of the soul when I think about an object is not the properties of the object but "an analog form of representation." But this makes no sense of the passages just cited, which seem to be emphasizing that what is in the soul is *identical* with the object in every respect except the matter.

Re (2): According to Caston, for Aristotle *sensation* is infallible. Indeed, Aristotle does say that "perception of the special objects of sense is always free from error," [49] and this is essential to the theory of reference Caston ascribes to Aristotle, because if sensation is sometimes erroneous, we cannot say that Aristotle thought that the "aboutness" of a sensation is simply its relation to its cause. Aristotle does qualify (*De an.* 3.3, 427b12) almost immediately, writing "Perception of the special objects of sense is never in error *or admits of the least possible amount of falsehood*" (*De an.* 3.3, 428b19). Concerning this pas-

---

47. Aristotle *De an.* 3.4, 430a3–5.
48. Caston, "Aristotle and the Problem of Intentionality," 261–262.
49. Aristotle *De an.* 3.3, 427b12.

sage, Caston writes, "Provided that we have a pure case of sensation, unadulterated by other mental processes, this qualification is inexplicable on Aristotle's account: sensation always corresponds to its cause."[50]

But what is "its cause"? Caston writes, "Although the odor or the color of a flower can affect our senses *per se,* the flower cannot *as a flower,*[51] and so is perceived only 'incidentally,' "[52] but the passage he cites does not mention the property of being a flower or any other property that we would class as "observable"; the property Aristotle says is perceived only incidentally is the property of being "the son of Diares." But let us accept Caston's interpretation. Then the problem arises: what singles out the color of the flower as *the* cause of the sensation, in the sense required for Caston's theory of reference (in the case of sensation, that the sensation is "about its cause") to be true?

First, Caston must claim that when I look at a rose, as it might be, I have one sensation of the red color and a different sensation of the shape of the rose. Even supposing that this is right (as an interpretation of Aristotle), it remains the case that each of these sensations has *the rose* as its cause in a perfectly ordinary sense that Aristotle would hardly wish to deny. So why is the sensation "about" the color or the shape and not about the rose? Is it supposed to be that I am never (or rarely) mistaken about the color of the flower but I am often mistaken about whether it is a rose? (In the case of roses I think that the reverse is true.) If sensations must be supposed to "single out" just *one* of their causes as *the* relevant one, this is a form of intentionality, and the appearance that we have *explained* the intentionality of sensations by saying that "sensation always corresponds to its cause" is illusory.

And, come to think of it, what *is* a sensation for Caston? It cannot be a direct contact with a property (the color or the shape), for that would imply that Aristotle does mean to claim that we are capable of having such properties in the thinking part of the soul. But it cannot simply be an "image" in a "corporeal substratum," as a phantasma is, for Caston is emphatic that we are not necessarily *conscious* of phantasmata (an inactivated memory trace is a phantasma); it is when the phantasma produces an "experience phenomenally like a perceptual experience"[53] that we have a memory, or an illusion, or a dream, or whatever. Presumably, then, sensations are "experiences" (sense

50. Caston, "Why Aristotle Needs Imagination," 53.
51. Aristotle *De an.* 2.6, 418a23–24.
52. Caston, "Why Aristotle Needs Imagination," 53.
53. Ibid., 48.

data?) for Caston. This is a modern "causal theory of perception" with a vengeance. But I repeat my skeptical question: is it Aristotle?

On the interpretation I propose, these problems vanish. A sensation is a perceptual taking in of a property that is externally manifested. That is its *nature*. When I have an illusion, I have "an experience phenomenally like a perceptual experience" in the sense that *it seems to me as if I were having a perceptual experience* (the same properties are called to mind), but not in the sense that there is a fundamental kind of particular, an "experience," that is in my mind. In short, Aristotle, like McDowell today, has a "disjunctive" account, one that does not postulate any "highest common factor" between an illusion and a veridical perception.[54] (I suspect that McDowell, who is a fine Aristotle scholar, *got* his account from Aristotle, in fact.) Thus the "infallibility" of sensation is a conceptual matter, not a problematic empirical claim. And the intentionality of sensation lies in the fact that it is a taking in of the very sensible quality that it is said to be "about." Of course, the presence of that quality is one of the causes of its being taken in; but that cause is singled out by *what sensation is*.

Re (3): Here is Caston's account of how the thinking part of the soul extracts universals from sensory experience:

> To arrive at this higher level of representation requires a different power in Aristotle's opinion, the power of conception or understanding which grasps part of a phantasma's content to the exclusion of others in a new mode—again, a form of transduction. It is unlikely it does this in the way imagined by later Aristotelians, by literally stripping away matter from the phantasma, and leaving the bare concept. To the extent that Aristotle himself says anything on the subject, he seems to have in mind, not so much the production of a separate entity, but a different way of handling the phantasma, by *ignoring* certain features. Different phantasmata, that is, *can be treated as equivalent,* insofar as they each have a certain part of their content in common.[55]

---

54. [2011]: Here I was too quick to identify Aristotle with "disjunctivism." My description of Aristotle's position was, I believe, correct, but this is not a "disjunctivist" position, because there obviously is a "highest common factor between an illusion and a veridical perception" in this account: the same forms are involved, or, as I put it, "The same properties are called to mind," and that is a highest common factor in precisely McDowell's sense.

55. Caston, "Aristotle and the Problem of Intentionality," 285.

Although the notion of *ignoring certain features of a phantasma* is indeed Aristotelian, it is noteworthy what Caston adds and what he omits from the passage he cites (*De mem. et remin.* 449b30–450a14). What he adds is his own identification of ignoring certain features of a phantasma with *treating different phantasmata as equivalent;* which is, of course, the British empiricist *substitute* for any use of the notion of attention to a universal, or, indeed, of a universal at all. Because the only content phantasmata have, on Caston's account, is the content of sensations, and sensations have only sensible qualities as content (or so Caston claims), it follows from his interpretation that the mind has only the resources available to it that Berkeley or Hume would have allowed. What Caston omits is the reference to abstraction (using an image that is quantitative to think of an object that is not quantitative by "thinking of it in abstraction from quantity"[56]—note that this says nothing about "treating different phantasmata as equivalent").

This concludes my defense of the "direct realist" reading of Aristotle against the only serious contemporary challenge that I know of.

---

56. Aristotle *De mem. et remin.* 450a5, Revised Oxford Translation. What the Greek says is that although we do not think of it (the object) as something having quantity, we present it to ourselves visually as a quantity.

35

Functionalism: Cognitive Science or Science Fiction?

There is an ancient form of "functionalism"—so Martha Nussbaum and I have argued—that can be found in Aristotle's *De anima*.[1] This is the view that our psyches can best be viewed not as material or immaterial organs or things but as capacities and functions and ways we are organized to function. In that wide sense of the term, I am still a functionalist. In this chapter, however, I will be considering a contemporary rather than an ancient way of

I have now [2011] only one disagreement with this paper, published in 1997: the "shudder quotes" around the words "cognitive science" in a number of places suggested that cognitive science consists of nothing but reductionist and scientistic errors, and that nothing in it deserves to be called "science." I would now say that although there are *instances* of scientism and some reductionist fantasies in present-day cognitive science, as in every human science, cognitive science is an exciting and many-sided field, and that even when a reductionist program "does not work," as my own computational functionalism did not work, we learn something valuable by investigating it. My present, much more optimistic view of the prospects for cognitive science and for what I call a "liberal" nonreductive functionalism are described in Chapter 3.

1. Cf. Hilary Putnam and Martha Nussbaun, "Changing Aristotle's Mind," in Martha C. Nussbaum and Amelie Rorty, eds., *Essays on Aristotle's "De Anima"* (Oxford: Oxford University Press, 1992), 27–56.

specifying what it is to be a functionalist, one I introduced in a series of papers beginning in 1960.[2]

The leading idea of this more recent view is that a human being is just a computer that happens to be made of flesh and blood, and that the mental states of a human being are its computational states. In this chapter I will try to explain why I was led to propose functionalism as a hypothesis, and why I no longer think that this contemporary sort of functionalism is correct.[3]

Different functionalists have tried to make this leading idea precise in different ways. In my own functionalist writings I exploited two ideas that I still find very important in the philosophy of science: the idea of "theoretical identification," and an idea that I introduced to go with it, the idea of "synthetic identity of properties."[4] The relation between our mental properties and our computational properties (or a subset of them) is, I suggested, just this kind of "synthetic identity" (here some functionalists, notably David Lewis, would disagree and insist that it must be some sort of conceptual or analytic identity), and functionalism itself, I maintained, should be viewed as an empirical hypothesis, on all four feet with the hypothesis that light is electromagnetic radiation, and not as a piece of conceptual analysis.

The reason that I was never attracted to the idea that functionalism is correct as a matter of analytic truth or even, in some looser way, as a clarification or "explication" of our ordinary psychological concepts is, in part, that in other areas of science we know that it is wrong to think that statements that make theoretical identifications of phenomena originally described in

2. Hilary Putnam, "Minds and Machines?" in Sidney Hook, ed., *Dimensions of Mind* (New York: New York University Press, 1960); Putnam, "Robots: Machines or Artificially Created Life?" *Journal of Philosophy* 61 (November 1964): 668–691; Putnam, "The Nature of Mental States," published as "Psychological Predicates" in W. H. Capitan and D. D. Merrill, eds., *Art, Mind, and Religion* (Pittsburgh: University of Pittsburgh Press, 1967), 37–48; and Putnam, "The Mental Life of Some Machines," in Hector-Neri Castañeda, ed., *Intentionality, Minds, and Perception* (Detroit: Wayne State University Press, 1967), 177–200. All of these are reprinted in Hilary Putnam, *Philosophical Papers*, vol. 2, *Mind, Language and Reality* (Cambridge: Cambridge University Press, 1975), 386–407, 429–440, 362–385, 408–428, from which I quote.

3. I explained at length why I gave up this view in Hilary Putnam, *Representation and Reality* (Cambridge: Cambridge University Press, 1988).

4. For a detailed discussion, see Hilary Putnam, "On Properties," in Nicholas Rescher, ed., *Essays in Honor of Carl G. Hempel: A Tribute on the Occasion of His Sixty-fifth Birthday* (Dordrecht: Reidel, 1970), 235–254; reprinted in Putnam, *Philosophical Papers*, vol. 1, *Mathematics, Matter and Method,* (Cambridge: Cambridge University Press, 1975), 305–322.

different vocabularies must be conceptual truths or must follow conceptually from non-question—begging empirical facts in order to be true. Light is electromagnetic radiation; but this is no "conceptual truth."

I introduced the notion of synthetic identity of properties because I wish to be able to say that not only is light passing through an aperture the same *event* as electromagnetic radiation passing through the aperture, but that the *property of being light* is the very same property as *the property of being electromagnetic radiation of such and such wavelengths*. In sum, I hold that properties can be synthetically identical, and that the way in which we establish that properties are synthetically identical is by showing that identifying them enables us to explain phenomena we would not otherwise be able to explain.

Applying this idea to the philosophy of mind, I proposed as a hypothesis that just as light has empirically turned out to be identical with electromagnetic radiation, so psychological states are empirically identical with functional states. Here is the hypothesis as I stated it at the time (for simplicity I stated it only for the case of pain, but I made clear that it was intended to hold for psychological states in general):[5]

1. All organisms capable of feeling pain are Probabilistic Automata.[6]
2. Every organism capable of feeling pain possesses at least one Probabilistic Automaton Description of a certain kind (i.e., being capable of feeling pain is possessing an appropriate kind of functional organization).[7]
3. No organism capable of feeling pain possesses a decomposition into parts that separately possess Probabilistic Automaton Descriptions of the kind referred to in (2).[8]
4. For every Probabilistic Automaton Description of the kind referred to in (2), there exists a subset of the sensory inputs such that an organism

5. Putnam, "Nature of Mental States," 434.

6. A Probabilistic Automaton is a device similar to a Turing machine except that (1) its memory capacity has a fixed finite limit, whereas a Turing machine has a potentially infinite external memory; and (2) state transitions may be probabilistic rather than deterministic. In this paper (see note 5) I assumed that the Probabilistic Automata in question were equipped with motor organs and with sensory organs.

7. A Description of a Probabilistic Automaton specifies the functional states of the Automaton and the transition probabilities between them.

8. Note that this rules out Searle's "Chinese room." See John Searle, "Minds, Brains and Programs," *Behavioral and Brain Sciences* 3, (1980): 417–457.

with that Description is in pain when and only when some of its sensory inputs are in that subset.

If this is an empirical hypothesis, however, then the questions one must ask are: How is the hypothesis to be empirically investigated? And what would the *verification* of such a hypothesis look like? My answer at the time was that to investigate the functionalist hypothesis, what we have to do is "just to attempt to produce 'mechanical' models of organisms—and isn't this, in a sense, what psychology is all about? The difficult step, of course, will be to pass from models of *specific* organisms to a *normal form* for the psychological description of organisms—for this is what is required to make (2) and (4) precise. But this too seems an inevitable part of the program of psychology."[9]

Notice that no argument was offered for the idea that the task of psychology is to produce "mechanical models of organisms" (in software, not hardware, terms, of course). Indeed, although I soon recognized that Turing machines and Probabilistic Automata could not possibly serve as such models, I held on for a long time to the idea that it is "an inevitable part of the program of psychology" to provide a "normal form" for those "mechanical models." By this I did not simply mean a normal form for the description of the psychology of human beings, although that now seems to me to be a completely utopian idea, but a normal form for the psychological description of an arbitrary organism.

So that is how we were supposed to investigate empirically the functionalist "hypothesis." We were supposed to construct "mechanical models" of species of organisms, including the human species, or at least the aspects that we take to constitute their psychological functioning; and then, by reflecting on the nature of these models, we were to pass to a normal form in which the "software description" of the psychological functioning of any physically possible organism could be written.

What about the question of empirical verification? This too seemed to me straightforward at that time. Consider how we verify that light is electromagnetic radiation of such and such wavelengths. What we do (here I still subscribe to a classic account) is to show that if we identify light with electromagnetic radiation of certain wavelengths, we can deduce the laws of unreduced optics—say, classical geometric optics or classical wave optics—to the extent that those laws were true (and, incidentally, to explain why they were not perfectly true as classically stated). In the same way, I thought, if we find a

9. Putnam, "Nature of Mental States," 435.

way to identify the properties we speak of in psychology—in ordinary language, or, say, clinical or behaviorist psychology—with computational properties (properties of our "software"), then we will be able to deduce the laws of these psychological theories, to the extent that they are true (and, incidentally, to show why and how they are not precisely true as presently stated).

However, as I have already mentioned, the normal form for the description of the psychology of an "arbitrary organism" cannot simply be the Turing machine formalism, for a number of reasons.

A "state" of a Turing machine is described in such a way that a Turing machine can be in exactly one state at a time. Moreover, memory and learning are not represented by states, in the Turing machine model, but by information printed on the machine's tape. Thus if human beings have any states at all that resemble Turing machine states, those states would have (1) to be states a human being can be in at any time, independent of learning and memory; and (2) to be totally instantaneous states of the human being, states that determine together with learning and memory (the contents of the "machine tape") what the next state will be, and thus totally specify the present psychological condition of the human being. Clearly, such states would neither be the familiar propositional attitudes nor the states postulated by any presently known psychological theory, be it clinical or behaviorist or cognitive or what have you. For other technical reasons, which I will not bore you with, the Probabilistic Automata formalism fares no better. But I claimed that it is not fatally sloppy to apply the notions of a functional description and of a "normal form" for functional descriptions to "systems" for which we have no detailed idea at present what such a description might look like—"systems" like ourselves. Although it is true that we do not now have a "normal form" in which all psychological theories can be written, we know, for example, that "systems" might be models for the same psychological theory without having the same physics and chemistry. A robot, a creature with a chemistry based on something that is not DNA, and a human being might conceivably obey the same or very much the same psychological laws. If they obeyed exactly the same psychological laws, they would be (in a terminology that I introduced in the series of papers I have been referring to) "psychologically isomorphic"; and psychologically isomorphic entities can be in the same psychological state without ever being in the same physical states. In short, psychological states are *compositionally plastic.*

So, in the functionalist hypothesis as spelled out by (1)–(4), it became necessary to replace the notion of a "Probabilistic Automaton description" by the notion of a "normal-form" description; and what I claimed was that we will know precisely what we mean by a "normal-form description" when we know what sort of description would be provided by an ideal psychological theory, where an ideal psychological theory is thought of as providing a "mechanical model" for a species of organism. This is what led me to think that it was an "inevitable part" of the "program of psychology" to provide mechanical models of species of organisms, and a normal form in which such models can be described.

The importance of this issue is this: if we deny that such a normal form can be provided, or that at least a "mechanical model" can be provided in the case of our own species, then the functionalist hypothesis cannot even be stated (not even if it is restricted to the human species). For what the project of functionalism (as opposed to, say, "eliminativism," that is, the denial that psychology, as opposed to neurology, has a legitimate subject matter) requires is not just that a computational description of the human brain and nervous system should be possible (an "eliminativist" may well think that that is true), but that such a description "line up" with psychology in the sense of providing computational properties that can be *identified* with psychological properties.

Psychology and Functionalist Speculation

The original idea of functionalism was that our mental states could be identified with computational states, where the notion of a computational state had already been made precise by the preexisting formalisms for computation theory, for example, the Turing formalism or the theory of automata. What I have just outlined is the manner in which the original idea of functionalism quickly became replaced by an appeal to the notion of an ideal "psychological theory." But this "ideal psychological theory" was conceived of as having just the properties that formalisms for computation theory possess.

A formalism for computation theory *implicitly defines* each and every computational state by the totality of its computational relations (e.g., relations of succession or probabilistic succession) to all the other states of the given system. In other words, the members of the whole set of computational states of a given system are *simultaneously implicitly defined;* and the implicit definition *individuates* each of the states in the sense of distinguishing it from all other computational states. But no psychological theory individuates or "implicitly defines" its states in this sense. Thus functionalism conceived of what it called an "ideal psychological theory" in a very strange way. No actual psychological theory has ever pretended to provide a set of laws that distinguish, say, the state of being jealous of Desdemona's fancied regard for Cassio from every other actual or possible propositional attitude. Yet this is precisely what the identification of that propositional attitude with a "computational state" would do. Thus functionalism brought to the study of the mind strong assumptions about what any truly scientific psychological theory must look like.

There is no reason to think that the idea of such a psychological theory (today such a theory is often referred to as "conceptual role semantics") is anything but utopian. There is no harm in speculating about scientific possibilities that we are not presently able to realize; but is the possibility of an "ideal psychological theory" of this sort anything more than a "we know not what"? Has anyone suggested how one might go about constructing such a theory? Do we have any conception of what such a theory might look like? Even if we had a candidate for such a theory, the question remains: how would *we* go about verifying that it does implicitly define the unreduced psychological properties? One hears a lot of talk about "cognitive science" nowadays, but one needs to distinguish between the putting forward of a scientific theory, or the flourishing of a scientific discipline with well-defined questions, from the proffering of promissory notes for possible theories that one does not know even in principle how to redeem.

Functionalism and Semantic Externalism

Ever since I wrote "The Meaning of 'Meaning'," I have defended the view that the content of our words depends not just on the state of our brains (be that characterized functionally or neurophysiologically), but also on our relations to the world, on the way we are embedded in a culture and in a physical environment.[10] A creature with no culture and no physical environment that it could detect outside its own brain would be a creature that could not think or refer, or at least (to avoid the notorious issue of the possibility of private language) could not think about or refer to anything outside itself. That, given our physiology and our environment, H_2O is the liquid we drink has everything to do with fixing the meaning of "water," I claim. *Au* is the substance that experts refer to as "gold"; and the cultural relations of semantic deference

---

10. Hilary Putnam, "The Meaning of Meaning" in *Mind, Language and Reality: Philosophical Papers,* vol. 2 (Cambridge: Cambridge University Press, 1975), 215–271.

between us laypersons and those experts has everything to do with fixing the reference of "gold" in our lay speech, I claim. Mere computational relations between speech events and brain events do not, in and of themselves, bestow any content whatsoever on a word, any more than chemical and physical relations do. But this implies that no mental state that has content (no "propositional attitude") can possibly be identical with a state of the brain, even with a computationally characterized state of the brain. Even if this is true, however, it has been suggested that one might abstract away from the content of a word in the sense I just used that notion (the sense in which content determines reference and, in the case of assertions, is what can be assigned the values "true" and "false") by simply *ignoring* all the external factors. The result of this abstraction is supposed to be a new notion of content, "narrow content" (the original notion being "wide content"); and it has been suggested that this notion of narrow content is the proper notion when our purpose is psychological explanation.

The advantage of this suggestion, from a functionalist point of view, is that "narrow content," by definition, has to do only with factors *inside* the organism; thus there is at least the hope that narrow contents might be identifiable with computational states of the organism, thus realizing a version of the original functionalist programs. But the suggestion has problems.

The key problem is totally obscured by the habit of tossing the term "narrow content" around in the literature as if the notion were really well defined. The problem is that we possess neither a way of individuating "narrow contents" nor a set of unreduced psychological laws involving "narrow contents" (unless the laws of "folk psychology" are supposed to be about "narrow contents"—a suggestion I find it hard to take seriously). But the very idea of a theoretical identification presupposes that the concepts to be reduced are already under some kind of scientific control (recall the case of optics or of thermodynamics). To introduce a set of concepts that at present figure in no laws (the "narrow contents" of our familiar propositional attitudes) and then immediately to begin talking of searching for theoretical identifications of these "narrow contents" with computational states of the brain (which, as we noted earlier, also have not been defined, because we have the problem of what formalism is being envisaged when one talks of "computational states" here) is to engage in a fantasy of theoretical identification. It is to mistake a piece of science fiction for an outline of a scientific theory that it only remains for future research to fill in.

In "The Meaning of 'Meaning' " I suggested that one might speak of "narrow contents" in the following way: For me to be in a mental state that has the "narrow content" "There is water on the table," or whatever the proposition p in question might be, is just for me to be in a total brain state such that (1) some person P with some language L might be thinking a sentence S that is *syntactically* the same as the sentence I am thinking; and (2) the sentence in the language L on the occasion of use by person P in question could be *translated* (preserving ordinary or "wide" content) as "There is water on the table" (or whatever p is in question). Thus if a Twin Earth speaker of Twin English whose "water" is actually XYZ and an Earthian whose "water" is H_2O are in the same brain state when they think the words "Water is on the table," we would say (on this proposal) that their words have the same "narrow content" even though they refer to different liquids as "water," because the Twin Earthian is in the same brain state as someone—say, myself—whose words "Water is on the table" have the *wide* content in question. Another proposal, advanced by Jerry Fodor, is to say that two thoughts have the same "narrow content" if the *wide content* of the thoughts in question (or of the corresponding utterances, either in natural language or in "mentalese") vary in the same way with environmental and cultural parameters. Like my proposal, this makes the notion of narrow content parasitic on the ordinary ("wide") notion of content. Neither proposal provides an *independent* notion of narrow content that could be the subject of psychological theorizing.[11] Ned Block has proposed that "narrow contents" might be identified with "conceptual roles."[12] But we lack an unproblematic conception of what we mean by the "conceptual role" of a sentence—especially if we are skeptical about the analytic/synthetic distinction—and it is also contentious to claim that "conceptual roles," if the notion *can* be made precise, in any way correspond to the contents of propositional attitudes.

Of course, one might decide to drop the notion of "narrow content" and say: "Very well, then. If mental states are individuated by contents that are themselves partly determined by the community and the environment, then let us widen the functionalist program and postulate that mental states are identical with *computational-cum-physical states of organisms plus communities*

---

11. Fodor, however, thinks that "wide content" can be defined in causal terms. See Jerry Fodor, *A Theory of Content and Other Essays* (Cambridge, Mass.: MIT Press, 1990). For a criticism of his proposal, see Hilary Putnam, *Renewing Philosophy* (Cambridge, Mass.: Harvard University Press, 1992), chap. 3.

12. Cf. Ned Block, "Advertisement for a Semantics for Psychology," *Midwest Studies in Philosophy* 10 (1985): 615–678. Cf. Putnam, *Representation and Reality,* 46–56, for a discussion.

plus environments." But how useful is it to speak of "computational-cum-physical states" of such vast systems?

The Issue of Utopianism

I have charged functionalism with utopianism, with being "science fiction" rather than the serious empirical hypothesis that, in my early papers, I hoped to provide. I have also suggested that bringing in such concepts as "narrow content" and "conceptual role" only drains the functionalist proposal of its original substance—it turns it into a case of using concepts that stand for we know not what as if they had serious scientific content. An example may help make it clear what I mean by this charge.[13]

Suppose that we encounter a primitive culture in which people are observed to say (or think) "Sheleg" when it snows. Folk psychology—and, we may assume, "the ideal psychological theory" as well, if there is such a thing—tells us that when it snows, people are likely to say and think "It is snowing." So it is certainly compatible with psychological theory that these people are saying (and thinking) that it is snowing when they say and think "Sheleg." But, of course, there are other possibilities. What a characterization of the assertibility conditions and the inferential relations (these are usually what people have in mind when "conceptual role" is spoken of) between these Sheleg sayings and thinkings and the other sayings and thinkings and doings of these people is supposed to accomplish is to provide us with a way of *individuating* the propositional content of "Sheleg" in the sense of distinguishing that content from all other possible contents.

But distinguishing a content from all other *possible* contents on the basis of any finite supply of facts about inferential relations and prima facie assertibility conditions is a tall order. Let us suppose that the members of this tribe have a religion according to which the one infallible sign of the anger of their gods is the falling of snow, and that what "Sheleg" actually means is that the gods are angry. If this is the case, then, of course, it will make a difference in how these natives talk. But can we say in advance just *what* difference it will make? Can we (or an "ideal psychological theory" that we could envisage being able to construct) survey all the possible differences it *could* make?

13. I take this example from Hilary Putnam, "Putnam, Hilary," in Samuel Guttenplan, ed., *Companion to the Philosophy of Mind* (Cambridge: Cambridge University Press, 1994), 507–513. I have borrowed a few sentences from that paper in this chapter.

Such a theory would have to be able to describe *the beliefs of a believer of any possible religion.* Or, to take a different example, consider the case of a tribe all of whose members are superscientists. They may be saying "Quantum state such and such" when it snows, or making a comment in a physical theory we do not have yet. A psychological theory that is able to individuate the contents of such sayings and doings would presuppose knowledge of a physical theory that we have not even imagined. In short, it looks as if an ideal psychological theory, a theory that would be able to determine the content of an arbitrary thought, would have to be able to describe the content of every belief of every possible kind, or at least every human belief of every possible kind, even of kinds that are not yet invented, or that go with institutions that have not yet come into existence. That is why I say that the idea of such a theory is pure "science fiction."

What finally pushed me over the antifunctionalist edge was a conversation I had one day with Noam Chomsky. Chomsky suggested that the difference between a rational or a well-confirmed belief and a belief that is not rational or not well confirmed might be determined by rules that are innate in the human brain. It struck me at once that it ought to be fairly easy to show, using the techniques Gödel used to prove his incompleteness theorems, that if Chomsky is right, then we could never discover that he is right; that is, if what it is rational or not rational to believe is determined by a recursive procedure that is specified in our ideal competence description, then it could never be rational to believe that *D is* our ideal competence description. And I was able to show that this indeed is the case without too much trouble.[14]

This argument does not apply directly to the present discussion, because what we are discussing here is not determining what is well confirmed but determining what the *contents* of thoughts are. The Gödel incompleteness theorems show that any description of our logical capacities that we are able to formalize is a description of a set of capacities that we are able to go beyond. But if this is true of our deductive logical and inductive logical capacities, why should it not be true of our interpretative capacities as well? Interpretation involves decisions that, interwoven as they are with our understanding of all the topics we can think about, are unlikely to be susceptible of *complete* formalization. It is true that we cannot *prove* that functionalism cannot be

---

14. Hilary Putnam, "Reflexive Reflections," *Erkenntnis* 22, no. 1 (January 1985): 143–154; reprinted in Putnam, *Words and Life,* ed. James Conant (Cambridge, Mass.: Harvard University Press 1985), 416–427.

made less vague than it presently is, but neither do we have any reason to believe that it can. In short, it seems to me that if there is an "ideal psychological theory," that is, a theory that does everything that the functionalist wants a "description of human functional organization" to do *(let alone* a "normal form for the description of the functional organization of an arbitrary organism"), then there is no reason to believe that it would be within the capacity of *human beings* to discover it. But it is worth pausing to notice that the *notion* of a "complete description of human functional organization" is itself a tremendously unclear notion. The idea that there *is* such a "complete description," even if a recognition procedure for it does not exist, surely goes beyond the bounds of sense.

Of course, those who are sympathetic to functionalism have not given up as a result of my recantation; there are a number of what we might call "postfunctionalist" programs on the market. One kind of postfunctionalist program seeks to avoid the difficulties inherent in the idea of "implicit definition by a theory" that was at the heart of classic functionalism[15] by relying *entirely* on external factors to fix the contents of thoughts. Dretske[16] and Stalnaker,[17] for example, try to define the content of thoughts, as well as of expressions in a language, by simply looking for *probabilistic relations* between the occurrences of thoughts and expressions and external states of affairs, bypassing entirely the question of the functional organization of the speaker, which would presumably come in at a later stage in the account. But both Loewer[18] and I[19] have argued that the information-theoretic concepts on which Dretske and Stalnaker rely cannot individuate contents finely enough.

---

15. This idea was central not only to my version of functionalism but also to the somewhat different version proposed by David Lewis in "Psychophysical and Theoretical Identifications," *Australasian Journal of Philosophy* 50, no. 3 (1972): 249–258, and in part 2 of Lewis, *Philosophical Papers,* vol. 1 (Oxford: Oxford University Press, 1983). In Lewis's view, we already *have* the ideal psychological theory required to implicitly define the content of an arbitrary thought: it is just folk psychology.

16. Fred Dretske, *Knowledge and the Flow of Information* (Cambridge, Mass.: MIT Press, 1981); Dretske, "Misrepresentation," in R. J. Bogdan, ed., *Belief* (Oxford: Oxford University Press, 1986), 17–36.

17. Robert Stalnaker, *Inquiry* (Cambridge Mass.: MIT Press, 1984).

18. Barry Loewer, "From Information to Intentionality," *Synthese* 70, no. 2 (1987): 287–316.

19. Cf. Hilary Putnam, "Computational Psychology and Interpretation Theory," in Rainer Born, ed., *Artificial Intelligence: The Case Against* (London: Routledge, 1987), 1–17; reprinted in Putnam, *Philosophical Papers,* vol. 3, *Realism and Reason* (Cambridge: Cambridge University Press, 1983), 139–154.

In response to this problem, Fodor proposes to rely not on information-theoretic notions, but instead on the notion of causality.[20] In *Renewing Philosophy* I argue (1) that the notion of causality Fodor employs itself presupposes intentional notions, and (2) that the assignments of contents that result if we look only at the causes of utterances are the wrong ones.[21]

The Relevance of Interpretive Practice

Before I close, there is one objection to my whole line of argument that I need to consider. It can be stated as follows: I have assumed that our ordinary practice of interpretation—what is often called "translation practice," although what is involved in interpretation is certainly much more than *translation*[22]—*is* the appropriate criterion for identifying the content of propositional attitudes. But is this not too "unscientific" a criterion for the purposes of a scientific psychology?

Let me begin by considering just one aspect of this large question. In "The Meaning of 'Meaning'" I argued that it would be in accordance with standard interpretive practice to say that the term "water" on Earth has the extension H_2O (give or take various impurities), and the term "water" on Twin Earth has the extension XYZ (give or take various impurities).[23] Because these are names of different liquids (much in the way in which "molybdenum" and "aluminum" are names of different metals), interpretive practice would conclude that believing that a lake is full of "twater" (Twin Earth water) is a different propositional attitude than believing that the lake is full of (Earth) water. And this is perfectly reasonable if the Earthians and the

20. Jerry Fodor, *A Theory of Content*. Note that Fodor needs to assume a distinction between contributory causal factors and *the* cause of an event. I regard this as an intentional notion because what is *the* cause of an event depends on the interests we have in the context; it is not something that is inscribed in the phenomena themselves.

21. Putnam, *Renewing Philosophy*, chap. 3.

22. More is involved in interpretation than just translation because, for one thing, in our own language we may be able to describe the extension of a term that we cannot translate into it (e.g., I can say in English "In Choctaw, *wakai* is the name for a kind of snail," but I may not know whether there is a word in English for that kind of snail). Moreover, paraphrase and even commentary are part of interpretation and are relevant to our identification of propositional attitudes.

23. Putnam, "The Meaning of 'Meaning'," 215–271. Actually the practice is more complex than this; for a discussion, see Putnam, *Representation and Reality*, 30–33, and Putnam, *Realism with a Human Face*, ed. James Conant (Cambridge, Mass.: Harvard University Press, 1990), 282.

Twin Earthians in question are scientifically sophisticated. But suppose that we are dealing with speakers who do not yet know about the difference between Earth water and Twin Earth "twater." Surely no psychologist would regard the fact(s) that water is H_2O and twater is XYZ as relevant to the subject in *this* case. For psychological purposes, should we not say that the Earth subject and the Twin Earth subject who both assent to the sentence "The lake is full of water" have the *same* "belief"? Most "cognitive scientists" would certainly answer yes.

But what exactly is the force of "for psychological purposes" here? If a psychologist finds that one of her subjects believes that molybdenum is a light metal and that another subject believes that aluminum is a light metal, must not the psychologist find out whether either subject has any beliefs that differentiate aluminum from molybdenum? Otherwise, has she not failed to determine whether the first subject's belief is "the same belief" as the second subject's "for psychological purposes"? Or is the fact that each subject believes that molybdenum is *called* "molybdenum" and aluminum is *called* "aluminum" *enough* to make these "different beliefs for psychological purposes"? Would the Earthian's belief and the Twin Earthian's belief become "different beliefs for psychological purposes" if Twin Earth English had a different *word* for the liquid? So if Twin Earth English has the word "twater" rather than "water," the belief becomes different *even if the liquid is the same.*

Compare the situation in evolutionary biology. Ernst Mayr has long urged that the evolutionary biologist employs what is essentially the *lay* notion of a "species," and that the idea of replacing that notion by a more "scientific" notion (or defining it "precisely") is misguided.[24] The nature of the theory of evolution, with its antiessentialism and its emphasis on variety, explains why there is not and cannot be a sharp line between a "species" and a "variety" (many populations are classified as "species" by one classification and as "varieties" by another). Perhaps it might seem more precise to replace talk of species by talk of "reproductively isolated populations," but then exactly what would count as a significant "population" (three rabbits in a cage?), and what as significant "reproductive isolation" (would purebred golden retrievers count as an object of evolutionary theory?)? If evolutionary biology, a science with far better credentials than "cognitive science," better serves the task of accounting for what we are interested in by sticking to ordinary imprecise ways

24. Ernst Mayr, *Evolution and the Diversity of Life* (Cambridge, Mass.: Harvard University Press, 1976).

of speaking, is it really necessary or desirable for cognitive psychology to depart from ordinary ways of individuating beliefs and other propositional attitudes?

But supposing that it sometimes is, we are still not driven to the fantastic supposition that we are in possession of or can usefully imagine a way of individuating propositional attitudes in utter independence from our normal standards and practices. If we want, for certain purposes, to ignore all facts about water and twater that are not known to the subjects in a particular experiment, we can decide to do so. But even if we identify the content of Earth English "water" (respectively, Twin Earth English "twater") with some salient set of beliefs about the liquid(s) that are known to the subject(s), as Akeel Bilgrami proposes in his *Belief and Meaning*,[25] *those* beliefs must ultimately be identified by the unformalized (and probably unformalizable) standards implicit in ordinary interpretive practice. We do not have to engage in science fiction by imagining that "cognitive science" suddenly delivers a way of individuating propositional attitudes wholly independent of the "whirl" of our spontaneous interests and reactions.[26] As Bilgrami argues, the result is not a notion of "content" that ignores the role of the environment in individuating beliefs; it is simply a notion that employs facts about the external referents of various terms *only when those facts are known to the subject.* This is still an "externalist" notion of content, even if it departs, in motivated ways, from the way content is (sometimes) individuated in translation.

In fact, this is not the first time that one of the moral sciences has been seduced (or at least tempted) by the dream of laws and concepts as rigorous as those of physics. Even before he hit on the term "sociology" for the subject he

---

25. Akeel Bilgrami, *Belief and Meaning* (Oxford: Blackwell, 1992).

26. In speaking of the "whirl" of our interests and reactions here, I am thinking of the following description of Wittgenstein's view: "We learn and teach words in certain contexts, and we are expected, and expect others, to be able to project them into further contexts. Nothing insures that this projection will take place (in particular, not the grasping of universals nor the grasping of books of rules), just as nothing insures that we will make, and understand, the same projections. That on the whole we do is a matter of our sharing routes of interest and feeling, sense of humour and of significance and of fulfilment, of what is outrageous, of what is similar to what else, what a rebuke, what forgiveness, of when an utterance is an assertion, when an appeal, when an explanation—all the whirl of organisms Wittgenstein calls 'forms of life.' Human speech and activity, sanity and community, rest upon nothing more, but nothing less than this. It is a vision as simple as it is difficult, and as difficult as it is (and because it is) terrifying." Stanley Cavell, *Must We Mean What We Say?* (New York: Charles Scribner's Sons, 1969), 52.

was proposing, Auguste Comte proposed that our social theories and explanations would reach such a state, and he confidently proposed that before long all our social problems would be solved by "savants" acquainted with the new social physics.[27] That the greatest sociologist of the subsequent century, Max Weber, would employ in his work such imprecise terms as "Protestantism" and "mass party"—terms that require interpretive practice for their application—would have seemed to Comte a betrayal rather than a fulfillment of his hopes for the subject. Twentieth-century positivists extended Comte's dream from sociology to history, insisting that eventually history would be systematized and made explanatory by the application of "sociological laws." Yet today we have evolutionary biologists such as Ernst Mayr and Stephen Gould arguing that evolutionary biology should not be expected to look like physics *because* evolutionary explanations are historical explanations, and, they say, we cannot expect history to resemble physics. Clearly, no one—or almost no one—still hopes that history will be able to dispense with notions that depend on what I have called "interpretive practice"—notions like the Renaissance or the Enlightenment, or, for that matter, the nation-state or ethnic conflict. I am convinced that the dream of a psychological physics that seems to be thinly disguised under many of the programs currently announced for "cognitive science" will sooner or later be realized to be as illusory as Comte's dream of a social physics.[28]

27. Cf. Auguste Comte, "Sciences and Savants," in Ronald Fletcher, ed., *The Crisis of Industrial Civilization: The Early Essays of August Comte* (London: Heinemann, 1974).

28. I do not, however, claim that the scientism I am criticizing comes directly from the idea of imitating physics. Two more direct sources are (1) the idea that our propositional attitudes will someday be explicated by such things as "belief boxes" and "desire boxes" in the brain (and their contents will be individuated by formulas in "mentalese"); and (2) more fundamentally, the idea that psychology will be absorbed in computer science—the very idea that was behind my own functionalist utopianism. With respect to (1), let me remark that the whole idea of "mentalese" depends on the idea that there could be a language (the "language of thought") with the property that the contents of its sentences are completely insensitive to context of use. If "mentalese" lacks this property, then "sentences in a belief box" cannot serve as a record of beliefs by themselves; two subjects may have the same sentences in their "belief box," if that metaphor makes any sense, and have different beliefs. On this, see Putnam, "Computational Psychology and Interpretation Theory." For an argument that no language has this kind of context insensitivity (and, I would add, we do not have any idea what a language with context insensitivity would be—yet another example of the constant appeal to concepts that look "scientific" but refer to we know not what in "cognitive science," see Charles Travis, *The Uses of Sense* [Oxford: Oxford University Press, 1981]).

36

How to Be a Sophisticated "Naïve Realist"

Why Were You Initially Drawn to Philosophy of Mind?

In the late 1950s an argument for dualism known as the "grain argument" was advanced in the philosophy of mind.[1] (The objection—that it is "unintelligible" to suppose that qualities whose "grains" are as different as those of neural properties and phenomenal properties are, in reality, identical—is an early ancestor of both Frank Jackson's "knowledge argument" and Tom Nagel's "what it's like to be a bat" argument.[2]) Although my "Minds and Machines"

New material now incorporated in section 5, "What Are The Most Important Open Problems in Contemporary Philosophy of Mind?," was formulated during a seminar I gave in 2009 in Tel Aviv, when I became convinced by Ned Block's paper "Inverted Earth" (*Philosophical Perspectives* 4 [1990]: 53–79), having already accepted his "Wittgenstein and Qualia" (*Philosophical Perspectives,* 21 [2007]: 73–115): in particular by his claim that there is a consilience of scientific and philosophical considerations supporting the idea that "qualia" (phenomenal characters of sensory experiences) are identical with brain events/states.

1. A good (later) account is Michael Green, "The Grain Objection," *Philosophy of Science* 46 (1979): 559–589.

2. Frank Jackson, "What Mary Knew," *Journal of Philosophy* 83. no. 5, (1986): 291–295; Thomas Nagel, "What Is It Like to Be a Bat?," *The Philosophical Review* 83, no. 4, (1974): 435–450.

was later seen as important—by myself, as well as by others, primarily because it suggested the functionalist account of mental states—the reason
I wrote it was to argue that *if the grain argument is right, it is available to a
robot as well,* and hence the form of "dualism" it is supposed to establish
could not be a threat to materialism.[3] Basically, I argued that the underlying
"dualism" is the "dualism" of *knowing about a psychological property via a
description* and *"knowing" about it by exemplifying it oneself,* and that this is
an inevitable dualism, even for machines. (I still think that this argument is
sound, by the way.)

Having dipped my toe in the water of philosophy of mind by writing that
paper, I was led to go on and reflect further on functionalism, the synthetic
identity of properties, and related topics. Another source of my interest in
philosophy of mind was a revulsion against behaviorism, particularly Norman
Malcolm's pseudo-Wittgensteinian version thereof, which attracted a lot of
attention at the time.

What Do You Consider Your Most Important
Contribution to the Field?

Proposing functionalism. Although I am no longer a *computer-program* functionalist, the idea that *our mental states are best conceived of as ways of functioning and exercises of those ways* still seems right to me, although not in the "internalist" (and reductionist) sense that went with the model of those states as
"the brain's software."

The reasons I gave up that "model" are three: (1) It cannot be the case that
there is a one-to-one mapping of such mental attributes as believing something, hoping for something, or desiring something onto precise kinds of
software, as functionalism hoped. If such states are "realizable" in software at
all, they are so in infinitely many different ways. We might call this *the computational plasticity* of mental states. (2) According to the "externalist" theory
of reference" I developed in "Is Semantics Possible?" and "The Meaning of
'Meaning,'" reference and meaning are not simply in our heads; meaning
and reference are "transactional," that is, they depend on both the organism
and the environment, and they cannot be simply read off from our brains
without looking at the kinds of interactions that take place among the brain,

3. Hilary Putnam, "Minds and Machines," *Philosophical Papers,* vol. 2, *Mind, Language
and Reality* (Cambridge: Cambridge University Press, 1975), 386–407.

the rest of the organism, and the environment.[4] If they are functional states in some sense (as I believe that they are), they are functional states with "long arms"; that is, they are *environment-involving* ways of functioning. (3) As a corollary of (1) and (2), the crucial notion of "sameness of content" between thoughts cannot be simply a matter of sameness of "program."

In any case, the question whether our minds/brains are best thought of as computers is important and exciting. Here, happily, is an area in which philosophers and scientists do talk to each other and recognize the profit in doing so. Also, although I now think that the computer-program functionalism I proposed was too simple, it still seems to me an excellent entering wedge into the philosophy of mind in our postcomputer age.

What Is the Proper Role of Philosophy in Relation to Psychology, Artificial Intelligence, and the Neurosciences?

To be a gadfly, of course.

Seriously, I learned from my teacher, Hans Reichenbach, that the most exciting task of philosophy of science is to combine clarification of the concepts of science with reflection on the implications of scientific theories, both proposed theories and theories that are not considered to be confirmed, for great metaphysical issues. That seems to me to be its role whether the science in question is physics or psychology or artificial intelligence or neuroscience.

Is a Science of Consciousness Possible?

I cannot answer this question with a simple yes or no. If the presupposition is that consciousness is a "mystery," I do not agree. I am still a "functionalist" in the wide sense that I think that having a functional organization of the kind humans typically have or sufficiently similar to what humans typically have (of course, that is really a "family-resemblance" affair) is all there is to having human consciousness (and analogously for chimpanzee consciousness, pussycat consciousness, and so on). The familiar objections (e.g., the "knowledge argument," the "zombie" argument, Kripke's argument in *Naming and Neces-*

---

4. Putnam, "Is Semantics Possible?" and "The Meaning of 'Meaning,'" in *Mind, Language and Reality*, 139–152, 215–271.

sity) fail to convince me. But, of course, there is a great deal to be *learned* about what sort of (environment-involving) functional organization that is.[5] For me, that is the real "problem of consciousness."

What Are the Most Important Open Problems in Contemporary Philosophy of Mind? What Are the Most Promising Prospects?

I can say only that the problem I find most *interesting* is the problem of perception. But everyone has his own pet problem. And there is an enormous amount of work to be done no matter what problem one finds interesting or what approach one favors.

Hilla Jacobson of Ben-Gurion University and I are currently working on a monograph on perception. Here is a sketch of our approach, which we call "transactionalism."

First, by way of background, the current approaches that Jacobson and I take most seriously are *intentionalism* (Dretske, Tye, and many others), *disjunctivism* (McDowell, Martin, and many others), and *phenomenism* (Ned Block, in particular). We see each of these approaches as having insights and oversights, and we believe that our approach will preserve the insights and avoid the oversights. One great advance in the discussion, in our opinion, is that none of the philosophers just mentioned think of perceptual experiences as objects in an "inner theater" that we "observe" with the aid of an inner perceptual capacity called "introspection." As Reichenbach long ago argued our perceptual experiences are states that we have, not objects that we observe[6]; and as he further argued, our awareness of our perceptual experiences does not issue in "incorrigible" reports.[7] These points are now widely accepted, although Reichenbach's anticipation of them seems to have been forgotten.

But there are further issues about which there continues to be serious disagreement. One concerns the question whether perceptual experiences and

---

5. Saul Kripke, *Naming and Necessity* (Cambridge, Mass.: Harvard University Press, 1980).

6. Hans Reichenbach, *Experience and Prediction* (Chicago: Chicago University Press, 1938).

7. Hans Reichenbach, "Are Phenomenal Reports Absolutely Certain?," *The Philosophical Review* 61, no. 2 (1952): 147–159. See also Hilary Putnam, "Reichenbach and the Myth of the Given," in Putnam, my *Words and Life* (Cambridge, Mass.: Harvard University Press, 1994), 115–130.

their "relatives" (illusions, hallucinations, and the like) are just "flat psychological surface,"[8] with no intrinsic connection to anything in the environment.
This is Ned Block's picture (the picture of these experiences as "mental
paint").[9] We, along with both the disjunctivists and the intentionalists, hold
that such a description, in addition to being inadequate to the phenomenology of such experiences, gives too much away to skepticism of both the Cartesian and the Kantian varieties.[10] In addition, intentionalists and disjunctivists disagree about whether there is a "highest common factor" present in
veridical perceptions and hallucinations/illusions. Both intentionalists and
disjunctivists defend "naïve realism," however, which they identify (mistakenly, in our view) with the idea that the description of the perceived public
property also exhausts the qualitative character ("phenomenal character") of
the experience, while Block argues (correctly, in our view) that it does not.

Rejigging Block's Thought Experiment

First let me explain Block's insight. Block's paper "Inverted Earth"[11] uses an
ingenious thought experiment to argue that it is perfectly conceivable that
identical objective colors (for example, the blue of the sky, the red of blood,
or the green of grass) could correspond to very different "color qualia" in the
case of different individuals. (To some extent, that this actually is the case is
supported by empirical evidence. For example, in a classic experiment much
discussed in the literature on color, Hurvich, Jameson, and Cohen found large
variation in the location of unique green.)[12] In Block's thought experiment

8. William James, "The Tigers in India," in James, *The Meaning of Truth* (New York:
Longmans, Green, 1909), 43–50 (extracts of a presidential address delivered at the American
Psychological Association and published in *Psychological Review*, 2 [1895]: 105–124), describes
conceptual representations in this way, although he had a "natural realist" view with respect to
perceptual experiences.

9. At my 85th Birthday Conference (held at Harvard and Brandeis universities, May 31 to June
3, 2011), Ned Block pointed out that although, on his view, qualia are flat psychological surfaces,
he agrees with Jacobson and myself that there is more to experience than just qualia, and he said
that, moreover, he sees no incompatibility between his phenomenism and our transactionalism.

10. On the difference between Cartesian and Kantian skepticism, see James Conant,
"Varieties of Skepticism," in Denis McManus, ed., *Wittgenstein and Skepticism* (London:
Routledge, 2004), 97–134.

11. Ned Block, "Inverted Earth," *Philosophical Perspectives* 4 (1990): 53–79.

12. L. M. Hurvich, Dorothea Jameson, and J. D. Cohen, "The Experimental Determination of Unique Green in the Spectrum," *Perception and Psychophysics* 4 (1968): 65–68. "Unique

the environment is supposed to be altered by moving the subject at an early age to Inverted Earth, a planet on which each object has "the complementary color of [the corresponding object] on Earth. The sky is yellow, grass is red, fire hydrants are green, etc."[13] The inhabitants also speak Color Inverted English ("red" means green, "blue" means yellow, and so on). But an Earthian subject, call her Alice, is fitted (without her knowledge) with color-inverting lenses and then moved (without her knowledge) to Inverted Earth. Alice then grows up on Inverted Earth and learns to speak Color Inverted English. Eventually her concepts became those of her new home planet: "blue" in Alice's idiolect now means "yellow." But the qualia that (for Alice, with the color-inverting lenses) correspond to the experience of seeing the sky of Inverted Earth on a sunny day are not identical with the objective color of the sky on Inverted Earth (yellow), as they are supposed to be by those philosophers of perception (the "intentionalists" and "disjunctivists" who claim that in normal perception our "qualia" just *are* the objective colors of the things we look at. Yet there is nothing physically impossible about this scenario. So, Block argues, the radical externalism with respect to the phenomenal character of experience defended by these philosophers cannot be right.

One obvious objection to Block's thought experiment is that it depends on deceiving the subject; the subject does not know that artificial devices (the color-inverting lenses) have been inserted in her body. And even the most radical "naïve realist" should not deny that mirrors, lenses, and other artificial devices can make it *seem* that we are "perceiving things as they really are" when we are not.

A second, less plausible objection would be to say that when Alice becomes fully "acculturated" on Inverted Earth, her color qualia will no longer be as they were when she was still a child on Earth; instead, when she looks at the (Inverted Earth) sky, Alice now has "yellow" qualia (i.e., the qualia she would have had looking at the Inverted Earth sky had she been born on Inverted Earth and never had the color-inverting lenses inserted). However, because of the lenses inserted into her eyes, the inputs to her visual cortex are exactly as they would be if she were a normal Earth person looking at a normal Earth sky. So this "way out" of the problem Block poses for the extreme "naïve realists" (that is, externalists with respect to the phenomenal) would

---

green," for a subject, is a shade that the subject does not report as looking either "blue green" or "yellow green," but just plain green.

13. Block, "Inverted Earth," 60.

require them to deny that the phenomenal character of our color experiences is supervenient on our brain state, and this seems incompatible with everything we have learned about the role of the brain in perception in the past few centuries. But the first objection still stands.

To meet the first objection, let me present a different version of the Inverted Earth thought experiment that I find completely convincing. Block's Inverted Earth thought experiment might be modified as follows: Let us imagine that instead of color-inverting lenses being inserted in an Earthian's eyes by scientists, evolution developed something equivalent in the eyes of native Inverted Earthians. (Perhaps it is advantageous to see the sky as blue and not as yellow because of the psychological effects of the two experiences.) Then the normal biological function of the brain mechanism that recognizes blue on Earth will be to recognize yellow on Inverted Earth.[14] (At least that is how we describe the situation in Earth English.) This is certainly a problem for Tye,[15] because he insists that the representational content is identical with the phenomenal character, and also that the representational content in the case of colors is a property (of the surface) that is dispositional but not relational (a disposition to affect light in certain ways under normal conditions). But yellow and blue, so understood (understood in terms of the opponent process theory Tye favors), are certainly different colors, and so the phenomenal character of the experience the Earthian describes as "experiencing blue" has to be different from the phenomenal character of the corresponding experience of the Inverted Earthian, on Tye's theory, even though (because of the naturally evolved color-inverting modules in the eyes that compensate for the difference in the "objective" colors) it would seem to be the same.

For reasons I shall shortly describe, Jacobson and I agree with Block on the psychological facts, though not with his conception of visual experiences as "mental paint," but we consider ourselves to be—like both disjunctivists and intentionalists—defending (and reinterpreting) "naïve realism." It should be

---

14. Note that the color-inverting lenses that evolution developed on Inverted Earth (in *my* Inverted Earth scenario) represent a biological modification of the eyes and optic nerve *external to the brain itself*. Inverted Earthians' brain states when they look at their yellow sky (which they call a "blue" sky) are identical with the brain states of various Earth people like ourselves when we look at our blue sky.

15. Michael Tye, *Ten Problems of Consciousness: A Representational Theory of the Phenomenal Mind* (Cambridge, Mass.: MIT Press, 1995); and Tye, *Consciousness, Color, and Content* (Cambridge, Mass.: MIT Press, 2000).

clear that it is not enough to say, "I am a naïve realist in the philosophy of perception." One has to say *how* one proposes to be a "naïve realist."

Let me now explain our criticism of the approaches just mentioned in a little more detail.

Disjunctivism

Like intentionalism, which I shall describe next, disjunctivism, a position introduced by J. M. Hinton in a 1967 article in *Mind* titled "Visual Experiences," has the ambition of reviving and reinterpreting naïve realism, that is, the view that the objects of veridical perception are the objects we see in our environment (in the case of visual perception), and not our own "sense data."[16]

At first blush, the claim of "disjunctivism" seems simple to state, although the position is obviously extreme from the standpoint of traditional philosophy. In a nutshell, disjunctivism claims that in the case of a *veridical* perception, for example, of a yellow wall, what I am directly aware of is the wall itself, and of its yellowness. I am not aware of "qualia," either on a phenomenalist conception of qualia or on a conception of them as identical with something inside my brain, and there is no "interface" between me and the wall. In particular, the claim that Russell (of *Problems of Philosophy*) would have expressed by saying that "the yellow I see" is not in or on the surface of a material object but is a "sense datum" in my "private space" is decisively rejected. Nowadays this is hardly controversial. Even Ned Block, while defending "qualia," explicitly attacks the "inner-arena" picture of our relation to those qualia. We need to dig deeper to expose the differences.

Not only do disjunctivists and intentionalists hold that veridical perception extends all the way to the environment itself, but they also hold that the qualities we perceive are simply the qualities of the objects we see (I confine attention to visual perception, which is the most discussed case). That is why I said before that "both intentionalists and disjunctivists defend naïve realism, however, which they identify (mistakenly, in our view) with the idea that the description of the perceived public qualities of the object we see (in the case of a veridical visual experience) exhaust the qualitative character ('phenomenal character') of the experience." This indeed distinguishes them

16. Here I refer to an example used by J. M. Hinton in "Visual Experiences," *Mind* 76, no. 302 (1967): 217–227.

from Ned Block, the "phenomenist," who argues that different perceivers not only may have but in fact do have quite different "qualia" when they look at, say, a green picket fence, and that none of these "qualia" can be identified with *the* color of the picket fence.

But the disjunctivist also rejects the claim of Tye and other intentionalists (self-styled "representationalists," although they also apply this term to disjunctivists) that the content of my perception *represents* the wall and its color, and that this "intentional content" is what I am directly aware of and not the wall itself. In sum, two sorts of interface are rejected: the sense data of Russell, Moore, Ayer, Price, and others, and the "intentional contents" of Dretske, Tye, and others.

What makes disjunctivism the most radical of the "naïve realist" positions is the denial that veridical experiences and illusions (even the perfect hallucination supposed to be produced by Nozick's famous "experience machine"[17]) *have anything in common.* Of course, if one were plugged into the experience machine, one would not be able to tell that she was not experiencing real colors, a real picket fence, or whatever. But according to Hinton and his followers (John McDowell, Mike Martin), that is not because the hallucination and the veridical experience have a "highest common factor." According to Hinton's classic paper, even referring to both the veridical perception of an object and the corresponding hallucination as "experiences" is misleading.

The disjunctivists would not deny, I believe, that I perceive something when I see a white wall that looks red because a red light is shining on it; they would say that I see the wall, and I see how it looks from here (something that is "objective" and could even be photographed), and that look is misleading. But, they claim, Macbeth did not perceive *anything* when he hallucinated a dagger. He was not even aware of a property of his experience, other than "being indistinguishable from seeing a dagger."[18] And his experience is not something that could be "the same" if it were veridical (if it were normally caused rather than being a hallucination).

A problem with this is that the disjunctivists simply assume that there could not be *scientific* reasons for identifying the *phenomenal* properties of experiences with brain states/events. This is especially clear in Hinton's

17. Robert Nozick, *Anarchy, State and Utopia* (New York: Basic Books, 1974), 42–45.

18. See Michael Martin, "On Being Alienated," in John Hawthorne and Tamar Szabo Gendler, eds., *Perceptual Experience* (Oxford: Oxford University Press, 2006), 354–410.

famous 1967 paper.[19] Once one argues (in good Blockian fashion) that systematic coherence (including the Inverted Earth argument) and explanatory power support such an identification, I see no reason to deny that persons whose relevant brain states are the same have experiences of the same phenomenal character, and if that is a "common factor," so be it.

The disjunctivists *do*, however, admit that there is something that happens (or would happen) in the brain both in the case of a veridical perception and in the case of a "perfect hallucination," and they even accept that that fact explains why I am deceived by the perfect hallucination. This seems to be an admission that the phenomenal similarity (although they would not put it that way) of the hallucination and the veridical experience is supervenient on the brain state: that is, *same brain state implies indistinguishable experiences.* Let us call this (ignoring possible howls of protest from the disjunctivist camp) *the supervenience of phenomenal character on internal neurophysiological state.* The disjunctivists also say, although not in that language, that *in the case of a veridical experience* the phenomenal character is determined by the property of the wall—its yellowness. In Alva Noe's metaphor, we "visually touch" the wall.[20] The external color itself is "involved" in the experience. Let us call this the supervenience of phenomenal character on the external property when the experience is veridical.

In the light of the physical and conceptual possibility of "Inverted Earth," there is a possible fatal tension between the two sorts of supervenience. Specifically, if the brain state covaries with *different* properties of the wall on Earth and on Inverted Earth, and the brain state determines the phenomenal character (up to indistinguishability, anyway), how can the phenomenal character be both the *same* (because the brain states are the same) and *different* (because the physical color of the wall is different)?

Intentionalism

According to intentionalism, perceptual experiences have *representational content.* (Most intentionalists hold that this is not the same as "conceptual content," but this is an epicycle I shall not enter.) They are thus *not* flat "mental paint"; they tell us how the environment is (or at least claim to do that,

19. Hinton, "Visual Experiences."
20. Alva Nöe, *Action in Perception* (Cambridge, Mass.: MIT Press, 2004), 96–100.

when they deceive us). Thus the perfect hallucination of the green picket fence and the veridical perception of the same *do* have a "highest common factor" according to intentionalists. However, as already said, intentionalists and disjunctivists both hold that the description of the perceived public qualities of the object we see (in the case of a veridical visual experience) exhausts the qualitative character ("phenomenal character") of the experience.

Most, if not all, of the intentionalists Jacobson and I know of, beginning with Dretske, hold that the content of a mental representation is the "information" it carries in something like Shannon's sense of "information."[21] Call this familiar version of intentionalism "reductive intentionalism." In principle, there is room for another version of intentionalism, a "liberal naturalist" version, which takes intentionality to be irreducible without "transcendentalizing" it.[22] Thus a liberal naturalist could agree that the intentionality of both words and perceptual experiences is something that can be scientifically as well as philosophically studied, something that depends on causal connections and on what Ruth Millikan calls "normal biological functioning," and something that many different sciences, as well as many sorts of conceptual inquiries, can contribute to our understanding of, while *denying* that one can reduce the intentional to the nonintentional.[23] Indeed, Jacobson and I would agree with this much of a "nonreductive intentionalist position."

What makes reductive intentionalism tremendously implausible is the difficulty in seeing what it means to say that "information" in, say, Dretske's sense (which is simply a matter of what a mental representation covaries with in the subject's environment) can be the same thing as "phenomenal quality."

This is not the difficulty that leads disjunctivists to reject intentionalism, however. According to disjunctivists, to concede that there is a highest common factor in the case of the hallucination and the corresponding veridical experience is to surrender to the skeptic. They claim that only by denying

21. Claude E. Shannon and Warren Weaver. *The Mathematical Theory of Communication* (Champaign-Urbana: University of Illinois Press, 1949).

22. For an articulation of the program of *liberal naturalism* and a defense of its advantages over contemporary forms of reductive or scientific naturalism, see Mario De Caro and David Macarthur, eds., *Naturalism in Question* (Cambridge, Mass.: Harvard University Press, 2004), and De Caro and Macarthur, eds., *Naturalism and Normativity* (New York: Columbia University Press, 2010).

23. Ruth Garrett Millikan, *Language, Thought and Other Biological Categories* (Cambridge, Mass.: MIT Press, 1984), 5.

that there is *any* highest common factor can we preserve the idea that veridical experiences *justify* and do not simply "trigger" our beliefs about them. In our view, this is a mistake. No philosophy of perception can claim to be "the only" way to answer the skeptic—not even if we agree that what we want is more than a "reliabilist" account of justification. On reflection, it seems to me that one might meet the skeptical worry by arguing that as long as our experiences *also* have a *functional* description, they can perfectly well enter into "the space of reasons." Of course, a lot more needs to be said.[24]

Phenomenism

Block, who is the principal representative of phenomenism, rejects the "inner-arena" picture—that is, the "qualia" he posits are not objects that we "observe" (although he does not go into sufficient detail about his alternative picture, in my view)—but otherwise his account resembles a classical sense-datum story. We are able to attend to qualia on his account, and they are inside our brains and nonrepresentational. In our view, the fact that a green object may look different to different subjects no more shows that the *capacity to have that look* is not a property of the object seen (say, the picket fence) than does the fact that the picket fence has a different look in bright sunlight and when a cloud passes over the sun. Looks, in my view, are *capacities that objects have* (and realizations of those capacities), not properties of a supposed "mental paint."

[Additional comment, 2009]: I no longer agree with this criticism of Block's view. I do still think that looks are capacities that objects have, but those capacities include the capacity to cause subjects to experience certain qualia under appropriate circumstances, and I agree with Block's claim that qualia are probably identical with brain events/states. Because qualia are what Block calls "mental paint," that means that I do not now see Block's "phenomenism" as *incompatible* with Jacobson's and my "transactionalism."

Transactionalism

Last but not least, let me very briefly sketch the alternative Jacobson and I favor.

---

24. See Chapter 3 in this volume, "Corresponding with Reality."

As already mentioned, one problem that we see with both disjunctivism and intentionalism has to do with the claim that the description of the perceived public qualities of the object we see (in the case of a veridical visual experience) exhausts the phenomenal character of the experience. *There are many counterexamples to this claim.* Here are three (for simplicity of exposition I confine attention to visual experiences, but similar points apply to the other sensory modalities):

1. The picket fence has a different look to someone who has astigmatism than it does to someone with "normal" vision.
2. In the case of vision, when they look at a white surface, most observers will observe a slightly different look when they view the surface with the left eye closed and when they view it with the right eye closed. (One look is "grayer.") This is not explained by appealing to parallax; the explanation is that there are slight but significant differences between the macular areas of the two eyes. (Try it yourself.)
3. Data cited by Block, Hardin, and others show that a given shade of green looks "pure green" to some observers and "yellow green" to others, and that there are reasons to think that neither observer misperceives the shade in question.

The existence of such cases illustrates that both disjunctivists and intentionalists are still in the grip of the "spectator" view of perception that Dewey, Gibson, and others have criticized; they fail to see the extent to which what we perceive depends on a *transaction* between ourselves and the environment, and hence they fail to see that the properties we perceive depend on our nature as well as the nature of the environment. (This is why we call them "transactional" properties.") Once this is recognized, there is room for an account that preserves what is right in talk of the "transparency" of perceptual experience—namely, the idea that in a successful perception we experience properties of the picket fence (or whatever), and not properties of our own minds or brains, while leaving room for recognition of subjective as well as objective factors in perception.

A second problem has to do with what both disjunctivists and intentionalists say about hallucinations and illusions. Because this is a huge topic, I will confine attention to just the case of what might be called "philosophical hallucinations," that is, the sort of hallucination imagined by myself when I imagined "brains in a vat" and by Nozick when he imagined an "experience machine."

Disjunctivists simply deny that there is such a thing as a "phenomenal quality" or even a common "intentional content" common to the hallucination and the corresponding veridical experience, while illusionists say that both exist (and are in fact identical), and the phenomenal quality is simply the "information" that reaches the brain.

My own view is that (environment-involving) "liberal functionalism" is the right way to think about this. A hallucination, *functionally* characterized, is a state in which it appears to the subject that the subject is perceiving, say, a green picket fence. Thus hallucinating a green picket fence is not *directly* a world-involving state (there does not have to be a picket fence or, indeed, anything at all the subject is seeing), but its functional characterization refers to such a state ("perceiving a green picket fence"). Because I still believe that Commander Data and I could both have the same mental states (a corollary of my original version of functionalism that I still subscribe to), it is not the case that hallucinating is a *neurologically* characterized brain state; it is, rather, a *functionally* characterized state of the organism.[25] It is part of this "functionalist" view that ("philosophical") hallucinations do have content (something radical disjunctivists deny); intentionalists are right in seeing the hallucination as involving the organism's being misinformed about the environment. The tricky question is the status of the hallucinated objects and qualities.

Quite simply, what I think is that the hallucinated white picket fence has the ontological status of a *fictional entity*. (What is that status? Pick your favorite story about such entities as "Hamlet.")[26] I do not mean to say only that perception involves an intentional content that is similar to a proposition expressed by a sentence with an empty name; rather, I want to point to the sense in which hallucinations are similar to intentional states that specifically concern *fiction*. When I get engaged with the fiction, Hamlet, who is not real, *seems real* to me. He seems to be presented to my mind in a way in which the

---

25. I would now [2011] say that this is right for nonqualitative mental states (e.g., knowing, believing, being aware that, and the other "propositional attitudes"). For qualia, I would now say that qualia are not identical with functional events/states but with neurological ones. If this is right, then Commander Data and I could have the same beliefs and fears, love the same person, and so on, but the question whether any relation of "identity" exists between the qualia of nonconspecifics (such as Commander Data and myself) is problematic. I would say, "Probably not."

26. Of course, real entities can have fictional counterparts; Napoleon was a real person and is also a character in *War and Peace*. Similarly, it is possible to have a hallucination about a real person or thing.

king of France surely is not, if it is just the intentional object of my reading Russell's famous sentence.[27] Likewise, in a perfect hallucination the orangish picket fence, which is not real, seems real to me. Hamlet and the picket fence seem real to me; it is not only that the proposition that they exist, or some specific predicative proposition about them, seems to me to be true.

How does transactionalism compare with intentionalism? Intentionalism treats all experiences, veridical and nonveridical alike, as, in effect, *representations*. According to intentionalists, the difference between a veridical experience and a nonveridical one is like the difference between a true newspaper article and a false one. In contrast, Jacobson and I think of veridical experiences as *externalistically* characterized states of the organism, states that essentially involve the object perceived. Hallucinations do not involve an object perceived; they do, indeed, only *represent* an object as perceived, but that is the case because they are, so to speak, defective states; they are malfunctions. Veridical perceptions *involve* the object; they do not merely "represent" it. Thus even if intentionalists were to agree that the status of the hallucinated picket fence is similar to the status of a fictional entity, their account of the nature of veridical perceptual experiences would still be very different from mine/ours.

How does transactionalism compare with disjunctivism? In agreeing that the qualities that are presented in a veridical perceptual experience of the picket fence are real (albeit "transactional") properties of the *fence* while, at the same time, denying that the "qualities" that are seemingly presented in a hallucination are properties of any real object, it goes a long way toward disjunctivism. I think, in fact, that it preserves what is right in disjunctivism.

I shall close by briefly considering two objections that are sure to occur to my readers.

First, it may seem that fictional characters are too "abstract" to be experientially presented, too abstract even to *seem* "real." In part this objection underestimates the extent to which an imaginative person can experience a fictional scenario as "real"; in my own case, there are portions of *The Wind in the Willows* that I read when I was about ten years old that I could swear that I *saw*. But in any case the degree to which a fiction seems "real" depends on the mode of presentation of that fiction, and hallucination is obviously a functionally different mode of presentation than, say, reading a novel.

27. Russell's sentence is, of course, "The present King of France does not exist." "On Denoting," *Mind* 14, no. 56 (1905): 479–493.

Second, and sharpening the first objection, it might be argued that although a fictional character can indeed seem fully real if the fiction is acted on the stage, the "reality" belongs to the actors and the scenery. So, this line of thought continues, there have to be "actors" and "scenery" in the case of a hallucination, and what could they be but our brain states?

Well, if you think that it is a priori that a functionally identical but physically differently realized person (e.g., Commander Data) could not have the same experiences you do, you may find yourself forced back to the idea that what you experience is your own brain states (in all cases, or only in the case of hallucinations?). Not only does this make it difficult to see how our experiences can justify beliefs about the external world (one of the insights of McDowell's *Mind and World*)[28], but it makes the sort of functionalism that I favor a priori false, which makes it extremely dubious, in my view. On the other hand, if what you claim is that what we experience (in the case of a hallucination) is a *functionally characterized state of our own brains,* then I would say that this view is compatible with my account if (but again this is a big "if") the right account of fictional entities is that they are functionally characterized states of the organism. But I will not decide here whether that is the right account of fictional entities.

28. John McDowell, *Mind and World* (Cambridge, Mass.: Harvard University Press, 2004).

Acknowledgments

Many of the papers reprinted here have been lightly edited to correct minor mistakes, avoid needless repetition and, in cases where the papers were originally given as talks, to remove indications of occasional address. Putnam has made minor additions or deletions to several papers for purposes of clarification. However, in cases where he has added a whole paragraph indicating a significant change in view, this has been indicated in the text by giving the date of the remark. In two cases, Chapters 27 and 36, substantial changes were made to the original papers by the incorporation of additional material from another source. Please see below for a more detailed explanation.

"Science and Philosophy." From *Naturalism and Normativity,* edited by Mario De Caro and David Macarthur, Copyright © 2010 by Columbia University Press. Reprinted with permission of the publisher.

"From Quantum Mechanics to Ethics and Back Again." Originally published in Maria Baghramian, ed., *Reading Putnam* (London: Routledge, 2012). Originates from a closing lecture to the "Putnam Fest" conference at University College Dublin, March 11–14, 2007.

"Corresponding with Reality." Unpublished. Based on the Prometheus Prize lecture to the American Philosophical Association (Eastern Conference), December 28, 2010.

"On Not Writing Off Scientific Realism." Unpublished. Based on a lecture given for the 50th Anniversary Celebration of the Boston Colloquium for Philosophy of Science, April 15, 2010.

"The Content and Appeal of 'Naturalism.'" Originally published in Mario De Caro and David Macarthur, eds., *Naturalism in Question* (Cambridge, Mass.: Harvard University Press, 2004), 59–70.

"A Philosopher Looks at Quantum Mechanics (Again)." Originally published in *British Journal for the Philosophy of Science* 56, no. 4 (2005): 615–634.

"Quantum Mechanics and Ontology." Originally published in *Analysis and Interpretation in the Exact Sciences: Essays in Honour of William Demopoulos*, Frappier, Melanie; Brown, Derek; DiSalle, Robert, Eds., The Western Ontario Series in Philosophy of Science Vol. 78 (January 2012). Based on a lecture given at a conference in honor of Bill Demopoulos at the University of Western Ontario, May 2–4, 2008.

"The Curious Story of Quantum Logic." Unpublished. Based on a lecture given at Tel Aviv University in 2011.

"Indispensability Arguments in the Philosophy of Mathematics." Unpublished. Derived from a lecture given at the 40th Chapel Hill Colloquium in Philosophy, October 6–8, 2006.

"Revisiting the Liar Paradox." Originally published in Gila Sher and Richard Tieszen, eds., *Between Logic and Intuition: Essays in Honor of Charles Parsons* (Cambridge: Cambridge University Press, 2000), 3–15, where it appeared with the title "Paradox Revisited I: Truth." Copyright © 2000 Cambridge University Press. Reprinted with permission.

"Set Theory: Realism, Replacement, and Modality." Unpublished. Based on a paper given at a conference held at Princeton University in Paul Benacerraf's honor on May 12, 2007.

"On Axioms of Set Existence." Unpublished. Derived from an earlier paper published with the title "Axioms of Class Existence," in *Summaries of Talks presented at the Summer Institute in Symbolic Logic* (Cornell: Cornell University, 1957), sponsored by the American Mathematical Society (Princeton: Communication Research Division, Institute for Defense Analyses, 1960).

"The Gödel Theorem and Human Nature." Originally published in Matthias Baaz, ed., *Kurt Gödel and the Foundations of Mathematics: Horizons of Truth* (Cambridge: Cambridge University Press, 2011). Copyright © 2011. Reprinted with permission.

"After Gödel." A shortened and slightly revised version of a lecture to the conference "Models of Computation" held at Tel Aviv University in March 2004. Originally published under the same title in *Logic Journal of the IGPL* 14, no. 5 (October 2006): 745–759. The shortening consists in omitting material found in Chapters 11 and 13 of the present volume.

"Nonstandard Models and Kripke's Proof of the Gödel Theorem." Originally published in *Notre Dame Journal of Formal Logic* 41, no. 1 (2000): 53–58. Copyright © 2000 Yale University. Reprinted by permission of the publisher, Duke University Press.

"A Proof of the Underdetermination 'Doctrine.' " Unpublished.

"A Theorem of Craig's about Ramsey Sentences." Unpublished.

"The Fact/Value Dichotomy and Its Critics." Originally published in Naoko Saito and Paul Standish, eds., *Stanley Cavell and the Education of Grownups* (New York: Fordham University Press, 2011).

"Capabilities and Two Ethical Theories." Originally published in *Journal of Human Development* 9, no. 3 (2008): 377–388, www.informaworld.com.

"The Epistemology of Unjust War." Originally published in *Royal Institute of Philosophy Supplement* 81, no. 58 (2006): 173–188.

"Cloning People." Originally published in Justine Burley, ed., *Genetics and Human Diversity* (Oxford: Oxford University Press, 1999), 1–13.

"Wittgenstein and Realism." Unpublished. Derived from a talk given at Hamilton College in 2007.

"Was Wittgenstein Really an Antirealist about Mathematics?" Originally published in Timothy McCarthy and Sean Stidd, eds., *Wittgenstein in America* (Oxford: Clarendon Press, 2001), 140–194.

"Rules, Attunement, and 'Applying Words to the World': The Struggle to Understand Wittgenstein's Vision of Language." Originally published in Ludwig Nagl and Chantal Mouffe, eds., *Pragmatism or Deconstruction* (New York: Peter Lang, 2001), 9–23.

"Wittgenstein, Realism, and Mathematics." Originally published in French, with the title "Wittgenstein, le réalisme et les mathematiques," in Jacques Bouveresse, Sandra Laugier, and Jean-Jacques Rosat, eds., *Wittgenstein, dernières pensées* (Marseilles: Agone, 2002), 289–313.

"Wittgenstein and the Real Numbers." Originally published in Alice Crary, ed., *Wittgenstein and the Moral Life* (Cambridge, Mass: MIT Press, 2007), 235–250.

"Wittgenstein's 'Notorious' Paragraph about the Gödel Theorem: Recent Discussions." This essay was cowritten with Juliet Floyd and derives from a paper published in German, with the title "Wittgensteins 'berüchtigter' Paragraph über das Gödel-Theorem: Neuere Diskussionen," in Esther Ramharter, ed., *Prosa oder Beweis? Wittgensteins >> berüchtigte << Bemerkungen zu Gödel, Texte und Dokumente* (Berlin: Parerga Verlag, 2008), 75–97. It incorporates new material from "A Note on Steiner on Wittgenstein, Gödel and Tarski," *Iyyun: The Jerusalem Philosophical Quarterly* 57 (January 2008): 1–11.

"Wittgenstein: A Reappraisal." Unpublished.

"Skepticism, Stroud, and the Contextuality of Knowledge." Originally published in *Philosophical Explorations* 4, no. 1 (2001): 2–16.

"Skepticism and Occasion-Sensitive Semantics." Originally published, with the title "Skepticism," in Marcelo Stamm, ed., *Philosophie in synthetischer Absicht* (Stuttgart: Klett-Cotta, 1998), 239–268.

"Strawson and Skepticism." Originally published in Lewis Hahn, ed., *The Philosophy of P. F. Strawson* (Chicago: Open Court, 1998), 273–287.

"Philosophy as the Education of Grownups: Stanley Cavell and Skepticism." The last two sections of this chapter have been substantially revised. The original version was published in Alice Crary and Sanford Shieh, eds., *Reading Cavell* (London: Routledge, 2006), 117–128.

"The Depths and Shallows of Experience." Originally published in J. D. Proctor, ed., *Science, Religion, and the Human Experience* (Oxford: Oxford University Press, 2005), 71–86.

"Aristotle's Mind and the Contemporary Mind." Originally published in Demetra Sfendoni-Mentzou, ed., *Aristotle and Contemporary Science* (New York: Peter Lang, 2000), 7–30.

"Functionalism: Cognitive Science or Science Fiction?" Originally published in David Martel Johnson and Christina Erneling, eds., *The Future of the Cognitive Revolution* (Oxford: Oxford University Press, 1997), 32–44. By permission of Oxford University Press, Inc.

"How to Be a Sophisticated 'Naïve Realist,'" Derived from an interview published in Patrick Grim, ed., *Mind and Consciousness: 5 Questions* (London: Automatic Press/VIP London, 2009). New material, now incorporated in the section titled "What Are the Most Important Open Problems in Contemporary Philosophy of Mind? What Are the Most Promising Prospects?," was given as a talk at a seminar in 2009 in Tel Aviv. Published by permission from Automatic Press/Vince Inc. Press.

Index